McGRAW-HILL SERIES
IN BIOENGINEERING

CONSULTING EDITORS

Professor S. J. Mason
Department of Electrical Engineering
Massachusetts Institute of Technology

and

Professor Eugene Yates
Department of Physiology
Stanford University

A SYSTEMS APPROACH

TO BIOMEDICINE

A SYSTEMS APPROACH

TO BIOMEDICINE

William B. Blesser

Professor of Bioengineering
Polytechnic Institute of Brooklyn

McGRAW-HILL BOOK COMPANY

New York   St. Louis   San Francisco   London
Sydney   Toronto   Mexico   Panama

R
856
B55

A SYSTEMS APPROACH TO BIOMEDICINE

Library of Congress Catalog Card Number 68-8658

05893

1 2 3 4 5 6 7 8 9 0 M A M M 7 6 5 4 3 2 1 0 6 9

To Lee, Danna, and Bonnie

PREFACE

The absence of the term "engineering" in the title of this work is not an oversight; it was a deliberate omission. The first three titles considered for the text were "Bioengineering," "Engineering in Biological Systems," and "Systems Engineering in Biomedicine." Each of these was rejected in turn as none really reflected the intent of the book. The difficulty was the word "engineering" and the incorrect picture it brings to mind. Engineering is thought of as a field; the intent here is to present a concept or viewpoint which, though often used in engineering, need not be restricted to that field. Engineering generally implies the application of ideas and concepts for construction, building, synthesis; the intent here is to utilize ideas and concepts for analysis and design of experiments and to develop insight about the operation of a system. In a sense, then, the discussion presented does not necessarily relate to engineering (or for that matter to biomedicine); instead the material can be viewed as an exposition of a philosophy that can be applied to any field.

The specific field chosen to illustrate this "systems approach" is biomedicine because it is the author's opinion that living systems offer the most exciting challenges in today's technology. In this area the reader can find subject matter to suit his every taste: from the well-organized laws in genetics to the unpredictability of human behavior, from the use of man-made systems in living systems to the use of living systems in man-made systems, from studies of component parts to studies of assembled systems, from the problem of the aquanaut to the problems of the astronaut. Here truly is an area where the imaginative reader can find a wide spectrum of interests.

In order to describe the level of the text it might be well first to describe its history. The original ideas for the text were developed during a course which the author presented to the medical personnel of the Pulmonary Function Laboratory at Downstate Medical Center Brooklyn, New York at the request of Dr. H. Lyons, Director of the Laboratory. The physiological background of these students was obviously excellent, but their mathematical background had suffered "disuse atrophy," and the initial plans for course presentation had to be modified. The material could not be just "presented" but had to be carefully related to the students' physiological background. In this first course the author must admit that he probably learned more than the students. In a second course, this time presented to engineering students at the Polytechnic Institute of Brooklyn, a reverse situation was encountered. The mathematical facility of the students was well developed; the inability to relate their background to an area well removed from their basic training was, however, a very real obstacle. Again course presentation had to be modified. After a number of such alternate presentations it came as a surprise to realize that the same introductory material was applicable to both groups. The groups differed only to the extent that the speed of presentation and attention to details had to be tailored to meet the needs of the students. The basic material could be the same, and it is this basic introductory material which forms the core of this text.

Although the book was written primarily to introduce systems concepts to biomedical personnel, it can also be used to introduce systems concepts to other groups. Thus, engineering and nonengineering students may find this material useful as initial preparation for more ambitious studies in both biomedicine and engineering.

Relatively little background is required of the reader. An attempt has been made to keep the physiological examples realistic yet simple. The mathematics required for analysis is in most cases developed in the text. Calculus is desirable only to the extent that the reader understand, at least conceptually, what a derivative and an integral mean. Additional background is, of course, helpful, but no advanced mathematical capability is required.

It is difficult to thank adequately all who have helped prepare and complete this manuscript. I might start by thanking all the students who were exposed to the unfinished manuscript; their patience (beyond the call of duty), understanding, and, above all, their constructive criticism went a long way toward making this project possible. Some specific students who deserve special mention include Dr. H. Lyons of Downstate Medical Center, Dr. H. Goldberg of Albert Einstein Medical Center (Philadelphia), Dr. E. Glassman of Long Island Jewish Hospital, and the late Dr. H. Gadboys of Mt. Sinai School of Medicine. These men worked actively to have the "Systems Approach to Biomedicine" presented as a course at their respective institutions and then, in spite of their heavy work loads, they faithfully attended classes until the completion of the course.

In addition, Drs. E. Meilman, C. Brill, M. Bleicher, and J. Truxal should be mentioned for their fine editorial comments about the technical aspects of the manuscript. I would like to thank the latter particularly for his continued encouragement, interest, and enthusiasm and specifically for the material which he contributed for Chapter 10 of this book.

This expression of gratitude cannot be complete until I include my appreciation for the efforts of Mrs. Rosalind Chester, who read an impossible handwriting and translated it into legible type on ditto mats. The total process would have been nerve-shattering for a less capable individual. Without Mrs. Chester's help and steadfast cooperation this manuscript might never have been completed.

To close, I must thank my wonderful wife and children for bearing with me through these trying months. Their forbearance and patience contributed in no small measure to the completion of this work. I am looking forward to becoming reacquainted with them.

WILLIAM B. BLESSER

CONTENTS

CHAPTER **5**   Introduction to Analysis of Systems with Combined Properties

CHAPTER **6**   The Transfer Function

CHAPTER **7**   The Impedance Concept

CHAPTER **8** Periodic Signals

CHAPTER **9** Transient Oscillations and Resonance

CHAPTER **10**   Feedback

CHAPTER ONE

INTRODUCTION

1.1 SOME PRELIMINARY REMARKS

It is difficult to know how to start a book which, we hope, will integrate
two fields that appear to be only superficially related. Systems and
biomedicine are not considered to be fields with a common foundation.
They stem from two sciences that have been traditionally considered
only distantly related, and they were developed for dissimilar reasons.
Systems science evolved from the engineering and physical sciences;
it relies heavily on mathematical analyses and models to aid in the design
of functioning systems. Biomedical science developed as a branch of
the life sciences and is primarily an experimental field. The mathematical
relationships used in biomedicine are, for the most part, empirical;
analyses, when employed, are used not for design but rather to understand how an existing system functions. In spite of these differences,
however, workers in both fields have shown that there is common ground
between the systems and biomedical sciences. The application of systems
techniques and concepts to biomedical problems in particular has been
developed to a large degree. Articles and papers which utilize block

diagrams and computer simulations can now be found regularly in biomedical publications and in various other journals devoted to the life sciences. There is hardly a technical meeting dealing with life-science problems that does not now include some phase of systems techniques applied to biomedicine.

Although emphasis at this time is on systems concepts applied to biomedicine, efforts are now under way also to utilize concepts derived from biomedical research to aid in the design of engineering systems. Studies of the bat's navigation system have caused designers of navigation and guidance systems to take a more critical look at their present designs. The pattern-recognition capability of the frog serves as a physiological model for moving-target-identification systems. The pattern-recognition capability of human beings is under study to determine how an effective reading machine can be devised and automatic language translation can be effected. These are but a few areas where research and analysis of physiological systems are used to suggest and improve designs of physical systems. Although this aspect of the biomedical-systems interchange has not yet been exploited to any great degree, it is expected that research along these lines will be increased in the near future.

As the cooperative efforts between the systems and biomedical scientists increase, it can be expected that systems techniques and concepts will be utilized to an even greater degree than in the past. It will become increasingly important, therefore, for modern researchers to be familiar with some of these systems viewpoints—to use these techniques if possible or at least to understand their limitations so that they can communicate with other researchers who use them. The material presented in this book is to introduce and expose workers in the life sciences to some of the modern systems concepts. In particular, the book has been developed to provide the biomedical scientist with an insight into and an understanding of the utilization of models, system analysis, and analog simulation. If the book does no more than introduce the concepts and provide some familiarity with the nomenclature, it can still be a useful study as it will keep open the communications channels between the systems and biomedical scientists. It is hoped, of course, that this study will do more and that it will help students and researchers to develop an understanding and a facility with systems concepts and the associated analytical techniques.

1.2 THE MODEL AND THE ANALOG

Before we begin formal discussions of the development of analogs and the analysis of "models," we shall first attempt to differentiate between these terms. This is no easy task, as the differences (if they exist) are very often academic and depend upon who was educated where. To those familiar with the terminology, there is generally little confusion between the terms; however, this is probably due to experience and exposure rather than precise, agreed-upon definitions. To those without experience the terms can be very confusing, and some attempt should be made to indicate the concepts associated with them.

The best way to show how analogs, models, and systems are related is through the use of a simple illustration. The diagram in Fig. 1.1 may prove helpful. The system to be examined is the human forearm when it is subjected to a static weight load (w). The actual system is not trivial and by no means simple but for illustrative purposes we shall assume that only a very simple study is required. In particular, we shall assume that only the relationship between the vertical muscle force and the static weight load w (for the horizontal forearm) is of interest. The variety of ways in which this relationship may be established and represented is shown in Fig. 1.1. From the way the term "model" is employed we see that it is associated with any representation (conceptual or physical) that describes the behavior of the system with the same variables as used in the basic system. The physical model is visually identifiable with the system it is supposed to represent. The model has a similar function and appearance as the actual system, and there is correspondence between the elements of the model and the physical system. The closer the correspondence between the elements, the more accurate is the model. The "graphical model" is simply a graph (or a set of graphs) that relates the variables of interest. In essence, the graphical model is merely a sophisticated way of describing the plot of two variables. This model may be obtained from measurements made directly on the physical system or from measurements made on the physical model or from the mathematical model (to be described next). In any case, since all models relate to the same physical system, all must be interrelated and, depending upon the degree of approximation, each is derivable from the other.

The mathematical model is another way of describing the mathematical expressions that relate the variables of the system under investigation. This model can be determined by an examination of the graphical model derived by direct measurements, it can be determined from the graphical model derived from the physical model, or it may be determined directly by analytic consideration of the actual system or the physical model. The mathematical models derived by these procedures may not be exactly the same. In fact, it would be very surprising if they were the same. The best one could realistically expect would be that all the mathematical models be reducible to a common simplified form when various simplifying approximations are used.

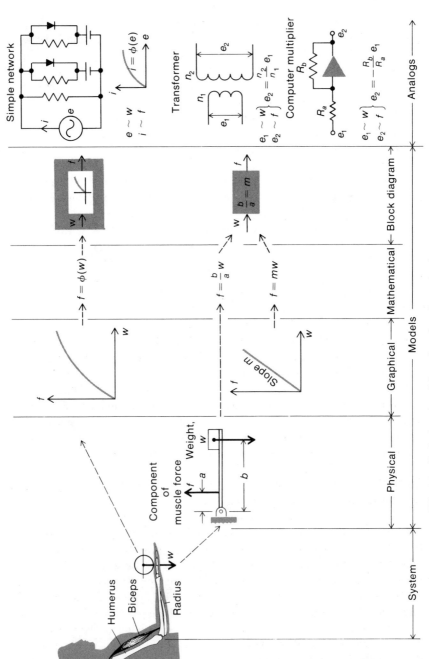

FIG. 1.1 *Evolution of models and analogs for a simple system.*

The block-diagram model is a pictorial representation of the pertinent mathematical relationships that govern the system. It describes in diagrammatic form how an excitation signal is related to a resultant response. For more complex systems the block diagram shows the interrelation between a set of mathematical equations, and it shows how the various signals "flow" through the systems to produce the various responses. Because of its latter function, the block diagram is sometimes called a signal-flow diagram. (Actually, however, the signal-flow diagram is generally drawn in a slightly different manner and is ordinarily limited to linear systems.) Although it is possible to manipulate block diagrams to provide for simplified analytic results,* it is often just as easy (and sometimes easier) to manipulate the mathematical expressions they represent. The utility of the block-diagram model therefore lies not in its function as an analytic tool but rather in the conceptual aid it provides and the analog-computer simulation that it suggests. The latter points will be more clearly understood as the subject matter is developed in later chapters.

As mentioned earlier, the term "model" is generally used when a system representation involves variables of the same dimensions as the system under study. The model may be a physical device (e.g., a lever) or it may be a conceptual formulation such as a graph or mathematical equation. In any case, the variables of a model are dimensionally equivalent to the variables of an actual system. The analog, on the other hand, generally is a physical device that is not basically identifiable with the actual system and is described by variables that are not dimensionally equivalent to those of the actual system. The analog and the actual system are the same only to the extent that the equations that govern the variables in both systems are the same. In Fig. 1.1 electrical analogs are used to represent the physiological system. In the topmost analog representation a voltage (e) represents the applied weight force (w), and current flow (i) represents the resultant muscle-force component (f). The electrical circuit is arranged so that current varies with voltage in the same manner as vertical muscle force varies with applied weight. A plot of e versus i would thus have the same shape as the graphical model of the physiological system. The lower two circuits in Fig. 1.1 may also be considered

---

* A table for block-diagram transformations may be found in Ref. 1, p. 252. A rather detailed discussion of how to construct a signal-flow diagram from a set of equations and how to manipulate the diagram to get results is given in Ref. 2, chap. 2. A more recent text which includes this information is Ref. 4. These references are mentioned here to indicate the wealth of information on systems concepts. They are, however, advanced in some respects, and it may be wise to defer consultation of these texts until more background has been developed.

analogs of the physiological system. In these circuits, however, voltages at different points on the circuit (e_1 and e_2) represent the applied weight (w) and the muscle-force component (f). Although only electrical analogs are shown in Fig. 1.1, there are many other possible analogs. The only requirement is that the analog and the actual system be mathematically similar. Electrical analogs are the most popular as they are the most versatile. Equations are generally easy to simulate with electrical analogs, and system parameters can easily be varied in such analogs. The manner by which these circuits may be developed will be discussed in later chapters.

We close this chapter by noting that there are no hard and fast rules regarding the use of the terms "model" and "analog"; as might be expected, the terminology is often mixed. To avoid confusion, here the terms will be used as described. In general, however, if the terms are taken in context the meaning will be clear.

Problems

1.1   The force-displacement characteristics of a tendon tested in vitro are shown in Fig. P1.1. The force f applied is a static load in the sense that it does not change continuously but varies in a stepwise fashion (as shown in Fig. P1.1b).

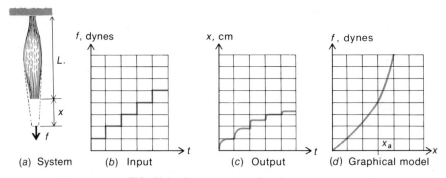

(a) System   (b) Input   (c) Output   (d) Graphical model

FIG. P1.1   *Representation of tendon test.*

The resultant displacement for each application of load is recorded in Fig. P1.1c. The input-output data are combined into a single graphical model as shown in Fig. P1.1d.

(a) Propose a mathematical model and block-diagram model for the full extension.

(b) Propose a mathematical model and block-diagram model if it is known that within the range of practical operation x does not exceed x_a.

(c) Sketch the graphical model of a simple extension spring. Under what circumstances might the spring be used as a physical model of a muscle?

(*d*) It may be possible to use a combination of springs to represent the muscle characteristics shown in Fig. P1.1*d*. Indicate by diagram how this may be done.

1.2 The analog-computer circuits shown in Fig. P1.2* have the input-output characteristics indicated. When the input signal e_1 is increased suddenly, the output of circuit *A* rises gradually to a new level; the output of circuit *B* rises and then falls. Circuit *C* adds the signals delivered to the input terminals; thus if e_a

* Information on analog computers and computers in general may be found in Refs. 5 and 6. Although they may be rather advanced at this stage of development, they indicate some applications of computers in medicine. References 7 to 9 are somewhat more basic and may be more useful for introductory background (see also Refs. 6 and 7 in Chap. 2). Two very useful introductory handbooks may be obtained from analog-computer manufacturers: One is "Applications Manual for Computing Amplifiers for Modelling, Measuring, Manipulating and Much Else!" issued by Philbrick Researches, Inc., Dedham, Mass.; the other is "Analog Computers—Operators Handbook" issued by Electronic Associates, Inc., Long Branch, N.J.

As there is adequate literature on the subject of analog-computer simulation and related topics, such material will not be formally presented in this text. However, simple problems will be included in the various chapters to expose the reader to some of the concepts associated with analog simulation. This, coupled with the elementary teaching program (see steps 19 to 35 in Appendix A), should be sufficient to provide an introduction to the use of analog computers for systems studies.

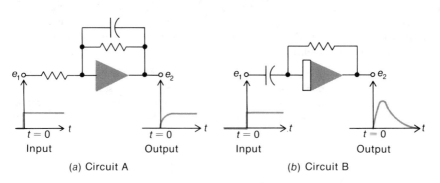

(*a*) Circuit A   (*b*) Circuit B

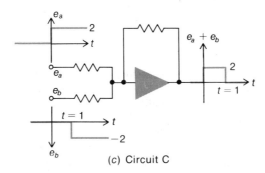

(*c*) Circuit C

FIG. P1.2 *Analog-computer circuits.*

and e_b are applied to the input of the circuit, the output will be $e_a + e_b$. (In each case the actual delivered signal will be the negative of the curves shown. This can be ignored here.)

(a) Which circuit can be used to simulate the test of Prob. 1.1?

(b) In a glucose-tolerance test a patient is given a dose of glucose (a form of sugar), and blood samples are taken periodically to determine how the sugar concentration in the blood varies. Sketch a graph representing the *amount* of glucose administered in terms of time. Sketch a graph that represents the sugar concentration at any time. (This can be obtained from any good physiology book; one can, however, easily imagine the shape of such a curve, as sugar cannot be stored continuously.) Which circuit should be used to represent this test?

(c) In a glucose-tolerance test a patient is given a dose of glucose at the beginning of the test, and 10 min later he is given another dose equal to the first. How could circuit C be used (in combination with others if necessary) to estimate the expected sugar concentration? (This presupposes that the additional glucose does not significantly change the patient's response.)

(d) The iron lung is a mechanical respirator sometimes used to induce breathing when a person is incapable of breathing for himself. The lung is an airtight rigid structure; a patient is enclosed in the chamber with only his head exposed. Changes of pressure in the chamber cause the patient's chest cavity to expand and contract. This expansion and contraction induce breathing. If the internal pressure of the iron lung is increased in a stepwise fashion, which analog most nearly represents (1) the air-flow intake, (2) the volume of gas inspired, (3) the velocity of the diaphragm?

1.3 Shown below is a diagrammatic sketch of the heart. Of particular interest in this case is a representation of the mitral valve. The valve is essentially a mechanical "check valve" which permits flow when the atrial pressure (p_a) exceeds ventricular pressure (p_v) and which prevents flow when the ventricular pressure exceeds the atrial pressure.

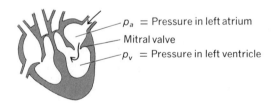

p_a = Pressure in left atrium

Mitral valve

p_v = Pressure in left ventricle

(a) Draw a mechanical model of the system.

(b) From the mechanical model, propose a graphical model of the system (pressure vs. flow). Include both conditions $p_a > p_v$ and $p_a < p_v$.

(c) A graphical model of the physiological system (the mitral valve) may be determined from published data. The simultaneous recordings of atrial pressure, ventricular pressure, and ventricular volume shown in Fig. P1.3 can be used to construct such a model. Using the data shown, plot the pressure drop across the

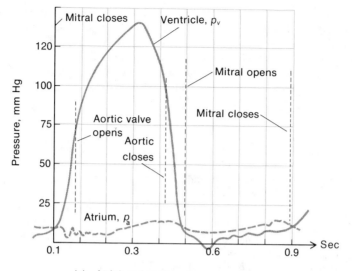

(a) Atrial and ventricular pressures

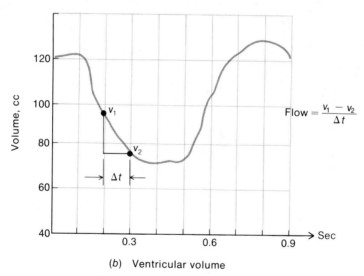

(b) Ventricular volume

FIG. P1.3 *Cardiac pressure and volume curves vs. time. (From J. C. Wiggers, "Circulatory Dynamics," Grune & Stratton, Inc., New York, 1952.)*

valve vs. flow through the valve (flow is obtained from the slope of the volume curve; i.e., flow = change of volume per unit time = dv/dt). Note the data are not "perfect"; therefore only approximate results can be expected.

(d) Compare the graphical model of the mitral valve with the graphical model of the mechanical system and of an electrical diode. Indicate why these systems may be considered analogous and which variables are analogous.

REFERENCES

1. Del Toro, V., and S. R. Parker: "Principles of Control System Engineering," McGraw-Hill Book Company, New York, 1960.
2. Truxal, J. G.: "Automatic Feedback Control System Synthesis," McGraw-Hill Book Company, New York, 1955.
3. Wiggers, J. C.: "Circulatory Dynamics," Grune & Stratton, Inc., New York, 1952.
4. Mason, S., and H. Zimmerman: "Electronic Circuits, Signals and Systems," John Wiley & Sons, Inc., New York, 1960.
5. Ledley, Robert S.: "Use of Computers in Biology and Medicine," McGraw-Hill Book Company, New York, 1965.
6. Stacy, Ralph W., and Bruce D. Waxman: "Computers in Biomedical Research," Academic Press Inc., New York, 1965.
7. Stice, J. E., and B. S. Swanson: "Electronic Analog Computer Primer," chap. 4, Blaisdell Publishing Co., Inc., New York, 1965.
8. Johnson, Clarence L.: "Analog Computer Techniques," 2d ed., McGraw-Hill Book Company, New York, 1963.
9. Soroka, Walter W.: "Analog Methods in Computation and Simulation," McGraw-Hill Book Company, New York, 1954.

CHAPTER TWO

SOME INITIAL CONCEPTS

2.1 BLOCK DIAGRAMS AND SIGNALS

In order to develop some preliminary background on signals and their relation to block diagrams, we first examine the role of signals in laboratory research. With very few exceptions, laboratory investigations involve the acquisition of data on the response to an applied stimulus. The experimenter excites a system with a known stimulus (or measures what he believes to be a stimulus) and simultaneously observes the response. From these stimulus-response signals the researcher hopes to gain insight and understanding about the system under study. A diagrammatic description of the outlined procedure is shown in Fig. 2.1. This diagram can be interpreted as a simple pictorial representation of the verbal statement "A stimulus is applied to a system or device; the system shows a response as evidenced by some output signal." This is the most elementary form of a block diagram.

The term "signals" means a time record of measured phenomena. Some representative examples are shown in Fig. 2.2. As can be seen, some of these may be spontaneous or self-generated signals which are

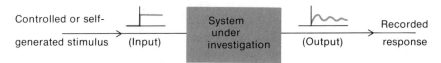

FIG. 2.1 *Block-diagram representation.*

inherently associated with the system. Intrapleural pressures during respiration and electrical signals from the heart are signals of this type. Either of these may, however, be considered "input" signals if there is reason to believe that these excitations cause other measurable effects. Intrapleural pressure, for example, may be considered an "input" that produces air flow during the respiratory cycle (Fig. 2.3a); electrical impulses from the myocardial cells may be considered the "input" that activates the heart muscle to produce cardiac flow and causes current flows that are detected as EKG signals (Fig. 2.3b).

In most situations, laboratory research consists of measuring input and output signals under controlled conditions. The most effective way to conduct such tests is to limit input excitations so that a single input-output relationship can be examined. Under these ideal circumstances, cause-and-effect relationships are much more easily established. Although

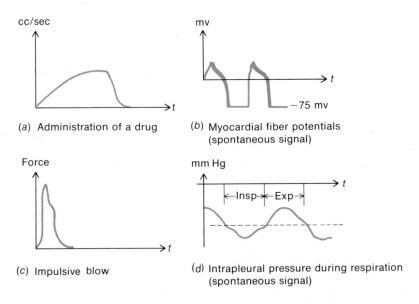

(a) Administration of a drug

(b) Myocardial fiber potentials (spontaneous signal)

(c) Impulsive blow

(d) Intrapleural pressure during respiration (spontaneous signal)

FIG. 2.2 *Typical signals associated with investigations of living systems.*

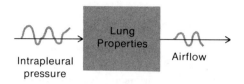

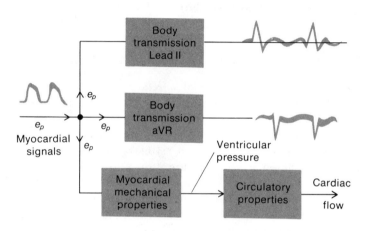

(b)   Heart-pacemaker activity
        as an input

FIG. 2.3   *Spontaneous signals as inputs.*

this ideal condition is not always attainable, it is closely approached in many situations. We shall start with the assumption that ideal conditions are closely realizable; we shall assume, therefore, that it will be possible to focus attention on single input-output relationships.

The block diagram of Fig. 2.1 indicates that there are three basic quantities involved in any investigation: the stimulus, the system, and the response. By defining any two of them, it should be possible to describe the third. Thus, if the stimulus and response are known or are measured variables, it should be possible to estimate the properties of the system; if the stimulus and the system properties are known, it should be possible to describe the form of the response. Both approaches are important in studies that involve living systems. The first includes the kind of problem most often encountered in laboratory research investigations. This may be considered the problem of systems analysis based upon stimulus-response studies. (Engineers often call this the identification problem.) The second approach includes problems that are most often encountered in design. This could be considered the systems synthesis (or design) approach. The ejection seat of the jet pilot or the reclining couch of the astronaut is an example of this aspect of systems studies. The thrust profile is the known stimulus; human acceleration endurance provides a limit on the acceptable response. The system must then be designed accordingly.* Both approaches to systems studies will be examined. The particular procedure used will depend upon the problem under discussion.

2.2 CHARACTERIZATION OF A SIMPLE PHYSIOLOGICAL SYSTEM

We start with an examination of a relatively simple situation where we desire to characterize the properties of an exposed trachea (Fig. 2.4). An animal is anesthetized and the pulmonary system temporarily paralyzed so that the air delivery can be controlled externally. The purpose of the investigation is to obtain an estimate of how flow $(\dot{q})$† through the trachea is related to the pressure drop (p_d) across the trachea. The instrumentation is arranged so that flow and pressure drop can be recorded instantaneously. The experimental setup shown in Fig. 2.4 is for illustrative purposes only; from a practical point of view a number of objections might logically be raised. One of them deals with the manner by which tracheal air flow is measured. The air delivered to the trachea

---

* Once a system has been "synthesized" (on paper or in the shop), it is ordinarily checked by an analysis of the input-output characteristics. In fact, one might say that a system is rarely ever directly synthesized. Rather, a system is postulated, analyzed, and then modified based upon the analysis. In effect, a "synthesized" system is "analyzed" into reality.

† Volume will generally be designated by the letter q; flow (which is described by the change of volume per unit time) will thus be designated by the notation $\Delta q/\Delta t$ or dq/dt or $\dot{q}$ (read cue dot). To see how volume change per unit time relates to flow, we use water flowing into a container as an illustration (see the following figure). The change in water level reflects the change in volume in the container; this is evidently related to the flow into the container or the flow from the pipe. In particular, the rate of change of volume in the container is a measure of the flow into the container.

Flow $= \dot{q} \longrightarrow$

Volume $= q$

Change of volume $= \Delta q =$ (flow) Δt

Rate of change of volume $= \dfrac{\Delta q}{\Delta t} =$ flow

To assign an appropriate sign for the flow term $\dot{q}$ we might consider flow into the container as positive; flow from the container would then be considered negative. In general, the sign considerations are arbitrary and are more easily established by inspection than by the use of rigid rules. This will become clear as various illustrative problems are examined.

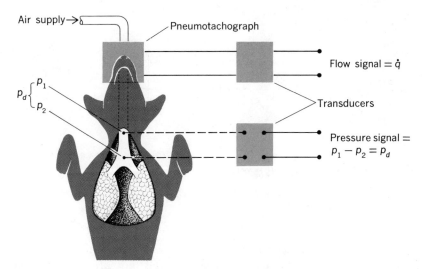

FIG. 2.4 *Signal acquisition for tracheal studies.*

is not at the same temperature as the air in the trachea. Under these circumstances, flow measurements (as determined by a pneumotachograph) do not necessarily reflect air flow in the trachea. This difficulty could be eliminated by including appropriate corrections or, alternatively, the delivered air could be preheated and saturated so that it is at body-temperature-standard-conditions (BTSC).

A more significant objection to the experimental setup is that flow and pressure drop do not specifically relate to the trachea. The characteristics of the trachea are under study, yet flow is measured at the mouth. In order to show that the latter measurement represents flow through the trachea, it would be necessary to show that the air is relatively incompressible for the pressure ranges used and that, within a reasonable approximation, tracheal flow and flow through the mouth are closely equal.

The realistic problems and limitations just mentioned emphasize that data may be (and, in general, are) acquired under less than ideal conditions. The measured data must then also be less than "perfect." Therefore the data must be viewed as an indication of trend rather than exact relationships.

With these views in mind, we now examine the signals obtained from the experimental arrangement shown in Fig. 2.4. The pressure drop $(p_1 - p_2)$ vs. time (t) signal is shown in Fig. 2.5a, and the flow $(\dot{q})$ vs. time (t) signal is shown in Fig. 2.5b. These signals define the operating characteristics of the system (the trachea) under the specified experi-

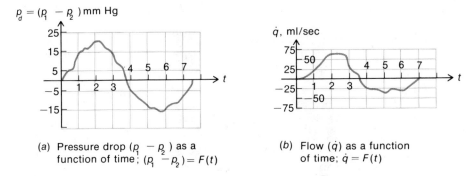

(a) Pressure drop $(p_1 - p_2)$ as a function of time; $(p_1 - p_2) = F(t)$

(b) Flow $(\dot{q})$ as a function of time; $\dot{q} = F(t)$

FIG. 2.5 *Pressure and flow signals obtained from the experimental arrangement shown in Fig. 2.4 (illustrative data).*

mental conditions.* In order to present characteristics in a more concise fashion (so that we can see how pressure and flow are related) time is eliminated as a parameter and pressure is plotted against flow, as in Fig. 2.6. The plot of Fig. 2.6 is the "graphical model" of the trachea or, as it may sometimes be called, the "graphical characteristics."

No new concepts have been developed in the foregoing discussion. Basically only terminology has been introduced, and the correlation of input-output data with the graphical model has been indicated for a particular system.

* These signals should be considered representative examples of a single breathing cycle. It is assumed that they typify the waveforms that would appear during steady respiration. Pressure drop is positive when p_1 is greater than p_2; flow is positive when it is directed from point 1 to point 2 (or when air is delivered from the air supply to the mouth).

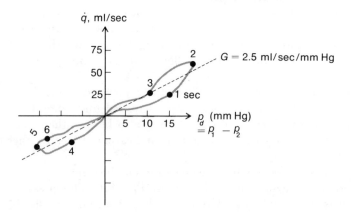

FIG. 2.6 *Pressure-flow characteristics of the trachea.*

If we desired, and were patient enough, we probably could derive a mathematical expression to fit the graphical characteristics shown in Fig. 2.6. This would give rise to *one* mathematical description or mathematical model of the trachea. At best, this would involve much tedious and time-consuming work and, in view of experimental limitations, the results might be questionable. A less "exact" though very much simpler approach would be to develop a very crude mathematical model by grossly approximating the graphical model. This prodecure would not be "right" but it cannot as yet be said to be completely "wrong," and it might give insight into the form of the true characteristics. The justification of the procedure is that it is simple and rapid and may provide insight—not necessarily that it is "right." To develop another, much simpler mathematical model, the graphical model, in the measured range, can be crudely represented by a straight line (shown as a dashed line in Fig. 2.6). With this gross approximation the mathematical model is written as

(2.1)   Flow is directly proportional to pressure drop

(2.2) $$\dot{q} = Gp_d = G(p_1 - p_2)$$

where G is a proportionality constant. Evidently G is the slope of the straight-line approximation in Fig. 2.6. (In this case, G is about 2.5 ml/s/mmHg.)

The mathematical model gives no more information than the graphical model; in fact, it often gives less information. However, the convenience it may afford very often compensates for the possible loss of information. To illustrate this point, we might go a step further with the last illustration and perform similar tests on the right and left bronchi to study distribution of air flow to the lungs. By performing such tests we obtain the graphical models of the main bronchi as shown in Fig. 2.7b and c (hypothetical data). From these data it should be possible to estimate the flow to both lungs for various given conditions. We should, for example, be able to estimate bronchial flows when p_1, p_3, and p_4 are specified; yet it is not at all clear how we could get this information from the graphical models.

As a specific case, let us assume that we require the various flows when p_1 is 790 mmHg, p_4 is 780 mmHg, and p_3 is 760 mmHg. It is to be noted that p_2 is not specified; yet p_2 is required in order to use the graphical characteristics of the tracheal and bronchial elements (all pressure drops in Figs. 2.6 and 2.7b and c depend upon p_2). The only way to use the graphical models is to employ a method of trial and error. A value for p_2 could be assumed; since p_1, p_3, and p_4 are given, $p_1 - p_2$, $p_2 - p_3$,

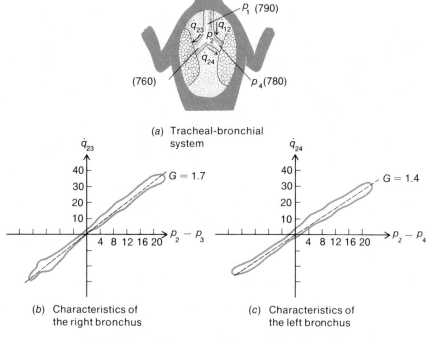

(a) Tracheal-bronchial system

(b) Characteristics of the right bronchus

(c) Characteristics of the left bronchus

FIG. 2.7 *Bronchial system and characteristics of bronchi.*

and $p_2 - p_4$ can be used to find $\dot{q}_{12}$, $\dot{q}_{23}$, and $\dot{q}_{24}$. We would then have to check to see if the sum of the bronchial flows ($\dot{q}_{23}$ and $\dot{q}_{24}$) is equal to the total tracheal flow ($\dot{q}_{12}$). That is, if the correct value for p_2 is chosen, the flows will be related by

$$(2.3)^* \qquad\qquad \dot{q}_{23} + \dot{q}_{24} = \dot{q}_{12}$$

As a specific "try," we assume that p_2 is 785 mmHg. Then

$$(2.4) \quad p_1 - p_2 = 790 - 785 = 5 \qquad \therefore \qquad 5 < \dot{q}_{12} < 15 \qquad \text{(Fig. 2.6)}$$

$$(2.5) \quad p_2 - p_3 = 785 - 760 = 25 \qquad \therefore \qquad \dot{q}_{23} \approx 32 \qquad \text{(Fig. 2.7b)}$$

$$(2.6) \quad p_2 - p_4 = 785 - 780 = 5 \qquad \therefore \qquad 6 < \dot{q}_{24} < 12 \qquad \text{(Fig. 2.7c)}$$

From Eqs. (2.5) and (2.6) we see that the sum of the bronchial flows is $38 < (\dot{q}_{23} + \dot{q}_{24}) < 44$. This total flow should be roughly equivalent to

---

* This flow equation should be self-evident by inspection. It states that flow to a junction is equal to flow from the junction and is essentially a restatement of the law of the conservation of mass. The corresponding law for electrical networks is Kirchhoff's law. A brief introduction to the development of circuit equations using Kirchhoff's law is given in Appendix A.

the tracheal flow ($\dot{q}_{12}$) as given by Eq. (2.4). As can be seen, the flow relations do not correspond; the equality required by Eq. (2.3) is not even approximately satisfied. The assumed value for p_2 was therefore incorrect. Another value for p_2 must be assumed and the calculations repeated. This is not an invalid procedure; it sometimes is fairly straightforward and rapid. Often, however, it is clumsy and tedious. It would be desirable to find some other technique to get at least some first estimate of an accurate value for p_2.

The use of the approximate mathematical model provides a very simple technique in this case. The graphical models are expressed as simple mathematical equations; the equations are then solved to obtain a first estimate for p_2. This limits p_2 to the vicinity of the "true" p_2. If more refined answers are required, at least we have a reasonable first estimate for p_2 and can then rapidly converge on an exact result by trial and error. To illustrate the technique described we first approximate the graphical models by straight lines as shown in Figs. 2.6 and 2.7b and c. The mathematical expressions for the lines are

(2.7) $$(p_1 - p_2)G_{12} = \dot{q}_{12}$$

(2.8) $$(p_2 - p_3)G_{23} = \dot{q}_{23}$$

(2.9) $$(p_2 - p_4)G_{24} = \dot{q}_{24}$$

where in all cases the G's are the slopes of the approximate straight lines and can be measured directly from the graphs. From Eqs. (2.7) to (2.9) it is now fairly easy to estimate p_2 if p_1, p_3, and p_4 are given, because by Eq. (2.3) we have

(2.10) $$(p_1 - p_2)G_{12} = (p_2 - p_3)G_{23} + (p_2 - p_4)G_{24}$$

Since p_2 is the only unknown, we can now solve for p_2:

(2.11) $$p_2(G_{23} + G_{24} + G_{12}) = G_{23}p_3 + G_{24}p_4 + G_{12}p_1$$

or

(2.12) $$p_2 = \frac{G_{23}p_3 + G_{24}p_4 + G_{12}p_1}{G_{23} + G_{24} + G_{12}}$$

By substituting known values and those obtained from the graphs, we get

(2.13) $$p_2 = \frac{1.7 \times 760 + 1.4 \times 780 + 2.5 \times 790}{1.7 + 1.4 + 2.5} \approx 778 \text{ mmHg}$$

With an approximate value for p_2 available, it is now possible to approximate the various flows by using the flow equations (2.7) to (2.9). In doing so we find

$$(2.14) \qquad \dot{q}_{12} \approx 30 \text{ ml/s}$$

$$(2.15) \qquad \dot{q}_{23} \approx 30.6 \text{ ml/s}$$

$$(2.16) \qquad \dot{q}_{24} \approx -2.8 \text{ ml/s}$$

There is a discrepancy here: $\dot{q}_{23} + \dot{q}_{24}$ is not exactly equal to $\dot{q}_{12}$. This is due to the approximations used for the calculated values of pressures and the general rounding off of numerical values. In this case, however, the results would be acceptable as the whole procedure is approximate and the discrepancy is relatively small. It is of interest to note that with the values chosen the flow in the left bronchus is negative. In this case, flow is from the left lung rather than to it. This is not obvious from the graphs and would probably have required a considerable search before this result would have been seen.

As pointed out above, if more exact answers are desired, the solved value for p_2 can be used as a first guess in Figs. 2.6 and 2.7, and a trial-and-error procedure can then be employed, as previously outlined. If one is interested in approximate values for the flow distribution, the ratio $\dot{q}_{23}/\dot{q}_{12}$ or $\dot{q}_{24}/\dot{q}_{12}$ can be formed by dividing Eq. (2.8) by (2.7) or (2.9) by (2.7). In any case it can be seen that the flow depends on the pressure and the system characteristics, which, in the approximation, are represented by G.

The preceding discussion provides a justification for "linearizing" graphical data when gross trends or approximate analyses are acceptable. The technique is not suggested as a panacea nor is it implied that the procedure will always yield valid results. It is contended, however, that when it can be used it helps to establish trends and allows the use of various computational and conceptual aids such as computers and analogs.

It will be seen later that linearization may not be so much an approximation as has been implied, particularly when used for small ranges of single variations. It will also be seen that "piecewise-linear approximations" (linear straight-line segments over small ranges) can be used to represent very nonlinear systems. Here, however, linearization is simply an expedient procedure to permit relatively easy calculations.

2.3 MODELS AND ANALOGS

The concepts of physical models and analog representations can be introduced here if the linearizing technique is accepted as a first-order approximation. However, it should be kept in mind that, although the concept is introduced on the basis of approximation, refined techniques often permit very close simulations of actual systems, particularly when analog and digital computers are used.

The basic idea behind the concept of simulation is that, if two systems have graphical similarity for certain aspects of behavior, either can substitute for the other in a particular phase of investigation. If the dimensions of both systems are the same, it is called a model simulation; if the dimensions are different, it is an analog simulation.

A proposed physical model of the trachea is shown in Fig. 2.8a; an electrical-analog representation is shown in Fig. 2.8b. The associated graphical characteristics and mathematical models of each are included. In both cases the graphical characteristics are closely represented by straight lines. If the straight-line representation is accepted for the tracheal characteristics, then either of these devices can be used to represent the trachea under the experimental conditions previously stipulated. This implies that, for a given pressure drop, the flow in the tube can be used to represent the flow in the trachea or as a measure of flow in the trachea. Similarly, if voltage and pressure are considered analogous variables, then, for a given voltage (pressure) drop, the current through the resistor can be used to represent the flow in the trachea (or, again, as a measure of tracheal flow).

The slope of the voltage-current curve is identified as the reciprocal of electrical resistance (see Fig. 2.8b). Reciprocal electrical resistance is generally designated as conductance. The slope of the pressure flow characteristics of the trachea can thus be considered the pneumatic conductance of the trachea. We shall examine this concept further in Chap. 3.

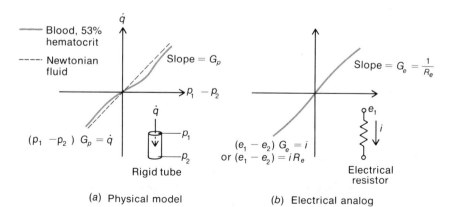

FIG. 2.8 *Physical model and electrical analog of the trachea.*

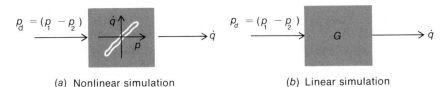

$$P_d = (P_1 - P_2)$$

(a) Nonlinear simulation   (b) Linear simulation

FIG. 2.9   *Block-diagram models of the trachea.*

The trachea may be represented in block-diagram form. A qualitative description of a system, using the block-diagram representation, was shown in Fig. 2.1. In order to include quantitative data the diagrams in Fig. 2.9a and b are often used. Figure 2.9a includes the total graphical characteristics of the system within the block (in this case the trachea). This representation may appear to be unwieldy and unnecessary. It will be seen later, however, that such "nonlinear" blocks may sometimes be simulated electrically. Then Fig. 2.9a could be considered an analog-computer simulation of the trachea. The pressure difference $(p_d = p_1 - p_2)$ would be a voltage signal delivered to a "computer block"; the block would "operate" on the input signal to deliver an output signal that would represent flow.

Figure 2.9b shows a linearized block-diagram simulation of the trachea. The symbol in the block represents the assumed mathematical properties of the system. As in the previous case, the notation in the box indicates the "operational effect" that the system has on the input signal. For the linearized representation of the trachea, the "operator" is the constant G. Since the linearized mathematical statement relating flow and pressure is

$$(2.2) \qquad\qquad \dot{q} = G(p_1 - p_2)$$

in this case "operates" implies simply multiplication. In later sections it will be seen that the term "operates" has broader significance.

One might well wonder why there is a need for a model or analog when the graphical characteristics are already available. In the case of the trachea, for example, the flow for any pressure drop can easily be determined by referring to the pressure-flow curve; for approximate mathematical calculations the linearized mathematical model can be used. Even for this simple case, however, the simulations can serve two useful purposes: One is the study of how a component of a system may affect the overall operation of a system; the other is to learn more about the system being studied by noting how its behavior differs from that of a known system. Both these applications will be discussed in the following sections.

2.4 SIMULATIONS USED FOR SYSTEMS STUDIES

From the discussion in Sec. 2.3 it is easy to see that each component of the tracheal-bronchial system (Fig. 2.7a) can be represented by a rigid tube or an electrical resistor. The simulation of the total system is evidently an assembly of the component elements. Figure 2.10 shows both the "pneumatic model" and an electrical analog of the system. The pressures needed to drive the pneumatic model are indicated by the symbols p_1, p_3, and p_4. In the electrical analog, these pressure excitations are represented by the voltage sources labeled e_1, e_3, and e_4. [Generally *constant*-voltage sources are represented by a battery symbol; variable, controlled-voltage sources are represented by the standard symbol shown. (See Appendix B.)] By making the proper measurements on these simulations, one should be able to predict approximately the various responses of the tracheal-bronchial system.

In order to be able to use either of these simulations, the physical components of the system must be appropriately specified. In the electrical system, for example, "just any resistance" cannot be used; the values of the resistors must be properly chosen. This introduces the problem of scaling (or calibration, as it might be called). Scaling problems are always encountered when simulations are used, and an introductory discussion here may prove worthwhile, for in this instance the situation is simple and not complicated by other considerations.

Scaling implies an appropriate choice of ratios between analogous

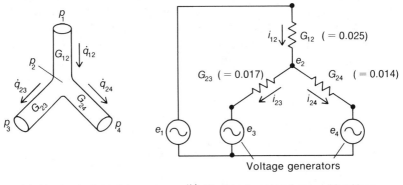

(a) Mechanical model   (b) Electrical analog:1 ma = 10 ml/sec

FIG. 2.10 *Analog of tracheal-bronchial system.*

variables. If we use the electrical analog of the trachea as an example, voltage and pressure are analogous and current and flow are analogous. (See Figs. 2.6 and 2.8.) The relation governing the electrical system is very simply

(2.17) $e = iR_e$ e in volts (V), i in amperes (A), R_e in ohms (Ω)

For the physiological system we have

(2.18) $p = \dot{q}R_p{}^*$ p in mmHg, $\dot{q}$ in ml/s, R_p in mmHg-s/ml

If we use a one-to-one correspondence between variables, then a 1-V drop in the electrical system will represent 1-mmHg pressure drop in the physiological system. Similarly, 1 A represents a flow of 1 ml/s, and the ohmic resistance (R_e) must be equal to the flow resistance (R_p). This would be the most straightforward simulation. We note, however, that for the trachea the numerical value of the flow conductance is 2.5 (see Fig. 2.6). We should therefore use an electrical resistance of 1/2.5 or 0.4 Ω. It is possible to obtain resistors of such low values, but they are not common nor are they practical in this case as the simulation resistor would probably be lower than the wires and the joint connections between resistors (as at junction e_2 in Fig. 2.10). For practical reasons, then, it would be desirable to use a much higher electrical resistance. A more reasonable value for R_e would be about 400 Ω, that is, about 1,000 times the flow resistance.

To understand how this might affect analogous quantities, we have but to realize that if flow resistance is increased by a factor of 1,000 then current flow per unit volt is reduced by a factor of 1,000. This means, therefore, if we use a 400-Ω resistor, each *milli*ampere (mA) of current represents 1 ml/s when 1 V is considered equivalent to 1 mmHg pressure. Alternatively, if 1,000 V [1 kilovolt (kV)] is considered equivalent to 1 mmHg then 1 A of current is equivalent to 1 ml/s. These ideas are shown diagrammatically in Fig. 2.11 and are expressed in equation form by dividing Eqs. (2.17) and (2.18).

(2.19) $$\frac{e}{p} = \frac{i}{\dot{q}} \frac{R_e}{R_p}$$

If R_e is 1,000 times R_p, then Eq. (2.19) becomes

(2.20) $$\frac{e}{p} = \frac{i}{\dot{q}} \times 1{,}000$$

If 1 V is equivalent to 1 mmHg, then e/p is unity and 1 A (i) represents

* Where R_p is defined simply as $1/G_p$, that is, the reciprocal slope. R_p could be considered the tracheal resistance to air flow.

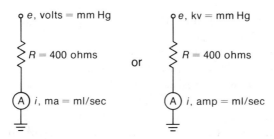

FIG. 2.11 *Scaled values in an analog simulation.*

a flow of 1,000 ml/s, or 1 A is equivalent to 1 ml/s. If 1 A is equivalent to a flow of 1 ml/s, then $i/\dot{q}$ is unity and 1 V represents $(1/1000)$ mmHg, or 1 kV is equivalent to 1 mmHg.

The simulation of the tracheal-bronchial system shown in Fig. 2.10b has been scaled so that the flow-to-current ratio $(\dot{q}/i)$ is 10,000:1. By using this value we can avoid the need for low-resistance values and excessive-voltage sources. The values for the electrical resistors are 100 times greater than the corresponding physiological values $[R_e/R_p$ in Eq. (2.19) is 100]. It is assumed that 0.01V, or 10 mV, is equivalent to 1 mmHg. The ratio e/p in Eq. (2.19) is therefore 1/100. With these values the ratio of $\dot{q}/i$ becomes 10,000:1; that is, 1/10 mA [100 microamperes (μA)] of current is equivalent to a flow of 1 ml/s. Once we have chosen appropriate values for the components, we can then use the simulation to study the responses of the system to various imposed stimuli. If, as an example, we choose to make p_1 790 mmHg, p_3 760 mmHg, and p_4 780 mmHg (see the problem of Sec. 2.2), then e_1, e_3, and e_4 in Fig. 2.10b would be adjusted to be 7.90, 7.60, and 7.80 V, respectively. The flows in the trachea and bronchi can now be directly determined by current-flow measurements in the corresponding branches. This study would, in essence, represent a simultaneous solution of Eqs. (2.3) and (2.7) to (2.9). The nodal voltage e_2 would represent the p_2, and the branch currents would represent the flows.

By using this same simulation we can also investigate more sophisticated situations. We could, for example, propose that p_3 and p_4 should be continuously variable (as in breathing). For this situation we could vary e_3 and e_4 in the manner prescribed for the variations of p_3 and p_4. Once again flows could be directly measured. As a more complex situation, we might also ask what effect would be expected if there was a continuous variation of the tracheal resistance (as might occur during a breathing cycle). To include this aspect, we would replace the constant resistor G_{12} with a variable resistor and then vary the resistor in a prescribed manner while e_3 and e_4 were simultaneously varied. In this case, too, we could still determine the various branch flows by measuring the proper branch currents.

A mechanical simulation (e.g., with a rigid tube or possibly with a flexible tube) presents essentially the same problem as the electrical simulation. The tube flow resistance must be adjusted so that it is equal to or proportional to the resistance of the physiological system. By appropriate choice of parameters (diameter, length, roughness) or by the use of a resistor valve the desired equivalency can be obtained.

The question of which simulation is better cannot be answered directly. The electrical-analog simulation is ordinarily easier to construct, as variable components are easily obtained; it is sometimes difficult, however, to achieve sufficiently characteristic behavior. The model simulation generally maintains visual correspondence between systems, and material properties can sometimes be chosen so that behavioral correspondence is close. However, the practical difficulties of construction often make this approach less desirable.

The simulations shown in Fig. 2.10 are fairly simple but they illustrate the basic concepts involved. More complicated systems require more elegant representations, such as computer simulations. The computer simulation is different from simple model and analog simulations only in that, in the former, the system and computer have mathematical similarity whereas in the latter the system and simulation have both mathematical and component similarity.

The form that a computer simulation should take can be indicated by constructing the appropriate block diagram for the system. Figure 2.12c shows a total assembly for the system. It should be noted that this diagram includes an operational symbol for addition or subtraction (see Fig. 2.12a). This symbol represents a device delivering an output signal that is the difference between two input signals. The utility of the symbol is shown in Fig. 2.12b. The latter figure is essentially that of Fig. 2.9b redrawn so that p_1 and p_2 can be delivered as two separate signals. The difference between these signals is delivered to block G where multiplication by G is performed. The computer simulation shown in Fig. 2.12c consists of an appropriate assembly of the operational symbols shown in Fig. 2.12a and b. Each of the operational symbols and blocks in Fig. 2.12 may be replaced by an electrical device that will perform the required operation.* These devices will deliver voltages that will represent

---

* See Appendix A. Steps 2 to 29 in particular include some background related to the realization of a few simple mathematical operations. Introductory information relating block diagrams and analog-computer elements can also be found in Ref. 5, chaps. 6–8. The author has not had the opportunity to study this text completely; on the basis of a brief examination, however, it appears that much of the material presented there would serve as good supplementary reading to reinforce the concepts of systems as applied to nonbiological fields.

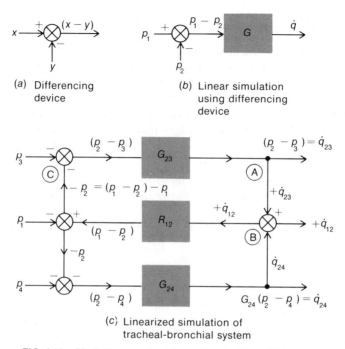

(a) Differencing
device

(b) Linear simulation
using differencing
device

(c) Linearized simulation of
tracheal-bronchial system

FIG. 2.12 *Block-diagram simulation of tracheal-bronchial system.*

the numerical values of the quantities shown. At this stage of development, however, we do not consider a block diagram to be an electrical circuit. For the purposes of our discussions, it will be sufficient if the diagram is considered simply a schematic representation of the pertinent equations governing the operation of a system. In this sense it serves the same function as the electrical analog of Fig. 2.10b (see, however, Prob. 2.6).

The computer simulation and the analog-circuit representation are significantly different in one respect: Whereas the latter has physical and mathematical similarity to the actual (tracheal-bronchial) system, the former has only mathematical similarity. In the analog circuit, for example, the branch-current flows actually represent the air flows through corresponding branches of the physical system; the sum of the currents leaving a junction of branches (node e_2, for example) is equal to the sum of currents entering the junction, comparable to what one would expect in the physical system. In the block diagram this component similarity does not exist. The branches all represent numerical values, not physical quantities. The fact that the numerical values may be determined by voltage measurements is incidental. Summations at a node take place only when such an operation is indicated by the presence of a "summer." If no operational process is indicated, the numerical values of information leaving a node are equal to the information entering the node. Thus all branches in Fig. 2.12 connected to the node labeled A have the value $\dot{q}_{23}$. On the other hand, the value of all branches leaving a summer is equal to the sum of the entering quantities. Thus at the summer labeled B (Fig. 2.12) all branches leaving the summer are designated as $\dot{q}_{12}$, where $\dot{q}_{12}$ is taken to be equal to the sum of the values entering the summer (which in this case is $\dot{q}_{23} + \dot{q}_{24}$; the signs adjacent to the input branches indicate the sign sense of the incoming signals).

Although this is a very brief introductory discussion, it should be possible for the reader now to check the diagram of Fig. 2.12c and see that this diagram is simply a schematic representation of Eqs. (2.3) and (2.7) to (2.9). Thus node B represents the operation indicated by Eq. (2.3); Eq. (2.8) is represented by the diagram between nodes C and A, etc. When appropriate voltage signals are applied at p_1, p_3, and p_4, the system will deliver voltages that represent the branch flows as indicated.

As mentioned earlier, one cannot say which simulation is better. The aim of simulation is to be able to use it as a "study substitute" for a real system; the more versatile the simulation and the closer the representation, the more useful it can be. Mathematical representations are sometimes the most versatile as mathematical solutions are general* and tend to indicate overall trends. However, the closer the realization, the more complex the mathematics becomes; furthermore, in complex situations, the solutions may be impossible to obtain (except by approximation). Although the model and analog simulations do not give general answers, "answers" for specific situations are ordinarily possible. Correspondence between system and simulation is often maintained so that interpretation is relatively straightforward. These advantages are retained to a large degree even when the simulation is complex and representation approaches the real system. The choice between analog vs. model simulations often depends on individual preference, although, in general, electrical simulations are more versatile because parameters are more easily and economically varied to suit particular needs.

2.5   UTILITY OF MODELS AND ANALOGS IN RESEARCH STUDIES

One of the advantages of the model concept is that it can be used as an aid in systems studies that involve the interaction between components

---

* Mathematical solutions can be completely "algebraic" without the need of numerical values. Numbers can be assigned in the final solution to cover a large variety of possible changes in variables. In model and analog simulations the choice of components and parameter variations is limited by practical considerations.

of the system. This application was briefly described in the preceding section. Here we examine the concept of the model as a research tool. Although this aspect of modeling may seem self-evident, it is worthwhile to point out just how the model is used in this capacity. We do this by approaching the question backward and first ask what would be expected of a "perfect model." Evidently such a model would have identical characteristics with the real system under all operating conditions. To build the "perfect model," then, one must know everything about the real system. In striving toward a perfect-model representation one is thus forced to examine systematically the gross aspects and then the more subtle properties of the system. This implies that a model cannot be developed by analysis alone. The process must consist of both analysis and experimental work, each supplementing the other. By examining the *difference* between a postulated model and the real system, the understood portion of the real system is, in effect, "subtracted out" and attention can then be focused on the more subtle properties of the system.

It is not easy at this point to illustrate the comments of the last paragraphs by example. We can, however, use the model derived for the trachea at least to justify broadly the views presented. It will be recalled that a graphical model of the trachea (repeated in Fig. 2.13a) was first derived from experimental data acquired in the laboratory. It would be presumptuous to assume that these characteristics are truly representative. The uncertainty of the experimental data, recording difficulties, etc., should cause the observer to be wary of reading too much into the data. If a model representation is to be attempted, it would be wise to simulate only the gross aspects of these characteristics. A straight-line model (Fig. 2.13b) or possibly a piecewise-linear model (Fig. 2.13c) would probably be adequate under these circumstances. Without worry-

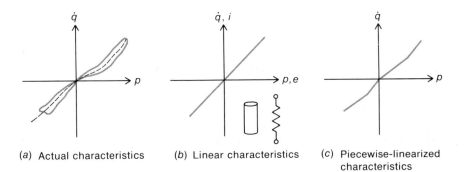

(a) Actual characteristics   (b) Linear characteristics   (c) Piecewise-linearized characteristics

FIG. 2.13   *Tracheal approximations.*

ing about parameter values, we can examine the significance of the models in Fig. 2.13. In Fig. 2.13b the conductance (G) is considered constant throughout the ranges of pressures used. If the corresponding mechanical and electrical models are deemed representative, we might ask what governs the conductance of a pipe or a wire resistance. Elementary physics and intuition indicate that these conductances increase with flow area but decrease with length. As an approximate mathematical relation, we write

(2.21) $$G_{\text{elect}} \propto \frac{A}{L} \propto \frac{D^2}{L}$$

(2.22) $$G_{\text{pneum}} \propto \frac{D^4}{L}$$

In both cases the conductance depends at least on diameter and length. Regardless of which of these relations is considered representative of the system under study, both state that if conductance is to remain constant one might expect, at least as a first estimate, that length and diameter are relatively constant. The actual graphical characteristics* seem to indicate that the conductance (G) varies with pressure. This would then imply that the diameter and/or length also varies with pressure. The question that must now be resolved is: Are the curved tracheal characteristics due to limitations of instruments and instrument techniques or are changes in tracheal dimensions responsible for the curved characteristics? To determine which of these viewpoints is closer to the truth, further experimentation is necessary. The criterion, which can be used to establish which viewpoint is more valid, would be to measure tracheal diameter simultaneously with flow and pressure measurements.†

Although the above is not a profound example, it indicates that the model can guide the next phase of investigation. We now go further and show, at least heuristically, how the experimental data can indicate a need for model revision. If the experiment discussed above were performed, it would show that there is, indeed, a diameter change in the trachea as the pressure drop changes. This would imply that the model in Fig. 2.13c is probably a closer representation than the straight-line model. This, however, creates a dilemma. If area changes with pressures, then the actual system has "stretchability," i.e., elastic properties.

---

* And the piecewise-linear model of Fig. 2.13c.

† From the above discussion, a careful experimenter would probably also look for a means of measuring "tracheal length" so that these data could be utilized to establish more definitively the relative significance of each change. There is experimental evidence that tracheal dimensions do change. Some animal tests have been performed and data have been published; see Ref. 4.

These properties do not show up in any of the simulations; that is, they do not show the possibility of storage. More precisely, the models indicate that the input and output flow must always be equal. For the actual system, however, some of the input flow may fill the stretched element and some may pass through. Input and output flow may thus not always be equal. The experimental data have therefore simultaneously cleared up one problem and introduced another. The model as it stands is obviously incomplete; another effect must be included. We cannot, as yet, include this effect with our present level of background. However, it is clear that to modify the trachea model a new "model element" must be developed to account for the phenomenon of stretchability. This element will be discussed in a later section.

To conclude this section, we summarize some of the ideas presented. Although much space was devoted to a discussion of simulations, the purpose here was not to develop models and analogs for their own sake but rather to provide a technique to gain insight and understanding of system functions. The sequence employed is generally to obtain input-output data for the system under study. From these data, graphical and mathematical models may be derived. Other devices or systems with similar graphical and mathematical models can then be used as model or analog simulations. The simulations are useful when interactions and coupling between systems are to be examined. More important, they indicate not only what is known about a system but also what is unknown.

SUPPLEMENTARY DISCUSSION

Problem 1.3 dealt with the model representation of the mitral valve of the heart. In view of the association of the mechanical variables pressure and flow with the electrical variables voltage and current and in view of the implications of the questions posed in Prob. 1.3, it should now be evident the electrical diode may serve as an analog of the mitral valve. Here we shall review and expand upon some of the ideas introduced by Prob. 1.3.

The pressure-flow characteristics of the mitral valve may be constructed from simultaneous recordings of flow through the valve and pressure across the valve. The latter information is readily available from pressure recordings in the left atrium and the left ventricle. A representative set of such curves is shown in Fig. 2.14. From these curves, it is possible to obtain a plot of the pressure drop across the valve $(p_a - p_v)$ vs. time. Such a plot is shown in Fig. 2.14b.

It is somewhat more of a problem to obtain the flow through the valve. However, time recordings of ventricular volume are available, and flow information can be obtained from them. A representative plot of ventricular volume is shown in Fig. 2.15a. During the period of aortic closure, blood enters the ventricle but no blood is removed from the ventricle; during this period, therefore, changes

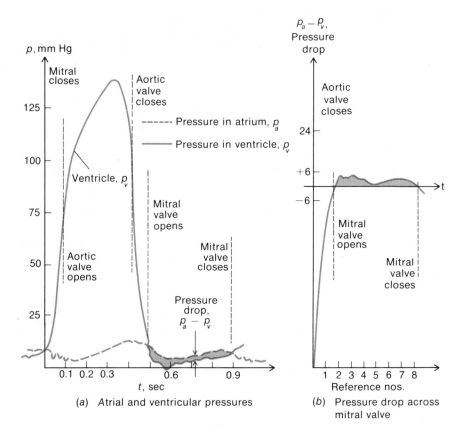

(a)   Atrial and ventricular pressures

(b)   Pressure drop across mitral valve

FIG. 2.14

of ventricular volume must reflect flow into the ventricle (see the first footnote on page 14). The flow through the mitral valve can thus be estimated by assuming

(2.23)   Flow = ventricular volume change per unit time

or

(2.24) $$\dot{q} = \frac{dq_v}{dt}$$

where q_v is the ventricular volume at any time t. The instantaneous slope of the ventricular volume vs. time curve can thus be used to construct the flow vs. time curve for the valve. This curve is shown in Fig. 2.15b. (A smooth curve was drawn between discrete points; there would be more irregularity if more points were used.)

The pressure-flow characteristics of the valve can now be constructed by

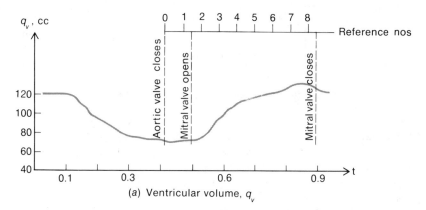

(a) Ventricular volume, q_v

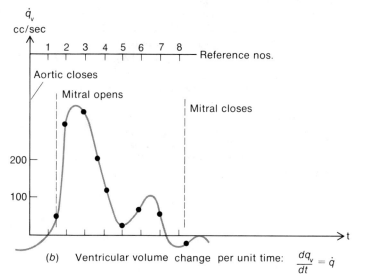

(b)   Ventricular volume change per unit time: $\dfrac{dq_v}{dt} = \dot{q}$

FIG. 2.15

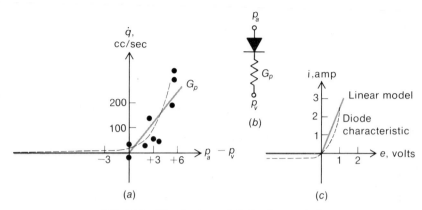

FIG. 2.16 *Mitral valve and diode characteristics.*

plotting simultaneous values of $p_a - p_v$ and $\dot{q}$ taken from Figs. 2.14b and 2.15b. Such a plot is shown in Fig. 2.16a; the dashed curve was drawn as an "eye approximation" through the plotted points.*

Even in this approximate form, the basic property of the valve shows up rather clearly. Flow takes place only in the section of the plot where $p_a - p_v$ is positive. When the pressure drop is negative there is practically no flow through the valve. This is significantly different from the characteristics of the pipes and tubes previously discussed, where flow could be either positive or negative. This unidirectional property of the mitral valve is found in many other devices, the most familiar of which are the mechanical check valve and the electrical diode. A diagrammatic representation of the equivalency of these devices is shown in Fig. 2.17; the mitral valve and the check valve permit flow in the direction shown

* There are mathematical techniques to locate a curve through the points with greater accuracy. This would be rather meaningless here, however, as these points were obtained for part of *one* cardiac cycle. For more accurate representation, the procedure should be repeated over many cycles, and then curve-fitting techniques for a larger number of points could be employed. For the purposes of illustration, however, it will be assumed that the construction in Fig. 2.16 is valid.

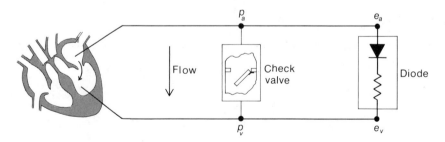

FIG. 2.17 *Mechanical and electrical analog of mitral valve.*

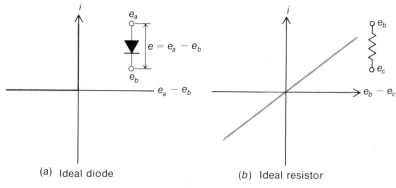

(a) Ideal diode   (b) Ideal resistor

FIG. 2.18

when $p_a > p_v$ but flow ceases when $p_a < p_v$. In a similar manner, the diode permits current flow when $e_a > e_v$; flow is reduced practically to zero when the voltage is reversed. The graphical characteristics of the diode are shown in Fig. 2.16c. The correspondence between the valve and the diode is obvious by inspection; flow and current are evidently analogous quantities, as are pressure and voltage.

From the graphical characteristics of the valve and the diode it can be seen that both devices have resistance when they permit flow (or conduct). This characteristic is more pronounced in a straight-line representation of the curves, as shown by the solid lines in Fig. 2.16a and c. The positive slope of pressure-flow or voltage-current characteristics may be considered an indication of the inherent resistance of these devices. The diode (or check valve) property and the associated resistance of such devices are sometimes represented by the schematic diagrams shown in Figs. 2.16b and 2.17. The triangular arrow and bar represent a "perfect" or "ideal" valve (or diode). Such a device would have no internal resistance; its graphical characteristics would be that shown in Fig. 2.18a. When the pressure (voltage) across the device is negative, no conduction takes place; when the pressure (voltage) exceeds zero, flow takes place freely in any amount delivered with no associated flow resistance. The resistor associated with the ideal diode represents the internal opposition to flow that would be generated by a practical device (the curves in Fig. 2.18a and b combine to give the solid curve shown in Fig. 2.16c).

It is easy to see how the diode might be used to simulate some pathological conditions of a mitral valve. If a normal mitral valve is represented by the diagram in Fig. 2.16b, additional resistors can be used to simulate a stenotic valve (a valve that cannot fully open) or a leaky valve (one that cannot close completely) or a combination of both. Figure 2.19 shows three graphical models of valves with an associated circuit analog. Model A is that of a normal valve. What pathological conditions would B and C represent? What would be the graphical and circuit representations for a leaky, stenotic valve?

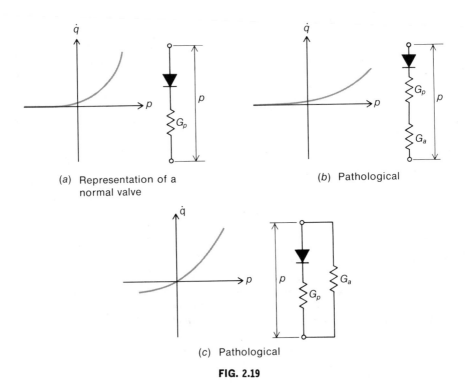

(a) Representation of a
     normal valve

(b) Pathological

(c) Pathological

FIG. 2.19

Problems

2.1 A diagnostic test on the kidney system is performed by steadily injecting inulin (a dye) until the blood concentration reaches some fixed level. A block diagram to represent the system is shown in Fig. P2.1.

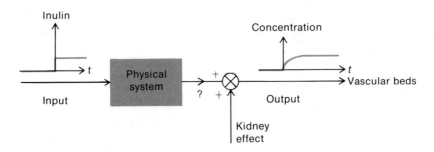

FIG. P2.1

(a) Does the input signal represent the weight of dye delivered or the rate of dye delivery?

(b) What must the "kidney effect" be in order for the concentration to reach a fixed level?

(c) If the kidney operates at reduced efficiency, how is the concentration level affected? Draw the resultant response curve.

(d) If the kidney does not operate at all, what will be the shape of the blood-concentration curve?

2.2 Pupillary response to light flashes is to be determined. Propose a block diagram that describes the experiment to be performed. Show a plot of the input as a single flash of light and a plot of pupillary response if the pupil diameter first increases and then decreases relatively slowly. Show the block diagram when a series of flashes are applied. Include input and output waveforms.

2.3 Three alveoli surround a pulmonary capillary arteriole, as shown in Fig. P2.3a. The flow through the capillary is fast enough so that a steady supply of O_2-deficient blood is always adjacent to the alveolar wall. The flow of O_2 to the blood is governed by the concentration gradient between O_2 in the alveoli and capillary (the concentration is related to the partial pressures). The gradient-flow characteristics are shown in Fig. P2.3b (hypothetical data).

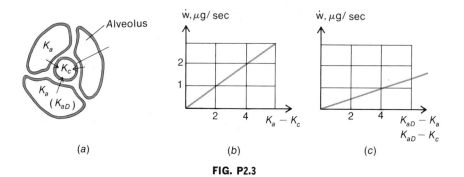

FIG. P2.3

(a) If the concentrations in the alveoli are the same, how would you obtain the flow to the capillary?

(b) If one of the alveoli is slightly defective (or the airways to it are defective) the O_2 concentration in this alveolus may be reduced (K_{aD}). Then O_2 might diffuse from one alveolus to another through the alveolar walls and then to the capillary. If the diffusion between alveoli is described by Fig. P2.3c, by what factor will the overall O_2 flow to the capillary be reduced? An analog circuit of this situation will prove helpful.

2.4 Figure P2.4 is a simplified representation of the upper respiratory tract. The pressures at both bronchi are the same ($p_{bR} = p_{bL}$) and are assumed to vary according to the plot shown. Plot the flow from each lung and the total flow from the mouth. If the bronchial resistances are doubled, what would be the effect on flow? If the tracheal resistance is doubled, how would this affect flow? If the right bronchus has half the resistance of the left bronchus but p_b at the right bronchus is twice that at the left bronchus, will the total flow be affected?

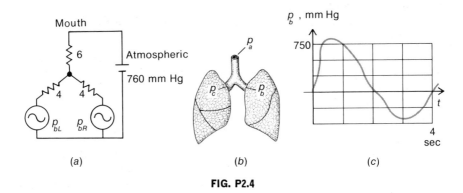

(a)   (b)   (c)

FIG. P2.4

2.5   Figure P2.5 shows a segment of a typical capillary network.

(a) Draw a circuit analog of the system (see Appendix A, step 12).

(b) If the pressures at points A, B, C, and D are known and the resistances of all branches are assumed equal, write the three nodal equations needed to determine the pressures at the unknown nodes.

(c) Determine the pressures at the unknown nodes in terms of p_A, p_B, p_C, and p_D.

(d) From the directions of flow indicated in the artery and the vein, it is evident $p_A > p_B$ and $p_D < p_C$; it is evident, also, that p_A and p_B are both greater than p_D and p_C. With these conditions, is there a resistance condition possible such that the identified capillary tube will not conduct flow?

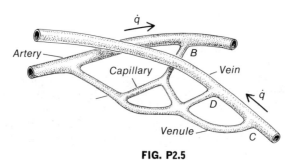

FIG. P2.5

2.6   The analog-computer circuits in Fig. P2.6 perform the operations indicated. (Note the values of the resistors; see also Appendix A.)

(a) Develop a computer configuration that can be used to solve the equation $\dot{q} = 20(p_1 - p_2)$.

(b) Develop a computer configuration that can be used to solve the equation $\dot{q}_1 = \dot{q}_2 + \dot{q}_3$.

(c) Develop a computer configuration to simulate the block diagram of Fig. 2.12.

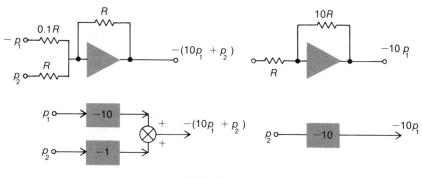

FIG. P2.6

REFERENCES

1. Comroe, J. H., Jr., R. E. Forster II, A. B. DuBois, W. A. Briscoe, and E. Carlsen: "The Lung—Clinical Physiology and Pulmonary Function Tests," 2d ed., The Year Book Medical Publishers, Inc., Chicago, 1962.
2. Lueg, R. E., and E. A. Reinhard: Circuit Analysis Techniques, in "Basic Electric Circuits," chap. 2, International Textbook Company, Scranton, Pa., 1967.
3. Angus, R. B., Jr.: "Electrical Engineering Fundamentals," chap. 5 (deals with circuit analysis), Addison-Wesley Publishing Company, Inc., Reading, Mass., 1961.
4. Kilburn, Kaye H.: Dimensional Changes of Bronchi in Apnic Dogs to Airway Pressure, Gases and Drugs, *J. Appl. Physiol.*, March, 1960.
5. Shearer, J. L., A. T. Murphy, and H. H. Richardson: "Introduction to System Dynamics," Addison-Wesley Publishing Company, Inc., Reading, Mass., 1967.
6. Milhorn, H. T., Jr.: "The Application of Control Theory to Physiological Systems" (analog computer programming), chap. 5, W. B. Saunders Company, Philadelphia, 1966.

CHAPTER THREE

SYSTEM PROPERTIES: RESISTANCE

3.1 INTRODUCTION

In the preceding chapter, we tacitly assumed that the trachea had flow-resistive properties. This assumption was based not upon any direct development or definitions but rather upon a comparison with an analogous system. The characteristics of the trachea were compared with those of a known "resistive" element; since there appeared to be some correspondence between the characteristics, it was concluded that the trachea also had resistive properties. This defines resistance by a "back-door approach," and it may lead to misinterpretation and confusion. It is almost always possible to compare graphical characteristics of a system with those of a "resistive" element. If only this criterion were used, all system properties could be considered resistive. If we note, for example, that the force-displacement characteristics of a spring are essentially linear (Hooke's law), we might consider a spring a "mechanical resistor"; a spring has the same characteristics as an electrical resistor if a force-current and displacement-voltage analog is used. This does not seem correct intuitively, particularly from a physical point of view. Mechanical resistance is associated with frictional effects and energy loss by generation of heat. Springs are not used to develop friction or heat; they are used to absorb energy (e.g., shock absorbers). Before an attempt to attribute any specific property to a system, one should first try to decide qualitatively what the property implies and then what parameters govern this property. In this chapter and the next we shall examine the relationships between the two, general-system properties and the associated graphical characteristics.

3.2 THE RESISTIVE PROPERTY

The concept of resistance appears in many fields although the parameters used to define the term vary from field to field. There is a thread of consistency associated with most of these definitions and we shall ultimately include this element of consistency in a generalized interpretation of the resistive property. At this point, however, we begin by examining the specialized concept of resistance as associated with energy or power loss. In many physical systems, resistance is essentially synonymous with friction, and it indicates energy (or power) loss in the system because of irretrievable heat loss. This implies that in fluid systems the power delivered to the inlet of the system is greater than that delivered from the outlet. If we use a rigid pipe as an example (Fig. 3.1), the power (energy per unit time) delivered to the pipe is the product of the variables, "flow" ($\dot{q}$) and "inlet pressure" (p_1),* and the power delivered from the

---

* This is easily confirmed by the use of "unit analysis." Pressure is defined by force per unit area; area is the product of two length dimensions. Therefore, pressure is defined by force/(length)². Similarly, flow is defined by volume per second or (length)³/s. The product is thus

$$\frac{\text{Force}}{(\text{Length})^2} \times \frac{(\text{length})^3}{s} = \frac{(\text{force})(\text{length})}{s}$$

but (force) $\times$ (length) is work and work per unit time is power. Therefore $p \times \dot{q}$ = power. A good discussion of dimensions and units is given in Ref. 1. See also Refs. 2 and 3.

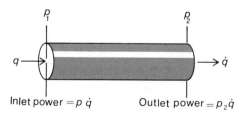

FIG. 3.1 *Power loss in rigid pipe.*

pipe is the product of flow and outlet pressure (p_2). The difference between the power supplied and the power delivered is evidently the transmission loss. Therefore,

$$(3.1) \qquad \text{Power loss} = p_1\dot{q} - p_2\dot{q} = (p_1 - p_2)\dot{q}$$

Thus we see that if a pressure drop exists between inlet and outlet there must be a power loss. Since there must be a pressure drop across the pipe to maintain a flow, there must also be a power loss in the process of fluid transmission. This loss is the energy required to overcome the frictional effects in the pipe or what may be called the flow resistance. It is clear now that, on the basis of this definition, the trachea has resistive properties. If we examine the original data for the trachea, even without linearization, the data seem to indicate a pressure drop must exist if flow is to be maintained and this drop must be increased if flow is to be increased. These are the characteristics of a system that has resistance.

The discussion in the preceding paragraph established one conceptual description of resistance: Resistance implies power loss; the p-$\dot{q}$ characteristics reflect this property. Nothing has yet been said about a quantitative description of resistance.* A quantitative definition of resistance should relate to the characteristic curves. If the curve is a straight line starting at the origin (as curve B in Fig. 3.2) the task is straightforward. When p and $\dot{q}$ are linearly related, the curve is defined by its slope. That is, for a straight-line p-$\dot{q}$ characteristic, the equation can be written

$$(3.2) \qquad \dot{q} = mp$$

* For some applications the qualitative description of resistance is completely adequate, particularly if the p-$\dot{q}$ relationship of a system is available. The curve tells all that is needed about the pressure-flow relationship. In Fig. 3.2, for example, it is easy to see that, for any given flow, the pressure drop in C always exceeds that of B. The loss (and hence the resistance) in C is thus always greater than that in B. Similar statements can be made about tubes A and B. In fluid studies, one seldom sees a specific definition of resistance. Instead, the concept of friction or head loss or drag is used. In physiological studies of the circulatory and pulmonary systems, however, resistance is a specifically defined concept.

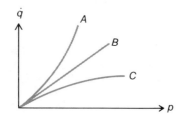

FIG. 3.2 *The p-$\dot{q}$ characteristics of flow-resistive systems.*

where m is the slope of the characteristic curve. Since the slope defines the curve, it can also be used to define the resistive property of the system. Relatively speaking, a large slope indicates a small pressure drop; therefore, a large slope also implies relatively small power loss or resistance. Consequently, resistance and slope are reciprocally related, and a meaningful definition for resistance is

$$(3.3) \qquad R = \frac{1}{\text{slope of characteristic curve}} = \frac{1}{G}^{*}$$

For linear characteristics, then, resistance as defined by Eq. (3.3) is essentially equivalent to a description of the p-$\dot{q}$ characteristics and can be used as a substitute in analysis. This implies that, instead of analyzing systems by graphical means, one can use the equivalent of the graph (G, in this case) and analyze mathematically.

3.3 LINEAR–RESISTANCE ANALYSIS

Before we begin a discussion of how flow resistance is defined when the p-$\dot{q}$ characteristics are not linear, we shall examine how the concept can be useful when a number of theoretically linear-resistive elements form a system. In Fig. 3.3b, c, and d three different tubular networks are shown; each is composed of three "separate" tubes (A, B, C) of known p-$\dot{q}$ characteristics. For simplicity, it is assumed that in all the networks the three tubes have the characteristics shown in Fig. 3.3a. As a particular problem we wish to know what pressure drop to expect in each case if a flow is specified. For a group of such simple networks and characteristic curves, the problem could be handled by using only the curves. For more complex networks this procedure would be clumsy and tedious even with linear characteristics. However, an analytic approach is straightforward even when the networks become more complex. To use the analytic approach we use the concept of resistance to relate pressure and flow. For the case shown in Fig. 3.3b we have

$$(3.4) \qquad p_1 - p_2 = \dot{q}_A R_A = p_A$$

$$(3.5) \qquad p_2 - p_3 = \dot{q}_B R_B = p_B$$

$$(3.6) \qquad p_3 - p_4 = \dot{q}_C R_C = p_C$$

---

* Since resistance and conductance (G) were previously defined as reciprocal functions (see Sec. 2.3), evidently large slope implies large conductance. Conductance can therefore be defined by

$$G = \text{slope of characteristic curve}$$

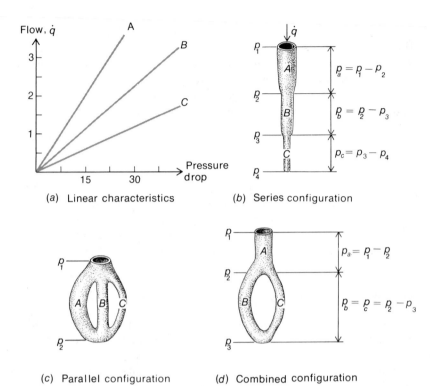

(a) Linear characteristics   (b) Series configuration

(c) Parallel configuration   (d) Combined configuration

FIG. 3.3 *Different configurations of tubular networks.*

The sum of the pressure drops is the total drop

$$(3.7) \qquad p_1 - p_4 = p_A + p_B + p_C = \dot{q}_A R_A + \dot{q}_B R_B + \dot{q}_C R_C$$

Finally, since $\dot{q}_A = \dot{q}_B = \dot{q}_C$ (the flow is the same in all tubes),

$$(3.8) \qquad p_1 - p_4 = \dot{q}_A (R_A + R_B + R_C)$$

Equation (3.8) gives the required pressure drop in terms of flow and the elemental resistances. As a by-product of the result, the total resistance for the composite system would be defined as $(p_1 - p_4)/\dot{q}_A$ so that for the series arrangement of Fig. 3.3b the total resistance is

$$(3.9) \qquad R_T = R_1 + R_2 + R_3$$

This simple example can be represented analogously by an electrical network of series resistors as shown in Fig. 3.4a.* The result given

---

* If flow and pressure are analogous to current and voltage, respectively, then, from the discussion in Chap. 2, the flow resistance is analogous to electrical resistance.

by Eq. (3.9) agrees with the well-known relation between series resistors and total resistance. In a similar manner, the network of Fig. 3.3c is clearly analogous to the electrical network shown in Fig. 3.4b. Without completing the analysis, it should be evident that the total flow resistance is related to the component resistances by

$$(3.10) \qquad \frac{1}{R_T} = \frac{1}{R_A} + \frac{1}{R_B} + \frac{1}{R_C}$$

or, in terms of conductances,

$$(3.11) \qquad G_T = G_A + G_B + G_C$$

For the network of Fig. 3.3c we can therefore immediately write

$$(3.12) \qquad p_1 - p_2 = \frac{\dot{q}_A}{G_A + G_B + G_C}$$

Finally, the networks of Figs. 3.3d and 3.4c are also obviously analogous so that simply by inspection we may write the total resistance as

$$(3.13) \qquad R_T = R_A + \frac{1}{G_B + G_C}$$

The pressure drop $p_1 - p_3$ is then determined by $\dot{q}R_T$.

As the problem just outlined was based upon the use of theoretical, linear-resistive elements, it might be considered merely an academic exercise. Yet, since many pressure-flow curves can, at least in part, be represented by linear segments, linear analysis may have practical applications. If, for example, the pressure-flow characteristics of the elements in the

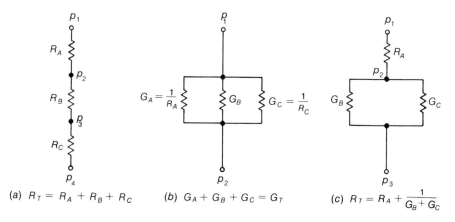

(a) $R_T = R_A + R_B + R_C$ (b) $G_A + G_B + G_C = G_T$ (c) $R_T = R_A + \dfrac{1}{G_B + G_C}$

FIG. 3.4   *Electrical analogs of networks in Fig. 3.3.*

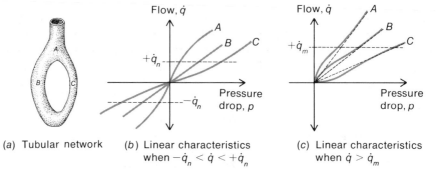

(a) Tubular network   (b) Linear characteristics when $-\dot{q}_n < \dot{q} < +\dot{q}_n$ (c) Linear characteristics when $\dot{q} > \dot{q}_m$

FIG. 3.5 *Linear characteristics in a limited range.*

previous problem were given by the curves in Fig. 3.5*b* (instead of those in Fig. 3.3), it should be obvious that a linear analysis would give accurate results if it were known that the flow was not expected to exceed $\pm\dot{q}_n$. In general, a simple linear analysis gives valid results when the actual pressure-flow characteristics are collinear with straight lines radiating from the origin. Thus a linear analysis would also be valid in the problem outlined if the characteristics were given by the curves in Fig. 3.5*c* and the flow was known to exceed $+\dot{q}_m$.

3.4 STATIC AND DYNAMIC RESISTANCE

It was indicated earlier that a meaningful definition of resistance should relate to the characteristic curves. When the curves are collinear with lines emanating from the origin, the reciprocals of the slopes of the lines serve as a meaningful and useful definition of resistance. By using this definition, analysis of system assemblies is relatively straightforward; analog simulation is feasible and uncomplicated. These advantages can, to some extent, be retained for nonlinear characteristics if these curved characteristics are grossly described by radial straight-line approximations (e.g., line OA in Fig. 3.6) or if it is assumed the curved characteristics are made up of a series of straight-line segments (such as BC or DE in Fig. 3.6). The first straight-line approximation gives rise to the concept of static resistance; the second, to that of dynamic resistance. If an investigation involves a relatively small section of the pressure-flow characteristics, surprisingly accurate analytic results can be obtained when these straight-line approximations are used to define the curve characteristics.

Both static and dynamic resistance are defined by the slopes of straight lines. Static resistance is the reciprocal slope of the radial line from the origin to some point on the characteristic curve; dynamic resistance is the reciprocal slope of a straight-line approximation to the characteristic curve for some small operating range. Evidently both these resistances have meaning only if an operating point or operating range is specified. In Fig. 3.6, if it is known that the tube will operate over a small pressure variation with an average pressure drop of about 40 mmHg, the static resistance is defined by the slope of the radial line from the origin to the 40-mm Hg point. Specifically,

$$(3.14) \qquad R_{\text{stat}} = \left. \frac{p}{\dot{q}} \right]_{p_{\text{av}}=40} = \frac{40 \text{ mmHg} - \text{min}}{1 \text{ ml}} = 40$$

For the same range of operation, the dynamic resistance can be defined by the slope of the tangent line BC constructed at the 40-mmHg point of the characteristic curve. Thus, for the curve in Fig. 3.6,

$$(3.15) \qquad R_{\text{dyn}} = \left. \frac{\Delta p}{\Delta \dot{q}} \right]_{p_{\text{av}}=40} = \frac{60 - 29}{2.5 - 0} = \frac{64}{5} = 12.8$$

Two questions of semantics should be resolved at this point. The first is the use of the terms "static" and "dynamic," and the second is the graphical interpretation of dynamic resistance. The adjectives "static" and "dynamic" have long been used to describe the resistance associated with the operation of an electron tube. These terms were (and still are) used in the same sense as we use them here: Static resistance is the ratio of total voltage (or pressure) to the total current (or flow) and is defined as E/I [or as Eq. (3.14)]; dynamic resistance is defined as the ratio of small incremental changes of the variables [as in Eq. (3.15)]. In recent

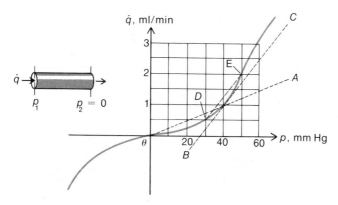

FIG. 3.6 *Characteristics of flow through a rigid tube over a large range of pressures (flow behavior of a real fluid is sometimes called an Ostwald curve; see Ref. 9).*

years, another terminology has become popular. Instead of static and dynamic resistance, the terms "total" and "incremental" resistance are used. Total and static resistance are thus interchangeable terms, as are dynamic and incremental resistance. Although the newer terminology is somewhat more descriptive, both designations are acceptable and both are widely used.

The graphical description of dynamic (or incremental) resistance requires some additional discussion. As described by Eq. (3.15), dynamic resistance refers to the best straight-line fit between two operating points on the characteristic curve. If the operating range in Fig. 3.6 is given as $30 < p < 50$ the line DE would probably be used to define the dynamic resistance. Another interpretation is possible if only the average operating value is specified and the operating range is considered small. Under these circumstances, the definition of dynamic resistance as the reciprocal slope of the tangent constructed at the average operating point would be acceptable. If the operating range is small, these two interpretations give rise to the same result. For most situations with which we shall be concerned, this will be the case and, unless otherwise stated, we shall use the tangency point to illustrate the concepts associated with dynamic resistance.

The utility of the concepts of static and dynamic resistance will be demonstrated in the following sections. Before we do so, however, it should be mentioned that the concept of static resistance is relatively well established in studies of blood circulation. The static circulatory resistance is generally expressed in terms of peripheral-resistance units (PRU). One PRU is defined as a 1-mmHg pressure drop for a flow of 1 ml/s. Under normal resting conditions the pressure drop between systemic arteries and veins is about 100 mmHg; the flow is about 100 ml/s. The static resistance (or the total peripheral resistance) is then about 1 PRU. Under abnormal conditions this value may vary from about 0.3 to 4 PRU. Plots of total peripheral resistance vs. pressure are often found in the literature.* Such plots are conceptually useful as they immediately show the variation of resistance within some operating range of pressure. It should be noted, however, that this same information would be directly available from a pressure-flow diagram.

3.5 SIMULATION AND ANALYSIS USING STATIC RESISTANCE

In order to demonstrate how static resistance can be used in simulation and analog studies we again examine the network of Fig. 3.3d but we now

* See, for example, Ref. 4, p. 347; also Ref. 5, p. 35.

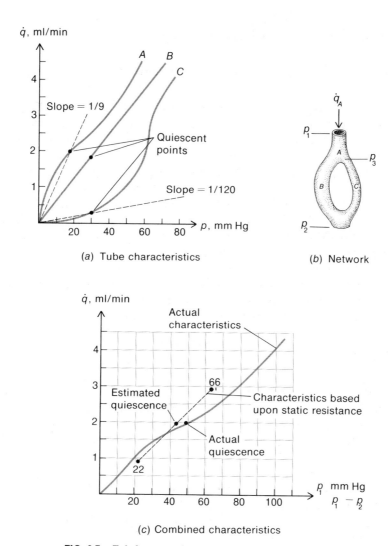

(a) Tube characteristics

(b) Network

(c) Combined characteristics

FIG. 3.7   *Tubular network with nonlinear characteristics.*

assume the pipe segments are given by the nonlinear characteristics shown in Fig. 3.7a instead of those shown in Fig. 3.3a. We also make the assumption that the total flow is known to be greater than 1 ml/min but less than 3 ml/min (that is, $1 < \dot{q}_A < 3$). The average total flow is thus about 2 ml/min. Since the total flow passes through tube A, the latter operates at an average flow of 2 ml/min. From Fig. 3.7a it can be seen that at this operating point the static resistance of tube A is

$$(3.16) \qquad R_A = \frac{p}{\dot{q}}\bigg]_{\dot{q}_A = 2} \approx \frac{18}{2} = 9 \text{ mmHg-min/ml}$$

In order to obtain static-resistance values for tubes B and C, some estimate of the quiescent operating point of these tubes must be made. To do so we note that the pressure drop across tubes B and C must be the same. Since the total flow through these tubes cannot exceed the flow in A, an estimate of the quiescent operating points of the tubes can be made by examining the tube flows when tube pressure drops are the same. From Fig. 3.7a we see that at a pressure of 30 mmHg the flow in B is about 1.8 ml/min and in C is about 0.25 ml/min. The flow to the two tubes is thus approximately 2 ml/s, which is equal to the flow from A at quiescence, as it should be. (This establishes the quiescent pressure drop across the assembly. The drop across A is about 18 mmHg and that across B and C is 30 mmHg. The total drop at quiescence is thus about 48 mmHg.) The static resistance at the quiescent operating points of the tubes can now be evaluated. From the curves, we get

$$(3.17) \qquad\qquad R_B = 15 \text{ mmHg-min/ml}$$

$$(3.18) \qquad\qquad R_C = 120 \text{ mmHg-min/ml}$$

Before using the static-resistance values given by Eqs. (3.16) to (3.18) we once again examine carefully what these values imply. The static resistance essentially defines the reciprocal slope of a line from the origin to an operating point on the graphical characteristics (see dashed lines in the plot of Fig. 3.7c). Since static resistance is defined by radial lines emanating from the origin, we can assume that, in the vicinity of the operating points, each tube can be represented by linear electrical resistors with values given by Eqs. (3.16) to (3.18). The tubular network of Fig. 3.7b can thus be approximately represented by the analog network shown in Fig. 3.8.

We can now estimate the total expected pressure drop in the operating range by direct measurement or, alternatively, we can calculate the drop by first calculating the total resistance and then using $p = \dot{q}R_T$. Using the latter procedure, we have

$$(3.19) \quad R_T = R_A + \frac{1}{G_B + G_C} = 9 + \frac{1}{\frac{1}{15} + \frac{1}{120}} = 22 \text{ mmHg-min/ml}$$

At a flow of 1 ml/min the expected pressure drop is thus

$$(3.20) \quad p = \dot{q}R_T = 1 \text{ ml/min} \times 22 \text{ mmHg-min/ml} = 22 \text{ mmHg}$$

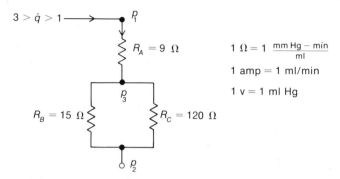

FIG. 3.8 *Simulation of tubular network of Fig. 3.7b using static resistance. $1\ \Omega = 1\ mmHg$-min/ml; 1 A = 1 ml/min; 1 V = 1 mmHg.*

At a flow of 3 ml/min the expected drop is

(3.21) $$p = \dot{q}R_T = 3 \times 22 = 66 \text{ mmHg}$$

Equations (3.20) and (3.21) in effect state that, as flow varies between 1 and 3 ml/min, the pressure drop varies linearly between 22 and 66 mmHg. In fact, the latter equations imply that, in the range of $1 < \dot{q} < 3$, the p-$\dot{q}$ characteristics of the composite network can be approximately represented by a radial straight line of slope $G_T = \frac{1}{2}2$. The dashed line in Fig. 3.7c thus represents the approximate p-$\dot{q}$ characteristics of the total system. For comparison, the exact characteristics of the composite network (determined graphically*) are also shown in Fig. 3.7c. It should be noted that at a flow of 1 ml/min the true pressure drop is 20 mmHg instead of the predicted 22 mmHg (an error of 10 percent); at a flow of 3 ml/min the actual value is about 80 mmHg instead of the predicted value of 66 mmHg (an error of about 20 percent). These are not very large errors, particularly in view of the crude flow approximations that were used. It is of interest to note that even with the crude approximations the theoretical and actual characteristics agree fairly well somewhat below midrange. In general, an intersection between the theoretical and actual characteristics can be expected. Therefore, somewhere within the stipulated range, one can expect tha theoretical results to be exactly correct. Evidently, the smaller the range variation the more accurate are the results.

---

* For this simple arrangement, graphical analysis is not very difficult. In fact, in this case it might be the easier technique. For complex assemblies, however, graphical analysis can be very involved.

3.6 SIMULATION AND ANALYSIS USING DYNAMIC RESISTANCE

As indicated earlier, dynamic resistance is defined by the slope of a tangent to the p-$\dot{q}$ characteristics. Since such tangent lines do not in general pass directly through the origin, simple electrical resistors, by themselves, cannot ordinarily be used to simulate characteristics that are approximated by dynamic resistance. This can best be illustrated by examining a specific situation.

The rigid tube of Fig. 3.9a is assumed to have the curved characteristics shown in Fig. 3.9b. An electrical simulation about the quiescent point $\dot{q}_q = 1.8$ ml/min and $p_q = 45$ mmHg is required. The dynamic resistance of the tube at the quiescent point is determined by the slope of the tangent line. From Fig. 3.9b it is seen that this resistance has been

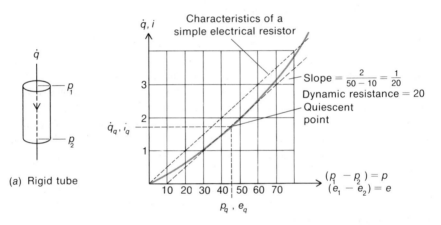

(a) Rigid tube

(b) Characteristics of a rigid tube

$$\text{Slope} = \frac{2}{50 - 10} = \frac{1}{20}$$

Dynamic resistance $= 20$

Quiescent point

$(p_1 - p_2) = p$
$(e_1 - e_2) = e$

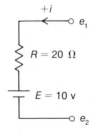

(c) Dynamic simulation

$R = 20\ \Omega$

$E = 10$ v

FIG. 3.9 *Pressure-flow simulation using dynamic resistance.*

evaluated to be 20 mmHg-min/ml. If the scaling ratio between electrical and fluid parameters is 1 A to 1 ml/s and 1 V to 1 mmHg, the simulation would require the use of a 20-Ω resistor. However, a simple resistor cannot be used by itself, as the characteristics of linear electrical resistors generally pass directly through the origin (as shown by the dotted line of Fig. 3.9b). To shift the linear characteristics of a simple resistor 10 units to the right as required by the dynamic-resistance line (note the x intercept of the tangent line), a 10-V battery is included in series with the resistor, as shown in Fig. 3.9c. We can demonstrate that this circuit is a valid representation of the p-$\dot{q}$ characteristics around the quiescent point by showing that the circuit equation defines the tangent (dynamic-resistance) line. If we use the circuit in Fig. 3.9c, the voltage on one side of the resistor is e_1; on the other side it is $e_2 + 10$.* The flow through the resistor is therefore

$$(3.22) \qquad i = \frac{e_1 - (e_2 + 10)}{20} = \frac{(e_1 - e_2) - 10}{20}$$

or with corresponding flow variables

$$(3.23) \qquad \dot{q} = \frac{(p_1 - p_2) - 10}{20}$$

When $p_1 - p_2$ is 45 mmHg, the flow is 1.8 ml/s, as it should be. Equation (3.23) is evidently the equation for the line passing through the q (quiescent) point and tangent to the actual characteristics; the simulation of Fig. 3.7c thus represents the pressure-flow characteristics of the tube of Fig. 3.9a around the quiescent point $\dot{q} = 1.8$, $p = 45$.

   In spite of the fact that dynamic-resistance lines do not pass through the origin, dynamic resistances can be combined in the standard fashion in order to determine the total dynamic resistance of a network. Thus the total dynamic resistance of the network in Fig. 3.7b is given by Eq. (3.19) except the values of the resistances are now somewhat different. As will be shown in the following paragraphs, such an analysis will in general yield a more accurate estimate of the network pressure-flow characteristics than that obtained by using static resistance.

   To show how dynamic resistances may be combined and how the results should be interpreted, we once again examine the system of Fig. 3.7. In this analysis, however, we use the dynamic resistances at the quiescent points instead of the static resistances. The tube characteristics are redrawn in Fig. 3.10, and the dynamic-resistance lines are drawn

---

* This may be more clearly understood if e_2 is considered a pressure and the battery E a small pump. The pump raises the pressure level from the input port to the output port; the output pressure is therefore $e_2 + 10$.

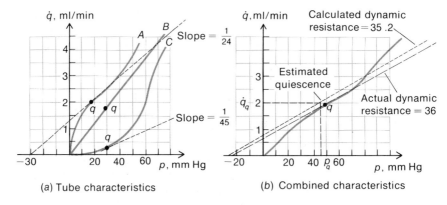

(a) Tube characteristics   (b) Combined characteristics

FIG. 3.10   *Dynamic resistances of network shown in Fig. 3.7b.*

tangent to the appropriate quiescent points. From the curves in Fig. 3.10a the dynamic resistances* are evaluated to be

$$(3.24) \qquad\qquad r_A = 24$$

$$(3.25) \qquad\qquad r_B = 15 \qquad (\text{or } g_B = 0.067)$$

$$(3.26) \qquad\qquad r_C = 45 \qquad (\text{or } g_C = 0.022)$$

In the manner outlined earlier we can now construct an electrical analog that will simulate the characteristics of the total system around the operating point. A diagram of the analog is shown in Fig. 3.11a. The battery values are determined by the x intercepts of the dynamic-resistance lines. (See Fig. 3.10a. The negative voltage implies only that the battery is connected in reverse.) We can now determine how the system will perform by direct measurement, or we can determine the expected performance by calculation.

Rather than analyze the circuit of Fig. 3.11a immediately, however, we shall first use some hindsight and intuitive reasoning to establish what such an analysis would be expected to show. First we must remember that the ultimate goal in this instance is to formulate some idea of the pressure-flow characteristics of the total circuit when information about the elements is available. Since for this simple case the p-$\dot{q}$ characteristics of the composite assembly are already available (see Fig. 3.7c or 3.10b) we might use this information to establish how the circuit of Fig. 3.11a relates to the actual characteristics. To develop this correspondence, we ask first what is the significance of the dynamic resistance

* In general, lower-case letters are used to represent dynamic resistance and conductance.

of the composite network. The fact that we are considering a composite network rather than an individual element does not alter the definition of dynamic resistance. The dynamic resistance of the total network is still the slope of the tangent to the system characteristics. For the composite characteristics shown in Fig. 3.10b the dynamic resistance is evaluated to be 36 mmHg-min/ml at the quiescent operating point $\dot{q} = 2$ ml/s, $p = 50$ mmHg. By the techniques previously developed we can now construct an electrical analog of the total system, using the dynamic resistance of the composite network. This analog simulation is shown in Fig. 3.11c.

We are now in a position to compare the simulation derived by an assembly of elements (Fig. 3.11a) with that derived from the actual characteristics of the composite assembly (Fig. 3.11c). Obviously the two circuits should be directly related. It should, in fact, be evident that by proper manipulation the circuit of Fig. 3.11a must be transformable into that of Fig. 3.11c. By comparing the two circuits we should intuitively

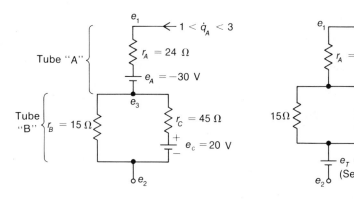

(a) Analog simulation using dynamic resistance

(b) An equivalent representation with a single battery

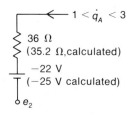

(c) Composite simulation

FIG. 3.11 *Simulation of Fig. 3.7b using dynamic resistance.*

expect that the first step in the transformation would be to combine all the resistors directly and to replace the batteries by a single equivalent battery.* The equivalent representation is shown in Fig. 3.11b. From this circuit, it is now very clear that the dynamic resistance of the composite assembly can be determined by combining the dynamic resistances of the components in a standard fashion. The calculated dynamic resistance for the total assembly would thus be

$$(3.27) \qquad r_T = r_A + \frac{1}{g_B + g_C} = 24 + \frac{1}{0.089} = 35.2$$

This result agrees surprisingly well with that taken directly from the actual characteristics of the composite assembly. What is more significant, however, is how closely the composite dynamic-resistance line represents the true pressure-flow characteristics in the stipulated operating range. In Fig. 3.10b the calculated dynamic-resistance line is drawn through the estimated quiescent point. It can be seen that, in the vicinity of the operating range, the dynamic-resistance line as calculated from the component dynamic resistances could, with good accuracy, be used to represent the pressure-flow characteristics of the composite system.

In the foregoing paragraphs it was intuitively established that the dynamic resistance of a composite network could be calculated directly from the dynamic resistances of the component elements. We shall now justify this procedure analytically. One way to approach this is to use the circuit in Fig. 3.11a to find the equation relating total flow to overall pressure drop and to compare this equation with that derived from the circuit of Fig. 3.11c. In effect, we are using Fig. 3.11a to obtain an equation for the dynamic-resistance line of the total system at quiescence and comparing this with the equation of the dynamic-resistance line taken from the characteristics of the total system. Concentrating first on the circuit of Fig. 3.11a, we find the total flow can be expressed by

$$(3.28) \qquad \dot{q}_A = [e_1 - (e_3 + e_A)]g_A = (e_1 - e_3 - e_A)g_A$$

where e_3 is the entrance pressure at tubes B and C and g_A is the dynamic conductance of tube A. To eliminate e_3 in Eq. (3.28) we express e_3 in terms of other known parameters. To do so, we note that the flow in A must be equal to the sum of the flows in tubes B and C. Therefore

$$(3.29) \qquad (e_1 - e_3 - e_A)g_A = (e_3 - e_2)g_B + (e_3 - e_2 - e_C)g_C$$

By manipulating and collecting terms, Eq. (3.29) can be rewritten so that

$$(3.30) \qquad e_3 = e_1 \frac{g_A}{g_S} - e_A \frac{g_A}{g_S} + e_2 \frac{g_B + g_C}{g_S} + e_C \frac{g_C}{g_S}$$

* This will be more rigorously established in a discussion to follow.

where

(3.31) $$g_S = g_A + g_B + g_C$$

When Eq. (3.30) is substituted into Eq. (3.29), the final result after manipulation is

(3.32) $$\dot{q}_A = \underbrace{\frac{1}{r_A + 1/(g_B + g_C)}}_{\text{Slope} = g_T} (e_1 - e_2) - \underbrace{\left(e_A g_A \frac{g_B + g_C}{g_S} + e_C \frac{g_C g_A}{g_S} \right)}_{\substack{y \text{ intercept} = e_T g_T \\ \text{equivalent battery} = e_T}}$$

Equation (3.32) now defines the dynamic-resistance line of the composite network (and is the equation that would be directly obtained if the circuit in Fig. 3.11b were used for analysis; e_T is the equivalent battery).

The comparable equation taken from the circuit of Fig. 3.11c (or Fig. 3.11b) is

(3.33) $$\dot{q}_A = (e_1 - e_2 - e_T)g_T = g_T(e_1 - e_2) - e_T g_T$$

where g_T is the slope of the composite characteristics at quiescence (or the dynamic conductance) and e_T is the x intercept of the dynamic-resistance line. By comparing Eqs. (3.31) and (3.32) we find

(3.34) $$\frac{1}{g_T} = r_T = r_A + \frac{1}{g_B + g_C}$$

and

(3.35) $$e_T = \left[e_A \frac{g_A(g_B + g_C)}{g_S} + e_C \frac{g_C g_A}{g_S} \right] r_T = e_A + \frac{g_C}{g_C + g_B} e_C$$

Equation (3.34) shows that the total dynamic resistance can be determined from the dynamic resistance of the component elements and that dynamic resistances can be combined as if they were simple linear resistors.

As a by-product of the analysis, the x intercept (equivalent battery) for the composite system can also be calculated from the values associated with the component elements. Equation (3.35) provides the required relationship. If appropriate numerical values are used in this equation (that is, $e_A = -30$ and $e_C = +20$), e_T is -25, which agrees fairly well with the value of -22 as determined directly from the graphical characteristics. It is of interest to note, however, that once the quiescent point and system dynamic resistance are determined there is no need to know the x intercept of the dynamic-resistance line. The quiescent point and the slope are all that is needed to provide a picture of the pressure-flow characteristics in the vicinity of the operating point. This implies that when the quiescent operating point is known the battery values are of little significance in an analysis. It also implies that, for analysis, only the dynamic resistances of the component elements are needed; the battery values (the x-intercept points) need not be determined. The significance of this simplification cannot be fully appreciated at this point but, as will be seen later, it does much to simplify analysis and aid conceptual visualization.

3.7 PIECEWISE–LINEAR APPROXIMATIONS

It is relatively easy now to see how the concept of dynamic resistance can be used to simulate curved characteristics even over a large operating range. Since any curve can be described by the tangents to the curve, any pressure-flow curve can be simulated by an appropriate set of dynamic resistances. When curved characteristics are approximated by straight-line segments, the process is called piecewise-linearized approximation.

To illustrate the procedure we again examine the characteristics

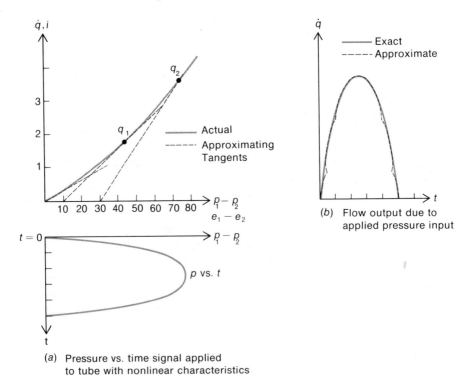

(a) Pressure vs. time signal applied
to tube with nonlinear characteristics

(b) Flow output due to
applied pressure input

FIG. 3.12 *Piecewise-linearized approximation for true characteristics.*

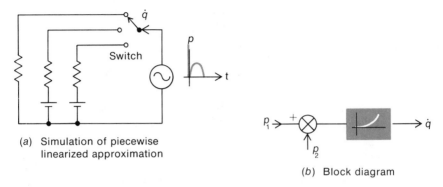

(a) Simulation of piecewise
linearized approximation

(b) Block diagram

FIG. 3.13 *Diagrams for simulation of p-q̇ characteristics of Fig. 3.12a.*

shown in Fig. 3.9*b* (redrawn in Fig. 3.12*a*). Instead of specifying that a simulation is required about a single point, however, we assume here that the simulation is to be for a pressure operating range of $0 < (p_1 - p_2) <$ 80 mmHg. In order to develop such an analog representation we approximate the actual curve by tangential line segments (in this case three segments as shown in Fig. 3.12*a*). The points of tangency (q_0, q_1, q_2) may now be considered the hypothetical midpoints for various operating ranges. Thus, when the pressure difference is between ± 28 mmHg, the midpoint of operation is at the origin; when the drop is greater than 28 mmHg but less than 63 mmHg, q_1 is the midpoint of operation, and for pressure drops greater than 63 mmHg the point is at q_2.

A simulation of the piecewise-linearized approximation is now straightforward. Each tangent line represents a different dynamic resistance. As the limiting value of the pressure drop is reached (28 mmHg for the first resistance) the second resistance is switched into the circuit and the first is switched out, etc. A schematic diagram of a possible circuit is shown in Fig. 3.13*a*. (See Prob. 3.4.)

To show to what extent the linearized approximation is an approximation, we examine the differences in the response predicted by the linearized characteristics and the actual characteristics. A variable-input pressure signal is applied across the tube (see the waveform below the characteristic curve in Fig. 3.12*a*). If we use the actual characteristics, the resultant flow is shown by the solid curve in Fig. 3.12*b*. If the linearized approximation is used (or flow measurements are made by using the circuit of Fig. 3.13*a*) the predicted flow is given by the dashed curve in Fig. 3.12*b*. The correspondence between the two curves is obviously fairly good. A very much closer approximation is possible if more hypothetical

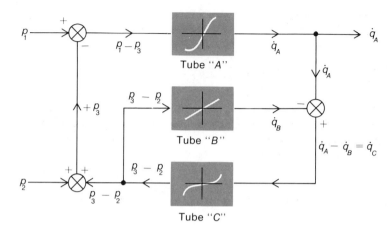

FIG. 3.14   *Block diagram of system in Fig. 3.7b.*

operating points are used (i.e., if more tangential approximations are made). Evidently, then, the more dynamic resistances used in the circuit simulation, the closer is the representation.

The piecewise-linearized simulation of Fig. 3.13a is the most elementary form of one type of function generator. In commercial units, secant segments are used instead of tangential segments to approximate nonlinear curves.* The slopes of the segments are variable by external adjustment, as are the battery voltages. This provides sufficient versatility so that a large variety of nonlinear characteristics can be closely approximated. Switching is automatically effected through the use of diode elements (electrical "check valves"; see Prob. 3.4).

In block-diagram form, the function-generator circuit in Fig. 3.13a is represented by the block shown in Fig. 3.13b. This diagram implies that, for each value of $p_1 - p_2$, the output flow $\dot{q}$ is determined by the graphical characteristics simulated by the function generator.

It is easy to see now how the total system in Fig. 3.7b can be completely simulated. The characteristics of each tube are represented separately by a function generator. The total tube assembly can then be represented by the analog-computer simulation shown in Fig. 3.14. This tube simulation can now be used in the simulation of a larger system of which the system in Fig. 3.7b is a part. In this manner more complex systems can be developed and the interaction between the component parts of the system can be studied.

---

* Although we have defined dynamic resistance in terms of tangent lines, the same definition is applicable if secant lines are used to approximate the characteristics over a small pressure range. (See the comments in Sec. 3.4.)

3.8 DISTRIBUTED AND LUMPED SYSTEMS

Before entering upon a discussion of resistance in other systems we shall distinguish between the concepts of distributed and lumped resistance. The adjectives are almost self-explanatory: Lumped resistance is the cumulative-resistance effect encountered over the full length of an element, and distributed resistance is the resistive effect for each small segment along the element length.

To illustrate the relationship between the two, we examine the tubular element shown in Fig. 3.15a. The pressure-flow characteristics of the overall tube are shown by the heavy solid curve in Fig. 3.15b. Resistances evaluated from this curve evidently represent the total resistive effect across the whole tube. The resistances thus determined are the lumped-resistive effect. If we consider the tube to be made up of three separate segments* a, b, and c, and if the pressure-flow characteristics of

---

* The tube should be divided into a much larger number of segments to obtain a more accurate picture of how the resistive effect is distributed. The choice of three segments is for simplicity of illustration.

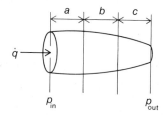

(a) Tubular element

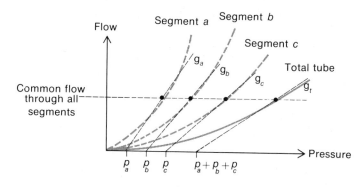

(b) Pressure-flow characteristics of segments

FIG. 3.15 *Lumped and distributed resistance of a tubular element.*

each segment are determined (dashed curves in Fig. 3.15b), each segment provides its own (and possibly different) resistive effect. It should be intuitively obvious that the lumped-resistive effect is equal to the sum of the resistance contributions of all the segments; in other words, the lumped effect is the sum of the distributed effects.

A circuit analog of a tube with distributed resistive effects can be developed if the dynamic resistances of tube segments are simulated. A simulation based upon a three-segment distribution is shown in Fig. 3.16a. The various resistance and battery values are determined from the characteristics of the segments. From the discussion in Sec. 3.6, it should be evident that the circuit in Fig. 3.16a can be redrawn as shown in Fig. 3.16b. In the latter figure, as expected, the dynamic resistance of the total tube is given by the sum of the dynamic resistances of the segments. That is,

$$(3.36) \qquad r_t = r_a + r_b + r_c = \sum_{x=a}^{x=c} r_x$$

If the slopes of the characteristic curves of all segments are the same at quiescence, then all the segmental dynamic resistances are the same. If there are n such segments, the total resistance is given by

$$(3.37) \qquad r_t = \sum_{x=1}^{x=n} r_x = nr$$

where r is the dynamic resistance of any one segment.

Generally the segments are taken to be of unit length (say, arbitrarily, 1 mm long). Then distributed resistance is expressed in terms of resistance per unit length. If the distributed resistance is constant over the total length, the total resistance is simply rL, where r is the unit-length resistance and L the total length. If the distributed resistance varies throughout the element length, the total resistance is obtained by integration:

$$(3.38) \qquad r_t = \int_0^L r \, dl = \sum_{x=0}^{\infty} r_x \, \Delta L_x$$

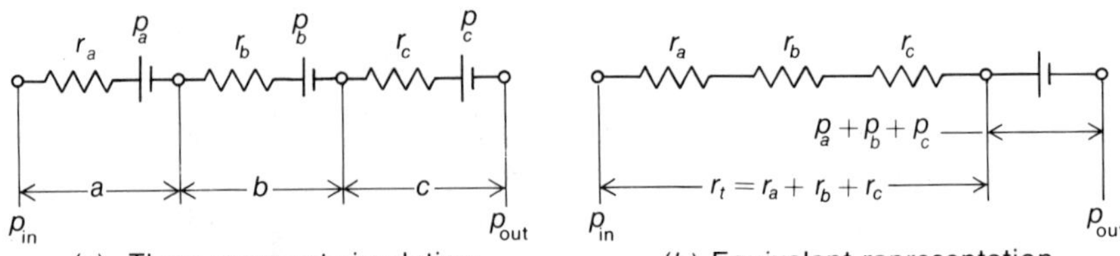

(a) Three-segment simulation   (b) Equivalent representation

FIG. 3.16 *Segmental representation of distributed resistive properties.*

The concept of distributed properties is of importance in studies of transmission effects through extended continuous media (such as fluid conduction along the aorta or force transmission through bone). In general, we shall not be concerned with distributed systems. If there is an obvious need to consider distributed effects, we shall assume that they can be described by a summation of discrete lumped systems.

3.9 RESISTANCE IN OTHER SYSTEMS

In Sec. 3.2 resistance was described as a system property that causes power loss. Although this interpretation of resistance has many applications, it is not sufficiently general to include other well-established concepts of resistance. Thermal resistance, for example, is a well-established property of thermal systems. However, thermal resistance is not conventionally considered a property that causes power loss. It is conceivable that there may be other systems where power is not a consideration, yet the concept of resistance has meaningful implications. If we are to include situations such as these, we must revise our concept of resistance.

Before we develop any general description of the resistive property we once again note that the concept of resistance is a convenience, not a necessity. As was indicated earlier, if one is willing always to work with graphical data, resistance need not be specifically defined (see footnote on page 42). However, it was seen that, when a specific definition for resistance is utilized, graphical data may be manipulated and interpreted more efficiently. In the same vein, a *generalized* description of the resistive property is not always needed. If one expects always to work with a single system (e.g., fluid or electrical), resistance can be specifically defined for each system and there need not be any relationship between these definitions. On the other hand, in order to work with combined systems and to resort to analogies, some general definition of resistance will facilitate manipulation and interpretation.

In order to develop guidelines for a general definition of resistance, we first examine some of the accepted definitions of this property. The resistive property is a well-established concept in electrical, fluid, and thermal systems. In electrical systems resistance is defined as the ratio of voltage drop to current; in fluid systems it is defined as the ratio of pressure drop to fluid flow; and in thermal systems it is defined as the ratio of temperature drop to heat flow. In all cases the measured parameters (voltage, flow, etc.) are considered to be constant with time. A comparison of the three definitions of resistance is shown in Table 3.1.

The ratios that define the resistive property are given in column 4; the units of these ratios are given in column 5. The similarity between the definitions of resistance is self-evident. The pattern that emerges is

TABLE 3.1 *A Comparison of Resistances in Various Systems*

1	2	3	4	5	6	7	8
System	Variables measured to detect R	Model	Evaluation of static R	Units of R	Basic laws	Early investigators	Resistance related to physical properties
Electrical	Voltage (e) Current (i)		$(e_1 - e_2)/i$	$M^2L^2Q^{-2}T^{-1}$	$e_1 - e_2 = \left(\rho\,\dfrac{x}{A}\right) i$	Ohm	$\rho\,\dfrac{x}{A}$
Fluid flow	Pressure (p) Flow ($\dot{q}$)		$(p_1 - p_2)/\dot{q}$	$ML^{-4}T^{-1}$	$p_1 - p_2 = \left(\dfrac{8\mu}{R^2}\dfrac{x}{A}\right) \dot{q}$	Poiseuille	$\dfrac{8\mu}{R^2}\dfrac{x}{A}$
Thermal	Temperature (T) Heat flow ($\dot{H}$)		$(T_1 - T_2)/\dot{H}$	$M^{-1}L^{+2}\theta T^{+3}$	$T_1 - T_2 = \left(\dfrac{1}{v}\dfrac{x}{A}\right) \dot{H}$	Fourier	$\dfrac{1}{v}\dfrac{x}{A}$
Mechanical	Velocity (V) Force (F)		$(V_1 - V_2)/F$	$M^{-1}T^{+1}$	$V_1 - V_2 = \left(\dfrac{1}{\mu}\dfrac{x}{A}\right) F$	Newton	$\dfrac{1}{\mu}\dfrac{x}{A}$
Diffusion	Concentration (k) Solute flow ($\dot{w}$)		$(k_1 - k_2)/\dot{w}_1$	$L^{-3}T^{+1}$	$k_1 - k_2 = \left(\dfrac{1}{D}\dfrac{x}{A}\right) \dot{w}$	Fick	$\dfrac{1}{D}\dfrac{x}{A}$

that resistance is described by the ratio of a distinctive pair of variables. The numerator of the ratio is a variable that might be described as a "relative" variable; the denominator can be described as an "absolute" variable. The units of the ratios are different for each system but in all systems the ratios include time units to an odd power (see column 5).

We now have the basic ingredients for a general definition of the resistive property. This property is defined by a pair of characteristic variables forming a ratio (of the form shown in column 4) that includes the dimensions of time to an odd power. The variables that define the system property may be described as a "relative" (or *across*) variable and a "transfer" (or *through*) variable. Pressure drop *across* a tube is an illustration of a relative or an across-variable; pressure at one end is measured relative to (or with respect to) the other end. Temperature drop *across* a wall is also a relative measurement. These variables are sometimes described as gradients (e.g., pressure gradient, temperature gradient, etc.). Fluid flow, current, and heat flow are obvious examples of "through-variables." These variables may also be conceived of as a measurement of "transfer of effect." On this basis, force would also be considered a through-variable (the transmission of a force effect *through* a rope or rod is not an unfamiliar experience). Through-variables are sometimes described as measurements of "flux." Thus the term heat-flux or current-flux measurements is frequently used.

As a brief illustration of how this concept of resistance is used, we examine the fourth system shown in Table 3.1. In mechanical systems viscous-frictional forces are generated when there is relative motion between mating members of the system. At the knee joint, for example, the motion of the femur with respect to the tibia is accompanied by frictional forces opposing the motion. The relative velocity between the bones and the force associated with the motion constitute a pair of across- and through-variables. The ratio of the across- to the through-variable $[(V_1 - V_2)/F]$ has time units to an odd power. This ratio may therefore be considered a definition of the mechanical resistance of the knee-joint system.*

Another example of the new concept of resistance is found in diffusion systems. Fick's law of diffusion (fifth system in Table 3.1) states that a constant concentration gradient across a membrane $(k_1 - k_2)$ causes a

---

* There is some inconsistency in the literature regarding the definition of mechanical viscous resistance. In older publications this property is defined by the ratio force/velocity, that is, $F/(V_1 - V_2)$; in newer publications the ratio velocity/force, that is, $(V_1 - V)/F$, is favored. In both cases the measured variables are force and velocity (the variables that define power), and in either case the ratio yields units that include time units to an odd power. In keeping with the uniform notation proposed in the text we shall use the latter definition for mechanical resistance, that is, $(V_1 - V_2)/F$.

constant flow of solute $(\dot{w})$ through the membrane. The ratio of concentration gradient (the across-variable) to the flow of solute (the through-variable) has time units to an odd power. Therefore the diffusion resistance of the membrane may be defined by $(k_1 - k_2)/\dot{w}$.

The concept of resistance has a dual function. On the one hand, it is used for the synthesis of systems or for the analysis of assembled systems. This utilization was illustrated in previous sections. Resistance may also be used to establish a characteristic, identifying property of a system material. To illustrate the latter utilization, we note for quiescent operation that the system variables that define resistance are related by the same type of equations (see column 6, Table 3.1*). The parenthetic expressions in these equations evidently represent the system resistance. The terms within the parentheses (repeated in the eighth column) show how the various resistances are used to determine the physical constants of the system materials.

Electrical resistance, for example, is assumed to be a function of the conducting path (length x and cross-sectional area A) and a function of the conducting medium (the resistivity ρ). By evaluating the electrical resistance $[(e_1 - e_2)/i]$ for a known conducting path, the resistivity of the conducting material may be evaluated. Thermal conductivity (v), membrane diffusivity (D), and fluid viscosity $(\mu$ and $\eta)$ may be determined in a similar manner. In the last case, resistance measurements include the effect of "surface" roughness as well as viscosity and system geometry. Here, therefore, resistance and the physical constant may not be simply related, and an additional relationship (or another procedure) may be required to evaluate the physical constant. In any case, the point to be established here is that resistance measurements relate to a specific system configuration (geometry, dimensions, roughness, etc.) and a material property. From a knowledge of the system configuration and measurements of resistance, the physical constants associated with the system materials can be evaluated.

From a graphical point of view, resistance is related to the reciprocal slope of the graphical characteristics generated by an appropriate pair of through- and across-variables. If we consistently plot the across- (or relative) variable on the x axis and the through-variable on the y axis, the reciprocal slope of a tangent to the curve is obviously a definition

---

* Probably the most striking feature of Table 3.1 is the column labeled "Basic laws." The perfect uniformity of these laws can hardly go unnoticed, yet the laws relate to five different systems, and the original statements of these laws were made by five different people. The symmetry between these expressions might be interpreted as an indication of the uniformity of the physical world or of the uniform pattern used to think of the physical world. In any case, simple expressions such as these form the foundation of the more elaborate theories found in the natural sciences.

of the dynamic resistance of a system. (The slope must, of course, have time units to an odd power.)

Static resistance is defined by the reciprocal slope of a radial line from the origin to a point on the across- vs. through-characteristics. A diagrammatic representation of these ideas is shown in Fig. 3.17; the defining equations are

(3.39)   Static resistance $= R = \dfrac{x_a}{y_t}$ (ratio has time units to an odd power)

(3.40)   Dynamic resistance $= r = \dfrac{dx_a}{dy_t}$ (ratio has time units to an odd power)

Since resistance and conductance have been described as reciprocal functions, the slopes of linear segments (rather than reciprocal slopes) are used to define the generalized property of conductance. Thus

(3.41)   Static conductance $= G = \dfrac{y_t}{x_a}$ (ratio has time units to an odd power)

(3.42)   Dynamic conductance $= g = \dfrac{dy_t}{dx_a}$ (ratio has time units to an odd power)

The generalized graph of Fig. 3.17 provides an easy way to develop analogs between systems. Once it is established that the slope of a characteristic curve represents the conductive property of a system, the x_a (or across-variable) may be considered analogous to electrical voltage and the y_t (or through-variable) may be considered analogous to electrical current. An electrical resistor could then serve to represent the resistance of the system under investigation. In the next section we shall develop the electrical analog of a simple thermal system where thermal resistance is replaced by an analogous electrical resistor.

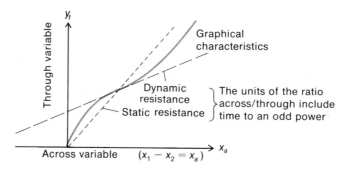

FIG. 3.17 *Generalized definition of static and dynamic resistance.*

3.10   THERMAL RESISTANCE

As an illustration of a system in which the resistance concept can be used even though power loss is not a consideration, we examine a simplified version of a physiological thermal system. We cannot examine the detailed aspects of such a system at this time as this would involve an extensive discussion of thermal-regulation mechanisms.* However, by making some simplifying approximations, we can gain insight into the relationship between heat flow and temperature in a living system.

Figure 3.18a shows a rough schematic of the system under investigation. A subject generates internal heat continuously and loses this heat continuously to his surroundings. During steady-state conditions the subject must lose the same amount of heat he generates or his temperature will steadily rise (or fall). The physiological system may be compared to a physical heating system: Internal combustion (metabolism) causes heat energy to be generated at a given rate; the internal temperature of the system remains higher than the surrounding temperature; the generated heat leaves the system by passing through the walls of the

* Some insight into what is involved in a physiological temperature-regulating system may be found in Ref. 6.

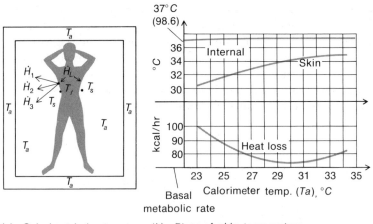

(a) Calorimetric heat and temperature measurements

(b) Plots of skin temperature, internal temperature and heat loss vs. ambient temperature (Ta)

FIG. 3.18   *Thermal considerations in human systems. [Part (b) is modified from J. D. Hardy and E. F. DuBois, The Significance of the Average Temperature of the Skin, in "Temperature— Its Measurement and Control in Science and Industry," Rheinhold Publishing Corporation, New York, 1941.]*

system. If the internal-heat production is constant, a steady flow of heat will pass from the high-temperature interior to the cooler exterior. To study the relationship between heat flow and temperature changes, simultaneous recordings of these variables must be obtained.

It is possible to obtain heat-flow and temperature information by considering either quantity an independent variable (i.e., a stimulus) and the other a dependent variable (i.e., the response). As temperature changes are easier to regulate, temperature is chosen as the controlled or independent variable (the input or stimulus); the heat flow is then considered the response.

Body-temperature changes may be produced by varying the ambient temperature. If the ambient temperature (T_a) in Fig. 3.18a is changed in a controlled manner, these changes would be reflected by a corresponding change in the surface temperature of the skin. This in turn would cause an associated change in heat flow and/or internal temperature. By measuring heat flow, internal temperature, and skin temperature while ambient temperature is altered, we should then have sufficient information to initiate a quantitative analysis of the thermal properties of the physiological system.

It is easy to state how the experiment should be performed; it is not so easy, however, to decide how to make appropriate measurements. We know, for example, that surface temperature of the skin would be a desired measurement; we do not know, however, where to make such a measurement. After a little thought, it becomes evident no one measurement can properly represent the surface skin temperature; this should be the average of several surface measurements taken at different points. To be practical, however, we must limit the number of surface measurements, and in general a few thermocouples (placed at appropriate points) are sufficient to provide a representative average. In one determination, nine surfaces (such as brow, thigh, chest, etc.) were monitored to provide a workable value for skin temperature. Each surface temperature was weighted by a factor to indicate the surface percentage of the total area. The average temperature was then the sum of the nine weighted temperatures.*

A similar difficulty is encountered in an attempt to define the internal temperature $(T_I$ in Fig. 3.18a). The problem is not as complex in this case, however, as the internal temperatures are relatively constant; therefore only a few measurements are necessary to establish an average value for the internal temperature. These can be obtained by thermocouples located within the subject; oral or rectal instruments could be

---

* A more complete description of this procedure and of other thermodynamic measuring techniques can be found in Ref. 7.

used for this purpose, or possibly a "pill" with telemetering equipment could be swallowed.

To determine the heat transmitted from the system interior to the system exterior, either the rate of heat production or the rate of heat loss can be measured. These quantities will be the same if steady-state operation is achieved. Steady-state operation implies, of course, that heat production remains reasonably constant during any one observation. To ensure that this condition prevails, the subject remains in a resting position during the test (on his back, possibly even asleep), and his food intake prior to the test is controlled (heat is then said to be generated at the basal-metabolic rate).

Heat produced and/or lost by the subject can be measured by various calorimetric techniques. In one procedure, the subject is enclosed in a chamber where oxygen consumption and carbon dioxide and water-vapor production can be monitored. From these figures it is possible to determine metabolic heat production. (The oxygen consumed and the carbon dioxide and water vapor produced indicate oxidation of carbohydrates. By knowing the thermochemical equations associated with the exothermic carbohydrate-oxidation reaction, it is possible to estimate the rate of heat production.) Some calorimetric chambers are also equipped with a circulatory water jacket; the heat delivered by the subject is absorbed by the water, and the temperature rise of the water is a measure of the subject's heat loss.

Having stated the conditions under which the necessary data can be obtained, we now examine experimental data. Figure 3.18b shows some typical plots of skin temperature (average), internal temperature (average), and heat flow (heat loss) all plotted as a function of the ambient calorimetric temperature T_a. As can be seen, the stimulus T_a produces changes in heat loss and skin temperature. The internal temperature remains essentially constant, however.

As we are primarily interested in the temperatures and heat flow associated with the physiological system, we can eliminate T_a as a variable and consider only the relationship between heat flow and body temperature. This information is available in Fig. 3.18b; it is not, however, displayed in a convenient fashion, and in Fig. 3.19 the information is replotted so that the temperature difference across the skin is shown as one variable and the heat flow is shown as the other. In keeping with the convention suggested in Sec. 3.9, temperature difference (the across-variable) is plotted as the abscissa and heat flow (the through-variable) is the ordinate. This presentation shows the skin properties in a clearer fashion. The resultant curve (labeled A in Fig. 3.19) is evidently a plot

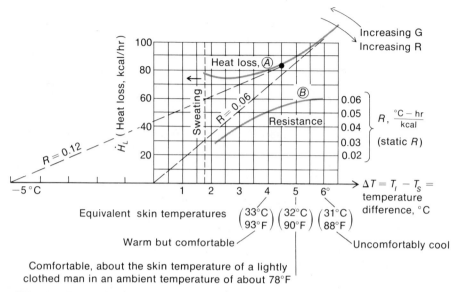

FIG. 3.19 *Data of Fig. 3.18b replotted to show thermal characteristics in terms of heat loss and temperature difference.*

of the across- vs. through-characteristics of the skin, and it can represent the thermal characteristics of the skin.

The curve in Fig. 3.19 indicates that heat loss increases as the *skin* temperature drops (i.e., as $T_I - T_s$ becomes greater or as T_s becomes lower). This is a reasonable and expected result; most materials behave in this fashion. This behavior does not, however, distinctively describe the thermal characteristics of the physiological system. For such purposes the static thermal resistance (i.e., the skin "insulation") provides a better picture of the thermal behavior of the system. If we use the definition of static resistance given by Eq. (3.39), for example, and plot R versus $T_I - T_s$ (curve B^* in Fig. 3.19) we find that the thermal resistance increases as temperature drop increases. This gives a better picture of the physiological thermal system; the body can "adjust" so that when cooled its resistance to heat loss increases (i.e., it tends to retain heat), and when warmed its resistance to heat loss decreases (it tends to lose heat). As a basis for comparison, for the temperature differences shown, the resistance can drop from a value of about 6×10^{-2} °C-h/kcal to 2.5 ×

* To obtain this curve we form the ratio temperature difference/heat loss for each value of the abscissa (temperature difference). Thus, when the temperature difference is 4°C, the heat loss is 80 kcal/h; the resistance at 4°C is therefore 4/80 or 0.05°C-h/kcal.

10^{-2} °C-h/kcal, a change of almost 60 percent from the highest value shown. In contrast, the insulation value of most physical materials decreases with increased temperature difference within a comparable range of temperatures.

In most literature, thermal properties are generally given in terms of conductance rather than resistance. For the previous example, then, the conductance of the system varies from about 17 to 40 kcal/°C-h. In order to account for individual differences, these values are normalized on the basis of skin-surface area. Then the term conductance per unit area is generally used instead of total conductance. This is more closely related to the material property than to the physical dimensions of the system and is designated as the conductivity of the material (see column 8, Table 3.1). The above values for conductance were based upon an average surface area of about 2 m². The variation of conductivity would therefore be given as 8.5 to 20 kcal/h/m²/°C.

The foregoing discussion is merely a very elementary study of the physiological thermal system. It answers one simple question (how temperature drop and heat loss are related) and poses many others. The interested investigator might now ask: Why the radical change in conductivity? Is this a passive property of the material? Does dry skin have the same properties? How does blood flow relate to the conductivity? etc. Some of these questions can be quantitatively answered by researches in the literature, and some open up new areas for investigation. Each bit of data permits a wider understanding of the interrelation between operating mechanisms of the system and delineates the part of the system not yet understood or adequately investigated.

3.11 SIMULATION OF THERMAL SYSTEMS

Electrical simulation of the thermal system discussed in the preceding section is relatively straightforward. If the temperature difference (the across-variable) is chosen to be analogous to voltage and if heat flow (the through-variable) is analogous to current, an electrical resistor can be used to represent thermal resistance (compare Figs. 2.8, 3.17, and 3.19). If the electrical simulation is to represent the physiological system around the quiescent skin temperature of 31.5°C ($T_I - T_s = 5.5$°C), the analog is particularly simple. At this quiescent point the thermal dynamic-resistance line passes through the origin (see Fig. 3.19); an electrical resistor (without a battery) is thus sufficient to represent the thermal dynamic-resistance line at this point. The simulation is thus simply a (temperature) source and a resistor (appropriately scaled) as shown in Fig. 3.20a. In order to permit more versatility, a variable-voltage source

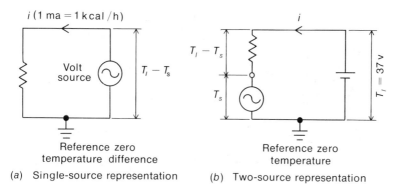

FIG. 3.20 *Electrical simulation of physiological thermal system at $T_s = 31.5°C$.*

is included in the diagram instead of a fixed battery. This allows for a variation of the across-variable $T_I - T_s$ around the quiescent point $T_I - T_s = 5.5°C$.

An alternative simulation for this simple system is shown in Fig. 3.20b. Instead of a single voltage source to generate $T_I - T_s$, two separate sources are used. A battery supplies a constant voltage to represent the constant internal temperature, and a variable source supplies a voltage to represent T_s. The across-variable is still $T_I - T_s$ but this value can now be varied by individually adjusting T_s (or T_I if a variable source replaces the battery). The representation in Fig. 3.20b is somewhat more versatile, as will be seen in a discussion to follow.

If a dynamic simulation is desired for skin-temperature variations between $31°C < T_s < 33°C$ [$4 < (T_I - T_s) < 6$], a single dynamic resistance at about $32°C$ ($T_I - T_s = 5°C$) could be used, or alternatively one could use two piecewise-linear approximations in this range. The latter simulation is shown in Fig. 3.21a, where e in volts represents temperature and i in milliamperes represents flow. When $5° < (T_I - T_s)$ $< 6°C$, the switch is in position A and the diagram is essentially the same as that of Fig. 3.20a. When $4° < (T_I - T_s) < 5°C$, the switch is in position B where a new dynamic resistance is used. The diagram for connection B is established from the characteristics at the quiescent point $T_I - T_s = 4.5°C$. The tangent line for the dynamic resistance at this point is given by

$$(3.43) \qquad \dot{H}_L = m[(T_I - T_s) - E_0]$$

where m is the slope of the line and E_0 is the x intercept. Since the slope is

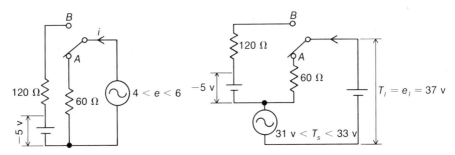

(a) Simulation with a single source   (b) Simulation with two driving sources

FIG. 3.21 *Piecewise-linearized simulation of steady-state thermal system in a human being.*

defined as the conductance (or reciprocal resistance), we may write

$$(3.44) \qquad \dot{H}_L = \frac{(T_I - T_s) - E_0}{r_{\mathrm{dyn}}}$$

From Fig. 3.19 it is seen the dynamic resistance is 0.12°C-h/kcal and E_0 is -5. Equation (3.44) then becomes

$$(3.45) \qquad \dot{H}_L = \frac{(T_I - T_s) + 5}{0.12}$$

Equation (3.45) is simulated by the circuit of Fig. 3.21a when the scaling factor for the resistance is 1,000 and the switch is in position B. Specifically, if the switch is in position B, the equation for the circuit in Fig. 3.21a is

$$(3.46) \qquad i = \frac{e - (-5)}{120} = \frac{e + 5}{120}$$

Equations (3.45) and (3.46) are essentially the same except for the factor of 1,000. Current in milliamperes thus represents flow in kilocalories per hour, and voltage in volts represents temperature differential $T_I - T_s$. The sources can now be separated in the manner previously described. The representation then becomes that shown in Fig. 3.21b, where T_I and T_s can be separately varied.

We conclude this discussion of thermal resistance with a brief presentation of how the model for an unclothed subject is modified by the addition of clothing. The across-variable for clothing is evidently the difference between skin temperature and ambient temperature. The heat lost through the clothing is the through-variable. These define the thermal resistance (or insulation) of clothing. The thermal resistance depends on a great number of factors (temperature, humidity, velocity, etc.). For convenience and standardization the armed services make use of a special unit to define clothing insulation (see Ref. 8). This unit, called the "clo," is based upon the comfort requirements of a clothed subject and is approximately equivalent to a thermal resistivity of 0.18°C-h-m²/ kcal. For a surface area of about 2 m², the resistance becomes 0.09°C-h/ kcal. The insulation resistance of a 1-clo suit ("ordinary" street clothes) is thus about one and one-half times the resistance of the physiological system.

If we now assume that the insulation value of clothing remains reasonably constant within a temperature range of interest, clothing may be electrically represented by a constant resistor. (By using such a simulation we are in effect assuming that the temperature-flow character-istics of clothing may be represented by a straight line through the origin. The reciprocal slope of the characteristics is the clothing resistance, which may be expressed in clo units.) To simulate the physiological system with clothing we now have only to combine the two models. Figure 3.22 shows the composite representation of a man clad in a 1-clo suit. The 90-Ω resistor is the scaled value of a thermal insulation equiv-alent to 1 clo (0.09°C-h/kcal). It is to be noted that heat from the physiological system passes directly through the suit (as it should). The skin temperature on one side of the suit and the ambient tempera-ture on the other side are the across-variable components of the suit (i.e., the across-variable is given by $T_s - T_a$). We can now realize why the physiological model of Fig. 3.21b is sometimes more convenient than that of Fig. 3.21a. In order to include the effect of clothing, T_s must be separately defined. The model of Fig. 3.21a does not include this feature.

We can now use the simulation of Fig. 3.22 to determine how skin temperature varies when an individual, clad in a 1-clo suit, is exposed to variable ambient temperatures. To use the simulation we simply

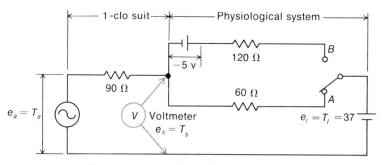

FIG. 3.22 *Model of man dressed in a 1-clo suit.*

vary T_a and read the corresponding values of T_s with a voltmeter. We should note, however, that the 60-Ω resistor is a valid representation of thermal dynamic resistance only when the skin temperature is less than 32°C (see Fig. 3.19). Therefore so long as T_s reads less than 32 V, the switch should be in position A. When T_s exceeds 32 V the switch should be moved to position B.

Instead of making actual measurements on the circuit of Fig. 3.22, we can use it to derive an analytic relationship between the ambient and skin temperatures (T_a and T_s, respectively). When T_s is less than 32°C, the switch is in position A and we have

$$(3.47) \qquad \frac{T_I - T_s}{60} = \dot{q} = \frac{T_s - T_a}{90}$$

(i.e., flow from the skin = flow through the suit). Therefore

$$(3.48) \qquad T_a = \frac{5T_s - 3T_I}{2} \qquad \text{valid for } T_s < 32°C$$

or

$$T_s = \frac{3T_I + 2T_a}{5}$$

From Eq. (3.48) we can determine corresponding values of T_s and T_a when $T_s < 32°C$ and $T_I = 37°C$. For comfort, T_s should be very slightly below 32°C (89.5°F). As calculated from Eq. (3.48), T_a would then be 24.5°C (76°F). A "resting" subject dressed in street clothes is thus comfortable when the ambient temperature is about 24°C (75°F). Should the ambient temperature drop to 21°C (69.8°F), we would find from Eq. (3.48) that skin temperature would change to about 31°C (88°F) and the subject would feel uncomfortably cool. The simple straight-line

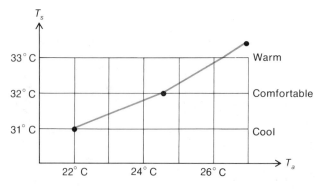

FIG. 3.23 *Skin temperature vs. ambient temperature for man clothed in a 1-clo suit.*

relationship between ambient and skin temperatures as given by Eq. (3.48) is shown in Fig. 3.23.

If skin temperature rises above 32°C, the 60-Ω resistor is no longer a valid representation of the thermal resistance, and the switch in the circuit of Fig. 3.22 must be moved to position B. The equation relating T_s and T_a now becomes

(3.49)
$$\frac{T_I - (T_s - 5)}{120} = \dot{q} = \frac{T_s - T_a}{90}$$

Therefore

(3.50) $$T_s = \frac{3T_I + 4T_a + 15}{7}$$ valid for $T_s > 32°C$

Equation (3.50) indicates that at an ambient temperature of 27°C (80.5°F) the skin temperature will rise to 33.4°C (92°F). At this skin temperature a person in a 1-clo suit (street clothes) is still comfortable though rather warm (as might be expected during a warm summer day). The linear relationship between T_s and T_a as given by Eq. (3.50) is also shown in Fig. 3.23.

Problems

3.1 Fluid power was defined by the product of flow and pressure (see Sec. 3.2).

(a) The curves shown in Figs. 2.14 and 2.15 include sufficient information to permit an evaluation of the power developed by the ventricle during systole. Obtain a plot showing the power output of the ventricle as a function of time.

(b) What additional data would be required to obtain the power loss due to flow through the aortic valve?

(c) Is it possible to determine the power loss across the mitral valve from the data shown in Figs. 2.14 and 2.15? If so, obtain a plot of power loss vs. time. If not, determine what further information is necessary, and indicate where or how such information may be obtained.

(d) Power loss across an electrical resistor is given by I^2R or E^2/R. During conduction the valve resistance may be roughly determined from the solid line in Fig. 2.16a. Use this resistance to estimate the average power lost because of flow through the mitral valve.

3.2 A section of a vascular bed is shown in Fig. P3.2. The vessels A, B, C, and D have approximately linear pressure-flow characteristics.

(a) Under what circumstances can flow through E be zero?

(b) Suppose the tubes shown in Fig. P3.2 have the graphical characteristics shown in Fig. 3.7 (tube D has the same characteristics as tube B in Fig. 3.7). Can the flow through E ever be zero? If not, why not? If so, at what value of total pressure drop $(p_1 - p_2)$?

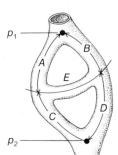

FIG. P3.2

3.3 Two tubular segments are connected in series, as shown in Fig. P3.3. The characteristics of one tube are described mathematically by the relationship $p = \dot{q}^2$. The other tube is described by $p = 5\dot{q}$. If the flow through the tube assembly varies between 0 and 3 ml/s:

   (a) What is the static resistance of each tube at the average flow? What is the static resistance of the series combination?

   (b) Obtain a mathematical expression that describes the total pressure-flow characteristics of the two tubes in series. What is the static resistance of the combination at average flow?

   (c) What are the dynamic resistances for each tube and for the combination at average flow?

   (d) Determine the dynamic resistance of the series combination from the expression determined in (b).

   (e) Which resistance (static or dynamic) would one use to estimate the power dissipated by each tube?

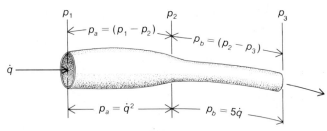

FIG. P3.3

3.4 In the supplementary discussion of Chap. 2 it was shown that the characteristics of a diode may be represented by the curve of Fig. P3.4a. The implications of the diagram are: Conduction will take place if $e_1 > e_2$; if $e_1 < e_2$ there will be no conduction.

   (a) If e_2 is 10 V what must e_1 be if conduction is to take place?

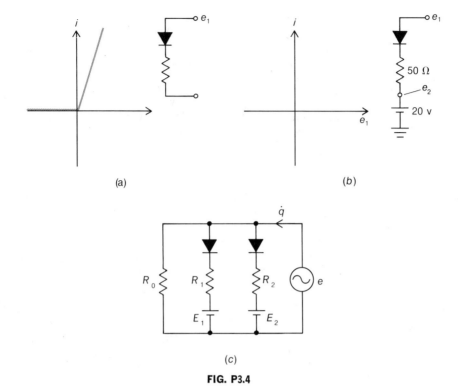

(a)

(b)

(c)

FIG. P3.4

(b) If the battery in Fig. P3.4b is 20 V, at what voltage will the diode begin to conduct?

(c) For the diode-battery configuration shown in Fig. P3.4b, draw the i versus e_1 characteristics on the axes shown (note abscissa is e_1, not $e_1 - e_2$).

(d) Figure P3.4c is a function-generator circuit which can be used to replace the circuit of Fig. 3.13a. The diodes and batteries are used to effect automatic switching as the applied voltage is increased. Determine values for the R's and E's to provide a piecewise-linearized approximation of the p-$\dot{q}$ characteristics shown in Fig. 3.12a.

3.5 Magnetic systems have not been included in Table 3.1. By analogy, obtain a definition for "magnetic resistance." Check the units of your definition to ascertain whether the criterion established for resistance is satisfied.

3.6 Stagnant air has high thermal resistance, and most thermal systems have stagnant surface films.

(a) For the system shown in Fig. 3.18, determine the thermal resistance of the air around the physiological system.

(b) Draw an electrical analog of the system which includes the air-film resistance.

(c) The air-film resistance is often considered to be dependent on the velocity of air flow around the system; velocity increases cause a resistance decrease. Show that it is possible to keep skin temperature roughly constant without sweating even if ambient temperature is increased so long as air velocity can also be increased.

3.7 The function generator included in the analog-computer circuit shown in Fig. P3.7 is needed to describe the temperature-flow characteristics of a physiological system.

(a) Identify the curve in Chap. 3 which should be used to adjust the input-output characteristics of the generator.

(b) Heat is lost from a physiological system primarily by three processes: convection, radiation, and vaporization. The first accounts for 15 percent of the heat lost, the second for 65 percent, and the third for 20 percent. Specify a proper set of R's in the circuit (R_1 to R_6) so that these heat losses may be determined.

(c) The relationship describing radiant heat lost in kilocalories per hour is given by $\dot{H}_r = K(\Theta_s^4 - \Theta_a^4)$; the coefficient K depends on the geometry and

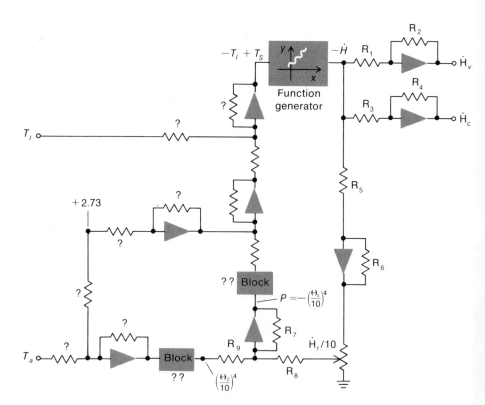

FIG. P3.7

material of the "radiator" and the enclosure. For this situation it may be taken as 1×10^{-6}. Θ_s and Θ_a are the absolute temperatures of the radiator and enclosure (skin and calorimeter, respectively, in this case; these temperatures would be described mathematically as $\Theta = T + 273$, where T is measured in degrees Celsius). To simplify the scaling problem, the radiation equation may be written $\dot{H}_r = K[(\Theta_s/10)^4 - (\Theta_a/10)^4] = 10^2[(\Theta_s/10)^4 - (\Theta_a/10)^4]$. Determine R_7, R_8, R_9, R_{10}, and R_{11} so that point P delivers a signal equivalent to $(\Theta_s/10)^4$.

(d) The complete computer circuit is designed to accept the internal temperature (T_I) and the ambient temperature (T_a) as input signals and to deliver signals that represent the skin temperature and heat loss by convection, radiation, and evaporation. Where can the skin-temperature reading be obtained? What values (or operations or function generators) are required to complete the diagram? (Note questioned elements.)

REFERENCES

1. Riggs, D. S.: "A Mathematical Approach to Physiological Problems," The Williams & Wilkins Company, Baltimore, 1963.
2. Abramson, H.: "Dimensional Analysis for Medical Students," Josiah Macy, Jr., Foundation, New York, 1950.
3. Langhoar, H. L.: "Dimensional Analysis and Theory of Models," John Wiley & Sons, Inc., New York, 1951.
4. Guyton, A. C.: "Medical Physiology," 2d ed., W. B. Saunders Company, Philadelphia, 1961.
5. McDonald, D. A.: "Blood Flow in Arteries," Edward Arnold (Publishers) Ltd., London, 1960.
6. Crosbie, R. J., J. D. Hardy, and E. Fessender: Electrical Analog Simulation of Temperature Regulation in Man, in J. D. Hardy (ed.), "Temperature—Its Measurement and Control in Science and Industry," Reinhold Publishing Corporation, New York, 1963.
7. Glasser, O. (ed.): "Medical Physics," vol. I, The Year Book Medical Publishers, Inc., Chicago, 1947.
8. Gagge, A. P., et al.: Physiology of Flight, ATSC, Engineering Div, Aero Medical Laboratory, Wright Field, Dayton, Ohio, *AAF Manual 25-2*, 1945.
9. Stacy, R. W., D. T. Williams, R. E. Worden, and R. O. McMorris: "Essentials of Biological and Medical Physics," McGraw-Hill Book Company, New York, 1955.

CHAPTER FOUR

SYSTEM PROPERTIES: STORAGE

4.1 THE PROPERTY OF STORAGE

To describe system properties in a concise fashion, we shall use only two general property classifications. The first of these, the resistive property, was discussed in the preceding chapter. It will be recalled that resistance was then defined in terms of the characteristics of the measured variables. In review: The resistive property of a system is defined by the ratio of an across-variable to a through-variable when this ratio includes time units to an odd power. In equation form this can be expressed as

(4.1)   Resistance = across/through

Conductance = through/across

(units $M^a L^b \theta^c T^m$, m odd)

The second property classification that we shall use to describe systems is called the storage property. This property is related to the ability of a system to store material (mass) or energy. ("Storage," of course, implies that recovery is possible.) Examples of physiological systems that demonstrate storage capability are the lungs, which store gas; the vascular beds, which store blood; the muscles, which store potential energy; the tissues, which store heat; etc.

In a conventional sense the storage property is generally associated only with those system elements having energy-storage capability. A spring, for example, is considered to be a system storage element because it stores potential energy; the mass of a fluid is considered to be the system storage element that serves as a storage repository for the kinetic energy of the moving fluid, etc. There are systems elements, however, that do not store energy yet obviously serve a storage function. Dye stored in the blood or tissues during clearance studies provides one such illustration. Gases stored in membranes during a diffusion process are another. In order to extend the conventional concept of the storage property so that it may include storage functions such as those just described, we do not use energy to define the storage property of a system. Instead, we relate this property to the measured variables (in the manner used to define resistance) and define storage as the ratio of a pair of characteristic variables that include time to an even power (zero is, of course, considered an even power). As yet, the form of the ratio has not been specified. For reasons that will become evident later, we shall use the same ratio to define storage as was used to define conductance. Specifically, then, storage is defined by the ratio of a through-variable to an across-variable when this ratio includes time to an even power. To express this concept in equation form we write

(4.2)   Storage = through/across   (units $M^a L^b \theta^c \cdots T^m$, m even)

It will be seen that by using this broader interpretation of storage [as defined by Eq. (4.2)] it is possible to describe a variety of systems where energy is not a prime consideration.

To illustrate briefly the implications of Eqs. (4.1) and (4.2), we examine the system shown in Fig. 4.1. An expansible tube (a blood vessel, a rubber tube, or a bronchiole) is shown under test. The ends of the tube are fixed; the tube is subjected to a steady (though changeable) pressure difference across the ends. For each pressure drop across the tube there are a cor-

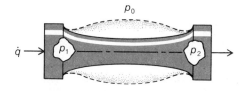

FIG. 4.1   *Expansible tube under test.*

responding flow through the tube and a corresponding volume of fluid in the tube. If the variables of interest during the test are the flow and the pressure differential, the test will define the conductive properties of this system. The ratio of the through-variable ($\dot{q}$) to the across-variable ($p_1 - p_2$) has the dimensions $M^{-1}L^{-2}T^{+3}$, which satisfies the criterion stated by Eq. (4.1). If, on the other hand, we consider the volume in the tube and the pressure differential to be the test variables of interest, then the test will define the storage properties of the system. The ratio of volume to pressure difference has the dimensions $M^{-1}L^{+4}T^{+2}$, which satisfies the criterion stated by Eq. (4.2). It may be noted that if the test variables of interest were the pressure differential and the diametral change of the tube then the ratio of variables would again satisfy Eq. (4.2). (The dimensions would be $M^{-1}L^{+4}T^{+2}$.) These variables would thus define another storage property of the system. In later sections we shall examine some of the various storage properties associated with systems. At this point we can observe that, even for a simple system, resistance and storage may both be present and there may be more than one kind of resistance and/or storage property in the same system.

The definitions for systems properties given by Eqs. (4.1) and (4.2) may be interpreted graphically. Figure 4.2 shows a set of generalized property coordinates. The ordinate is an assigned through-variable (y_t), and the abscissa is an assigned across-variable (x_a). The units of the through-variable include time to the $(m + n)$th power, and those of the across-variable include time to the nth power. The ratio of the variables (through/across) yields time to the mth power. When m is odd, the graphical characteristics define a conductive property of the system; when m is even, the characteristics define a storage property. This representation does not add any further information; it serves only to

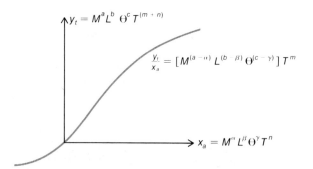

FIG. 4.2 *Generalized graph for both storage and resistance. The ratio of the variables yields time to the mth power (T^m). When m is odd, the graph defines resistive properties; when m is even, the graph defines storage properties.*

emphasize that the properties conductance and storage are defined by the slope of the graphical characteristics generated by a pair of through- and across-variables.

Another point should be made here regarding the data presentation in Fig. 4.2. Even though the variables y_t and x_a may have been obtained as a function of time, the plot does not contain time as an independent variable. This implies that time-dependent measurements (or dynamic measurements; see Fig. 2.5) are not needed to describe resistance or storage. In fact, it will be seen later that only in special cases can dynamic measurements be used to describe these properties. It should be assumed then that the properties defined by Eqs. (4.1) and (4.2) are based upon time independent of "static measurements," where "static measurements" imply that a constant level of input produces a constant level of output. A constant pressure difference that produces a constant flow through a tube is an example of a static measurement. If *either* resistance or storage is predominantly present in a system, dynamic or static measurements can be used to detect and quantify the property. However, if a combination of these properties exists (as in most systems) then only static measurements yield graphical characteristics that can be easily identified. When it is not possible to obtain static data (i.e., the assumed input and output are not time-independent, e.g., pulmonary blood flow and pressure) the data must be carefully interpreted before the separate properties can be evaluated. Such complex situations will be examined in later chapters.

As a brief summary of the foregoing paragraphs: System properties can be recognized by the criteria stated in Eqs. (4.1) and (4.2). Numerical values for the system properties can be obtained from the across vs. through graphical characteristics (the properties are related to the slopes of the curves). The graphical characteristics should be obtained by static measurements. If dynamic measurements are used for the through- and across-variables, the graphical characteristics will ordinarily reflect both the storage and resistive properties. Such curves cannot be used to determine the resistive or storage properties directly.

4.2 SYSTEMS WITH VOLUME STORAGE

We now examine some systems in which the property of storage predominates. This is an academic examination in the sense that the systems to be discussed are known to be storage systems; in an actual investigation this information is not available, and the purpose of the investigation is, in fact, to obtain such information. However, to gain experience and understanding of what the property of storage implies, we shall first examine the characteristics of known storage systems. We shall later use this knowledge to help identify system properties.

We begin with an examination of the volume-storage capability of various physiological elements. By volume-storage capability we mean simply that the element can hold a volume of fluid. A simple container would thus qualify as a volume-storage element. In physiological systems volume-storage capability is exemplified by the various hollow elastic organs. These include the heart, the bladder, lung passageways (particularly at the lower extremities of the bronchial tree), blood vessels, etc. In this first portion of the discussion we deal with hollow elastic elements having a single entrance port during some period of operation. The heart and bladder belong in this classification; the alveolar sacs of the lungs (or, considered as a composite assembly, the lungs) are also in this category. In a later section we shall examine the storage capability of elastic elements that continuously conduct flow. These include tubular elements such as the blood vessels and the bronchioles at the lower end of the bronchial tree.

As an illustration of a single-port elastic element in which the volume-storage property predominates, we examine the system shown in Fig. 4.3. The hollow elastic element shown could represent any one of the single-port storage elements of a physiological system (or it could be a model of such systems, e.g., a rubber balloon). The experimental arrangement shown is such that the internal pressure of the element can be varied and the volume change can be determined simultaneously. The upper piston serves to "pump up" the elastic member; the internal pressure is determined by the gauge reading p_i. The lower piston serves simultaneously to maintain the chamber pressure (p_0) at some fixed value and to register the volume increase of the elastic member. To perform the required measurements, a displacement is applied to the upper piston; the lower piston is then withdrawn until chamber pressure p_0 returns

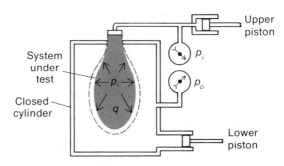

FIG. 4.3 *System with volume-storage capability.*

to the fixed quiescent value. The pressure across the elastic element is given by the gauge readings $p_i - p_0$, and the volume increase of the element can be determined from the linear displacement of the lower piston. By repeating this procedure we can generate a set of static measurements relating the pressure across the element to the volume delivered to the element. If we assume that, at zero-pressure differential (that is, $p_i = p_0$), the volume in the hollow elastic member was zero (this in effect assumes that the elastic element has no rigidity of its own), the pressure-volume characteristics of the elastic chamber would be approximately as shown in Fig. 4.4.

The pressure-volume curve in Fig. 4.4 has a characteristic S shape typical of elastomeric materials, which include as a class many of the elastic tissues in physiological systems. Thus the pressure-volume curves for the bladder, heart, lungs, etc., all show some form of this S-shape characteristic. The balloon under test is known to be a storage element and, as would be expected, the characteristics shown in Fig. 4.4 satisfy the criterion stated by Eq. (4.2). In most literature this volume-storage capability is specified by the term compliance.

We are now essentially in the same position as we were when the pressure-flow characteristics of the trachea were available. Either we can use the characteristics shown in Fig. 4.4 to describe the storage property of a hollow elastic element, or we can obtain a mathematical model to describe the pressure-volume behavior,* or finally we can describe the

---

* Such a relation has been derived and is given by the Langevin equation:

$$p - p_0 = A \left(\frac{r_0}{r}\right)^2 \left\{ L^{-1}\left(\beta \frac{r}{r_0}\right) - \left(\frac{r_0}{r}\right)^2 L^{-1}\left[\beta \left(\frac{r_0}{r}\right)^2\right] \right\}$$

where A and β are functions of temperature, chain links of carbon atoms, the initial shell thickness, and initial shell radius (r_0). L^{-1} is defined as the inverse Langevin (L), where $L(u) = \coth u - 1/u$. A brief description of the derivation of this is given in Ref. 2.

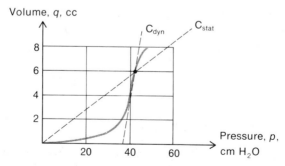

FIG. 4.4 *Pressure-volume characteristics of a balloon. (From A. L. King and R. W. Lawton, The Elasticity of Soft Body Tissues, Sci. Monthly,* **71:***258 (1950). Volume scale was changed from relative volume to absolute volume by assuming the initial volume was zero.)*

volume-storage property in the same manner as was used to describe resistance, i.e., by linearized approximations. The latter description is somewhat more useful from a simulation point of view and is the one that will be used in this text.

The discussion at this point can be almost a duplication of the earlier discussion of resistance. We may now define linear compliance by

$$(4.3) \qquad \text{Compliance} = C = \frac{q}{p}$$

or

(4.4) Volume-storage capability = compliance
= slope of pressure-volume curve

For nonlinear characteristics we can use the concepts of static and dynamic compliance and define these as

(4.5) C_{stat} = slope of radial line from origin to a point on pressure-volume curve

(4.6) C_{dyn} = slope of tangent at some point on pressure-volume curve
$$= \frac{dq}{dp}$$

Figure 4.4 can be used to illustrate the concepts of linear, static, and dynamic compliance. If the pressure on the element is not expected to exceed 20 cm H_2O, the characteristic curve can be considered linear, and the compliance is given by

$$(4.7) \qquad C = \frac{\frac{1}{2}}{20} = 0.025 \text{ cm}^3/\text{cm } H_2O$$

If the element is expected to operate within a small range around 43 cm H_2O pressure, the static compliance is given by

$$(4.8) \qquad C_{stat} = \frac{6}{43} = 0.14 \text{ cm}^3/\text{cm } H_2O$$

and the dynamic compliance is

$$(4.9) \qquad C_{dyn} = \frac{6}{5} = 1.2 \text{ cm}^3/\text{cm } H_2O$$

4.3 ELECTRICAL ANALOG OF COMPLIANCE

The property of storage has its counterpart in other systems, as does the property of resistance. In electrical systems two devices exhibit the storage property: The capacitor is capable of storing electric charge, and the inductor can store (concentrate might be a better word in this case) magnetic field within its core. Either of these devices can be used for analog representation of storage. The choice of which electrical device to use for a simulation depends only upon how voltage and current are used as analogous variables. This unfortunately can lead to confusion; e.g., in older literature the inductor is used to simulate compliance; in more recent publications compliance is simulated by a capacitor. In the interests of consistency, only one analog representation will be developed here; this analogy will depend upon across-variables always being associated with voltage (or terms related to voltage) and through-variables with current (or terms related to current). This may not eliminate entirely the possibility of dual representation but for most practical purposes it will lead to consistent results.

As pointed out in the previous paragraph, electrical analogs for storage elements will be developed on the basis of analogous variables. If we use the analogies developed in the discussion of resistance, the capacitor is the more natural analog for compliance. To point this out, we note that by Eq. (4.4) compliance may be defined by

$$(4.4) \qquad \text{Compliance} = \frac{q}{p} = \frac{\text{volume stored}}{\text{pressure differential}}$$

where p is evidently an across-variable and q may be considered a through-variable (or at least related to the through-variable flow). Since volume stored is really accumulated flow, it can be represented by

$$(4.10) \quad q = \int \dot{q}\, dt \qquad (\text{rate of volume delivery} \times \text{delivery time})$$

Equation (4.4) can now be written

$$(4.11) \qquad \text{Compliance} = \frac{q}{p} = \frac{\int \dot{q}\, dt}{p}$$

If we retain the analogies developed in the discussion of resistance, the flow $(\dot{q})$ and current (i) are analogous variables, and pressure (p) and voltage (e) are analogous variables. We therefore write

$$(4.12) \qquad \text{Compliance} = \frac{\int \dot{q}\, dt}{p} \sim \frac{\int i\, dt}{e} = \frac{\text{charge (coulombs)}}{\text{volts}}$$

The ratio of electrical variables in Eq. (4.12) defines electrical capacitance:

$$(4.13) \qquad \text{Capacitance} = C_e = \frac{\int i\, dt}{e}$$

The analogy between electrical capacitance and physiological compliance is clearly established when we compare Eqs. (4.12) and (4.13).

Electrical capacitance is sometimes defined by

(4.14)
$$C_e \frac{de}{dt} = i$$

[which is obtained by rewriting Eq. (4.13) as $C_e e = \int i \, dt$ and then differentiating]. The analogous physiological equation would be $c(dp/dt) = \dot{q}$. The latter relation could also serve as a definition for compliance though it is not as often used as that given by Eq. (4.4). We shall have occasion to use this relationship later when combined systems are examined.

An electrical capacitor is a linear device when used for a limited range; capacitors can therefore serve as electrical analogs of linear com-

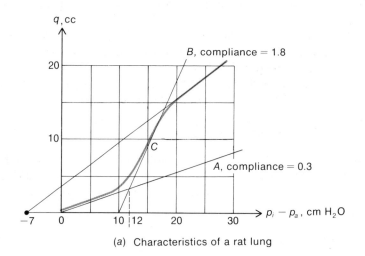

(a) Characteristics of a rat lung

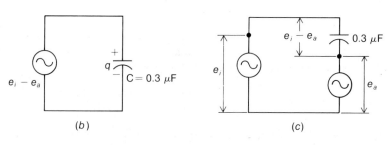

(b)

(c)

e_i (in volts) $\sim p_i$ (in cm H_2O) internal pressure
e_a (in volts) $\sim p_a$ (in cm H_2O) external pressure
1 μF capacitance $\sim$ 1 unit compliance
10^{-6} coul charge $\sim$ 1 cc

FIG. 4.5   *Volume-storage element and simulation. [Part (a) is taken from Ref. 1.]*

pliances or analogs of representations of compliance (e.g., dynamic compliance). The capacitor shown in Fig. 4.5b could thus be used to simulate the compliant system of Fig. 4.5a if it is known that the pressure differential is not expected to exceed 10 cm H$_2$O. In Fig. 4.5b a single "pressure" supply is used to generate the pressure differential across the element; in Fig. 4.5c two supplies are used to generate the internal and external pressures. The latter procedure is comparable to that in Fig. 3.20b when two separate temperature sources were used to generate the across temperature variable.

If the range of operation for the system in Fig. 4.5a is between 0 and about 18 cm H$_2$O, a piecewise-linear simulation can be used to represent the system. The analog in Fig. 4.5b or c is coupled with another branch to extend the range to 18 cm H$_2$O, as shown in Fig. 4.6a or b. When the pressure differential exceeds 12 cm H$_2$O, the switch operates to exclude the lower compliance and include a larger compliance. The equation governing operation when the switch is in position A is

$$(4.15) \qquad (e_i - e_a)C_1 = q \qquad e_i - e_a < 12$$

This is essentially the equation of the line OA in Fig. 4.5a. The equation governing operation when the switch is in position B is

$$(4.16) \qquad [(e_i - e_a) - 10]C_2 = q \qquad e_i - e_a > 12$$

Equation (4.16) defines the line BC in Fig. 4.5a.*

We shall, at this point, include the block-diagram representation of a compliant element. If the pressure-volume characteristic of the element is a straight line, the element is linear. The characteristic may then be described by the linear equation

$$(4.17) \qquad (p_{int} - p_{amb})C = q$$

The block diagram corresponding to Eq. (4.17) is shown in Fig. 4.7a. If the characteristic is nonlinear, the element can be schematically

* It should be mentioned once again that the diagrams in Fig. 4.6 are primarily for illustrative purposes. One can get the proper operating equations from the diagram [Eqs. (4.15) and (4.16) describe the volume stored as a function of pressure], but in a practical situation one would not use a switch to construct the analog circuit; a diode would more likely be used to effect switching. (See Prob. 4.3.)

It is possible to construct the analogs in Fig. 4.6 but some modification would be needed if the charge delivered is to represent the volume stored. When the switch is flipped from A to B, 3.6 units of charge has already been stored in C_1; an additional 3.6 units of charge will be delivered to C_2 when $e_i - e_a = 12$ if C_2 is empty. To avoid delivery of this extra charge, C_2 should have 3.6 charge units before switching. In effect, then, C_2 will have an initial voltage of 2 V. However, this is an academic point to preclude the possibility of engineering misinterpretation. In a practical case, switching would be handled by a diode.

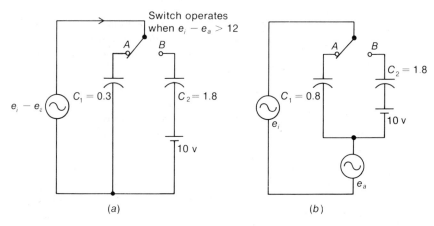

FIG. 4.6   *Piecewise-linear simulation of nonlinear compliance.*

represented by the diagram in Fig. 4.7*b*. A piecewise-linearized representation of the latter diagram is shown in Fig. 4.7*c*. The diagram of Fig. 4.7*c* is essentially a simulation of Eqs. (4.15) and (4.16). At this early stage, the block diagram of Fig. 4.7 can be considered only an academic concept

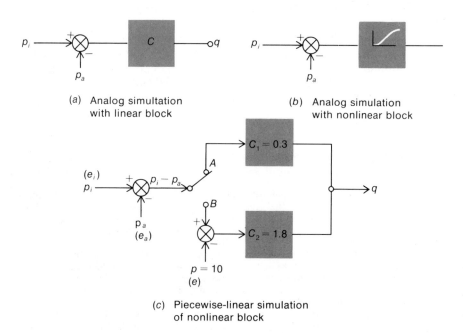

FIG. 4.7   *Block-diagram simulations of characteristics shown in Fig. 4.5a.*

which serves to illustrate how nonlinear elements may be represented by appropriate linear elements. Such diagrams can be useful when analog simulations are to be developed. See Prob. 3.7.

4.4 COMBINED HOLLOW ELASTIC ELEMENTS

As an illustration of a system where hollow elastic elements exist in combination, we examine the lung system within the thoracic cage.* This total system consists essentially of a hollow elastic element enclosed within another compliant element, as shown in Fig. 4.8a and schematically in Fig. 4.8b. The wall of the cage cavity is expanded by muscular action. The reduced pressure in the pleural space then permits outside air pressure to expand the lungs. Although muscle operation induces active breathing, "breathing" is still possible even if the muscles are inactive. Artificial respiration, forced ventilation, and the use of the iron lung are all illustrations of respiration without muscle action. We might therefore conceive of the respiratory system (at least for considerations of compliance) to be simply one compliant element enclosed within another, as shown by the model in Fig. 4.8b. The inner compliant element

* We dwell on gas "flow" systems here in order to avoid a discussion of inertia or massiveness. This aspect of systems will be discussed in a later section.

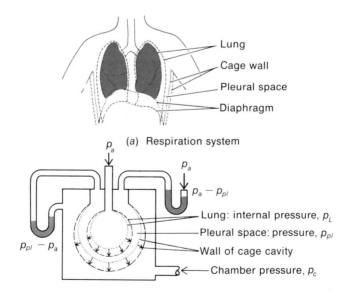

(a) Respiration system

(b) Simplified model of respiratory system

FIG. 4.8 *Compliant elements in series.*

represents the lung; the outer one represents the chest cage. The total system is mounted in a closed chamber. By changing the chamber pressure p_c, the system can be induced to "breathe." (One might consider this a simulation of artificial respiration or breathing as induced by iron-lung operation.)

The model shown in Fig. 4.8b may be compared with Fig. 4.3. If the cage and lung are considered a single element, the diagrams are identical. The internal pressure p_L in Fig. 4.8b is equivalent to p_i in Fig. 4.3, and the chamber pressure p_c in Fig. 4.8b is equivalent to p_0 in Fig. 4.3. The total pressure-volume characteristics of the complete system in Fig. 4.8 are thus obtained by measuring the pressure differential $p_L - p_c$ and the stored volume simultaneously. The pressure-volume characteristics of the *component* compliant elements are obtained in a similar manner except that the pressure differential $p_L - p_{pl}$ is used for the inner element (the lung) and $p_{pl} - p_c$ is used for the outer element (the cage).

To show how the total compliance of the model in Fig. 4.8b is related to the individual compliances, we assume that the ambient pressure (p_a) remains constant and that the chamber pressure (p_c) is varied. The pressure differentials $p_a - p_{pl}$, $p_{pl} - p_c$, and volume changes q are then determined for each value of p_c. (It is assumed q is measured by some means not shown.) From such measured information the pressure-volume characteristics of each element can be obtained. We assume that these characteristics are as shown in Fig. 4.9. Curve A represents the characteristics of the inner element (i.e., the lung); curve B represents the characteristics of the outer member, i.e., the cage wall.* From curves A and B a curve can be constructed to represent the pressure-volume characteristics of the total system. To do so we must realize that the volume stored in the inner element is the same as that in the outer element. Thus, if 500 cm³ of gas is delivered to the inner element, the volume contained by the outer element (and in fact by the whole system) must have been increased by 500 cm³. In short, then, the volume increase of A is common to that of B, which is common to that of the whole system. The pressure drop across the whole system for a given storage volume is equivalent to the sum of the pressure drops across each element. We can easily see this by simply adding the respective pressure drops. That is,

$$(4.18) \qquad (p_L - p_{pl}) + (p_{pl} - p_c) = p_L - p_c$$

* For the purposes of this preliminary discussion, the gas storage is assumed to be zero when no pressure differential exists. This presumes that the compliant elements have no "body" or rigidity of their own and will shrink to exclude all air if the internal pressure is not greater than the atmospheric pressure. This artificiality is used only to simplify the initial discussion and to permit the pressure-volume curves to pass through the origin. A more realistic situation is examined when the static-lung characteristics shown in Fig. 4.13 are discussed. See also Prob. 4.2.

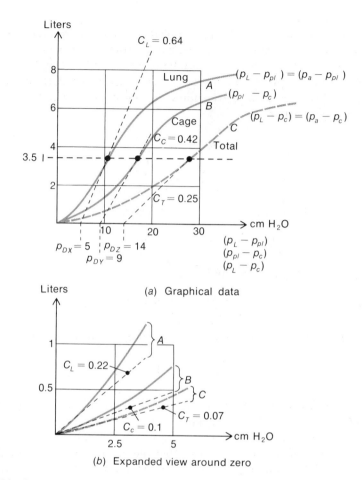

FIG. 4.9 *Pressure-volume characteristics of two hypothetical compliant elements in series.*

To obtain the graphical characteristics of the total (composite) system, therefore, we simply determine the pressure drop across each element for equal storage. The sum of the drops is the pressure drop across the composite system for this value of storage. Thus in Fig. 4.9a we find that at 3.5 l the inner system has a pressure drop of 11 cm H_2O (curve A); the outer system has a drop of 16 cm H_2O at this same volume (curve B). The composite system therefore has a total pressure drop of 27 cm H_2O when the stored volume is 3.5 l. By continuing this procedure, the total curve for the composite system may be generated. This curve is shown as a dashed line in Fig. 4.9a (curve C).

We can determine the relationship between the compliance of the composite system (C_T) and the compliances of the individual elements $(C_L$ and $C_c)$ if we examine first the characteristics curves in the vicinity of zero. In this area the curves may be represented by straight lines (see Fig. 4.9b); the slopes of the lines are the compliances of the individual elements. The equations of the lines for the inner and outer elements may thus be written

$$(4.19) \qquad\qquad p_L - p_{pl} = \frac{q_L}{C_L}$$

$$(4.20) \qquad\qquad p_{pl} - p_c = \frac{q_c}{C_c}$$

The equation for the composite characteristics is

$$(4.21) \qquad\qquad p_L - p_c = \frac{q_T}{C_T}$$

By using Eq. (4.18) we can now write

$$(4.22) \qquad\qquad \frac{q_T}{C_T} = \frac{q_L}{C_L} + \frac{q_c}{C_c}$$

Since the volumes are all equal $(q_T = q_L = q_c)$ we have finally

$$(4.23) \qquad\qquad \frac{1}{C_T} = \frac{1}{C_L} + \frac{1}{C_c}$$

We can quickly verify this result by numerical values. From Fig. 4.9b we find $C_c = 0.22$ and $C_L = 0.1$. These values, when used in Eq. (4.23), yield a value for C_T of 0.07. This agrees with the value for C_T obtained from the slope of the curve representing the composite characteristics.

The relationship derived and finally stated in Eq. (4.23) is the total electrical capacitance of two capacitors in series. As might have been expected, therefore, a compliant element enclosed within another compliant element can be represented analogously by two capacitors in series.* With a little experience the latter point becomes obvious by inspection, and one can immediately construct the analog directly from the model, as shown in Fig. 4.10. The plates of the capacitor are assumed to be on either side of the compliant element; one plate is connected to a source that represents the inside pressure of the compliance; the other is connected to a source representing the pressure outside of the compliance.

The relation given by Eq. (4.23) was derived for the special case

---

* Capacitors are in series when the sum of the voltage drops across each capacitor is equal to the voltage drop across the combination and the storage in all the capacitors is the same. See Probs. 4.1 and 4.2.

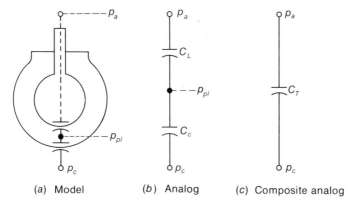

(a) Model   (b) Analog   (c) Composite analog

FIG. 4.10   *Analog of compliances in series.*

when the linearized approximations for the pressure-volume curves pass through the origin. However, it is valid even when this restriction is removed. To justify this claim, we repeat the analysis, using a different quiescent point. If, for example, the quiescent is $q = 3.5$ l then from Fig. 4.9 the linearized approximation of the pressure-volume curves for inner and outer elements may be written

(4.24) $$[(p_L - p_{pl}) - p_{DX}]C_L = q$$

(4.25) $$[(p_{pl} - p_c) - p_{DY}]C_c = q$$

where p_{DX} and p_{DY} are the x intercepts of the linearized approximations (and may be evaluated at 5 and 9 cm H_2O, respectively; see Fig. 4.9a). The C's are the slopes of the curves or the dynamic compliances of the inner and outer elements. ($C_L = 0.64$, $C_c = 0.42$ at $q = 3.5$ l.) By rearranging Eqs. (4.24) and (4.25) and adding we get

(4.26) $$(p_L - p_{pl}) + (p_{pl} - p_c) = \left(\frac{1}{C_L} + \frac{1}{C_c}\right) q + (p_{DX} + p_{DY})$$

or

(4.27) $$p_L - p_c = \left(\frac{1}{C_L} + \frac{1}{C_c}\right) q + (p_{DX} + p_{DY})$$

Equation (4.27) should now describe the linearized approximation of the *composite* pressure-volume curve around the quiescent point $q = 3.5$ l. If the linearized approximation of the composite curve were determined directly from Fig. 4.9, we would write

(4.28) $$[(p_L - p_c) - p_{DZ}]C_T = q$$

or

$$(4.29) \qquad p_L - p_c = \frac{1}{C_T} q + p_{DZ}$$

By comparing Eqs. (4.27) and (4.29) it is seen that

$$(4.30) \qquad \frac{1}{C_T} = \frac{1}{C_L} + \frac{1}{C_c} \qquad \left(\frac{1}{0.25} = \frac{1}{0.64} + \frac{1}{0.42} \right)$$

and

$$(4.31) \qquad p_{DZ} = p_{DX} + p_{DY} \qquad (14 = 5 + 9)$$

These results can be quickly verified by using numerical values derived from the actual curves as shown by the parenthetic equations following Eqs. (4.30) and (4.31). Equation (4.30) in effect states that the total dynamic compliance is equivalent to the series combination of the inner and outer dynamic compliances.*

Simulation of the series combination of compliant elements about the quiescent point $q = 3.5$ l is essentially the same as that shown in Fig. 4.10 except that batteries are now included to account for the fact that the linearized approximation no longer passes through the origin. The diagrams in Fig. 4.11 thus serve as analog simulations of the model shown in Figs. 4.8b and 4.9.

The illustrative example given in the preceding paragraphs is somewhat artificial in that it was based upon the model shown in Fig. 4.8b and the hypothetical characteristics of Fig. 4.9. More realistic data on the actual physiological system can be obtained by use of some special

* The concept that series or parallel combinations of dynamic elements can be combined as if they were simple electrical components was developed earlier when the total dynamic resistance was calculated from component elements. (See Sec. 3.6.)

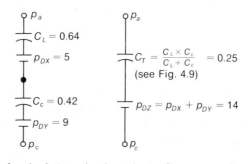

FIG. 4.11   *Analog simulation using dynamic compliances around a quiescent point.*

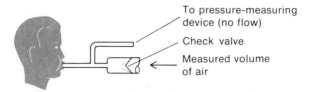

FIG. 4.12   *Method to obtain pressure-volume characteristics of a human subject.*

techniques. A relatively simple way to obtain data for the total system was first described by Rohrer in about 1916.*

A subject inhales a known volume of gas through a tube which allows flow in only one direction (see Fig. 4.12). The volume of inhaled gas can be determined by spirometric measurements or by integrating the flow measurements. As the subject relaxes his respiratory muscles in an attempt to exhale (exhalation through the tube is blocked by a check valve), he builds up a pressure which is registered by some pressure-measuring device at the mouth. Since there is no expiratory flow, there is no flow drop, and the pressure at the mouth is essentially the same as that in the lung. By using suitable corrections for pressure and temperature, the gas volume in the lung is available and a pressure-volume point of the characteristic curve can be obtained. By repeating the procedure for various volumes of gas, a pressure-volume curve can be generated. A representative plot is shown by the curve labeled $(p_L - p_a)$† in Fig. 4.13, where p_L is the absolute lung pressure and p_a is the body-surface pressure, assumed here to be atmospheric. Within the range of normal breathing it is seen that the compliance of the total system can be taken as 0.1 l/cm H_2O (which is about "normal").

* F. Rohrer, Der Zusammenhang der Atemkrafte und ihre Abhangigkeit vom Dehnungs-zustand der Atmungsorgane, *Arch. Ges. Physiol.*, **165**:419–444 (1916). A complete description of this and other techniques can be found in Ref. 3.

† In most literature this is known as the "pressure vs. vital capacity" curve, where "vital capacity" is the volume from maximum expiration to maximum inspiration (see Fig. 4.13b). The curve shown is based upon a vital capacity of about 4.8 l. In order to construct such a curve, the residual volume (gas remaining after maximum expiratory effort) must also be known. This volume cannot be obtained directly but can be obtained by the nitrogen-washout technique: The subject inhales pure oxygen which mixes with the residual air in the lung that remained from previous inhalations. The expirate thus contains unused oxygen, carbon dioxide, and some residual nitrogen from the lungs. The expirate is collected, and the nitrogen content is measured. After repeated inhalations of pure oxygen, the expirate contains no measurable amount of nitrogen; the total nitrogen which has been collected and measured provides a measure of the residual volume of the lung. This value was taken to be about 1.2 l. The details of such measurements and a more complete discussion of respiratory mechanics may be found in Ref. 3 and in J. H. Comroe, Jr., R. E. Forster, II, A. B. DuBois, W. A. Briscoe, and E. Carlsen, "The Lung—Clinical Physiology and Pulmonary Function Tests," 2d ed., The Year Book Medical Publishers, Inc., Chicago, 1962.

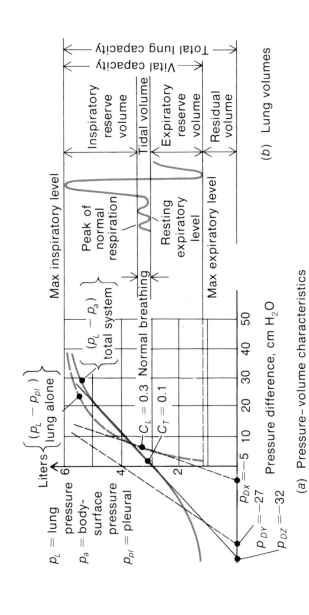

(a) Pressure–volume characteristics
of the respiratory system

(b) Lung volumes

FIG. 4.13 *Static-lung characteristics.*

In order to generate a curve for the lung characteristics above, pleural-pressure measurements are required. The pressure in the pleural cavity is not easily obtained; instead, an esophageal balloon is generally used to reflect pleural pressures.* By measuring lung pressures as previously outlined and pleural pressures by an esophageal balloon, the dashed curve in Fig. 4.13a can be generated. As can be seen from the curve, in the range of normal breathing this subject has a lung compliance of about 0.3 l/cm H_2O (slightly above "normal"; normal is generally taken to be roughly 0.2 l/cm H_2O).

From the available information we could directly construct the pressure-volume curve for the rest of the respiratory system (i.e., the thoracic cage), or we might analytically determine the curve in the vicinity of normal breathing by the procedure previously outlined. By the latter procedure the compliance of the rest of the system is found by Eq. (4.23). Using the appropriate values, we have

(4.32)
$$\frac{1}{0.1} = \frac{1}{0.3} + \frac{1}{C_{th}}$$

where C_{th} is the compliance of the thoracic cavity. By performing the required algebra, C_{th} is found to be 0.15 l/cm H_2O. This establishes the slope of the p-q curve of the thoracic cavity in the vicinity of breathing. The x intercept is simply $p_{DZ} - p_{DX} = -27$ cm H_2O (see Fig. 4.13a). The p-q curve is shown as a dashed line in Fig. 4.13a; the curve is valid in the vicinity of $q = 3$ l.

The dynamic-compliance analog for the system first discussed is shown in Fig. 4.14a. This simulation is inadequate as we cannot include the effect of different external body pressures. As most women in girdles will testify, lung storage is appreciably changed when the girdles are removed. The compliance of 0.15 therefore includes not only that of the chest cage but all factors associated with the respiratory system except that of the lungs.

In order to separate these factors, we note from Fig. 4.14b that between the pleural cavity and the surface of the body there are at least two possible paths through compliant elements. One is directly through the rib cage; the other is through the diaphragm into the abdomen and then through the abdominal wall. The total compliance of the respiratory

---

* A latex balloon at the end of a catheter (a long hollow tube) is threaded through one nostril and located in the esophagus. The balloon is inflated so that it is bounded by the esophageal wall. Changes in pleural pressure affect the diametral dimension of the esophagus, which in turn causes changes in balloon pressure. This pressure change is transmitted through the catheter and is picked up externally by a pressure transducer. Additional information can be found in Ref. 4.

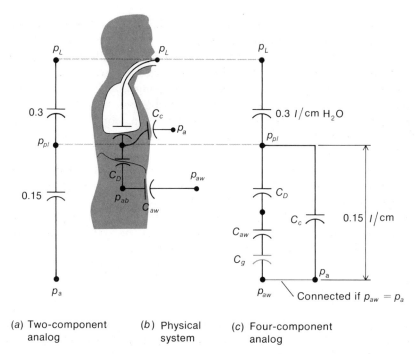

(a) Two-component analog   (b) Physical system   (c) Four-component analog

FIG. 4.14 *Simple analog extended.*

system is thus made up of the compliance of the abdominal wall (C_{aw}) in series with that of the diaphragm (C_D), where the latter combination is in parallel with the compliance of the rib cage (C_c), and the combination of the latter three are in series with the compliance of the lungs.

A composite simulation of the overall system with the above-mentioned elements is shown in Fig. 4.14c. The pressure on the abdominal wall and that on the chest are separated so that they can be individually varied. (The separate points would be connected if $p_a = p_{\mathrm{aw}}$.) Included also in the diagram, in dashed representation, is the compliance of any external garment C_g (such as a girdle) which might be worn on the abdominal wall. As can be seen from the diagram, if this garment has low compliance (little stretch) then for all practical purposes the left branch of the diagram is open-circuited and as such it will contribute little to the overall storage capability of the system.

Before the simulation of Fig. 4.14 can be used, values must be assigned to the various elements. The compliances of the diaphragm, abdominal wall, and rib cage are not individually known although the collective compliance of the three can be determined. Techniques are described* whereby the contributions of the diaphragm and abdominal wall can be theoretically evaluated. From this information the contribution of the rib cage can then also be determined.

The simulation also suggests a possible method of evaluating the combined compliance of the diaphragm and abdominal wall. Since these elements are in series, the storage due to the diaphragm and to the wall is the same (if air inclusion in the abdomen can be ignored). By measuring the distension of the abdominal wall (by, say, a series of mercury-filled elastic tubes) the storage contribution of the wall and of the diaphragm can be assessed. Since $p_{pl} - p_a$ is measurable, the compliance of the diaphragm-wall combination can be determined.

When numbers have been assigned to the various elements, the analog simulation may then be used to study some of the static properties of the system. We might, for example, examine how a percentage change in the various compliances affects total storage and which of them appears to be most significant†; the effect of various external pressures also can be examined. Any conclusions derived from the circuit should then be verified by experiment and discrepancies noted so that the simulation can be modified. In this manner, the model can indicate where understanding of the system operation is incomplete.

* See Ref. 3.

† A mathematical estimate could also be made. To do so, we note that, for the total system

$$(p_L - p_a)C_T = q$$

For a fixed pressure drop, the percent change in storage is equal to the percent change in compliance. That is,

$$\frac{dC_T}{C_T} = \frac{dq}{q}$$

Since

$$\frac{1}{C_T} = \frac{1}{C_L} + \frac{1}{C_{\mathrm{Daw}} + C_c}$$

where C_{Daw} is the combined compliance of the diaphragm and abdominal wall, dC_T/C_T can be determined as a function of the other compliances. Specifically,

$$\frac{dC_T}{C_T} = \frac{C_T}{C_L}\frac{dC_L}{C_L} + \frac{C_T C_{\mathrm{Daw}}}{(C_{\mathrm{Daw}} + C_c)^2}\frac{dC_{\mathrm{Daw}}}{C_{\mathrm{Daw}}} + \frac{C_T C_c}{(C_{\mathrm{Daw}} + C_c)^2}\frac{dC_c}{C_c}$$

for $C_T \approx 0.1$, $C_L \approx 0.3$, $C_c \approx 0.04$, $C_{\mathrm{Daw}} \approx 0.1$. (These values are estimates taken from the curves shown in Fig. 10, chap. 13, of Ref. 3.) Therefore

$$\frac{dC_T}{C_T} = 0.33\frac{dC_L}{C_L} + 0.92\frac{dC_{\mathrm{Daw}}}{C_{\mathrm{Daw}}} + 0.37\frac{dC_c}{C_c}$$

The latter result implies that the percent change in storage is most critically affected by compliance changes in the diaphragm-abdomen combination. The result does not seem unrealistic but of course it should be verified by experiment or substantiating data from the literature.

4.5 CYLINDRICAL ELEMENTS

Volume-storage capability is evidently not restricted to hollow spherical elastic elements. Any hollow element that can stretch is capable of storage. Thus, elastic tubes such as arteries, arterioles, bronchi, bronchioles, etc., all qualify as volume-storage elements.* Cylindrical storage elements warrant a separate discussion, however, in that it is possible to sustain a steady flow in such elements even after storage has taken place. In order to establish some preliminary ideas we first examine such elements under no-flow conditions. At the end of this section we shall examine these elements when they sustain a flow.

The discussion of volume storage in tubular elastic elements under no-flow conditions is not materially different from the discussion in Sec. 4.4. The pressure-volume characteristics of the system element define its volume-storage capability, and this capability is described by compliance. The diagram of Fig. 4.15 shows one possible arrangement for acquiring the pressure-volume characteristics of a segment of a tubular element. A known volume of fluid can be introduced into the reservoir. From the new reservoir level, the amount of fluid entering the element can be determined. At the other end of the tube the fluid leaving the element can be directly observed. The difference between the input and output q is the volume stored. The pressure in the tube with respect to ambient pressure is given by the corrected manometer reading (p_m). The manometer reading is thus the differential pressure between the inside and outside of the tube [($p_i - p_0$), often called the transmural pressure], and the pressure differential is the across-variable. The volume stored is the through-variable.

---

* A good discussion of the general subject of elastic elements in physiological systems is given in Ref. 5.

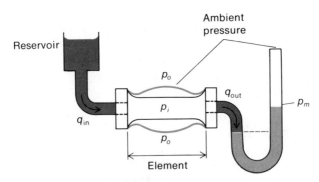

FIG. 4.15 *Arrangement to obtain pressure-volume characteristics of a tubular elastic element.*

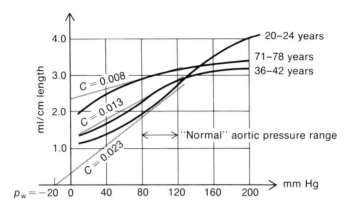

FIG. 4.16 *Pressure-volume characteristics of the aorta. [Modified from P. Hallock and I. C. Benson, Studies on the Elastic Properties of Human Isolated Aorta, J. Clin. Invest., **16**:595 (1937).]*

Figure 4.16 shows the storage characteristics for human aortas for various age groups. As noted, the volume storage is given in terms of per centimeter of length. The compliance determined from these curves thus refers to an elemental segment of the aorta. From the curves it can be seen that within the normal operating range the compliances decrease as the age group increases. In the 20- to 24-year-old group the dynamic compliance in the normal operating range is 0.023 ml/mmHg/cm length; in the 36 to 42 age group the value decreases to about 0.013, and in the oldest group shown the value is lowest at about 0.008. The stretchability of the aorta apparently decreases with age.

The simulation of the storage property of a single segment of the aorta (i.e., a 1-cm segment) is now straightforward. The dynamic compliance is represented by a capacitor; a battery can be used to set the proper operating level. The diagram in Fig. 4.17a shows the simulation of an aortic segment of the 20- to 24-year group in the range 80 to 120 mmHg.

In order to simulate an aortic length of 3 cm (chosen arbitrarily) we assume that this longer segment is composed of three shorter (1-cm) segments, as shown in Fig. 4.17b. It is further assumed that each of these 1-cm segments may be represented by the simulation in Fig. 4.17a. The resultant electrical analog of the 3-cm segment is then a combination of the 1-cm simulations. The total representation is shown in Fig. 4.17b. The capacitors are all connected at p_i since, for the no-flow condition, p_i (the internal pressure of the tube) is the same all along the tube. The separate sources, p_{01}, p_{02}, etc., are to allow for different external pressures

p_o , tissue pressure

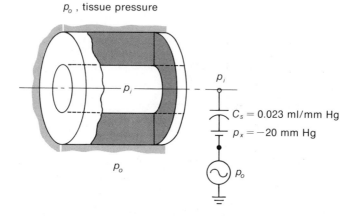

(a) Simulation of a 1–cm aortic segment

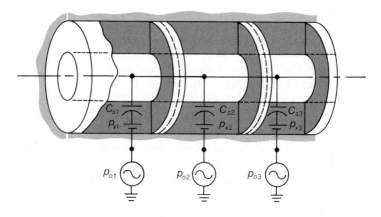

(b) Simulation of a series
of aortic segments

FIG. 4.17 *Simulation of aortic segments.*

along the aortic length. If these are all measured with respect to the same reference (say atmospheric pressure) they too may be connected as shown. [Such sources can also be used to represent the possibility of muscular action along a compliant element (see Prob. 4.5).] The different values for the capacitors and batteries are to allow for the possibility of different pressure-volume characteristics for segments along the aortic length.

If we refer to the discussion of dynamic resistance or of dynamic compliance, we see that the batteries in Fig. 4.17b can be replaced by a single equivalent battery. The simulation then reduces to the circuit in Fig. 4.18a. Should the tissue pressure be relatively constant for the tube length under consideration, the sources p_{01}, p_{02}, etc., are all equal. The circuit simulation can then be further simplified as shown in Fig. 4.18b. Finally, the parallel combination of capacitors can now be combined into a single equivalent capacitor, and we have the simple representation of Fig. 4.18c. For short lengths of tube the compliance does not materially change from section to section. The compliances C_{s1}, C_{s2}, etc., are therefore nearly equal. For simpler representation we could consider the sum of the compliances as nC_s, where n is the number of segments and C_s is the compliance of a single segment.

In the foregoing discussion, the volume-storage property of a tubular elastic element was described on the basis of constant internal pressure along the element length. This description is meaningful for the imposed condition of no flow. It is, however, unrealistic since the prime function of a tubular element is to conduct flow and a pressure drop must accompany such flow. To describe properly the storage property of a cylindrical elastic element, one should examine the element under the conditions of flow and account for the variability of transmural pressure.

It is possible to describe the pressure-volume characteristics of an elastic tube with flow by mathematical techniques. However, the mathematical treatment is rather complex and not within the scope of this book. We can nevertheless still obtain a meaningful simulation of such a system if we make some intuitive guesses as to what such a simulation would involve. Consider, for example, the elastic tube of Fig. 4.19 when steady

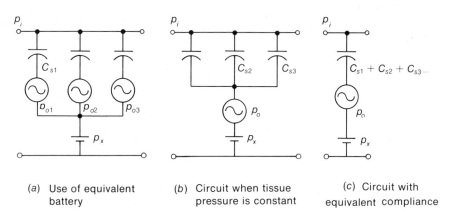

(a) Use of equivalent battery

(b) Circuit when tissue pressure is constant

(c) Circuit with equivalent compliance

FIG. 4.18 *Simplification of simulation shown in Fig. 4.17b.*

flow is maintained through the tube. For such operation, inlet and outlet pressures (p_{in} and p_{out}) remain constant (though $p_{in} > p_{out}$). The pressure at points along the tube axis (for example, p_1, p_2, p_3) steadily decreases as distance from the inlet increases, as shown by curve a in Fig. 4.19b. These pressures, too, are constant with time. For this introductory discussion it will be assumed that the ambient pressure (p_0) remains constant. The fluid volume in the tube is, of course, governed by the geometric dimensions of the element.

Should we now increase the inlet pressures to a new steady value (p_{in}) but keep p_{out} and p_0 constant, the pressure curve would rise (see curve b in Fig. 4.19b). The average pressure in the tube has obviously increased. This increase is accompanied by an increase in the physical dimensions of the tube. The fluid volume in the tube has evidently increased. It becomes clear, then, that the storage property of the tube (i.e., its compliance) could be approximately described by a plot of $(p - p_0)_A$ (average pressure drop across the tube) versus q (stored volume in the tube). From such a plot the compliance would be evaluated by the procedure outlined earlier. For static compliance, the slope of the line from the origin to the quiescent point would be used; for dynamic compliance the slope of the tangent at the quiescent point would be used. If we choose to work with the dynamic compliance, the electrical simulation of the tube would consist of a capacitor and a battery, as shown in Fig. 4.20a. However, the transmural pressure across the element is now $p_{av} - p_0$.

The simulation in Fig. 4.20a is not significantly different from that in Fig. 4.18c except that now the internal pressure is taken to be the average pressure across the ends of the elastic tube. If a "short" segment length is used, we can assume that this average pressure is given by $(p_{in} + p_{out})/2$. This in effect assumes that the pressure drop is linear along a small tube segment (i.e., the curve in Fig. 4.19b is approximately

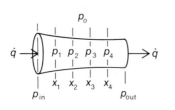

(a) Elastic tube with flow

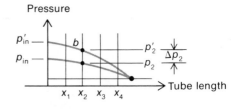

(b) Pressure distribution along a tube with flow

FIG. 4.19 *Effect of flow through a tubular elastic element.*

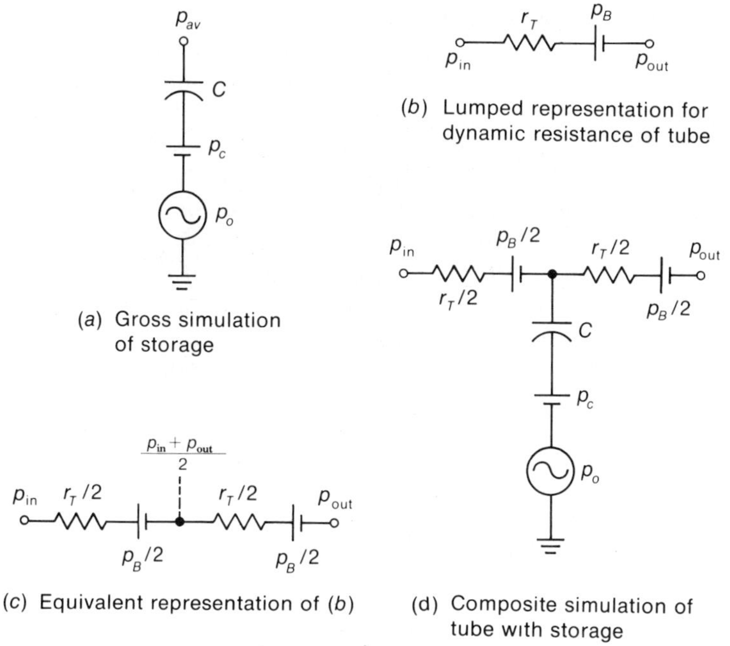

(a) Gross simulation of storage

(b) Lumped representation for dynamic resistance of tube

(c) Equivalent representation of (b)

(d) Composite simulation of tube with storage

FIG. 4.20 *Simulation of storage with flow.*

a straight line), and this in turn assumes that the tube fluid resistance may be represented by a dynamic resistance, as shown in Fig. 4.20b. (See also the discussion in Sec. 3.8.) If we divide the dynamic-resistance effect into two equal portions, as shown in Fig. 4.20c, we then have a discrete point between the resistors that represents the average pressure. It should now be intuitively evident that if we wish to include the storage effect of an elastic tube with flow we can do so by connecting a capacitor to this average pressure point. The resistor would provide the required pressure drop along the tube length; the capacitor would represent the storage of the tube. A diagram of the resultant representation is shown in Fig. 4.20d.

The simulation in Fig. 4.20d can be considered only an overall lumped representation of the distributed properties of an elastic tube. It will be seen later that even in this form it can serve as a useful representation of a tubular elastic component in a more complex system.

Should we desire a closer simulation of the elastic tube, we would consider the tube to be composed of an assembly of smaller segments. The simulation would then be the circuit in Fig. 4.21b. (The batteries are omitted for the sake of simplicity.) The dynamic resistances are evaluated as outlined in Sec. 3.8. If the tube is relatively uniform, the dynamic compliances can be determined from a pressure-volume curve for a single segment (Fig. 4.21c). Thus C_a is the slope of the curve p_a (which is the operating point of segment a). Similarly, C_b and C_c are the slopes at p_b and p_c, respectively. The different values of ambient-pressure sources (p_{0a}, p_{0b}, and p_{0c} in Fig. 4.21b) are to allow for a variation of ambient pressure along the external surface of the tube.

It is of interest to note that, if the ambient pressure is constant and all pressure measurements are made with respect to ambient, the distributed system is approximately represented by a combination of resistors and capacitors arranged in what is often called a "ladder network" (Fig. 4.22a). The more sections used in this network, the better is the representation of the distributed system. Ladder networks with an infinite number of identical sections are often used to represent the dis-

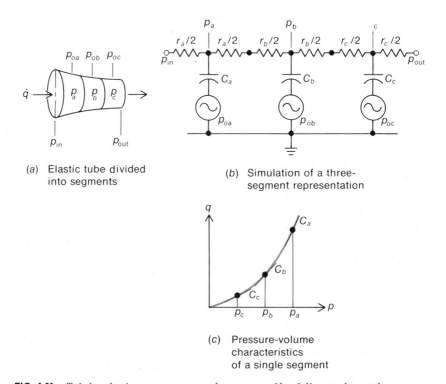

(a) Elastic tube divided
     into segments

(b) Simulation of a three-
     segment representation

(c) Pressure-volume
     characteristics
     of a single segment

FIG. 4.21 *Tubular elastic system represented as an assembly of discrete elemental segments.*

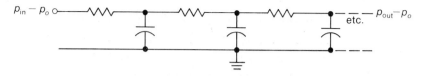

(a) Ladder-network representation

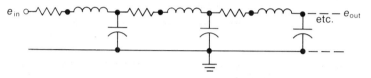

(b) Transmission-line representation

FIG. 4.22 *Fluid and electrical transmission-line analogy.*

tributed properties of an electrical transmission line. In fact, the equations that govern transmission-line operation are derived by considering the line an infinite set of ladder sections. In some transmission-line representations an inductive effect (another storage property) must also be included, as shown in Fig. 4.22b. It will be seen in a later section that this representation also has application to fluid flow through tubular elements. The inductor in this case is analogous to the inertia of the fluid.

4.6 STORAGE IN THERMAL SYSTEMS

As a second illustration of the property of storage we examine the heat-storage capability of thermal materials. The storage property in this case is related to the familiar concept of the "heat capacity" of materials. It may be recalled from elementary physics* that heat capacity is defined by the quantity of heat that must be added to (or removed from) a body to produce a temperature change in the body. In equation form this may be expressed as

$$(4.33) \qquad\qquad C_H = \frac{\Delta H}{\Delta T}$$

where ΔH represents the change in heat content of a body and ΔT is the corresponding temperature change. We can identify heat capacity as a storage property of a thermal system if we compare Eq. (4.33) with Eq. (4.2). If we consider the heat content a through-variable (i.e., the

* See, for example, Ref. 7.

TABLE 4.1 *Heat Capacities of Cutaneous and Subcutaneous Tissues of the Pig* (cal/g/°C)

Epidermis	Dermis	Fat	Muscle
0.887	0.785	0.538	0.890
0.845	0.753	0.573	0.923
0.860	0.770	0.550	0.910

integrated heat flow to a body) and temperature the across-variable, the ratio given by Eq. (4.33) defines a thermal-system property. This ratio satisfies the criterion stated by Eq. (4.2); heat capacity is thus considered a storage property of a thermal system. [This conclusion is intuitively obvious. The name of the property, "heat *capacity*," virtually suggests that it is a storage property. The comparison with Eq. (4.2) is simply to emphasize the correspondence between this intuitive conclusion and the formal definition for the storage property as stated by Eq. (4.2).]

One way to evaluate the heat capacity of a material is to use Eq. (4.33) directly. The heat gained (or lost) for a sample of the material is determined; the corresponding temperature change is noted. The slope of the curve generated by a plot of heat content vs. temperature is the heat capacity of the material. A procedure to determine the change in heat content of a material is to drop a sample of the material at a known temperature into a water calorimeter. The final temperature of the water provides a measure of the temperature change in the sample and the heat gained (or lost) by the sample. The ratio of these quantities for small temperature changes can be taken as the dynamic-heat capacity of the sample; this ratio for large temperature changes is considered the static-heat capacity.

The specific heat* of sections of cutaneous and subcutaneous tissue of a pig was evaluated by Henriques and Moritz[†] in 1947. Average values for sample sections of three different animals are given in Table 4.1. These values are fairly representative for comparable tissue of other forms of mammalian life (including man). The variation of the heat capacity between the various tissues is due primarily to the water content; the higher the water content, the higher is the heat capacity. In fact, as an overall approximation for the specific heat of tissue, Stacy (see Ref. 2)

---

* As a review, it is appropriate here to differentiate between heat capacity and specific heat. The first term refers to the property of an object; the second refers to the property of a unit weight (or mass) of the object. A can of water may have a heat capacity of 50 cal/°C but the specific heat of water is 1 cal/°C/g. The specific heat is thus the heat capacity of 1 g of water.

† See Ref. 8.

uses the relation

(4.34) $SH = (1.0 \times H_2O_{(f)}) + (0.4 \times S_f)$

where $H_2O_{(f)}$ represents the percentage of water in the tissue and S_f represents the percentage of dry material in the tissue. If we assume that a human being has a 60 percent water and 40 percent solid distribution (as a rough average), the specific heat for human tissue would be about 0.76. This agrees fairly well with the accepted average of 0.83 cal/g/°C.

4.7 ELECTRICAL SIMULATION OF THERMAL STORAGE

It should be obvious at this point that thermal storage may be analogously represented by electrical-charge storage and that the property of heat capacitance may be represented by an electrical capacitor. To establish this correspondence in a more quantitative fashion we can use the same procedure used to show that compliance and capacitance are analogous systems elements (see end of Sec. 4.2). Thus if the through-variables heat flow and current are considered analogous quantities (that is, $\dot{H} \sim i$; see Sec. 3.11) then the total heat delivered to a mass may be represented by $\int i \, dt$. Since temperature (T) and voltage (e) are also analogous quantities, we can now write Eq. (4.33) as

(4.35) $$C_H = \frac{\int \dot{H} \, dt}{\Delta T} \sim \frac{\int i \, dt}{\Delta e} = C_e$$

Dynamic-heat capacitance is thus obviously analogous to dynamic electrical capacitance.

A charged capacitor may be used to represent an insulated thermal mass at constant temperature, as shown diagrammatically in Fig. 4.23. The voltage of the charged capacitor (e_1) measured with respect to "ground" (a zero reference) is analogous to the temperature of the insulated mass (T_1) measured with respect to some temperature refer-

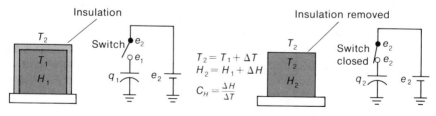

(a) Insulated mass at temperature T_1 (b) Mass exposed to temperature T_2

FIG. 4.23 *Electrical representation of thermal storage.*

ence. The quantity of charge stored in the capacitor is analogous to the amount of heat stored in the mass. The insulation around the mass ensures that no heat is lost and that the mass remains at a constant temperature T_1, in spite of the different ambient temperature T_2; the switch on the capacitor ensures that the capacitor retains its charge and that the voltage remains constant in spite of the nearby voltage source e_2. When the insulation around the mass is removed, heat is delivered to the mass and ultimately the mass reaches the same temperature as its surroundings (it does not generate its own heat). This is represented analogously by closing the electrical switch. Charge is delivered to the capacitor (i.e., current flows for a short period). The voltage of the capacitor increases until it becomes equal to that of the excitation source.

4.8   SIMULATION OF A THERMAL SYSTEM WITH COMBINED PROPERTIES

From the discussion in Sec. 4.7 we can easily see that under no-flow conditions the electrical simulation of a thermal system with lumped elements is simply an assembly of capacitors. Thus if a system is composed of a number of massy elements, as shown in Fig. 4.24, the electrical analog of the "no-flow" (equal temperature) thermal system is simply a parallel combination of electrical capacitors. The total heat capacity of the system is the sum of the individual heat capacities.

The no-flow condition imposed on the system of Fig. 4.24 reduces it to a trivial unreal system. In a more realistic situation one should include the possibility of heat flow from one shell to another or, what amounts to the same thing, the possibility of a temperature gradient across the shells. If the inner core, for example, is at higher than ambient temperature

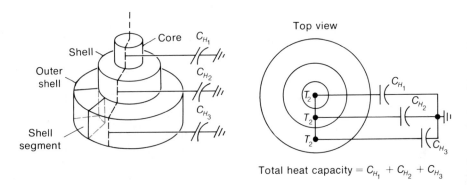

FIG. 4.24   *Heat capacity of an assembly of thermal-storage elements.*

(because of some form of heat generator such as the combustion of gases or metabolic processes), then heat will flow radially outward. We can easily see that, for such a situation, a parallel arrangement of capacitors would not be an adequate representation of the physical system. Capacitors represent only the heat-storage capability of the system. The temperature gradients that accompany heat flow indicate that the system also has flow-resistive properties. Therefore, to represent the thermal system more accurately both capacitors and resistors must be included in the simulation.

In order to understand how to simulate a thermal system with flow, we shall first review the procedure used to develop the lumped simulation of an elastic tube (see Sec. 4.5). It will be remembered that the tube was divided into small segments. Each segment was represented by a pair of resistors and a capacitor. One resistance was considered the opposition to flow delivered to the segment, and the other the opposition to flow delivered from the segment; the intermediate capacitor represented the storage within the tube. An appropriate assembly of such resistance-capacitance networks served to represent a length of elastic tube (Fig. 4.21). In an analogous manner a segment of a thermal system with undirectional flow is represented by a pair of resistors and an intermediate capacitor. The resistors represent flow opposition to and from the system; the capacitor represents heat storage within the system. An appropriate assembly of such networks serves to represent a thermal element.

To show how this procedure is used, we examine a segment of the outer shell of the system of Fig. 4.24. An expanded view of the segment is shown in Fig. 4.25a. It is assumed that heat flows radially from the inner face to the outer face (i.e., from T_1 to T_2). The resistance to flow from T_1 to the center of the segment (T_A) is represented by the lumped resistor r. The resistance to flow from the center of the segment to T_2 is represented by another lumped resistor r. It is assumed that T_A represents the average temperature of the segment and that the heat stored in the segment is adequately represented by the product of the heat capacity and average temperature. The capacitor connected at T_A thus represents the storage capability of the segment. (The direct analogy between the shell segment and the tube segment should be obvious.)

To construct the analog of the total shell we now simply connect the analog networks of the segmental components in an appropriate manner. If the total shell is divided into n segments, then n analog networks are needed to represent the system. These networks are joined at T_1 and T_2 (as shown in Fig. 4.25b) as the inner surfaces of all segments are at the same temperature, T_1, and the outer surfaces are at T_2. For a homo-

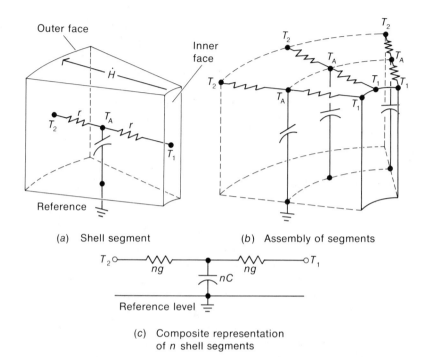

(a) Shell segment   (b) Assembly of segments

(c) Composite representation
of *n* shell segments

FIG. 4.25 *Simulation of a thermal system with radial heat flow.*

geneous shell material it could be expected that the average temperature
(T_A) in each segment is the same and that the resistance and capacitances
of each segment are the same. Under these circumstances all the "T_A
nodes" in Fig. 4.25*b* may be joined, and we then see that the resistances
and capacitances now form parallel networks. These may be combined
into single equivalent resistors and capacitors, and the result is the
simple network shown in Fig. 4.25*c*. This network now represents a
lumped analog representation of a thermal system (which can be compared with the lumped analog representation of an elastic tube).

The shell configuration in Fig. 4.24 has physiological interest as it
represents one type of model proposed for the study of the thermal
behavior of the human torso.* The inner core represents the "deep tissue"
such as the various internal organs. The middle shell includes subcutane-

---

* From J. Stolwicyk, Jr., unpublished lecture, Yale University, Pierce Foundation, New
Haven, Conn., 1962. Mentioned by J. H. Milsum (Ref. 9). Chapter 3 in Milsum's book is
devoted to a rather detailed discussion of a mathematical model of the human thermo-
regulating system. A fairly comprehensive explanation of a computer simulation of the
human temperature-regulating system is given in Ref. 10.

ous fat and muscle tissues; the outer shell represents the skin covering. A diagrammatic sketch of how the human torso may be represented by a model composed of three cylinders is shown in Fig. 4.26. The inner core is of relatively large diameter and of relatively constant temperature. The other two shells are very thin; together they probably do not exceed a thickness of about 2 to 3 cm. (In most areas of the body a constant temperature is attained after a penetration of about 3 cm.)

From the discussions of the preceding paragraphs it is easily seen how an approximate, lumped model of the physiological thermal system may be represented by an electrical analog. If we assume that heat flow is primarily radial, each shell is represented by a T network of the form shown in Fig. 4.25c. The T network for each shell is connected to the network of an adjacent shell at points that represent contacting surfaces (i.e., points of equal temperature). For n such shells we would generate a ladder network of the form shown in Fig. 4.22a. For the three shell representations suggested in the previous paragraph, the composite simulation is the network in Fig. 4.27. Some comments about this circuit are in order for, as can be seen, it includes only two T networks (for the skin and subcutaneous shells) instead of the expected three. The explanation is straightforward. Through internal measurements it is shown that the inner core is at relatively uniform temperature. This implies that

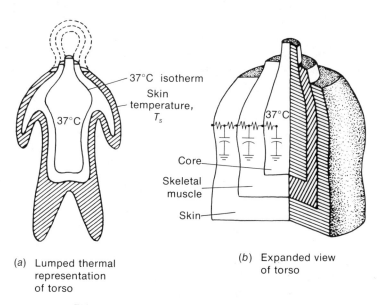

(a) Lumped thermal representation of torso

(b) Expanded view of torso

FIG. 4.26 *Three-cylinder model of human torso.*

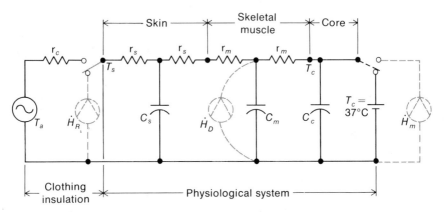

FIG. 4.27   *Electrical analog of a lumped model of physiological thermal system.*

for our purposes we can assume there are no temperature gradients in the inner core. Resistive effects in the inner core can therefore be neglected; resistive elements are thus omitted in the simulation for this thermal mass. On the other hand, the inner core is a massy element at some temperature; it can therefore store heat. A capacitor would therefore be required to represent the heat-storage capability of the inner core.

The circuits in Figs. 4.27 and 3.20*b* were developed to represent essentially the same system; yet by inspection we see that the circuits are obviously different. It is important to determine why these differences exist and how they may be reconciled because, as has been repeatedly pointed out, by examining the differences between systems (or different analogs of the same system) we gain insight into the operation of the systems. To start, we observe that, if the capacitors are neglected, the circuit of Fig. 4.27 reduces essentially to that of Fig. 3.20*b*. (The sum of the series resistors can be represented by a single equivalent resistor.) We suspect, then, that the circuit of Fig. 4.27 is a more general representation of the physiological system. We can justify this viewpoint when we remember that the circuit of Fig. 3.20*b* was developed with the assumption of "steady operation"; i.e., heat loss was evaluated only when temperature readings reached steady values (see the beginning of Sec. 3.10). No such constraint was included in the development of the circuit in Fig. 4.27. This circuit must therefore be the more general circuit valid for both steady (or dc) and transient operation. We can now intuitively realize that, when a simulation includes only resistive elements, the simulation either represents a system which in actuality does not have storage or it represents only the dc operation of a system with storage.

Some comments are also in order regarding the representation of the driving sources shown in Fig. 4.27. To explain these source representations, we digress in order to discuss excitation sources in general. (This is primarily a qualitative discussion. For a more complete discussion of source representations see Appendix B.) The use of the voltage source as an excitation has been frequently encountered. In Fig. 2.10 such a source was used to represent a pressure excitation; in Fig. 3.16 the voltage source was used to represent a temperature excitation. Pressure and temperature are across-variables as, of course, is voltage. It seems reasonable, therefore, that pressure and temperature can be represented by a voltage source, and it also seems quite reasonable to expect voltage sources can be used to represent across-variable excitations in general.

If we examine the function of a voltage source when it acts as an analog excitation, we realize that in most cases such a source must not only supply the across-variable but must also serve as a current (through-variable) source. We can easily justify this contention when we note that the voltage source T_I shown in Fig. 3.21 must supply not only a voltage (to represent the internal or core temperature) but also a current flow to represent the heat transferred (or the heat generated by metabolic processes). In a similar manner, if a voltage source represents a pressure supply in a fluid system, the source must not only supply voltage to represent pressure; it must also be able to supply current to represent flow. In general, then, sources must be devices that can supply both through and across excitations.

This conclusion must now give us cause to wonder why an excitation supply is not simply described as a source instead of a *voltage* source. What is implied by the descriptive adjective, voltage? At this point it may be worthwhile to separate need from availability. For simulation purposes, it is often necessary to supply or control a specific excitation waveform of one kind of variable or another; the uncontrolled variable simply adjusts to meet the needs of the system. Whether physical hardware is available which can do the required job is a separate problem. To simulate a temperature-controlled room (which maintains a constant temperature), for example, we might consider an excitation source which can maintain a constant voltage to represent room temperature. The current drawn from the source would represent the heat which must be removed from the room to maintain the fixed temperature. This current drain would vary, depending on the demands of the system; the voltage supplied, however, should remain constant, in spite of variable current requirements. (The need for the constant voltage source to simulate the constant controlled temperature should be evident.)

Whether sources are available (or can be built) which can operate at constant voltage within the required limits of current range is a question which must be separately resolved. A simple battery may not maintain the stable voltage output required during heavy current drain (see Appendix B) in which case more elaborate (and expensive) voltage sources must be used. For most applications, however, it is possible (within a specified range at least) to obtain a source which will supply a specified voltage waveform though current demand may vary.

It is easy now to extend the previous discussion to include the concept of current sources. For some simulations it is often necessary that the source deliver a controllable through-variable; the across-variable is then self-adjusting. If we use current as an analog for through-variables, then we can readily appreciate the need for electrical sources which will deliver a controlled current in spite of variable voltage requirements. Though electrical current sources are not as popular or familiar as voltage sources, they may be constructed in the laboratory. (See Appendix B and the discussion following Fig. 4.29.) Recently, commercial power supplies have become available which include rather versatile current sources.

Returning now to Fig. 4.27, we note that three current sources are included. These simulate the delivery of heat to the physiological system. The sources may be constant or variable, depending upon the conditions we wish to simulate. The source labeled $\dot{H}_m$ represents a generator that delivers current equivalent to the metabolic rate. By using a single current source, we in effect assume that the distributed metabolic processes may be adequately represented by a heat source located within the deep tissues of the body. For many purposes this approximation may be sufficient. In some cases, however, a more distributed representation may be required. When there is additional muscular activity due to exercising (or shivering, as induced by artificial cooling, i.e., hypothermia), for example, the local metabolic processes generate additional heat in the tissues nearer the skin surface. To represent this additional distributed mechanism, the current source labeled $\dot{H}_D$ is included (in phantom section) in Fig. 4.27. Finally, to represent the delivery of a controlled heat flow *to* the system, the source $\dot{H}_R$ is included in the diagram. This source would be used to simulate the application of heat to the physiological system, as might be experienced during diathermy treatments or exposure to infrared (or other heat-producing) radiative sources.

In general across- and through-variable sources do not drive the same point. One is sufficient to supply the necessary excitation. Thus in Fig. 4.27 the inner core could be excited by either a voltage source or a current source, not both. The choice of which one to use depends upon the information available and the simulation desired. If, for example, the core temperature for a particular test is known and this test is to be studied, then a controlled voltage source would be used in the simulation. (For this situation the current drawn from the voltage source could be measured to estimate the metabolic rate.) If, on the other hand, the metabolic rate is known, then a controlled current source would be used for excitation (and the voltage at the source could be used to estimate core temperature). This will be more clearly understood after a discussion of the specific example presented in the next section.

4.9 ANALOG STUDY OF A THERMAL–SYSTEM RESPONSE

The utility of the analog developed in Sec. 4.8 and the concept of source excitations can be better understood if we examine them in the context of a specific illustration. We cannot use this simulation to examine all the contingent complexities and subtleties of the physiological thermal system. The analog is far too simple for such a study. However, we shall see that, simple though it may be, the model of Fig. 4.27 can be used to investigate some of the general aspects of the thermal-system transient response for some simple though realistic situations.

As an illustration of how the analog may be used in a specific investigation we shall examine the temperature response of the physiological system when it is subjected to change of external and internal operating conditions. In particular, we shall examine the response of a human subject when he is clad in clothing designed to protect him against the extreme-temperature environments of outer space or nuclear blasts. First, we shall examine the initial temperature response when the subject dons the insulated suit; then we shall examine the temperature response of the subject when he performs tasks that increase metabolic activity.

In order to utilize the circuit of Fig. 4.27 to perform the suggested studies we must first decide upon the sources to excite the system and practical resistors and capacitors for the circuit elements. To resolve the former problem we make the simplifying assumption that for the first study the subject is completely passive; he will not dress himself but will instead be dressed by others. Under these circumstances muscle activity will be minimal. For the initial investigation we assume therefore that $\dot{H}_D$ may be neglected as a separate source and that muscular activity may be lumped with the other metabolic processes of the core. As a further simplification of the representation, we assume that total heat transfer between the subject and his environment can be adequately (or equivalently) described by heat flow through an insulation resistance (r_c). Radiated heat to and from the subject may therefore be neglected as a separate transfer mechanism and the radiative heat source $(\dot{H}_R)$ shown in Fig. 4.27 may thus also be omitted from the diagram.

Finally, to decide what source should be used to excite the core, we note that, for healthy subjects at rest, the metabolic rate stays relatively constant for a rather wide range of skin-temperature variations.* Although the internal temperature or core temperature is also relatively constant for a wide range of temperature variations, this condition is achieved only after long operating times when dc or steady operating conditions finally prevail. Transient changes in body temperature may certainly be expected, particularly when the total body surface is exposed to changes in ambient temperatures. Under these circumstances, a current (heat) source rather than a voltage (temperature) source would be the more appropriate and simpler core excitation. If the suggested changes are included in the circuit of Fig. 4.27, the result will be that shown in Fig. 4.28. The voltage (temperature) source T_a is retained so that the ambient conditions may be varied as desired.

The question that must next be resolved is what numerical values to use for the circuit elements. The system is not linear, and for large temperature variations the thermal resistance and capacitance cannot be expected to remain constant (see, for example, Sec. 3.10 and in particular Fig. 3.18). For an initial study, however, we assume average values for the circuit elements, with the understanding that this representation has limited applicability; i.e., it can be used only to study the gross operating characteristics of the system when large signal variations are applied.

To determine an appropriate average resistance that might be used in the simulation, we note that, within a range of temperatures that are tolerable by the human system, static thermal resistance varies from about 0.02 to 0.06°C-h/kcal (see Fig. 3.19). We hope to keep the physiological system within a tolerable temperature range, and so we accept an average value of 0.04°C-h/kcal for the overall thermal resistance during steady, quiescent operation. Since the overall resistance for steady operation is equivalent to the sum of the resistors in Fig. 4.28, $r_1 + r_2 + r_3$ must be equal to 0.04 thermal resistance unit.†

---

* Metabolic rate changes at a rate of about 10 percent per degree Celsius change of average body temperature (Ref. 12, p. 1046). Average body temperature may be approximated by

$$T_b = 0.3 T_s + 0.7 T_c$$

where T_s and T_c are the skin and core temperatures, respectively, and the factors 0.3 and 0.7 are weighting factors that reflect the relative significance of these temperatures. Evidently, if core temperature changes slowly (or not at all), average body temperature will not show large variations and hence metabolism will be relatively constant. This is a rough estimate; if more careful work is required, a more accurate evaluation of metabolic rate would be necessary.

† There are no standard thermal units for heat capacity and heat resistance (although the clo could probably be used as a unit for the latter property). For brevity of notation we

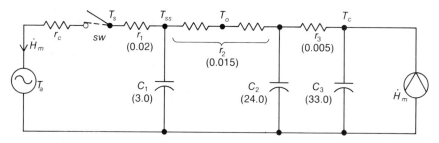

FIG. 4.28 *Simulation of an insulated physiological system with a steady metabolic rate* $(\dot{H}_m)$ *exposed to a variable ambient temperature* (T_a).

With no further information we could not even guess how to distribute the total resistance among the three resistances shown. However, it is known that internal body temperatures remain relatively constant and that temperature gradients become rather high near the surface of the physiological system. This would lead us to expect higher resistance values near the system surface. With this as a basis for judgment, we arbitrarily assume r_1 to be 0.02 unit, r_2 to be 0.015 unit, and r_3 to be 0.005 unit. This choice of values represents merely an educated guess. We can defend this procedure, however, because this is only an introductory investigation and not a detailed accurate study. Furthermore, since the electrical analog is so conveniently versatile, there is no need to specify uniquely the circuit components immediately. If more information becomes available so that a more accurate estimate is possible, the values of the circuit component can easily be changed.

To obtain some estimate for the capacitors to be used in the simulation, we take the average value for the heat capacity of physiological tissue to be 0.83 cal/g/°C. (See the end of Sec. 4.6.) A normal man weighs about 70 kg. The overall heat capacity can therefore be taken to be approximately 60 kcal/°C. This would mean that $C_1 + C_2 + C_3$ (in Fig. 4.28) must be equivalent to 60 kcal/°C. (See the discussion at the beginning of Sec. 4.8 and Fig. 4.24.) To distribute this thermal capacity, we note that the skin is about 5 percent of the body weight and the skeletal muscles are about 40 percent of the body weight.[*] From these figures we can estimate C_1 to be 3 kcal/°C, C_2 to be 24 kcal/°C, and C_3 to be 33 kcal/°C.

---

shall define our own units and consider a thermal capacitive unit (TCU) to be 1 kcal/°C and a thermal resistive unit (TRU) to be 1 °C-h/kcal. Thus heat delivered through 1 TRU at a rate of 1 kcal/h will cause a 1°C drop across the TRU; 1 kcal of heat delivered to 1 TCU causes a temperature change of 1°C in the TCU.

[*] See Ref. 11.

The numerical values for the various thermal circuit elements are parenthetically shown in Fig. 4.28. These values are impractical from an electrical point of view. If heat flow ($\dot{H}_m$) in kilocalories per hour is represented by current flow in amperes and temperatures in degrees Celsius is represented by volts, the values shown for the circuit elements are in ohms and farads, i.e., TRU = 1 Ω and TCU = 1 F. The resistors are impractically small and the capacitors are impractically large. The circuit shown in Fig. 4.28 would have to be scaled to include practical components. Scaling in this case is not as simple as for the purely resistive system previously encountered (see Sec. 2.4). The storage element introduces an additional complication. In Chap. 5 we shall see how a system "time-constant" can be used in such cases to help establish an appropriate scale. Here we resort to the less elegant—but no less effective—technique of "guessing" coupled with intuitive reasoning.

As a start we observe that, even if we could use the correspondence between variables as suggested in the preceding paragraph, we have already introduced a shift in time scale between the electrical and thermal systems. To show this we note first that, by our choice of analogies, the kilocalorie of heat automatically becomes equivalent to a coulomb of charge. (A coulomb of charge delivered to a 1-F capacitor causes a voltage change of 1 V. A kilocalorie of heat delivered to a heat capacitance of 1 kcal/°C causes a temperature change of 1°C. If volts and degrees Celsius are analogous and 1 F is equivalent to 1 TCU, then 1 kcal is equivalent to 1 C.) We note further that the choice of analogies also automatically establishes a correspondence between the heat flow in kilocalories per hour and the current flow in amperes (coulombs per second). [A 1-V drop is developed by a 1-A current through a 1-Ω resistor; a 1° temperature change (°C) is developed by a flow of 1 kcal/h through 1 TRU. Volts and degrees Celsius have already been chosen as analogous variables. If in addition we choose 1 Ω to be equivalent to 1 TRU, then 1 kcal/h automatically becomes equivalent to 1 C/s.] Since kilocalories and coulombs have already been established as analogous quantities, 1 s in the electrical system is equivalent to 1 h in the thermal system. This implies that events taking 1 h to occur in the thermal system will occur in 1 s in the electrical system. This compression of time can be a tremendous advantage as it permits performing many experiments on the simulation in the same time it would take to perform a single experiment in the real system.

To show some of the ideas developed in the preceding paragraph in equation form we compare the basic relations for storage in both systems:

(4.36) $q = C_e e$ q in coulombs, e in volts, C_e in farads

$H = C_T G$ H in kilocalories, T in °C, C_T in TCU

If we divide these equations, we have

(4.37)
$$\frac{q}{H} = \frac{C_e}{C_T}\frac{e}{T}$$

When the ratios on the right side of Eq. (4.37) are established, the storage correspondence between the systems is established. We now compare the basic relations for flow in both systems:

(4.38)
$$e = iR_e$$

$$T = \dot{H}R_T$$

By dividing these equations we again form analogous ratios:

(4.39)
$$\frac{e}{T} = \frac{i}{H}\frac{R_e}{R_T}$$

Since the ratio of the across-variable has already been established, the flow correspondence between the systems is established when the resistance ratio is specified. Thus the capacitive ratio establishes a storage scale, the resistive ratio establishes the flow scale, and together they establish the time scale.

To apply the above procedure, let us make 1 microfarad (μF) equivalent to 1 TCU and 1 V equivalent to 1°C. From Eq. (4.37), then, 1 microcoulomb (μC) is equivalent to 1 kcal. If we also make 1 megohm (1 megohm = 1 MΩ = $10^6\Omega$) equivalent to 1 TRU, then from Eq. (4.39) 1 microampere (1 μA = 1 μC/s) is equivalent to 1 kcal/h. From the two relations, 1 s in the electrical system is equivalent to 1 h in the thermal system. The scale relations just developed are shown in the third row of Table 4.2. The numerical values of the components are still not convenient, and by repeating the outlined procedure we arrive at the more convenient scale shown in the fourth row of the table.

We are finally in a position to use the circuit of Fig. 4.28 for the study suggested at the beginning of this section. The purpose of the study, it will be recalled, is to investigate the transient temperature response of the physiological system when quiescent conditions are altered. The actual circuit used to perform the first study is shown in Fig. 4.29. When the switch is at A and the circuit is in the steady state, it represents a passive subject dressed in a 1-clo suit (represented by a 90-Ω resistor; ambient temperature is 25°C, that is, 2.5 V). It is assumed that the subject generates approximately 100 kcal/h (10 μA) during the "resting" activity. This is simulated by the 10 μA current source shown by dashed representation in Fig. 4.29. [The current source actually used was a "high" voltage supply (100 V) driving a 9.7-MΩ resistor. This resistor is so

TABLE 4.2

	C_1	C_2	C_3	r_1	r_2	r_3	Time	Through	Across	Row
Thermal	3 TCU	24 TCU	33 TCU	0.02 TRU	0.015 TRU	0.005 TRU	1 h	1 kcal/h	1°C	1
Electrical	3 F	24 F	33 F	0.02 Ω	0.015 Ω	0.005 Ω	1 s	1 A	1 V	2
	3 F	24 F	33 F	0.02 MΩ	0.015 MΩ	0.005 MΩ	1 s	1 μA	1 V	3
	0.03 μF	0.24 μF	0.33 μF	0.02 MΩ	0.015 MΩ	0.005 MΩ	0.01 s	0.1 μA	0.1 V	4

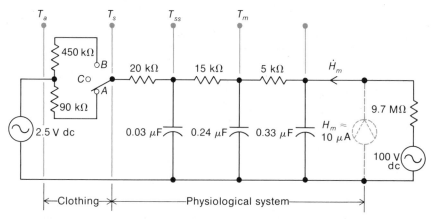

FIG. 4.29 *Practical simulation to study transient effects of insulated clothing.*

much higher than the rest of the circuit resistance that the steady current flow through the system could be taken as (100-2.5)/9.7 μA or approximately 10 μA. Throughout any one test, current flow varied from a high of 100/9.7 = 10.3 μA to a low of 100-2.5/10.2 = 9.6, which for our purposes can be considered relatively constant.]

To determine the transient response of the subject when he is attired in insulated clothing, the switch is flipped from A to B. This changes the clothing resistor from 90 to 450 kΩ, which in effect simulates a sudden, fivefold increase in the insulation value of the subject's dress (from 1 to 5 clo). The various temperature responses can be determined by voltage measurements at the various points. For a condition of 25°C constant ambient temperature, the measured transient response of the skin temperature (T_s), the subdermal temperature (T_{ss}), and the internal or core temperature (T_c) are shown in Fig. 4.30a. As might be expected, all body temperatures rise (and if allowed to continue would ultimately stabilize at a higher level, as shown by the dotted curves). The rise, though rapid, is not instantaneous; the heat capacitance of the tissue causes a transition time lag. This can easily be verified by repeating the test with the capacitors removed. The results of such a test are shown by the dashed lines in Fig. 4.30. From a practical point of view, it would not be feasible physiologically to wait until the temperatures stabilize. The subject would become very uncomfortable and would probably sweat profusely. The model does, however, give a crude estimate of what might be expected if the abnormal situation persisted.

When the switch in Fig. 4.29 is flipped to A we can "watch" the temperature changes that might be expected when the subject removes

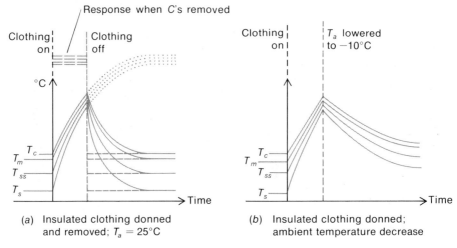

FIG. 4.30 *Transient temperature responses of a physiological system as predicted by the circuit of Fig. 4.29.*

the insulated clothing. These curves are also shown in Fig. 4.30a. It is to be noted that the temperature rises somewhat more slowly than it drops. The higher clothing resistance seems to slow the transient responses. It will be seen later that this should be expected as the higher clothing resistance causes a "longer time-constant" and hence a slower transient response.

Figure 4.30b shows the response curves that might be expected when the subject is clothed in a 5-clo suit and is then subjected to an ambient temperature of $-10°C$. As expected, body temperatures drop when the ambient temperature is lowered. In this case we note that at the beginning of each transient the rate of temperature change is about the same. The system "time-constants" are not changed when the system components are not changed. The concept of time-constant and some of its implications will be examined in Chap. 5 and in later chapters. It is interesting to observe that, with a 5-clo suit, a subject could remain comfortable "indefinitely" if he remains inactive and the external temperature is about $-10°C$ (18°F). In a 5-clo suit it is not likely that a person could do anything but remain inactive; clothing with this insulation value is very bulky, and simple movements become almost impossible in such inconvenient gear. We are therefore led to the conclusion that, as a practical matter, it is not possible to clothe an individual adequately so he can remain inactive and comfortable in an environment of less than $-10°C$ (18°F).

This may appear surprising; 18°F does not seem to be a very low temperature. Individuals have been exposed to much lower temperatures and have survived. However, for survival under such conditions either the persons were active for long periods and inactive for short periods or their apparel was artificially heated (e.g., by electrical means). The validity of these conclusions is obvious from the reaction of people waiting for a bus in winter weather. Few can stand still for more than 5 min, no matter how well dressed. This may be due to impatience but the primary motivation is to generate some additional internal heat. Regardless of dress, passive metabolism is not enough to keep the body temperature up when ambient temperatures drop below 18°F.

It will be interesting to compare the results obtained from the analog simulation with those of an actual test on a human subject. Some experimental data which can be used to make the desired comparison are available. Minard et al. (Ref. 13) studied the deep body and skin temperature responses of a subject suddenly insulated so that "the entire resting heat production was stored in the body by preventing its escape." To produce this condition the supine subject was clothed from head to foot in impermeable plastic garments with blankets above and below him. Saturated, heated air was delivered for breathing. We have not performed an analog study of this situation as yet, but it would be easy to do so. By flipping the clothing switch in Fig. 4.29 to point C, we in effect introduce infinite insulation (resistance) into the system and thereby interrupt current flow from the system. This would simulate the physiological conditions induced by Minard and coworkers.

The recorded responses of both systems are shown in Fig. 4.31. The curves in Fig. 4.31a represent the simulated temperatures as measured on the analog circuit. The curves in Fig. 4.31b represent the actual temperature measurements taken from the physiological system. The physiological temperature measurements, T_R (rectal temperature), T_E (tympanic temperature), and T_s (weighted skin temperature), correspond to the voltage measurements T_c, T_m, and T_s, respectively. As can be seen, two of the responses shown by the simulated circuit compare favorably with the response of the actual physiological system. When current flow is interrupted in the simulation circuit, T_s almost immediately rises to the level of T_m and thereafter both rise at a fairly rapid rate. When current flow is reestablished (by switching back to A) these temperatures fall at separate rates (although in both cases they fall at a more rapid rate than they rise) until they reach the same pretest equilibrium levels. The physiological system behaves in an almost identical fashion when heat flow is interrupted (insulation applied) and when heat flow is again permitted (insulation removed). The numerical values of the temperature

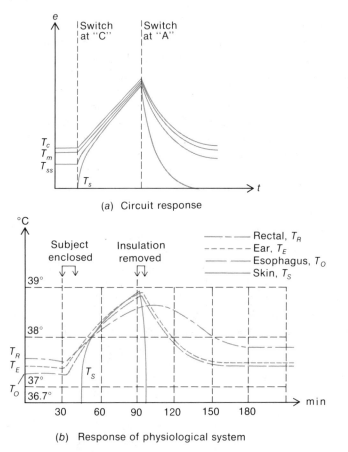

(a) Circuit response

(b) Response of physiological system

FIG. 4.31   *Transient temperature response of a subject enclosed in impermeable insulating garments.* [*Part (b) is adapted from Ref. 13.*]

levels attained and the system operating times were not checked closely though they appeared to be in approximate agreement. Good numerical correspondence could not be expected as the circuit parameters were not chosen particularly to match the characteristics of the physiological subject. The circuit, for example, was driven by a 10-μA current source. This is equivalent to a metabolic rate of 100 kcal/h. The metabolic rate of the human subject who was tested was about 80 kcal/h. Other minor numerical differences would easily account for differences in numerical results.

There is one glaring difference between the data acquired from the analog simulation and from the physiological system. In the simulated system, the core temperature and other body temperatures become very nearly equal as soon as flow is interrupted. Thereafter the rate of core-temperature rise is almost identical to the rise rate of other body temperatures. In the actual system, skin and esophageal temperatures become equal to the rectal temperature soon after flow is interrupted. Thereafter, however, skin and esophageal temperatures rise together at a faster rate than does the rectal temperature. The former two thus become and remain higher than the latter throughout the period of interrupted flow. This constitutes a fundamental difference not easily explained by component dissimilarity. A different choice of values for the components or the use of nonlinear components would not change the circuit response in the proper manner; the basic difficulty is the circuit configuration. The circuit is arranged so that current can only flow outward—from the core to the skin. This automatically means that the core voltage must always exceed the other voltages of the system; expressed in another way, the core voltage can never be less than the other circuit voltages. We see then that, if a single current (heat) source "centrally" located is used, the internal temperature response of the physiological system cannot be reproduced. The remedy now seems self-evident: If the distal nodal voltages are to exceed the core voltage, there must be current flow toward the core. To provide such current (heat) flow, a current source must be located at some position between the core and the external surface. In short, if the analog circuit is to provide a more valid simulation of the internal response of the physiological system, it should include an additional current source as illustrated by Fig. 4.32. The arrows radiating from the source indicate it can produce current flow to the skin, to the dermal-shell storage area, and to the inner core. The direction and the amount of flow depend upon the relative magnitudes of the two sources and external conditions.

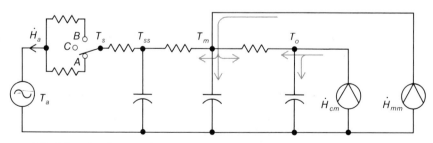

FIG. 4.32 *Simulation of Fig. 4.29 modified by addition of a distal current source.*

The location of the added source in Fig. 4.32 was chosen arbitrarily. We could just as well have located the sources at node T_{ss} or T_s. Any (or all) of these sources could have been used to provide for flow in the desired direction. If we include the current source only to make the circuit respond in a more realistic fashion, then, to some extent, we lose the advantage of the analog simulation. In order to use the simulation to its fullest advantage, not only should we use circuit changes to make the response more realistic but also to try to understand how such changes relate to the physiological system. In this case, the relationship between the required circuit change and the actual physical system is easily understood. In an earlier discussion, the current source in Fig. 4.27 was initially neglected, as it was assumed "muscular activity will be minimal"; from this assumption it was further assumed that "muscular activity may be lumped with the other metabolic processes of the core." It is easy to see how the current source required to make the circuit provide a more realistic internal temperature response may actually be considered the muscle heat source which was originally neglected. Evidently, then, even if muscular activity is minimal, the metabolic processes associated with skeletal muscles cannot be neglected or lumped with other processes when internal temperature responses are under consideration.* The additional current source required should therefore be one that simulates the heat generated by muscular metabolic processes. It should be located as a supply source to the storage element representing the muscle shell. The additional source required should therefore drive the T_m node, as shown in Fig. 4.32.

In order to use the newly developed circuit analog of Fig. 4.32, we must first decide upon the magnitudes of the excitation sources. The decision is not critical in this case, as the investigation is only an introductory one not designed for high accuracy; furthermore, if necessary, we can easily change any choice we may make. As a rough guess, therefore, we shall assume that the skeletal muscles generate 40 percent of the total metabolic heat (skeletal muscles constitute 40 percent of body weight). The core source ($\dot{H}_{cm}$) in Fig. 4.32 should therefore deliver 6 μA, and the muscle source ($\dot{H}_{mm}$) should deliver 4 μA. When the suggested modifications are included in the circuit and Minard's experiment is simulated once again, we find that the circuit and physiological responses are in closer agreement. (See Fig. 4.33.) There are still differences, and

* As far as "external" temperature responses are concerned, we shall see that it makes little difference whether all metabolic processes are lumped or not. Skin temperatures, for example, will remain relatively unaffected if the heat excitation is a single source or two sources located separately. This seems intuitively obvious. We would not expect the rearrangement of internal heat sources to affect external temperatures drastically. We would, however, expect that internal temperatures could be significantly affected by such a change.

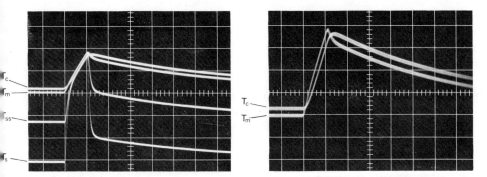

FIG. 4.33 *Response of the circuit of Fig. 4.32 when switching between A and C.*

in order to reconcile them further modifications would be required. However, the major aspects of the responses are similar, and this gives us confidence that we understand at least the gross operating characteristics of the physiological system. We observe that the change from one to two excitation sources did not drastically change the responses of T_{ss} and T_m, but the core response T_c has changed significantly. This agrees with the comment made earlier: The reorientation of the source excitation does not affect the external response to any great degree. However, the internal response is altered.

Having gained confidence in our circuit configuration, we can now use it to make a more quantitative comparison between the systems. We could do this by appropriately adjusting or analyzing the analog circuit and then comparing analog and physiological data. Although we are not yet in a position to make a detailed analysis of this circuit analog (even if we could, it might not be warranted at this stage of an investigation), there are some elementary analytic relations that can be used to obtain some quantitative results. As an illustration, we note from Fig. 4.34 (which is a modified version of the circuit in Fig. 4.32) that, when we interrupt the delivery of heat to the surrounding environment, all the generated heat is stored, and we may write

(4.40)   Rate of heat generated = rate of heat stored

or

$$\dot{H}_{mm} + \dot{H}_{cm} = \dot{H}_s + \dot{H}_m + \dot{H}_c$$

where $\dot{H}_{mm}$ and $\dot{H}_{cm}$ are the muscle and core metabolic rates and $\dot{H}_s$, $\dot{H}_m$, and $\dot{H}_c$ are the rates of heat storage in the "skin" shell, "muscle" shell, and core, respectively. Since the sum of the core and muscle metabolic rates is equivalent to the total metabolism ($\dot{H}_m$), the left side of Eq. (4.40) can

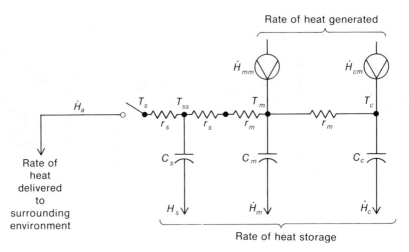

FIG. 4.34 *Modified schematic diagram of the thermal-analog circuit.*

be replaced by $\dot{H}_m$. The rate of heat stored in the skin may be given by $C_s(dT_{ss}/dt)$ or $C_s\dot{T}_{ss}$ [that is, $H_s = C_sT_{ss}$; therefore, $\dot{H}_s = C_s\dot{T}_{ss}$. See also the analogous relation Eq. (4.14)]. In a similar manner, the rate of heat storage in the muscle is given by $C_m\dot{T}_m$ and that in the core by $C_c\dot{T}_c$. By substituting these relationships in Eq. (4.40) we now have

(4.41) $$\dot{H}_m = C_s\dot{T}_{ss} + C_m\dot{T}_m + C_c\dot{T}_c$$

The time derivatives of the temperatures (that is, $\dot{T}_s$, etc.) are in the temperature changes that develop because of the steady increase of stored heat. These rates of temperature change may be interpreted as the slopes of the respective temperature vs. time curves that are obtained when there is a change in the amount of stored heat. From the response of the analogous circuit we can see that when the T_m and T_{ss} curves rise with a constant slope the T_c curve also rises at a constant slope and all three slopes are equal. If we look for the corresponding behavior in the physiological system, we note that the temperature-response curves of Fig. 4.31b become parallel (at $t \approx 50$ min) as predicted by the analog circuit. The parallelism of the actual-response curves is not as pronounced as that shown by the circuit analog. One of the reasons for this difference is that the period of excitation for the analog system was proportionately longer than that for the physiological system, or, expressed somewhat differently, the excitation period in the physiological test was not of sufficient duration to produce the definitive results evidenced by the circuit analog. From the available physiological data, however, it can be seen that the three temperature curves tend to approach a condition of constant and equal slopes.* Once we realize this condition can and does exist, we can rewrite Eq. (4.41) as

$$(4.42) \qquad \dot{H}_m = (C_s + C_m + C_c)\dot{T}$$

Equation (4.42) indicates a method by which an in vivo measurement of the total heat capacity of the physiological system can be obtained. By Eq. (4.42) this value is simply the ratio of the total metabolic rate to the slope of the temperature-response curve. If we use the values given in Minard's paper (Ref. 13), we find $\dot{H}_m$ is 92 kcal/h; from the response curves we find $\dot{T}$ is 1.44°C/h. The sum of the heat capacities is thus 92/1.44 kcal/°C or 64 kcal/°C. This agrees very nicely with the assumed value of 60 kcal/°C derived earlier by using the average value of 0.83 cal/g/°C for the heat capacity of tissue.

Although some additional observations and comparisons between the physiological and analog systems can be made, further examination at this time would be somewhat anticipatory. The discussion here was an introductory exposition of a (distributed) system which could not be considered either purely resistive or purely storage. More detailed examination of such systems will be presented in later chapters. We note here, however, that even with only an introductory background it is possible to use analog concepts to gain insight into systems with combined properties. The procedure is continuously to compare the responses of the analog (or model) with those of the physiological system. When there is a discrepancy, an attempt then is made to modify the analog (or model) to obtain a more adequate representation. In so doing the investigator is forced to focus attention on the intrinsic properties of the actual system.

---

* This parallel response of the temperature curves could have been predicted by heuristic or theoretical consideration of the circuit analog. By the first approach, we note from Fig. 4.34 that if $\dot{H}_{cm}$ and $\dot{H}_{mm}$ are constant we must intuitively expect that $\dot{H}_s$, $\dot{H}_m$, and $\dot{H}_c$ will each ultimately be constant. If $\dot{H}_c$ is constant, then $\dot{T}_c$ is constant and so is $T_c - T_m$. However, if the latter difference is constant, then T_c and T_m must increase at the same rate, or $\dot{T}_m = \dot{T}_c$. By similar reasoning, we can see that $\dot{T}_{ss}$ must be equal to $\dot{T}_m$. To express these ideas analytically at the T_c node, we write

$$\dot{H}_{cm} = C_c\dot{T}_c + \frac{T_c - T_m}{r_m} = \text{heat to storage} + \text{heat flow from core}$$

If $\dot{H}_{cm}$ and C_c are constant and $\dot{T}_c$ becomes constant, then by differentiating the last expression we get $\dot{T}_c = \dot{T}_m$. Thus, when $\dot{T}_c$ becomes constant, $\dot{T}_m$ becomes constant.

At the T_m node we write

$$\dot{H}_{mm} = \frac{T_m - T_{ss}}{r_s + r_m} + C_m\dot{T}_m + \frac{T_m - T_c}{r_m}$$

If $\dot{H}_{mm}$, C_m, $\dot{H}_{cm}$, and C_c are all constants, then, when $\dot{T}_c$ becomes constant, $\dot{T}_m$ and $T_m - T_c$ become constant. Differentiation of the last expression then yields $\dot{T}_m = \dot{T}_{ss}$.

4.10 STORAGE IN A MECHANICAL SYSTEM

A muscle is ordinarily thought of as an active system or a source generator (see Appendix B); its prime function is to generate force. However, if the muscle is "deactivated" (i.e., its force-generating mechanism remains inactive), it still retains (or shows evidence of) passive mechanical properties. An externally applied force, for example, will cause a deenergized muscle to stretch so that the characteristic force-displacement curve shown in Fig. 4.35b is generated. Such force-displacement characteristics

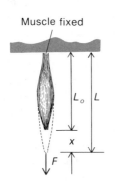

(a) Passive muscle under test

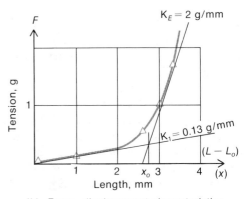

(b) Force–displacement characteristics

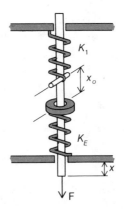

(c) Mechanical simulation

FIG. 4.35 *Passive force-displacement characteristics of muscle. [Part (b) is adapted from E. H. Sonnenblick, Implications of Muscle Mechanics, Federation Proc.,* **21**:*(November–December, 1962).]*

are common to many physiological tissues. Tendons and ligaments, for example, show this same general behavior of stretch under load. We can recognize that these characteristics define a storage property of tissue if the units of the variables involved are examined. The displacement $(L - L_0 = x)$ may be taken as a relative or an across-variable; the force (F) is the through-variable. The ratio of these, which is the slope of the force-displacement characteristics, has time units to an even power. The ratio F/x therefore defines a storage property of the system.

The storage property represented by the force-displacement characteristics may be recognized as the mechanical property of elasticity or springiness which was first studied quantitatively by Robert Hooke (1678). Hooke's law states that, for a limited range, deflection is proportional to applied force $(F \propto x)$. The constant of proportionality is generally given by K and is called the spring constant (or elastic coefficient, or stiffness coefficient, etc.*). Evidently the slope of the force-displacement characteristics shown in Fig. 4.35b defines the spring-constant of the muscle.

It seems entirely reasonable that a model simulation of a muscle should include a mechanical spring to represent the elasticity of the muscle. The simulation shown in Fig. 4.35c may thus be considered an elementary model of muscle under static (no motion) load. The upper spring has a spring-constant of K_1 ($= 0.13$ g/mm), which would be equivalent to the dynamic spring-constant of the muscle between the deflection $0 < x < x_0$ (where, in keeping with the convention previously established, the dynamic spring-constant is the slope of a tangent to the force-displacement characteristics). In order to extend the range of simulation, a second spring is included in the model so that beyond a deflection of x_0 ($x > x_0$) the spring K_1 and K_E act together to provide a combined spring-constant equivalent to K_2 (2 g/mm). We can easily see how to choose the value of K_E when we note from the graph that at deflections beyond x_0 the applied force is given by

$$(4.43) \qquad\qquad f = K_1 x_0 + K_2 \delta$$

where $K_1 x_0$ is evidently the force generated up to the deflection x_0 and $K_2 \delta$ is the additional force needed to provide an additional deflection δ beyond x_0. From the mechanical simulation in Fig. 4.35c, beyond x_0

---

* There is much confusion regarding the exact connotations of these terms. Here we shall denote the system property described by F/x as the springiness (or spring-constant) of the system, and we shall designate this property by K. For a more complete description of terms associated with muscle studies, see J. W. Remington, Introduction to Muscle Mechanics with a Glossary of Terms, *Federation Proc.*, **21**(6):(November–December, 1962).

the applied force can be described as

(4.44) f = (force to extend to x_0)

$\qquad\qquad$ + [force to extend beyond x_0 (that is, δ)]

$$= K_1 x_0 + (K_1 \delta + K_E \delta) = K_1 x_0 + (K_1 + K_E)\delta$$

By comparing Eqs. (4.43) and (4.44), we find that

(4.45) $$K_1 + K_E = K_2$$

Thus K_E is $K_2 - K_1$ or 0.87 g/mm.

This preliminary discussion illustrates two points which should be noted. The first is that one can simulate nonlinear characteristics by an appropriate arrangement of linearized *mechanical* elements in the same way as one might use linearized electrical elements. Linearized elements are combined by an appropriate "switching function" so that different straight-line segments are generated for different ranges of operation. The "switch" in this case is simply a mechanical connection. The second point that should be noted is that Eq. (4.45) is in essence a relationship that describes the effective springiness of linear springs having equal deflection under load. After spring K_1 in Fig. 4.35c meets the spring labeled K_E, both springs always have the same deflection. Such springs are considered to be in parallel, and the effective spring-constant for such configurations is the sum of the individual spring-constants. Thus in Fig. 4.36a, b, and c, the effective spring-constant (for an applied deflection) is given by $K_a + K_b$.

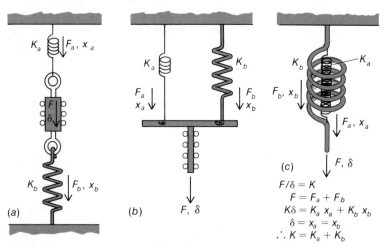

FIG. 4.36   *Springs in parallel.*

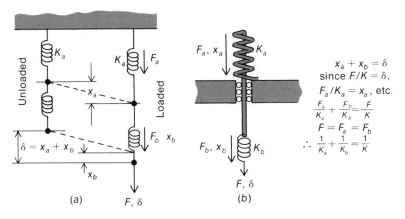

FIG. 4.37 *Springs in series.*

In Fig. 4.37 all the springs deflect different amounts although all carry the same load. Springs arranged in such a manner are considered to be in series. The deflection of the total system is the sum of the deflections of each spring. As a consequence, the reciprocal spring-constant of the total system is equal to the sum of the reciprocal spring-constants of the component springs. The development of this relationship is shown in Fig. 4.37.

4.11 THE ELECTRICAL ANALOG OF SPRINGINESS

To develop an electrical analog for the property of springiness we follow the same procedure used to establish the analog of compliance; we express springiness in terms of analogous variables and attempt to identify the electrical property that is described. Springiness is defined by the ratio of force to deflection:

$$(4.46) \qquad \text{Springiness} = K = \frac{f}{x} = \frac{\text{force}}{\text{deflection}}$$

If we retain the analogies established earlier (see Table 3.1), mechanical force is analogous to current and velocity gradient is analogous to voltage gradient. Since "distance" or deflection is the time integral of velocity ($x = \int v\, dt$), deflection may be considered analogous to the time integral of voltage, that is, $x = \int e\, dt$. Equation (4.46) can now be written

$$(4.47) \qquad K = \frac{f}{x} \sim \frac{i}{\int e\, dt}$$

The ratio of the electrical variables in Eq. (4.47) defines the reciprocal of

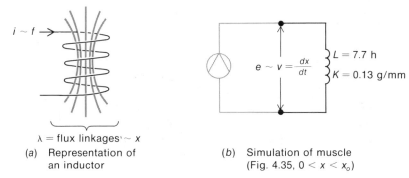

λ = flux linkages $\sim x$

(a)   Representation of
        an inductor

(b)   Simulation of muscle
        (Fig. 4.35, $0 < x < x_o$)

FIG. 4.38   *Electrical inductor representing mechanical property of springiness.*

electrical inductance. That is,

$$(4.48) \qquad L = \frac{\int e\,dt}{i} \qquad Li = \int e\,dt \qquad \text{or} \qquad L\frac{di}{dt} = e$$

Therefore, springiness in mechanical systems may be represented analogously by an inductor; the inductor is equivalent numerically to the reciprocal of the mechanical spring-constant ($L = 1/K$).

The electrical inductor is primarily a "magnetic device"; its operating characteristics depend upon a generated magnetic field. The flow of current through a coil of low-resistance wire (ideally, zero-resistance wire) sets up the magnetic field around the coil; the more the current, the more the magnetic field. The strength of the magnetic field is often expressed by lines of flux. The more the current through the coil, the more lines of flux link with the coil (i.e., the more the flux linkages; see Fig. 4.38a). Flux linkages and current are, in fact, related by the system property of inductance. If the flux linkages (given by λ) are proportional to current, the constant of proportionality is the inductance L, and $\lambda = Li$. By comparing this expression with Eqs. (4.46), (4.47), and (4.48) we see that flux linkages (λ) and displacement (x) are analogous quantities. If we examine the analogy further, we note that, for the spring, deflection changes when force changes. For the inductor, therefore, flux linkages change when current changes. Since the time rate of change of deflection is velocity ($dx/dt = v$), in the analogous inductor the time rate of change of flux linkages is voltage ($d\lambda/dt = e$). Therefore, for an ideal inductor (one with zero resistance), voltage across the inductor is generated only when there is a change of current through the inductor. The more rapid the current change through the coil, the higher is the voltage across the coil. (The arc produced when a switch interrupts current to a coil is an indication of the magnitude of the voltage that may be generated. The high rate of current change causes a voltage high enough to ionize the air.) These relationships may be quantitatively expressed by the equation $e = L(di/dt)$. This expression is often used as the defining expression for the units of an inductor. A 1-henry (H) inductor is considered to be a coil that will generate 1 V when the current change through the coil is 1 A/s.

The symbol for the inductor is a coil of wire (which is, appropriately enough, the same symbol used for a spring). Figure 4.38a is a schematic diagram of an inductor; in Fig. 4.38b the inductor is used to simulate the characteristics of the muscle shown in Fig. 4.35b. Since the dynamic inductance used represents the characteristics only up to $x = x_0$ (see Fig. 4.35b), the simulation is evidently not valid beyond this point. The current source is used to provide a simulated force excitation. So long as the source delivers a constant current, the flux linkages remain constant and the voltage remains zero. For any constant force, therefore, the spring deflection is constant and the end of the spring has no velocity, which is as it should be.

Electrical analogs of springs in parallel and springs in series are shown in Fig. 4.39a and b. Figure 4.39a is evidently the simulation of the parallel spring combinations shown in Fig. 4.36, and Fig. 4.39b is the simulation of the series spring combinations of Fig. 4.37. Since springiness and inductance have been shown to be reciprocal functions, for inductances in parallel the total reciprocal inductance is equal to the sum of the individual reciprocal inductances (see Fig. 4.39a); when inductors are in series the equivalent inductance is the sum of the individual inductances (see Fig. 4.39b).

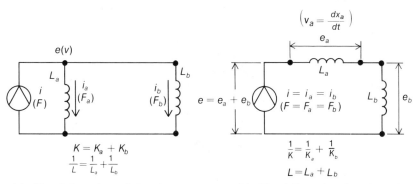

(a) Simulation of parallel springs   (b) Simulation of springs in series

FIG. 4.39 *Electrical analogs of spring combinations.*

It is conceivable that nonlinear springiness could be simulated by a proper choice of linear inductors and an appropriate switching sequence (in somewhat the same manner as nonlinear resistance and capacitance were simulated; see Sec. 3.7 and the end of Sec. 4.3). However, simulation of nonlinear functions with linear inductors is not very practical as the circuitry is clumsy and difficult to adjust. Linear and low-resistance inductors (generally called high-Q coils) are not standard stocked items and may not be available in sufficient variety to permit the needed flexibility. Furthermore, for this application the switching circuitry should depend upon current levels rather than voltage levels (switching is required when forces reach appropriate levels). Current-sensitive switching (i.e., by relays) is not as easy, or dependable, to work with as voltage-sensitive switching (i.e., diodes).

Although it is not practical to simulate nonlinear springiness with linear inductors, such functions may be simulated by computer techniques. The block-diagram representation is shown in Fig. 4.40. As mentioned earlier, these diagrams are simply schematic representations of the pertinent mathematical relationships that exist between the variables of interest. In Appendix A, however, it is noted that the blocks may be realized by certain configurations of operational amplifiers (see also Probs. 4.4 and 4.8).

The comments of the last two paragraphs do not mean that inductors cannot be used in circuit simulations. For operations over a limited range, inductors are practical and may be effectively used in the construc-

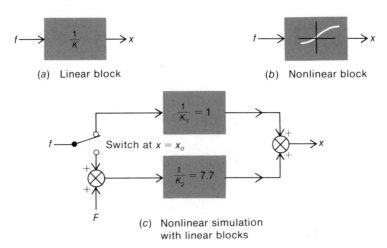

(a) Linear block

(b) Nonlinear block

(c) Nonlinear simulation
with linear blocks

FIG. 4.40 *Block diagrams to represent a system with springiness.*

tion of piecewise linearized simulations (for studies over small operating ranges). For sudies involving extended ranges of nonlinear operations, however, physical inductors are inconvenient and they cannot be practically used in actual simulations. Even in nonlinear situations, however, the concept of inductance—and, of course, springiness—provides valuable insight into the relationship between system responses and system properties. In Chap. 5 we shall examine some of the quantitative relationships that exist between various properties and responses.

4.12 FURTHER DISCUSSION OF NONENERGIZED MUSCLE TISSUE

We cannot leave the discussion of muscles with the viewpoint that such tissue may be described by a spring (i.e., by springiness) alone. Intuitively we realize that this representation must be incomplete as no means are included to account for frictional or resistive loss. Mechanical systems cannot in general be considered frictionless and muscle tissue is no exception.

Muscles (and elastomers* in general) have significant resistive properties; mechanical manipulation of such tissue (such as extension and release) is always accompanied by significant temperature increases within the material. The temperature rise is due to the action of some frictional mechanism within the material which converts a portion of the applied mechanical energy into heat energy. The recognition that elastomers must contain some frictional mechanism has led to two elementary model representations for elastomeric materials. These are given in Fig. 4.41. The first has been called a Voigt element or model, and the second, a Maxwell element or model. (Included also are the corresponding electrical analogs.) It should be noted that in both representations a spring represents the elastic properties (i.e., the springiness) of the tissues; in addition to the spring element, both models include an element to indicate that the system also has frictional properties. This frictional element is often described as a "damper" or "dashpot," and it represents "linear viscous friction." Before we discuss either of these representations (and show application to muscle tissue) we must examine briefly the concepts of friction and viscous friction as symbolized by the dashpot.

* Muscle tissue is generally included in a larger class of materials called elastomers. This class encompasses materials which are probably best described by indicating that rubber is an example of an elastomeric material. We have already encountered other body tissues that have been described as elastomeric materials. In Sec. 4.2 it was mentioned that the lungs, bladder, and other hollow viscera are elastomeric in nature. (It is of interest to note here that we used the capacitor as an electrical analog for the hollow viscera whereas an inductor was used to represent electrically the analog of a muscle, although both are elastomers. We shall examine the reason for these differences in representation in the next section.) Some discussion of elastomers and a good cross section of the general field of tissue properties can be found in Ref. 15.

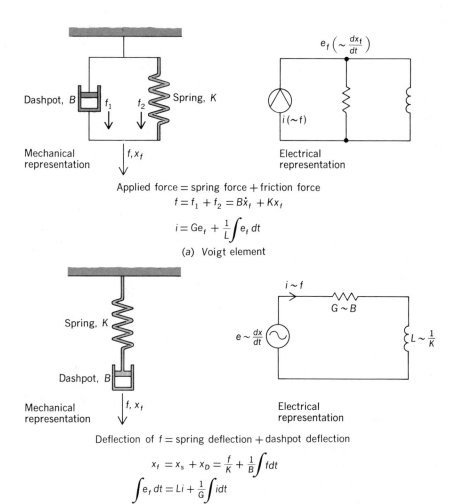

Mechanical representation

Electrical representation

Applied force = spring force + friction force

$$f = f_1 + f_2 = B\dot{x}_f + Kx_f$$

$$i = Ge_f + \frac{1}{L}\int e_f \, dt$$

(a) Voigt element

Mechanical representation

Electrical representation

Deflection of f = spring deflection + dashpot deflection

$$x_f = x_s + x_D = \frac{f}{K} + \frac{1}{B}\int f \, dt$$

$$\int e_f \, dt = Li + \frac{1}{G}\int i \, dt$$

(b) Maxwell element

FIG. 4.41 *Representation of elastomeric tissue including frictional effects.*

The most familiar concept of mechanical friction is that of the relatively constant force that opposes motion between two mating, non-lubricated surfaces. Such friction is thought to be related to the surface materials, surface condition, and pressure between the surfaces but relatively independent of the surface area and relative velocity between the surfaces. The characteristics of this concept of resistance are summarized in Fig. 4.42a.

Viscous friction is the friction force generated by the motion of a fluid. Such friction is related to the fluid viscosity, the amount of fluid in motion, and the velocity gradients in the fluid. Figure 4.42*b* indicates how viscous friction affects the motion of moving surfaces. Two surfaces (*A* and *B*) are separated by a fluid which serves as a lubricant (e.g., grease). The force needed to move one surface with respect to the other is the frictional force generated by the viscous friction in the fluid. Such friction depends upon the area of the wetted surface, the viscosity of the fluid, and the velocity of one surface with respect to the other. In equation form we write

(4.49)   Frictional force generated (f_f)

$\quad\quad$ = function of viscosity (μ), area (A), and relative velocity (v)

or

$$f_f = \phi(\mu, A, v)$$

Newton proposed that this frictional force be described by

(4.50) $$f_f = \mu A \frac{dv}{dx}$$

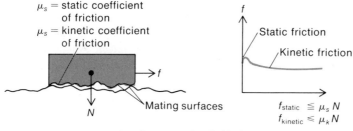

μ_s = static coefficient of friction
μ_s = kinetic coefficient of friction

$f_{\text{static}} \lessgtr \mu_s N$
$f_{\text{kinetic}} \leq \mu_k N$

(a)   Dry or coulomb friction

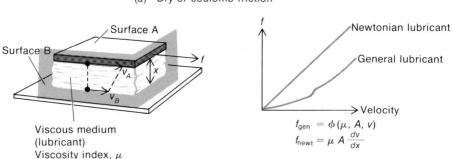

$f_{\text{gen}} = \phi(\mu, A, v)$
$f_{\text{newt}} = \mu A \frac{dv}{dx}$

(b)   Wet or viscous friction

FIG. 4.42   *Concepts of wet and dry friction.*

where dv/dx is the velocity gradient between the moving surfaces. When the motion is uniform and the surfaces are flat, this gradient may be expressed by $(V_A - V_B)/x$. Equation (4.50) then becomes

$$(4.51) \qquad f_f = \frac{\mu A}{x} (V_A - V_B)$$

Equation (4.51) is the relationship used to describe mechanical friction in Table 3.1.*

The steady forces needed to maintain a constant relative velocity between two portions of the system define the graphical characteristics of this viscous-resistive property of physical systems. The slope of the resultant curve defines the mechanical conductance (or damping coefficient, as it is sometimes called) of the system. The mechanical conductance is generally designated by the letter B, where B is analogous to the electrical conductance G. When the force-velocity curve is a straight line, the system evidently has a linear-resistive property, and the mechanical damping coefficient (B) is constant. Mechanical friction is then defined by $f = Bv$ (as shown in Fig. 4.43a). The dashpot symbol shown in Fig. 4.43 is generally used in mechanical-circuit diagrams to represent linear viscous friction; the physical dashpot (described in the following paragraph) is used to provide mechanical friction in actual mechanical simulations.

Physically, a dashpot may be described as a loose or "leaky" piston riding in a fluid-filled cylinder. As the piston is moved with respect to the cylinder, the fluid is "dragged" by the relative motion between the piston and cylinder surfaces. The fluid therefore generates friction drag forces

---

* This discussion is an oversimplified version of the general concept of friction. Friction has long remained an enigma. No general theory has yet been developed to explain the great variety of observed data. However, much research has been performed in recent years, and it seems likely that there will soon be a less empirical approach to friction studies.

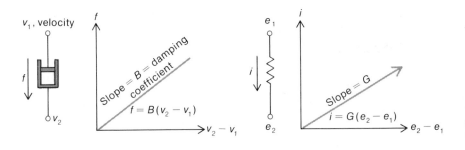

(a) Mechanical resistance   (b) Electrical resistance

FIG. 4.43 *Mechanical-electrical analogs of resistive property.*

to be exerted on both elements. Theoretically, if the dashpot is arranged so that the piston remains concentric with the cylinder and no pressure is generated in the fluid during piston motion, Eq. (4.51) may be directly applied. The relative velocity between the piston and cylinder is given by $V_A - V_B$; the area term A is the piston surface that mates with the cylinder; the distance x is the clearance between the piston and cylinder. We can easily see that the area term (A) is constant. If the dashpot is designed so that the clearance remains constant and if the viscosity is constant (a doubtful "if") then $\mu A/x$ is constant and Eq. (4.51) can be written $f = B(V_A - V_B)$, where B is the constant damping coefficient and is evidently equivalent to $\mu A/x$. Although this condition cannot be easily realized in actual practice, the miniature dashpots used in instruments often approach linear operation over a limited range. Other physical devices using this dashpot principle are door checks and automobile shock absorbers. By now it should be apparent that the mechanical dashpot is the mechanical analog of the electrical resistor. These analogous elements are compared in Fig. 4.43.

In almost all instances, physiological tissue contains fluid constituents. It seems reasonable to expect, therefore, that the mechanical friction in physiological materials is essentially of the fluid or viscous variety. In fact, elastomers and elastic tissue are often referred to as viscoelastic materials. Viscous friction is thus considered to be the primary component of the mechanical resistive property in a physiological system, and the force-velocity characteristics of the system are used to define this viscous, mechanical-resistive property.

We are now in a position to evaluate the model representations of tissue in Fig. 4.41. Both models recognize the viscoelastic properties of elastomeric tissue by virtue of the spring-dashpot (inductor-resistor) combinations used in the model simulations. The equations that govern the operation of these models are also given in Fig. 4.41. In succeeding chapters we shall examine how these equations may be used to determine the system responses to various stimuli and how such information can be used to gain insight into the physiological system. Before any analysis is made, however, the models should be examined to ascertain whether either of them is a reasonable representation of actual tissue. We start by noting that if smooth muscle is loaded, energized to contract, and then allowed to extend passively at constant load (i.e., isotonically) the response will be approximately that shown in Fig. 4.44a.*

For similar conditions of extension under constant load we would intuitively expect the model responses to be those shown by Fig. 4.44b and c. If the Maxwell model is allowed to respond isotonically it will

* See, for example, Emil Bolzer, Extensibility of Contractile Elements, in Ref. 15.

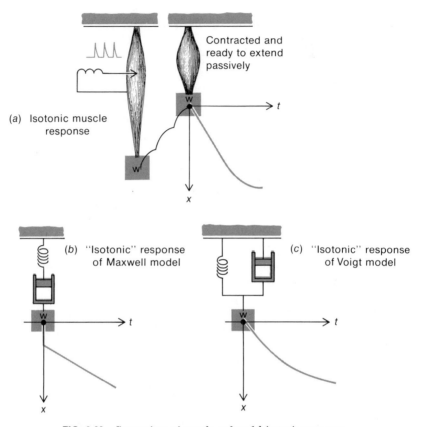

FIG. 4.44 *Comparison of muscle and model isotonic responses.*

immediately drop a short distance until the spring extends (the dashpot cannot instantaneously acquire a velocity); it will, theoretically, then extend ad infinitum at a constant rate. This response does not appear to conform with the response of the physiological system. Under isotonic conditions the Voigt model will initially fall at a constant rate. As the spring becomes stretched, the rate of extension is reduced until finally motion will entirely cease. This response is more in keeping with that of the physiological system, and we might accept the Voigt model as a preliminary representation of the physiological system. It is emphasized that this provides only a preliminary "guess" as to the model configuration. Further analysis, examination of data, and laboratory experimentation are required to confirm or deny the choice of model configuration and/or indicate the need for modifications to provide better representation.

4.13   DUAL REPRESENTATION OF THE STORAGE PROPERTY

In the preceding section we saw that an inductor can be used to represent the elasticity of tissue; yet in an earlier section a capacitor was used to represent this same general characteristic of tissue (i.e., its stretchability; see Secs. 4.2 and 4.3). This change in representation may at first appear to be an inconsistency. If we examine the basic reasons for the difference in representation, however, we realize that this change is not an inconsistency but simply a change in the choice of analogous variables.

To point this out in more specific terms, let us examine the situation shown in Fig. 4.45a. The elastic properties of a cat lung are to be studied

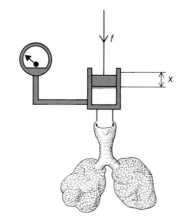

(a)   Experimental arrangement

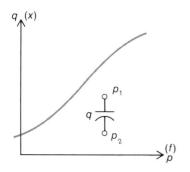

(b) Pressure–volume characteristics of saline filled lungs

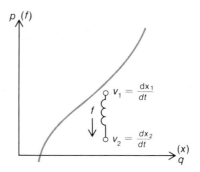

(c)   Alternate representation

FIG. 4.45   *Characteristics of saline-filled cat lung. [Part (b) is adapted from E. P. Radford, Jr., Recent Studies of Mechanical Properties of Mammalian Lungs, in Ref. 15.]*

by filling it with saline solution.* The procedure to be employed is indicated by Fig. 4.45a. After the lungs have been degassed, the trachea is connected to a piston-pump arrangement. When the piston is depressed, a known volume of saline solution is delivered to the lungs. The pressure required to deliver this volume is determined by the pressure gauge shown in the figure. By progressively indexing the piston, a set of values relating the variables pressure and volume may be obtained. These measured variables describe a storage property of the system. When the pressure-volume points are plotted, the resultant curve defines the graphical characteristics of this storage property. This curve is shown in Fig. 4.45b, where pressure is the across-variable and volume the through-variable. We have previously defined the slope of this curve as the compliance of the system, where compliance is one system property that reflects the elastic characteristics of the system. (In this case we might consider the curve to be a representation of tissue elasticity with the effects of surface tension excluded or at least minimized.) If we were to propose an electrical-analog simulation of this system we would probably associate pressure with voltage and volume with charge. Under these circumstances, tissue elasticity would be represented by an electrical capacitor.

In Fig. 4.45b it is of interest to note that volume is related to the piston displacement x by a scale factor A, the area of the piston (that is, $q = Ax$). In a similar manner, the pressure is also related to the applied piston force by a scale factor $1/A$ (that is, $pA = f$). If we change the scales of the coordinate axes, then the graphical characteristics remain unchanged but the defining variables are now x and f. Evidently then, if displacement (x) is related to current (as was volume) and if force (f) is related to voltage (as was pressure), the electrical analog of the system property is still a capacitor. On the other hand, if we relate the variable f to electric current and the displacement to voltage, then, on the basis of the discussion in Sec. 4.12 (see also Fig. 4.45c), we would associate this system property with an inductor. We see clearly that in the last analysis the real criterion that determines which device shall represent a system property is governed by the choice of analogous variables.

The comments in the preceding paragraph may be somewhat dis-

---

* The purpose of such a study is to learn how surface tension may affect the elastic properties of lung tissue. Air-filled lungs show hysteresis effects (different curves during emptying and filling). It is thought that these effects are due to the surface tension of the mucus-lined lung. To ascertain the importance of this effect, the lungs are filled with saline solution rather than air. The saline simultaneously washes out the mucus and minimizes the surface-tension effect. The pressure-volume curves so obtained thus reflect primarily the elastic properties of the lungs. See E. P. Radford, Jr., Recent Studies of Mechanical Properties of Mammalian Lungs, in Ref. 15.

couraging. After having established the concept of through- and across-variables and associating them with corresponding variables of other systems, we now find that (as was implied earlier) the concept is quite arbitrary. What we consider a through-variable from one point of view might well be taken as an across-variable when another viewpoint is used. Rather than being discouraging, this turn of events is very encouraging, as we cannot be "wrong" no matter which we choose; we are at liberty to relate analogous variables in any way we desire. This cannot be an unexpected "discovery." An analog is simply a convenient means of describing a situation; any description is acceptable so long as the analog yields results that duplicate the results of the real system. The only requirement one might make in developing an analog is that analogous variables be clearly indicated. In general, once a set of analogous variables has been selected, it is possible to retain a consistent notation throughout the simulation. In some situations, however, a change in notation may be desirable (such as pressure analogous to voltage but force analogous to current). In such cases it is necessary only that the diagrams, elements, and variables be clearly labeled to avoid confusion.

There is one general "rule" that we shall follow when circuit analogs are to be developed. Resistance properties of a system (or lossy properties, as they are sometimes called) will be electrically simulated in a circuit analog by an electrical resistor; storage properties will be represented by inductors and/or capacitors. We shall still retain the criteria by which we judge storage and resistive properties: When the ratio of two measured variables has time units to an odd power we shall consider that a resistive property is implied; when the ratio has time units to an even power we shall assume a storage property is implied. The designation of through and across-variables will be retained to some extent but this terminology is rather arbitrary. We shall now depend upon the association of analogous variables to determine whether a ratio defines an electrical resistance or conductance on the one hand or a capacitance or inductance on the other hand.

Problems

4.1 Two compliances are in parallel when the pressures across them are equal. In Fig. P4.1a, for example, the internal pressures of the compliant members are the same and the external pressures are the same. These compliances are in parallel and may be represented schematically by the capacitors shown in Fig. P4.1b.

(a) If the storage characteristics of the compliances are given by the curves in Fig. P4.1c, show that the static compliance of the parallel combination at any pressure is the sum of the separate compliances at that same pressure.

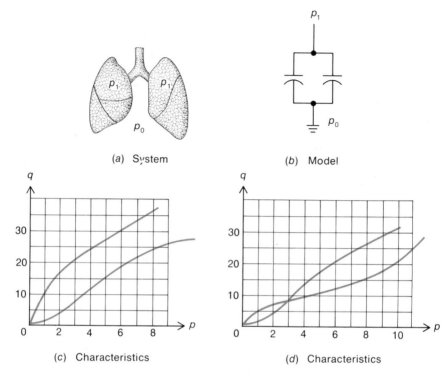

(a) System   (b) Model

(c) Characteristics   (d) Characteristics

FIG. P4.1

(b) Show that the dynamic compliance of the combination is the sum of the separate compliances.

(c) If the compliant members of Fig. P4.1a have the characteristics shown in Fig. P4.1d, what would be the shape of the static compliance vs. pressure curve? What would be the shape of the dynamic compliance vs. pressure curve?

4.2 From the discussion in Prob. 4.1 it should be apparent that the two lungs shown diagrammatically in Fig. P4.2b are in parallel. If the pressure-volume characteristics of one lung are as shown in Fig. P4.2c and both lungs are assumed to be the same, what are the pressure-volume characteristics of the combination?

Two compliances are in series if the volume stored in each compliance is the same and the pressure drop across the series combination is the sum of the pressure drops across each compliance. When the lungs are in the thoracic cage the system may therefore be represented schematically by the circuit shown in Fig. P4.2d. If the total pressure-volume characteristics of the combination are as indicated in Fig. P4.2c, construct the pressure-volume curve for the cage alone.

If the thoracic cage is considered primarily as skin stretched over a bony framework (Fig. 4.2a), it is easy to see the pressure-volume characteristics for the empty cage (cage alone without lungs included) will not pass through the origin; a storage space will be enclosed by the cavity even when the external pressure p_a and the internal pressure p_{pl} are equal (that is, $p_a - p_{pl} = 0$). When the lungs are in the chest-cavity storage space they cling to the chest wall and the thoracic cage contracts slightly. At resting expiratory level the lungs contain about 3 l of gas at atmospheric pressure. What is the pleural pressure at the time?

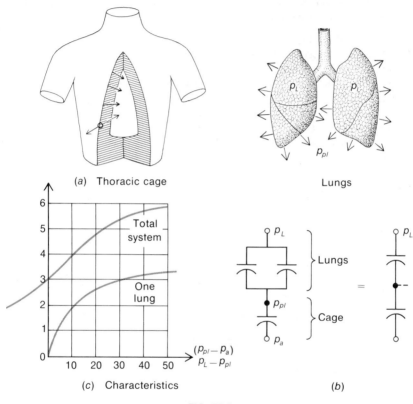

(a) Thoracic cage

Lungs

(c) Characteristics

(b)

FIG. P4.2

If a patient has a pneumothorax, the lungs cannot adhere to the chest wall everywhere as air has somehow become trapped between the lungs and the cage (i.e., in the pleural space). If 300 cm³ of air (at pleural pressure) is trapped in the pleural space, the storage space in the thoracic cage at resting expiratory level is the volume stored in the *lungs* plus 300 cm³. What will be the pleural pressure at resting expiratory level (i.e., when $p_L = p_a$) now, and what volume of gas will the lungs contain?

4.3 Figure 4.6 shows a piecewise-linearized simulation of the nonlinear storage element in Fig. 4.5a. The switch is included for illustrative purposes only. In a more practical simulation a diode would replace the switch.

(a) Show a simulation that includes a diode (see the supplementary discussion in Chap. 2 and Prob. 3.4).

(b) Use the simulation to derive an equation relating the stored charge to the applied signal.

(c) Diodes have inherent resistance. How will this affect the simulation if only steady-state operation is considered?

4.4 A physical system that has a nonlinear pressure-volume curve can be simulated with capacitors if the shape of the curve continually increases within the range of operation (i.e., the dynamic compliance increases). Thus the curve in Fig. 4.5a can be simulated by a set of parallel capacitors if proper switching and bias voltages (batteries) are used and the range of operation is from 0 to about 20 V (see Fig. 4.6 and Prob. 4.3). If a simulation is to include a range of operation in which the characteristic curve has increasing and decreasing slope (in mathematical terms, the second derivative may be both positive and negative) the simulation cannot be readily realized with parallel capacitors. In

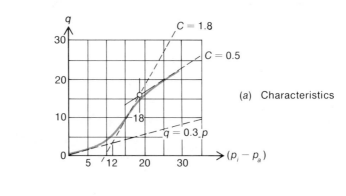

(a) Characteristics

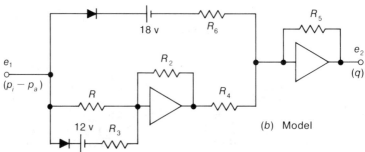

(b) Model

FIG. P4.4

such instances an operational-amplifier configuration can be used to simulate the given characteristic curve. One such circuit is shown in Fig. P4.4b. The operation of the circuit is such that neither diode circuit conducts if $e_1 < 12$; the output of the overall circuit is then $e_2 = +e_1(R_2/R_1)(R_5/R_4)$. If e_2 represents q, and e_1 represents $p_i - p_a$, and if the R's are properly chosen, this circuit will generate the first straight-line approximation to the pressure-volume characteristic shown in Figs. 4.5a and P4.4a.

When $12 < e_1 < 18$, the lower diode circuit conducts but the upper will not. Show that with a proper R_3 (based on the R's previously determined) the circuit will generate a straight-line approximation of the pressure-volume curve in Fig. 4.5a for the range $12 < (p_i - p_a) < 18$.

When $e_1 > 18$, all branches conduct. With the R's determined previously and an appropriate value for R_6, show that the circuit now generates a straight-line approximation of the $p_i - p_a$ versus q curve in Fig. 4.5a in the range $p_i - p_a > 18$.

4.5 A schematic diagram of the compliant elements in the respiratory system is shown in Fig. P4.5a. The voltage sources are included to simulate the artificially induced breathing of a polio patient in an iron lung. The source p_{1L} represents the pressure changes (above and below atmospheric) generated by the iron lung on the surface of the patient. The source p_a is atmospheric pressure.

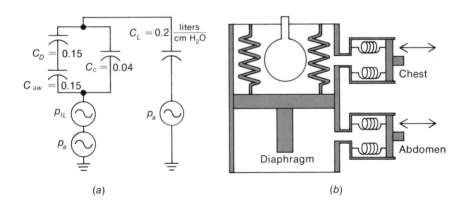

FIG. P4.5

(a) How much of a pressure change (p_{1L}) is required in order to deliver and remove 500 cm³ (i.e., a tidal volume) from the lung?

(b) In normal breathing, pressure differences are generated by various "respiratory" muscles (i.e., diaphragm, intercostal, and abdominal; see Fig. P4.5b). These muscle forces may be represented by equivalent pressure (or voltage) sources. Show where the various pressure sources due to muscle action should be located on the circuit schematic.

(c) If the abdominal muscles must work alone, how much equivalent pressure must they generate if they are to move a tidal volume of gas to and from the lung?

(d) Repeat part (c) if the intercostal muscles must work alone or if the diaphragm muscles must work alone.

(e) How would you define "respiratory compliance" in each of the cases cited in (c) and (d)?

4.6 In the model of Fig. 4.29 the metabolic rate is assumed to be constant. It is known, however, that metabolic rate increases with increasing temperature; the relationship is approximately a 10 percent increase in metabolism for each 1°C rise of average body temperature.* Mathematically this may be expressed as

$$\dot{H}_m = \dot{H}_{m0}(1 + 0.1\Delta T)$$

where $\dot{H}_{m0}$ is the metabolic rate at a "normal" temperature and ΔT is the change in average body temperature. In Fig. 4.29, for steady state when the switch is at A, ΔT is zero and $\dot{H}_{m0}$ is 100 kcal/h. Average body temperature may be approximated by skin and core temperatures. The relationship can be expressed by

$$T_{\text{av}} = 0.3T_s + 0.7T_c$$

Therefore

$$\Delta T = 0.3\Delta T_s + 0.7\Delta T_c$$

where ΔT_s is the change of skin temperature from quiescence ($\Delta T_s = T_s - T_{s0}$) and ΔT_c is the change of core temperature from quiescence. The analog circuit in Fig. P4.6 shows one configuration from which the change in skin temperature can be determined.

* Ref. 12, p. 1046.

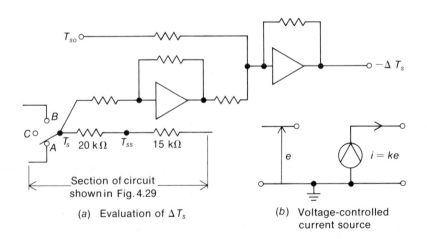

(a) Evaluation of ΔT_s

(b) Voltage-controlled current source

FIG. P4.6

(a) Show a circuit configuration from which the metabolic rate may be determined as a function of the change in skin and core temperatures.

(b) It is possible to build variable-current sources that will deliver current depending on an applied signal voltage. Such devices are called voltage-controlled current sources. A diagrammatic representation of such a device is shown in Fig. P4.6b. The signal derived in part (a) will be a voltage that represents the metabolic rate as a function of average body temperature. How can one devise a circuit that will deliver a current to represent the change in metabolic rate as the average body temperature changes?

(c) Modify Fig. 4.29 so that the delivered current varies with the skin and core temperatures.

4.7 To extend the forearm, the triceps brachial muscle is activated. The contraction of this muscle causes the radius and ulna in the forearm to swing back at the elbow around the humerus. This motion extends the biceps brachial muscle and brachialis and their associated tendons. In effect, then, the triceps contraction must generate enough force to extend the synergistic antagonists, the biceps and brachialis. A model of the system is shown in Fig. P4.7b. The springs shown represent the elasticity of the muscles and tendons (subscript m for muscle; subscript t for tendon).

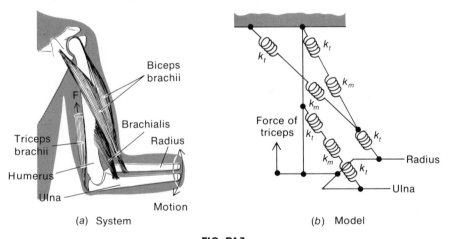

(a) System   (b) Model

FIG. P4.7

(a) What is the approximate equivalent spring-constant against which the triceps must initially operate?

(b) If the tendon spring-constants k_t increase 10 percent, what additional force will the triceps be required to generate?

(c) Does the spring effect increase or decrease as the forearm is extended?

4.8 A block diagram to simulate nonlinear spring characteristics is shown in Fig. 4.40. Show a more realistic simulation using operational amplifiers and diodes as suggested by Prob. 4.4.

REFERENCES

1. King, A. L., and R. W. Lawton: The Elasticity of Soft Body Tissues, *Sci. Monthly*, **71**:258 (1950).
2. Stacy, R. W., D. T. Williams, R. E. Worden, and R. O. McMorris: "Essentials of Biological and Medical Physics," McGraw-Hill Book Company, New York, 1955.
3. Fenn, W. O., and H. Rahn (eds.): "Handbook of Physiology," Respiration, Sec. 3, vol. I, American Physiological Society, Washington, D.C., 1964.
4. Mead, J., M. B. McIbray, N. J. Selverstone, and B. C. Kriete: Measurement of Intraesophageal Pressure, *J. Appl. Physiol.* **7**:491 (1955).
5. "Advances in the Biological and Medical Sciences," No. 10, Elastic Reservoir Theory Academic Press, Inc., New York, 1965.
6. Hallock, P., and I. C. Benson: Studies on the Elastic Properties of the Human Isolated Aorta, *J. Clin. Invest.*, **16**:595 (1937).
7. Halliday, David, and Robert Reswick: "Physics for Students of Science and Engineering," John Wiley & Sons, Inc., New York, 1962.
8. Henriques, F. C., Jr., and A. R. Moritz: Studies of Thermal Injury, *Am. J. Pathol.*, **23**:531 (1947).
9. Milsum, J. H.: "Biological Control Systems Analysis," McGraw-Hill Book Company, New York, 1966.
10. Crosbie, R. J., J. D. Hardy, and E. Fessenden: Electrical Analog Simulation of Temperature Regulation in Man, in J. D. Hardy (ed.), "Temperature: Its Measurement and Control," vol. III, Reinhold Publishing Corporation, New York, 1963.
11. Berensen, P. J., Prediction of Human Thermal Comfort in Oxygen-Nitrogen Atmospheres, and E. C. Wortz et al., Body Heat Storage in Full Pressure Suits, in Eric Burgess (ed.): "Physiological and Performance Determinants in Manned Space Systems," vol. 5, AAS Science and Technology Series.
12. Ruch, T. C., and H. D. Patton: "Physiology and Biophysics," 19th ed., W. B. Saunders Company, Philadelphia, 1965.
13. Minard, David, Louis Copman, and A. R. Dasler: Elevation of Body Temperatures in Health, in Harold E. Whipple (ed.), "Thermography and Its Clinical Applications," *Ann. N.Y. Acad. Sci.*, **121** (art. 1): (October, 1964).
14. Sonnenblick, E. H.: Implications of Muscle Mechanics, *Federation Proc.*, **21**:(November–December, 1962).
15. John W. Remington (ed.): "Tissue Elasticity," American Physiological Society, Washington, D.C., 1957.

CHAPTER FIVE

INTRODUCTION TO
ANALYSIS OF SYSTEMS
WITH COMBINED PROPERTIES

5.1 PRELIMINARY DISCUSSION

Before discussing how signal relationships (i.e., input-to-output signals) are affected by various combinations of system properties, we shall first review what has been done, why it has been done, and what we hope ultimately to achieve.

The premise has been that, if the response to a known stimulus can be obtained, the data can be used to provide information about the properties of the system under investigation. The ultimate purpose is to determine how input-output data may be used to define and study the properties of a system. From a research point of view, such information can be used to foster an understanding of the system and to point out where further research may be necessary. From a design point of view, the information can be used to determine how system response can be altered to modify the behavior in a desired manner.

In order to limit the problem of classifying system properties, we have separated these properties into two general categories: resistance and storage. Each of these was individually defined on the basis of known

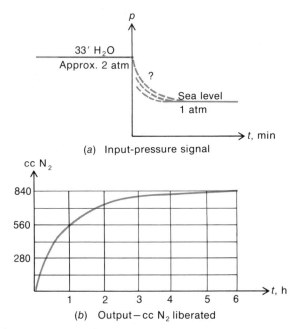

FIG. 5.1 *Rapid-decompression characteristics of a person returning to sea level after high-pressure exposure.* (*Adapted from H. G. Armstrong, "Principles and Practice of Aviation Medicine," Williams & Wilkins Company, Baltimore, 1952.*)

input-output data. The data used were chosen rather carefully; in particular, the properties were defined on the basis of static, or time-invariant, information. Given the *static* input-output characteristics of a system, its properties can be described in terms of resistance and/or storage.

Although it may sometimes be possible to obtain appropriate static data to determine the resistance and storage properties of a system, in many cases such data may not be easily available. In some cases they may, in fact, be unobtainable. To study the aspects of nitrogen liberation during decompression, for example, we must depend upon data such as those shown in Fig. 5.1.* The storage property of the system can be determined from the amount of nitrogen released per unit pressure

---

* A man returning to sea level from deep submergence must be decompressed to allow his system to readjust to normal atmospheric pressure. The process consists of a gradual reduction of pressure exposure until the excess nitrogen dissolved by the system at the higher pressures has been eliminated. The nitrogen liberated depends upon the pressure change to which the subject is exposed. The across-variable in this case is obviously the pressure difference, and the through-variable is the nitrogen eliminated.

after steady readings are obtained. (To define the storage property, we should obtain a set of values relating pressure change and nitrogen volume; the slope of the curve thus generated defines the "capacitance.") Evidently the system also has resistive properties, as there is a nitrogen flow during pressure change. However, at the present stage of development, we can define these properties only through the use of time-invariant pressure and flow information. (When such data are plotted, the slope defines the resistive property.) Another test must therefore be performed which will provide data relating steady flow to steady pressure.

Unfortunately, there is no easy experimental way to obtain such data, as a steady nitrogen flow cannot be maintained. We now know that the system has resistive properties but we have no way to define them. We realize the input-output data of Fig. 5.1 must contain information about the resistive property (as well as storage); what has not yet been developed is how such information can be extracted from the time-variant data. In this chapter we shall examine some of the relationships between signal information and combined properties in order to understand how these properties can be studied when static-signal information is not available.

5.2 STEP-RESPONSE OF A RESISTANT-COMPLIANT SYSTEM*

To learn how input-output data may be analyzed so that the properties of a system can be defined, we use a retrograde approach. Instead of using the stimulus-response data to define a system (Fig. 5.2a), we use a known stimulus and known system to define the response (Fig. 5.2b). By gaining experience in how various systems affect applied stimuli (as evidenced

* The term "step-response" is used to designate the response of a system to a sudden change of the applied stimulus from one fixed level to another fixed level (see Fig. 5.1a). Mathematically, the step-stimulus is considered an "instantaneous" shift from one level of excitation to another. From a practical point of view, however, "instantaneous" must be considered a relative term; at this point, we accept the intuitive description that step-stimulus implies a "rapid" shift in operating level. At the end of this chapter we shall attempt to describe "rapid" in a somewhat more quantitative manner.

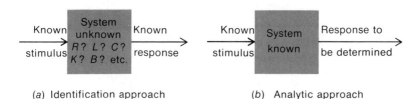

(a) Identification approach   (b) Analytic approach

FIG. 5.2 *Viewpoints related to investigation of a system.*

by the response) we can learn what system properties are reflected by what stimulus-response combinations. It is obviously not feasible to investigate every possible combination, but some relatively simple combinations can be used to cover a surprisingly wide range of possibilities.

As a first illustration of the procedure, we examine the system shown in Fig. 5.3a.* It consists of a rigid tube with known characteristics as given by Fig. 5.3b and a hollow compliant element with characteristics given by Fig. 5.3c. A pressure is to be applied at the entrance of the

* This might be considered a very approximate lumped representation of the tracheal-lung system. See Comroe et al., Ref. 1, Chap. 2.

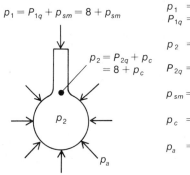

$$p_1 = P_{1q} + p_{sm} = 8 + p_{sm}$$

p_1 = Total applied pressure
P_{1q} = Quiescent value of applied pressure
p_2 = Total pressure in the compliant element
P_{2q} = Quiescent pressure in the compliant element
p_{sm} = Variation of applied signal above and below quiescence
p_c = Variation of response signal in the compliant element
p_a = Ambient pressure

$$p_2 = P_{2q} + p_c = 8 + p_c$$

(a) System configuration

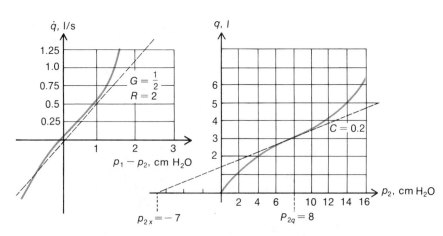

(b) Tube characteristics

(c) Compliant characteristics

FIG. 5.3   *Resistive-compliant system.*

tube, and we seek to determine either the pressure or volume response within the compliant element or the flow response through the tube. Should any one of these responses be determined, the other two can be found.

Before we make an analytic determination, we shall first make some intuitive guesses as to what can be expected. Let us assume the compliant element is slightly inflated because of a steady pressure applied to the entrance of the rigid tube ($P_{1q} = 8$ cm H_2O, Fig. 5.3a). The pressure in the compliant element is thus constant at $P_{2q} = P_{1q}$; the flow in the system is zero. These quiescent conditions are graphically depicted by the curves in Fig. 5.4 for the period $t < 0$. At $t = 0$, the system is subjected to a sudden increase in p_1 so that, for $t > 0$, p_1 is now $P_{1q} + p_{sm}$ (Fig. 5.4a). In more descriptive language one would say the input signal is a step-function of amplitude p_{sm}.

The pressure and flow responses are easy to imagine. Just before $t = 0$ the pressure in the compliant element is $P_{2q} = P_{1q} = 8$ cm H_2O.

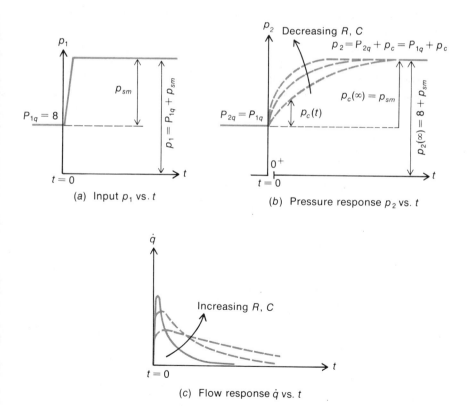

(a) Input p_1 vs. t

(b) Pressure response p_2 vs. t

(c) Flow response $\dot{q}$ vs. t

FIG. 5.4 *Pressure and flow response to a step input.*

Some time after the application of the step-function, the pressure in the compliant element will finally reach the applied pressure [as $t \rightarrow \infty$ the pressure p_2 will become $p_2(\infty) = 8 + p_{sm}$; see Fig. 5.4b]. In the time interval between $0^+ < t < \infty$ the pressure within the compliant element must have gradually increased from 8 to $8 + p_{sm}$ (dashed curves in Fig. 5.4b). How this increase took place is not yet known but it seems reasonable to suppose that the lower the resistance (or compliance, i.e., less stretch) the more rapidly p_2 will increase. Intuitively we realize that the rate of rise is intimately related to the resistance and compliance of the system.

We can make similar qualitative observations about the flow curve. Before the step-function of pressure is applied, the flow is zero; after a reasonable time interval the flow is again zero. In the time interval $0^+ < t < \infty$ the flow increased and then decreased. Again, by intuitive reasoning, we would expect that the lower the resistance the higher the maximum flow, and the lower the resistance and compliance the faster the flow reduces to zero (Fig. 5.4c). These intuitive observations seem self-evident, and any analytic solution would be expected to justify these conclusions.

To obtain a more quantitative statement about the relationship between the resistance, compliance, and response, we must decide upon the equations needed to describe the system. They can be determined by examining the original system shown in Fig. 5.3a or the analog simulation in Fig. 5.5a.* In either case, the flow through the tube can be written

$$(5.1) \qquad (p_1 - p_2)G = \dot{q} = [(P_{1q} + p_{sm}) - (P_{2q} + p_c)]G$$

Since P_{1q} and P_{2q} are equal, Eq. (5.1) can be written

$$(5.2) \qquad (p_{sm} - p_c)G = \dot{q}$$

The storage in the compliant element is given simply by

$$(5.3) \qquad (p_2 - P_{2x})C = q$$

Since the change of volume stored per unit time (i.e., the volume delivered per unit time) is equivalent to the flow into the storage element,

---

* It should be fairly obvious how the simulation was developed. The rigid tube is represented by a resistor (no battery is required as the resistance line passes through the origin; see Fig. 5.3b); the compliant element is represented by a capacitor with an appropriate battery (P_{2x}), as the compliance line at the quiescent pressure of $P_{2q} = 8$ does not pass through the origin (see Fig. 5.3c). The driving source or pressure supply is shown as two separate units: The battery represents the quiescent supply pressure; the "variable" source represents the "variable" applied signal (p_{sm}) which in this case is simply a step-function. The ambient pressure is assumed to be constant and is chosen as the reference pressure. This implies that all pressure measurements are made with respect to the ambient pressure. The flow after p_{sm} is applied is represented by $\dot{q}$.

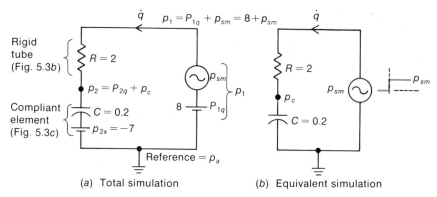

FIG. 5.5 Analog simulation of the system shown in Fig. 5.3a.

Eq. (5.3) can be written

$$(5.4) \qquad C\frac{d(p_2 - P_{2x})}{dt} = \dot{q} = C\frac{d(P_{2q} + p_c - P_{2x})^*}{dt}$$

Since P_{2x} and P_{2q} do not change (i.e., they are both constants), Eq. (5.4) can be rewritten as

$$(5.5) \qquad C\frac{dp_c}{dt} = \dot{q}$$

It is of interest to note that in Eqs. (5.2) and (5.5) no constant sources appear. These equations could have been obtained directly from the simulation in Fig. 5.5b. This circuit is the same as that in Fig. 5.5a except that all sources associated with quiescent operation have been eliminated. This agrees with the observation made earlier that quiescent values need not be included so long as their existence is recognized.

The relationship between Eqs. (5.2) and (5.5) should now be self-evident. Since the flow through the tube must be equal to the flow into the compliant element, we have

$$(5.6) \qquad (p_{sm} - p_c)G = C\frac{dp_c}{dt}$$

or, by rearranging,

$$(5.7) \qquad \frac{dp_c}{dt} + \frac{G}{C}p_c = \frac{G}{C}p_{sm}$$

* It will be helpful to keep in mind that this derivative form is essentially the limiting case of changes over finite time periods. The approximate form is

$$C\frac{\Delta(p_2 - P_{2x})}{\Delta t} = \frac{\Delta q}{\Delta t} \approx \dot{q}$$

Equation (5.7) is a simple first-order linear differential equation. (Only the first derivative of the variable appears; the coefficients G and C are constants in the range of operation.) The response p_c is related to the stimulus p_{sm} and the system parameters G/C. When the exact solution for p_c is written, we shall be able to see how the system parameters affect the response.

Rather than use conventional methods to effect an exact solution of Eq. (5.7), we shall first attempt an approximate numerical solution based upon the interpretation of the derivative as the ratio of a small finite difference. To do so, we interpret dp_c/dt as $\Delta p_c/\Delta t$, where Δp_c is the change of p_c in the time interval Δt. Thus, to determine the value of p_c at 0.4 s after the application of p_{sm}, Eq. (5.7) can be written

$$(5.8) \qquad \frac{p_c(0.4) - p_c(0)}{\Delta t(=0.4)} + \frac{G}{C} p_c(0.4) = \frac{G}{C} p_{sm}$$

where $p_c(0.4)$ is the value of p_c at the end of the 0.4-s time interval; $p_c(0)$ is the value of p_c at the beginning of the time interval so that $p_c(0.4) - p_c(0)$ is the change of $p_c(\Delta p_c)$ during the 0.4-s period; Δt is the 0.4-s period. If we use the values of G and C given earlier and if p_{sm} is assumed to be a step-function of the magnitude 2 cm of H_2O, we can solve Eq. (5.8) for p_c (0.4):

$$(5.9) \qquad p_c(0.4) = \frac{(G/C)\,\Delta t p_{sm}}{1 + (G/C)\,\Delta t} + \frac{p_c(0)}{1 + (G/C)\,\Delta t}$$

$$(5.10) \quad p_c(0.4) = \frac{(0.5/0.2)(0.4)(2)}{1 + (0.5/0.2)(0.4)} + \frac{0^*}{1 + (0.5/0.2)(0.4)} = 1.0 \text{ cm } H_2O$$

This method of finite differences can be used to generate a series of values for p_c. Thus to find p_c at 0.8 s $[p_c(0.8)]$, we could interpret Δp_c as the change in p_c in the interval $0.4 < t < 0.8$; that is,

$$(5.11) \qquad \Delta p_c = p_c(0.8) - p_c(0.4)$$

Therefore, Δt is still 0.4 s, and Eq. (5.7) can now be written

$$(5.12) \qquad \frac{p_c(0.8) - p_c(0.4)}{\Delta t} + \frac{G}{C} p_c(0.8) = \frac{G}{C} p_{sm}$$

wherefore

$$(5.13) \qquad p_c(0.8) = \frac{(G/C)\,\Delta t\, p_{sm}}{1 + (G/C)\,\Delta t} + \frac{p_c(0.4)}{1 + (G/C)\,\Delta t}$$

---

* $p_c(0)$ is the value of p_c at the instant that p_{sm} is applied $(t = 0^+)$. p_2, the pressure in the compliant element (see Fig. 5.3 or 5.5), has not yet had the chance to increase above quiescence so that $p_2 (0^+) = P_{2q}$ and p_c (the variable portion of p_2) must be zero at $t = 0^+$.

To evaluate $p_c(0.8)$ we must know $p_c(0.4)$. But this value was previously determined [Eq. (5.10)]. Equation (5.13) can thus be used to evaluate $p_c(0.8)$:

$$(5.14) \qquad p_c(0.8) = \frac{(2.5)(0.4)(2)}{1 + 1} + \frac{1 + 0}{1 + 1} = 1.5$$

The pattern is now self-evident. For every succeeding value of p_c we use the previously determined value and continuously solve for a new p_c. In each calculation we use an expression similar to that of Eq. (5.8). In fact, if n is used as any time instant and $n - 1$ as the previous instant, Eq. (5.8) can be more generally written as

$$(5.15) \qquad \frac{p_c(n) - p_c(n - 1)}{\Delta t} + \frac{G}{C} p_c(n) = \frac{G}{C} p_{sm}$$

or

$$(5.16)^* \quad p_c(n) - \frac{1}{1 + (G/C)\,\Delta t} p_c(n - 1) = \frac{(G/C)\,\Delta t}{1 + (G/C)\,\Delta t} p_{sm}$$

* This type of equation is generally known as a first-order difference equation. Such equations are often used in digital-computer simulations. In this case the program would be arranged so that p_{sm} and $p_c(0)$ would be fed to the computer to generate $p_c(0.4)$. Thereafter the computer would automatically use $p_c(0.4)$ to compute $p_c(0.8)$, etc. One outline of the procedure may be schematically represented by the block diagram shown below.

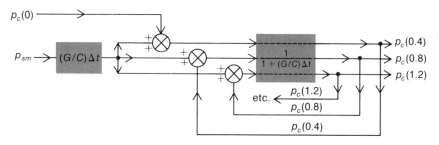

This finite-difference technique is not without pitfalls; the choice of Δt is not as completely arbitrary as the text would seem to imply. If the time interval Δt (or the sampling period, as it is sometimes called) is too long, this procedure does not yield a stable solution. Instead, the calculated curve will oscillate around the true curve with ever-increasing error. The criterion for the choice of Δt in this case depends on the term $(G/C)\,\Delta t$. To obtain a stable solution for this situation, this term should not exceed 2 [that is, $(G/C)\,\Delta t < 2$]; the finite-difference procedure in this case will therefore yield a stable solution if $\Delta t < 0.8$ (see Prob. 5.1). We shall not discuss digital programming in this text and therefore shall not be able to investigate the aspects of programming for stable solutions. It should be recognized, however, that this problem can exist. For more information on the subject, Ref. 2 may be consulted.

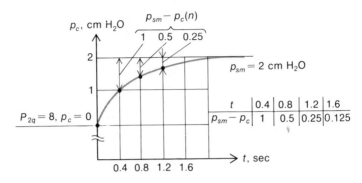

(a) Pressure response

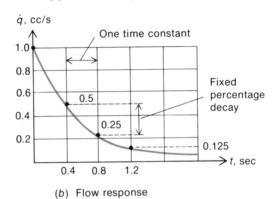

(b) Flow response

FIG. 5.6 *Approximation for the step-function response of system shown in Fig. 5.3.*

and finally, if numbers are used,

$$(5.17) \quad p_c(n) = \frac{(2.5)(0.4)(2)}{1 + (2.5)(0.4)} + \frac{p_c(n-1)}{1 + (2.5)(0.4)} = 1 + \frac{p_c(n-1)}{2}$$

An approximate solution for $p_c(n)$ based upon Eq. (5.17) is shown in Fig. 5.6a. The pressure in the compliant element of Fig. 5.3 rises gradually as expected; ultimately, it becomes equal to the applied-signal pressure p_{sm}. Shown also in Fig. 5.6 is the system flow response. If p_c is defined at every 0.4-s interval and p_{sm} is known, the pressure drop across the rigid tube is known. The flow through the tube as a function of the time increments can thus be easily determined [see Eq. (5.2)].

Although the solutions for pressure and flow just developed are approximate and graphical, they point out how the response and system parameters are related. First we make the general observation that the response is related to the ratio of G/C. All the equations from (5.7) to (5.17) contain this ratio, and as this ratio changes, so must the response. If, for example, G/C were 25 instead of 2.5, then Eq. (5.17) would be

$$(5.18) \qquad p_c(n) = \frac{20}{11} + \frac{p_c(n-1)}{11} \approx 2 = p_{sm}$$

and p_c would reach p_{sm} almost immediately.* On the other hand, if G/C were 0.25 instead of 2.5, then $p_c(n)$ and $p_c(n-1)$ would change very little for each time interval and it would take a very long time for p_c to reach p_{sm}. The ratio G/C thus governs the time-response of the system.

We make another observation regarding the ratio G/C. If we examine the expression $1 + (G/C) \Delta t$ we note that the first term is a unitless number. The second term must therefore also be a unitless number. Since Δt has the units of time, the ratio G/C must have the units 1/time; in other words, the reciprocal ratio (C/G) must also have the units of time.† In this case the ratio C/G would be considered the *time-constant* of the system. Obviously the time-constant of a system governs the time-response of the system and is intimately related to the system parameters. A more general statement of the significance of the time-constant will be given in later sections.

Having established that the ratio G/C directly governs the response, we now seek to determine how G/C can be evaluated from the response curves. To do this we should determine the exact analytic response by solving Eq. (5.7). We might, however, use the approximate solution developed here to indicate the trends to be expected. For example, in the table accompanying Fig. 5.6a, p_c shows a very organized increase. For every 0.4 s, the response approaches its final value at a fixed percent (50 percent for the approximate solution). Thus, if at 0.4 s p_c is 1 cm H_2O removed from its final value of 2 cm H_2O, at 0.8 s p_c is 0.5 cm H_2O removed, at 1.2 s it is 0.25 cm H_2O removed, etc. (see Fig. 5.6a). The flow response has these same properties. One could almost intuitively state the mathematical relationship governing this organized pressure response: After n 0.4-s intervals, the response p_c will be within $(\frac{1}{2})^n$

---

* Here the author falls into the same trap he warns against in the previous footnote. With a time interval of 0.4 and G/C chosen as 25, the product of these terms is 10; this is greater than the allowed limit of 2 for this product (see previous footnote). The point to be illustrated, however, is that the response depends upon the ratio G/C, and this conclusion is valid. If the reader needs to convince himself of this point, he can use G/C as 25 and then use $t = 0.075$. With these numbers, he will find the pressure p_c reaches a value of approximately 2 in 0.3 s (see Prob. 5.1).

† This can easily be verified by determining the units of C and G and forming the appropriate ratio.

of its final value; that is,

$$(5.19)^* \qquad\qquad p_{sm} - p_c(n \times 0.4) = (\tfrac{1}{2})^n p_{sm}$$

The decay factor of 50 percent (in the approximate case) is due directly to the expression $1/[1 + (G/C)\,\Delta t]$ and particularly the term G/C. When Δt is equal to the time-constant G/C, the term $(G/C)\,\Delta t$ is unity and $1/[1 + (G/C)\,\Delta t]$ becomes $\tfrac{1}{2}$. The relationship between the time-constant and the response should now be evident: The response approaches its final value at a fixed percent for each time interval equivalent to the time-constant. (For the approximate case the decay factor is 50 percent; it will subsequently be seen that for an exact analysis the factor is about 37 percent.) The time-constant can thus be determined from the response simply by finding the time interval during which the response approaches its final value with a specified percentage change. We shall use this fact after we determine the correct percentage change to be employed.

Throughout the foregoing discussion it was emphasized that the 50 percent factor is based upon an approximate analysis. The approximation stems from the choice of Δt. For example, the value of p_c at 1.2 s depends upon that at 0.8 s, which depends upon that at 0.4 s; obviously, if the

---

* This relation could have been directly determined by using Eq. (5.16) as a recursion formula. Thus

$$p_c(n \times \Delta t) = \frac{(G/C)\,\Delta t}{1 + (G/C)\,\Delta t}\, p_{sm} + \frac{1}{1 + (G/C)\,\Delta t}\, p_c[(n-1) \times \Delta t]$$

By using the same relation, we can determine $p_c[(n-1)\,\Delta t]$:

$$p_c[(n-1)\,\Delta t] = \frac{(G/C)\,\Delta t}{1 + (G/C)\,\Delta t}\, p_{sm} + \frac{1}{1 + (G/C)\,\Delta t}\, p_c[(n-2)\,\Delta t]$$

Substitution of the latter expression in the former gives

$$p_c(n\,\Delta t) = \left\{ \frac{1}{1 + (G/C)\,\Delta t} + \frac{1}{[1 + (G/C)\,\Delta t]^2} \right\} \frac{G}{C}\, \Delta t\, p_{sm} + \frac{1}{1 + (G/C)\,\Delta t}\, p_c[(n-2)\,\Delta t]$$

By repeated substitution we have

$$p_c(n\,\Delta t) = \left\{ \frac{1}{1 + (G/C)\,\Delta t} + \frac{1}{[1 + (G/C)\,\Delta t]^2} + \cdots + \frac{1}{[1 + (G/C)\,\Delta t]^n} \right\} \frac{G}{C}\, \Delta t\, p_{sm}$$
$$+ \text{ zero [as } p_c(0) = 0]$$

The coefficient of p_{sm} (in braces) is a geometric progression. When the sum of the progression is taken and used in the last expression, the result is

$$p_c(n\,\Delta t) = \left\{ \frac{-1}{[1 + (G/C)\,\Delta t]^n} + 1 \right\} p_{sm}$$

By transposing and using the appropriate numerical values we obtain the relation given by Eq. (5.19).

value of p_c at 0.4 s is slightly in error, all the other values are also. It is also obvious that the value of p_c at 0.4 s might very well be different if Δt were chosen to be 0.2 s rather than 0.4. This can be established very quickly by using the relation given by Eq. (5.19) except that $1/[1 + (G/C) \Delta t]$ is used instead of $\frac{1}{2}$. That is,

$$(5.20) \qquad \frac{p_{sm} - p_c(n \, \Delta t)}{p_{sm}} = \left[\frac{1}{1 + (G/C) \, \Delta t}\right]^n$$

or

$$(5.21) \qquad p_{sm} - p_c(n \, \Delta t) = p_{sm}\left[\frac{1}{1 + (G/C) \, \Delta t}\right]^n$$

or

$$(5.22) \qquad p_c(n \, \Delta t) = p_{sm} - p_{sm}\left[\frac{1}{1 + (G/C) \, \Delta t}\right]^n$$

When Δt is 0.2 at the first interval ($n = 1$, $t = 0.2$) we have

$$(5.23) \quad p_c(0.2) = +p_{sm} - p_{sm}\frac{1}{1 + (2.5)(0.2)} = 2\left(1 - \frac{2}{3}\right) = \frac{2}{3}$$

At the second interval ($n = 2$, $t = 0.4$) we have

$$(5.24) \quad p_c(2 \times 0.2) = p_c(0.4) = p_{sm} - \left(\frac{2}{3}\right)^2 p_{sm} = 2\left(1 - \frac{4}{9}\right) = \frac{10}{9}$$

We now see that, at 0.4 s, the value of p_c is somewhat greater than previously calculated (1.12 instead of 1.0; see Fig. 5.7) and that the decay factor at one time-constant is now 44 instead of 50 percent (when 0.4-s

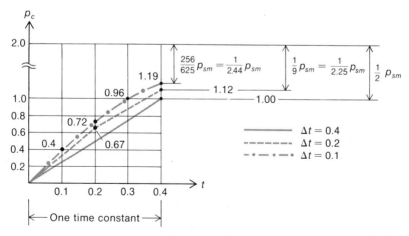

FIG. 5.7 *Response during an interval of one time-constant using $\Delta t = 0.4, 0.2, 0.1$.*

intervals were used). If we decrease the interval further to 0.1 s, the decay factor at one time-constant would be 41 percent. The decay factor at one-time-constant intervals thus continuously decreases (at a decreasing rate: 50 to 44 to 41 percent) as the time interval Δt decreases. If the interval is decreased to infinitesimally small values, the decay factor at one time-constant approaches the limit of about 37 percent, which is equivalent to $1/2.718$. The latter figure is the reciprocal of the natural number ϵ.*

We have now developed the exact relationship between the time-constant and the response and have in effect simultaneously solved the differential equation (5.7). With regard to the former contention, we now see that the response approaches its final limit by 37-percent increments for every interval of time equivalent to the time-constant.

$$(5.25) \qquad \frac{p_{sm} - p_c(t = \tau)}{p_{sm}} = \frac{1}{\epsilon} = \epsilon^{-1}$$

For two time-constants,

$$(5.26) \qquad \frac{p_{sm} - p_c(t = 2\tau)}{p_{sm}} = \epsilon^{-2}$$

To find a general expression at any time $(t = k\tau)$ we write

$$(5.27) \qquad \frac{p_{sm} - p_c(t = k\tau)}{p_{sm}} = \epsilon^{-k}$$

where k is now t/τ. Therefore

$$(5.28) \qquad \frac{p_{sm} - p_c(t)}{p_{sm}} = \epsilon^{-t/\tau}$$

or

$$(5.29) \qquad p_c(t) = p_{sm}(1 - \epsilon^{-t/\tau}) \qquad \tau = \frac{C}{G} = RC$$

Equation (5.29) is the general solution of the first-order differential equation when $p_c(0) = 0$ (that is, when initial conditions are zero).

For this simple, first-order system (systems governed by the first-order differential equation) all variables respond in an exponential fashion with the same time-constant. The flow, for example, must approach its final value with the same percentage decay as does the pressure rise in the compliant element. We can illustrate this point by calculating

---

* This can be developed mathematically by finding the limit of $[1 + (G/C) \Delta t]^n$ as Δt approaches zero. Now n is defined as the number of intervals required to reach one time-constant. Thus, if one time-constant (C/G) is represented by the symbol τ, n can be written $\tau/\Delta t$ and the term $[1 + (G/C) \Delta t]^n$ becomes $(1 + \Delta t/\tau)^{\tau/\Delta t}$. The limit of this last term approaches a constant value of approximately 2.718 $(= \epsilon)$ as Δt approaches zero.

the flow response from either Eq. (5.2) or Eq. (5.5). If we use the former, $\dot{q}$ can be written

$$(5.30) \qquad \dot{q} = G[p_{sm} - p_{sm}(1 - \epsilon^{-t/\tau})]$$

where p_c in Eq. (5.1) has been replaced by its equivalent value as given by Eq. (5.29). By expanding and collecting terms, we have

$$(5.31) \qquad \dot{q} = Gp_{sm}\epsilon^{-t/\tau}$$

As can be seen, the flow approaches a final value of zero, and it has the same percentage decay as p_c. (See Fig. 5.8.)

The volume in the compliant element must also increase exponentially. We can establish this by using two different approaches. From one point of view, the volume in the compliant element is defined by the pressure across the element and the compliance of the element (that is, $q = Cp_c$). Since p_c has already been defined by Eq. (5.29), we can immediately write

$$(5.32) \qquad q = Cp_{sm}(1 - \epsilon^{-t/\tau})$$

From another point of view, volume and flow are integrally related; i.e., the volume curve is generated by finding the area under the flow curve. Since the flow curve is a falling exponential, the area under this curve must reach a limiting value; the volume curve must therefore rise to a maximum level (Fig. 5.8b). To obtain an analytic expression for the volume curve, we could formally integrate the flow equation [Eq. (5.31)]; the result must obviously be Eq. (5.32).

The long mathematical discussion in this section may have beclouded our purpose. We briefly summarize what has been done and why it has been done. To show how step-response data can be used to determine system properties, we have used the reverse approach. Known system properties (resistive and compliant) were used with a step-input

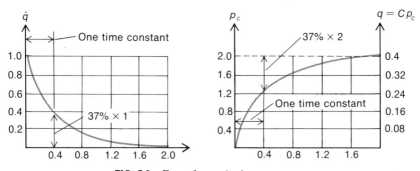

FIG. 5.8   *Exact flow and volume response.*

to determine the response. In using this approach, we established that, if a system is composed of (or can be represented by) a resistive and a compliant element in series, a distinctive step-response for small excitations can be expected. The shape of the response is governed by a time-constant, and the time-constant is related to the system properties. Given the step-response data, therefore, we now know that, if the response has a steady exponential decay toward equilibrium, some of the gross properties of the system can be described.

If this development provided insight to only resistive-compliant networks and step-response data, it would represent considerable work with relatively little reward. Many of the concepts developed here are more general and not necessarily restricted to resistive-compliant networks or to step-response. It will be seen later that any system with this exponential step-response must also have lumped properties that can be represented by a resistive and a storage element (not necessarily compliant). An even more general conclusion will be that such systems can be described by first-order differential equations. (Because of the latter connection, the exponential step-response is often described as a first-order step-response.) Since a differential equation has a unique solution for every excitation (when initial conditions are zero), any input should yield a predictable output. Ultimately, therefore, the step-response investigation should lead to a technique whereby any input-output data can be used to ascertain whether a system can be considered a first-order system (represented by a single resistive-storage combination) and, if it cannot be so represented, what representation will be needed.

5.3 AN EXAMINATION OF THE STEP–FUNCTION

Before we discount the idea of a step-function as simply a mathematical concept, we shall examine its significance. The step-function input is theoretically defined as a sudden change of the input stimulus from one operating level to another. Obviously, one cannot expect a stimulus to change from one value to another in zero time; some finite time is required for the change. The total pressure on a man ascending from deep submergence (Fig. 5.9a) or the variation in pressure to which he is subjected (Fig. 5.9b) cannot therefore be realistically represented by the solid curves shown. As a practical matter, one of the dashed curves should be used to represent the pressure signal. However, according to the time scale of the response data (Fig. 5.1b), response time is measured in hours. If the rate of ascent is rapid, say 1 ft/s (which is not very rapid), the pressure signal reaches a new operating level in about $\frac{1}{2}$ min. This represents such a

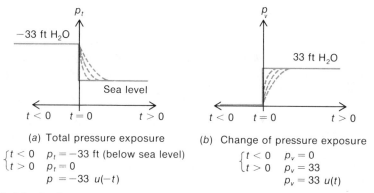

(a) Total pressure exposure

$$\begin{cases} t < 0 & p_t = -33 \text{ ft (below sea level)} \\ t > 0 & p_t = 0 \end{cases}$$
$$p = -33\ u(-t)$$

(b) Change of pressure exposure

$$\begin{cases} t < 0 & p_v = 0 \\ t > 0 & p_v = 33 \end{cases}$$
$$p_v = 33\ u(t)$$

FIG. 5.9 *Total pressure and pressure variation during an ascent from deep submergence.*

small fraction of the response time scale that it can safely be ignored, and the pressure signal can be considered a sudden change appropriately described as a step-function.

To describe the step-function mathematically, two separate expressions must be used: one to describe the level before the time of observation (i.e., the initial condition at $t < 0$) and one to describe the level during the observation period. The curve for the total pressure exposure in Fig. 5.9a (p_t) could thus be written

$$(5.33) \qquad p_t = 33 \text{ ft} \qquad \text{for } t < 0 \qquad p_t = 0 \qquad \text{for } t \geq 0$$

and the curve for the variation in pressure exposure (p_v in Fig. 5.9b) could be written

$$(5.33a) \qquad p_v = 0 \qquad \text{for } t < 0 \qquad p_v = 33 \text{ ft} \qquad \text{for } t \geq 0$$

The notation used in Eqs. (5.33) and (5.33a) is completely adequate though somewhat clumsy as two expressions are required to describe the total waveform. To simplify the representation, the special symbol for a unit-step-function, $u(t)$, is often employed. This symbol automatically indicates a step has occurred at time $t = 0$. In specific terms, the symbol $u(t)$ indicates a signal is zero when t is less than zero and unity when $t \geq 0$. In equation form this is written

$$(5.34) \qquad\qquad\qquad u(t) = 0 \qquad \text{for } t < 0$$

$$(5.34a) \qquad\qquad\qquad u(t) = 1 \qquad \text{for } t \geq 0$$

Therefore when we write $y = u(t)$ we automatically picture a step-signal of unit amplitude at $t = 0$ (see Prob. 5.2).

If we use this unit-step notation, we can describe the waveforms of Eqs. (5.33) and (5.33a) by single concise expressions. Equation (5.33), for example, can be written

$$(5.35) \qquad p_t = 33 - 33u(t)$$

When $t < 0$, $u(t)$ is zero; therefore $33u(t)$ is zero and $p_t = 33$. When $t \geq 0$, $u(t)$ is unity; $33u(t)$ is 33 and $p_t = 33 - 33 = 0$. In a similar manner p_v may be written

$$(5.35a) \qquad p_v = 33u(t)$$

When $t < 0$, $u(t)$ is zero and p_v is zero, as it should be. When $t \geq 0$, $u(t)$ is unity and p_v is 33 ft.

Although a true step-function is a mathematical concept, it can often be used to represent the realistic situations encountered in research experiments. Similarly, although the mathematical first-order step-response developed in Sec. 5.2 is not completely realistic, it can be used to approximate many physical and physiological responses. To be able to determine whether a step-response is indeed first-order (i.e., with an exponential approach to equilibrium), one should have some techniques to test the step-response curves for the first-order property. In the following section we shall examine some of the methods that can be used to establish whether or not a system has a first-order step-response.

5.4 AN EXAMINATION OF STEP–RESPONSE DATA

The method to determine whether a system actually has a first-order step-response was essentially outlined in Sec. 5.2. The data are examined to determine whether there is a relatively consistent time-constant throughout the response. Should this be the case, the response can then be considered that of a first-order system.

From the discussion of Sec. 5.2 we know that the step-response of a first-order system is governed by a fixed percentage decay for each time interval equivalent to one time-constant. After one time-constant the response is within 37 percent of its final value (or it will have gone 63 percent of its total excursion). After two time-constants it will be within $(0.37)^2$ (13.7 percent) of its final value (or it will have gone $100 - 13.7 = 86.3$ percent of its total excursion), etc. Therefore, if lines representing fixed percentages of the maximum value are drawn on the response curve (see Fig. 5.10) and if they intercept the curve at equal time intervals, the system has an exponential response as defined by Eq. (5.29). Each time interval, of course, represents one time-constant. In establishing that the system is first-order, the time-constant is automatically obtained and can be used as part of the information required to describe the system properties.

For a quick check, curve slopes are sometimes used to ascertain whether a system has a first-order step-response. The tangent line at

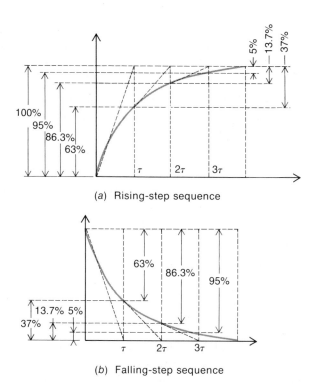

(a) Rising-step sequence

(b) Falling-step sequence

FIG. 5.10   *Use of the exponential-decay constant and curve shapes to check a step-response for first-order properties.*

any point on an exponential response intercepts the final value in an interval of one time-constant. Thus in Fig. 5.10a and b a tangent to the curve at $t = 0$ intercepts the final value at one time-constant; a tangent to the curve at one time-constant intercepts the final value at one time-constant removed,* etc. By drawing appropriate tangent lines, one can

* As an analytic justification we need only write the equation for a first-order response and then find the slope by differentiation. If we use Eq. (5.29) as an example,

$$p_c = p_{sm}(1 - \epsilon^{-t/\tau}) \qquad \frac{dp_c}{dt} = \text{slope} = \frac{p_{sm}}{\tau} \epsilon^{-t/\tau}$$

The slope at $t = 0$ is the ratio of the final value to the time-constant.   A tangent line at $t = 0$ thus rises to $p_c = p_{sm}$ in an interval of one time-constant. If we draw the tangent at any point, say where $p_c = f p_{sm}$ (f is any fraction), then

$$p_c = f p_{sm} = p_{sm}(1 - \epsilon^{-t/\tau}) \qquad \text{or} \qquad 1 - f = \epsilon^{-t/\tau}$$

The slope at any point is therefore

$$\text{Slope} = \frac{p_{sm}}{\tau}(1 - f) = \frac{p_{sm} - f p_{sm}}{\tau} = \frac{p_{sm} - p_c}{\tau} = \frac{\text{rise to equilibrium from any point}}{\tau}$$

A tangent at any point thus reaches the equilibrium level in one time-constant.

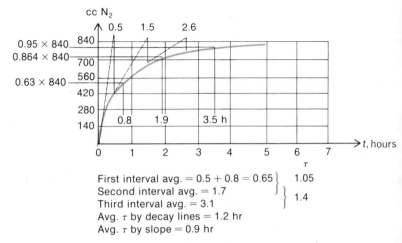

FIG. 5.11 *Check of nitrogen-liberation curve (Fig. 5.1b) for first-order properties.*

therefore determine whether or not the time-constant is relatively consistent throughout the response.

As an illustration of how these techniques may be employed, we examine the step-response curve in Fig. 5.1. This curve is repeated in Fig. 5.11. The percentage-decay lines and the tangent lines are drawn directly on the curve. As can be seen, the percentage-decay technique and the slope method both indicate that the curve is not really the step-response of a first-order system. The time-constant at the beginning of the curve seems to be about 0.65 h; after 0.65 h the curve appears to have a time-constant of about 1 h. Since the time-constant is not consistent throughout the response, this is not, strictly speaking, a first-order system. However, we are now considering a real physical system, not a mathematical one. For a real system there is always a question of accuracy of data and degree of accuracy required. If the data are not as accurate as they could be but accurate enough to provide a picture of gross effects, a first-order representation may be a reasonable approximation. If the data are known to be accurate, a first-order approximation might still be acceptable for some applications. In the latter case, however, if the effort is made to acquire accurate data, a more exact representation would probably be required. In order to develop a technique that may be used to describe the step-response data of Fig. 5.1 more accurately, a more sensitive method to test such data is needed. Such a technique will be developed in the following paragraphs.

The methods described for testing step-response data are rather simple and quick. These are their prime advantages. By using these tests one can rapidly decide whether or not a first-order representation seems reasonable. However, these tests are necessarily crude and may in themselves be inaccurate. Tangents to curves, particularly for physical data, are always difficult to construct. In the case of a decaying exponential it is almost impossible to draw tangent lines in the vicinity of equilibrium. A similar difficulty is encountered when the percentage-decay lines are drawn. The first few lines can easily be located. Thereafter, the spacing between the lines becomes so small that accuracy is impossible.

In order to develop a more sensitive test we take advantage of the exponential property of a first-order system and the exponential property of logarithms. Natural (or common) logarithms are merely an indication of the exponent of a specified base number (ϵ or 10; $\ln \epsilon^1 = 1$, $\ln 8 = \ln \epsilon^{2.08} = 2.08$; $\ln \epsilon^{-t/\tau} = -t/\tau$). Since the first-order step-response approaches equilibrium exponentially with time, the natural logarithm of the approach is linearly related to time, with $1/\tau$ as a proportionality constant. These ideas are more precisely expressed mathematically. If we use Eq. (5.29) as an example,

$$(5.29) \qquad\qquad p_c = p_{sm}(1 - \epsilon^{-t/\tau})$$

the approach to equilibrium is given by

$$(5.36) \qquad\qquad p_{sm} - p_c = p_{sm}\epsilon^{-t/\tau}$$

If natural logarithms are used, Eq. (5.36) becomes

$$(5.37) \qquad\qquad \underbrace{\ln\,(p_{sm} - p_c)}_{y} = \underbrace{\ln p_{sm}}_{b} \underbrace{- \frac{t}{\tau}}_{+\ mx}$$

Equation (5.37) indicates that plotting the natural logarithm of $p_{sm} - p_c$ (call this y) vs. the time t (call this x) should generate a straight line with a negative slope (m) of $1/\tau$ and a y intercept (b) of p_{sm}.

The implications of this development should now be obvious. If the step-response data are replotted so that the x axis is in units of time and the y axis is the logarithm of distance to equilibrium, the result will be a straight line if the step-response is that of a first-order system. The negative reciprocal of the slope of the line is the time-constant.

As an illustration of the procedure we again examine the nitrogen-liberation data of Fig. 5.1b. In Fig. 5.12a the data have been replotted as $N_{2max} - N_2$ versus t and in Fig. 5.12b the data are shown replotted as $\ln\,(N_{2max} - N_2)$ versus t. Also included in Fig. 5.12b is a logarithmic scale

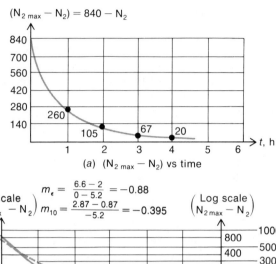

FIG. 5.12 *A plot of $N_{2\max} - N_2$ versus time in linear coordinates and semilog coordinates.*

using semilog paper, where the ordinate is automatically calibrated in logarithms.*

As might have been expected, the plot is not a straight line. It was already shown that the nitrogen-liberation curve is not strictly the step-response of a first-order system. We ask the same questions previously

* Semilog paper, on which the ordinate scale is calibrated in logarithms, can be conveniently used. However, commercial semilog paper uses a log scale to the base 10 (not to the base ϵ). Equation (5.37) should therefore be rewritten as

$$\overbrace{\log (p_{sm} - p_c)}^{y} = \overbrace{\log p_{sm}}^{b} - \frac{t}{\tau} \log \epsilon = \log p_{sm} - \overbrace{\left(\frac{t}{\tau}\right) 0.435}^{mx} \qquad \text{where } 0.435 = \log_{10} \epsilon$$

The negative reciprocal of the slope here is therefore $\tau/0.435$ or

$$\tau = \frac{0.435}{-\text{slope}}$$

posed: Are the data accurate? Is a good degree of accuracy required? etc. Since no general answers are possible, we shall examine the data with both points of view in mind. In this section we shall assume a gross approximation is possible. At the end of the next section we shall examine the data, assuming a better degree of accuracy is desired.

If a gross representation is acceptable, we can assume that the logarithmic plot in Fig. 5.12b may be represented by a straight line. The slope of the line defines the time-constant of the system. To decide what straight line can be drawn should not be a problem in these cases. The data will probably be close to a straight line to start with; otherwise there would be no justification for considering straight-line approximation. Also, since an overall approximation is acceptable it would seem unnecessary to use curve-fitting techniques. Under these circumstances, probably a simple estimate by eye would serve the purpose.

If we employ this procedure we find that the plot has a slope of 0.88 when natural logarithms are used; when common logarithms are used the plot has a slope of -0.395. In the first case the time-constant is evaluated to be

$$(5.38) \qquad\qquad \tau = \frac{-1}{-0.88} \approx 1.1 \text{ h}$$

In the second case the time-constant is given by

$$(5.39) \qquad\qquad \tau = \frac{-0.435}{-0.395} \approx 1.1 \text{ h}$$

The value obtained will vary somewhat depending upon the eye approximations used. It is surprising, however, that estimates by two individuals seldom differ by more than 20 percent. We have now, in effect, assumed that the nitrogen liberation can be considered an approximate first-order system with a time-constant of about 1.1 h.

We should review once more the implications of accepting the view that the nitrogen-liberation system has a first-order step-response. This means that physically the system has properties that are grossly described by a single resistive and a single storage element. Mathematically, the assumption has been made that the nitrogen-liberation step-response may be described as

$$(5.40)^* \qquad N_2(t) = N_{2\text{max}}(1 - \epsilon^{-t/\tau}) = 840(1 - \epsilon^{-t/1.1})$$

* These statements can be related to the concepts presented in Sec. 5.2. Equation (5.40), for example, can be directly compared with Eq. (5.29). In Sec. 5.7 we shall justify these concepts, using generalized through- and across-variables.

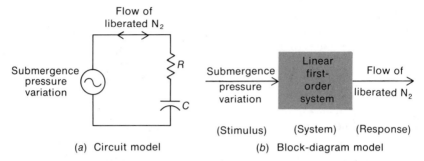

(a) Circuit model   (b) Block-diagram model

FIG. 5.13 *Model representations of nitrogen-liberation system upon ascent from deep submergence.*

and that the system may be described by the linear differential equation

(5.41)
$$R \frac{dN_2}{dt} + \frac{1}{C} N_2 = p(t)$$

where R ($= 1/G$) and C are the resistive and storage properties of the system, giving rise to the time-constant of RC. In analog form we assume the system can be represented by a series resistive-capacitive circuit as shown in Fig. 5.13a. In block-diagram form the representation is that of Fig. 5.13b.

It will be seen in the next section that, in spite of the gross approximations, the results have some utility. By using these approximate results, one can, for example, estimate the response that might be expected for different stimulus waveforms. If these results are used to pose some leading questions, such as what the resistive and storage properties imply, the approximate results may help to gain insight into the operating mechanism of the system.

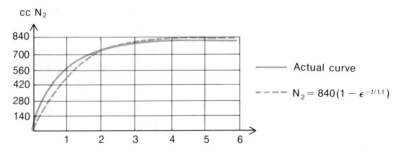

FIG. 5.14 *Comparison of actual nitrogen liberation curve with linear first-order step-response curve [Eq. (5.40); τ = 1.1].*

As a matter of interest, we should see how well or how poorly Eq. (5.40) represents the actual response. In Fig. 5.14 the actual-response curve is repeated; also shown is a plot of Eq. (5.40) with a time-constant of 1.1 h. The curves agree fairly well toward the end of the response time; there is, however, an appreciable error at the beginning of the response.

5.5 USE OF STEP–RESPONSE DATA TO ESTIMATE PERFORMANCE

In the preceding section it was noted that, if the first-order step-response is accepted as an approximation to given step-response data, a number of specific comments can be made about the system. These are all related and stem directly from the discussion of Sec. 5.2. In order to examine the significance of these comments and how they relate to the physical system, they are listed below; in each instance the nitrogen-liberation system is used to illustrate the points made. In review, then, if a system is stimulated with a step-stimulus and the response can be approximated by a linear first-order response, the following may be stated:

1. The system has a storage property that may be considered constant within the range of the applied step.* The storage property may be evaluated either from the ratio of the through- to the across-variable or from the time-constant. For the nitrogen-liberation system the ratio of the volume delivered (through-variable) to the pressure change of 33 ft H_2O (across-variable) defines the system storage property. The storage in this case is given by

$$(5.42) \qquad \text{Storage} = \frac{840 \text{ cm}^3}{33 \text{ ft}} = 25.4 \text{ cm}^3/\text{ft } H_2O$$

2. The system has a linear-resistive property in the range of observation. This property can also be evaluated either from the ratio of the variables or from the time-constant. Since the time-constant of the nitrogen-liberation system is 1.1 h and the through-variable storage is given by 25.4, the resistance is calculated to be

$$(5.43) \qquad R = \frac{\tau}{C} = \frac{1.1 \text{ h}}{25.4 \text{ cm}^3/\text{ft } H_2O} = 0.0434 \frac{\text{ft } H_2O}{\text{cm}^3/\text{h}}$$

3. The system may be represented by a resistance and a capacitor (or inductor) in series. (See Figs. 5.13a and 5.17.)

4. All responses of the system can be expressed in terms of the

---

* A more accurate statement is that the system time-constant remains constant within the range of excitation.

exponential function $\epsilon^{-t/\tau}$. In the case of the nitrogen-liberation system the volume of nitrogen liberated is expressed as

$$(5.44) \qquad N_2 = N_{2\max}(1 - \epsilon^{-t/1.1})u(t)$$

where $u(t)$, the unit step, indicates there is no response for $t < 0$. The response starts at $t = 0^+$. Since $N_{2\max}$ can be expressed in terms of the pressure and storage [Eq. (5.42)], we have

$$(5.45) \qquad N_2 = 25.4P(1 - \epsilon^{-t/1.1})u(t)$$

where P is the constant (step) pressure stimulus (33 ft H_2O in this case).

5. The system is governed by a first-order differential equation. This equation can be constructed from the response equation (or can be written from memory). To construct the equation, the response is differentiated and the result added to the response; e.g., by differentiating Eq. (5.45),

$$(5.46) \qquad \dot{N}_2 = \frac{25.4P}{1.1}\,\epsilon^{-t/1.1}u(t)$$

or

$$1.1\dot{N}_2 = 25.4P\epsilon^{-t/1.1}u(t)$$

By adding Eqs. (5.45) and (5.46), we get

$$(5.47) \qquad 1.1\dot{N}_2 + N_2 = 25.4Pu(t)$$

or

$$(5.48) \qquad \frac{1.1}{25.4}\,\dot{N}_2 + \frac{1}{25.4}\,N_2 = Pu(t)$$

6. Since the circuits (of Figs. 5.13a and 5.17) and the differential equation (5.48) are independent of the stimulus and depend only on the system, one can determine the response to any stimulus waveform (assuming the magnitude of the stimulus does not exceed the range of the initial step-function to ensure constancy of system properties). If, for example, a man is returned to sea level at a rate such that the pressure decrease is linear with time, we can write

$$(5.49) \qquad P = ktu(t) \qquad \text{or} \qquad P = kt \qquad t > 0$$

where k is the slope of the pressure curve. Equation (5.48) can now be written

$$(5.50) \qquad \frac{1.1}{25.4}\,\dot{N}_2 + \frac{1}{25.4}\,N_2 = ktu(t)$$

If k is specified, Eq. (5.50) can be solved by the standard techniques of differential calculus or by a transform method to be described in the next chapter. In any case, a solution of Eq. (5.50) is possible, and a plot of the response can be obtained.

Only the first two or three comments listed above deal with the internal properties of the system. The rest relate to the form of the response. The first few observations thus focus attention upon the system elements; the rest deal with the functioning of the system as a whole. The latter concepts are most useful when the composite system is under consideration; the first few concepts are most useful when seeking to understand or investigate the internal construction or operation of the system.

In this section we examine how a model derived from step-response data might be used to study the external behavior of the composite system. In the next section we shall examine how these same data may be used to gain insight about the internal functioning parts of the system.

Although the circuit model and differential equation need not accurately indicate the internal interaction between the functioning components of a system, they should accurately represent the system's external behavior. Once the model or equation is available, therefore, one should be able to predict the expected response for various stimulus waveforms, or, alternatively, to postulate an appropriate stimulus to provide a desired response. There are a variety of situations where it would be desirable to be able to predict responses to imposed stimuli. The astronauts' response to the acceleration profile expected during blast-off is one; the aquanauts' response to a variable-pressure exposure during a return to sea level is another. In the latter case a rapid pressure change during return to sea level can result in decompression sickness (the bends) which can cause anything from mild discomfort to death. In some underwater operations even mild discomfort may lead to tragic results.

To show how the equation or model may be used to study and shape a response curve, we again use the nitrogen-liberation system during decompression. This time we examine the system to propose an ascent program for a diver that would minimize the time for emergence (to use the diver's air supply efficiently) and yet not expose him to conditions that might cause decompression sickness. We start with the knowledge that an ascent from 33 ft (1 atm) can take place almost "instantaneously" with no ill effects (Fig. 5.1). For such an ascent the delivery rate of nitrogen can be estimated from Eq. (5.46) (or from the derivative of the liberation curve shown in Fig. 5.1b). A plot of the delivery rate vs. time

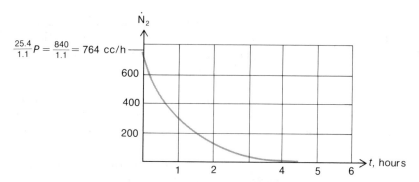

FIG. 5.15 *Nitrogen-liberation rate for ambient-pressure change of 1 atm.*

can thus be drawn, as in Fig. 5.15. From the curve it can be seen that the delivery rate starts at a maximum value of 764 cm³/h and diminishes steadily thereafter.

It would appear, then, that the body can eliminate nitrogen at least at a rate of 764 cm³/h with no ill effects. If we use this as a preliminary design figure, we now ask the question: What must the ascent curve be if emergence is from 66 ft H_2O and liberation flow is not to exceed 764 cm³/h?* We can now use the mathematical description of the system to good advantage. Equation (5.48) shows the system-flow response for any given pressure input. Our problem here is to find the pressure necessary to provide for a given flow. [From one point of view, the problem now is to find the input pressure (p) when the system and the output are specified; from another point of view, we can consider the stimulus to be flow and the output response to be the pressure profile.] Since flow is to be constant at a rate of 764 cm³/h, the delivered volume (N_2) at any time must be 764t. Direct substitution in Eq. (5.48) gives

$$(5.51) \qquad \frac{1.1}{25.4} \times 764 + \frac{1}{25.4} \times 764t = p = 33 \text{ ft} + 30 \frac{\text{ft}}{\text{h}} \times t$$

Equation (5.51) indicates that if flow is to be limited to 764 cm³/h the diver can ascend as rapidly as he likes for about 33 ft; thereafter he should ascend at a rate not to exceed 30 ft/h. The pressure curve is shown

---

* It should be noted that the proposed pressure signal is between 0 and 66 ft. The model representations we have derived are based upon a pressure signal between 0 and 33 ft H_2O. Therefore, if we use any of these model representations, we are assuming that the system properties do not show a radical departure from linearity with changed stimulus. With this assumption and the realization that human characteristics vary, it is clear that the results obtained cannot be used to propose a rigid operating schedule. Instead, the results should be viewed as an indication of the range within which an acceptable program might be developed.

in Fig. 5.16a, and the corresponding flow and volume curves are shown in Fig. 5.16b and c.

The above analysis gives rise to a slightly "oversafe" program. According to the Navy Standard Decompression Table (see Ref. 3), a man submerged at a depth of 70 ft for 180 min (bottom time; this determines the amount of nitrogen that will be stored in his system; see Prob. 5.3) can ascend as rapidly as he likes for 50 ft but should remain there for 21 min. He can then rise to 10 ft and remain there for 32 min, after which he may surface. (See "staircase" function; Fig. 5.16a.) As a total, therefore, the Navy suggests a decompression time of 53 min for a 70-ft submergence instead of the calculated 1.1 h for a 66-ft submergence. The flow curves based upon a 1.1-h time-constant and the Navy's program are shown by dashed lines in Fig. 5.16. The average "safe" flow is evidently somewhat higher than the "design" figure used here.

Aquanauts and sportsmen are acutely aware of the danger of decompression sickness. The demand has, in fact, alerted companies to consider the production of commercial-type computers and gauges which can be

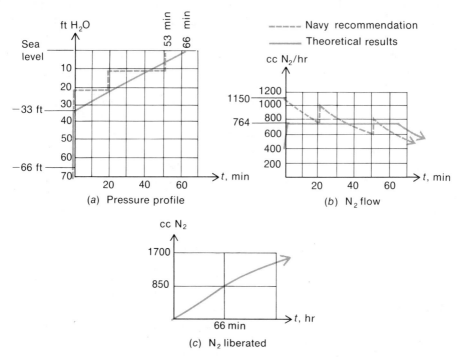

FIG. 5.16 *Ascent curves limiting liberation rate to safe values.*

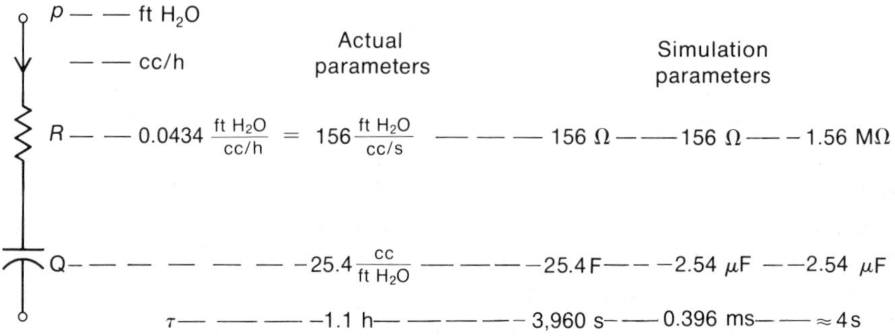

P — — ft H_2O

— — cc/h

Actual parameters

Simulation parameters

R — — 0.0434 $\dfrac{\text{ft } H_2O}{\text{cc/h}}$ = 156 $\dfrac{\text{ft } H_2O}{\text{cc/s}}$ — — — — 156 Ω — —156 Ω — — 1.56 $M\Omega$

Q — — — — — — — −25.4 $\dfrac{\text{cc}}{\text{ft } H_2O}$ — — — — −25.4 F — − −2.54 μF — —2.54 μF

τ — — — — −1.1 h— — — — — − 3,960 s— — 0.396 ms— — $\approx$ 4 s

FIG. 5.17 *Simulation of decompression during ascent from submergence.*

incorporated as part of standard diving equipment.* These devices would be used to calculate the oxygen requirements for the ascent to sea level and would warn divers when their oxygen supply is getting low. The problem of decompression sickness is not merely an academic one, particularly since underwater sports have become popular.

Circuit simulation can also be used to study the effect of different ascent programs, but this does not provide more information. The mathematical representation and the analog simulation are essentially complementary techniques, and, for particular problems, both should theoretically lead to a solution. In some problems, however, the mathematical approach involves such complex operations that solutions are not practically possible. With a simulation, however, a graphical solution to a variety of problems is almost always possible. The simulation technique has one obvious disadvantage: The analog must be built.

Figure 5.17 shows how a practical analog simulation of the nitrogen-liberation system may be developed. The storage property of the system is represented by a capacitor; the electrical resistor represents the resistive property of the system. No driving source is shown; either a voltage source can be used to simulate various pressure signals and the resultant current (nitrogen) flow measured, or a current source (a device that can deliver a regulated flow, something like a constant-delivery pump) can be used to simulate nitrogen flow and the required voltage drop (pressure signals) be measured.

To choose values for the electrical components we start by using a one-to-one correspondence between physical and electrical parameters. The value of the capacitor would then be 25.4 F. To obtain a value for the resistor we note that the electrical time-constant in seconds is given by C/G or RC (the product of ohms and farads yields time in seconds).

* See Refs. 4 and 5.

The system time-constant is 1.1 h, or 3,960 s; therefore the ohmic resistance is 3,960/25.4 or 156 Ω. If these values for the electrical components could be used, 1 V would represent a pressure change of 1 ft H_2O, and 1 A would represent a nitrogen flow of 1 cm³/s. The value for the electrical resistance is reasonable (resistors as low as 0.1 Ω to as high as 100 MΩ are easily available) but the value for the electrical capacitor is entirely impractical. Standard capacitors are in the range of 10^{-12} F [a picofarad (pF)] to about 100×10^{-6} F (100 μF). To be practical, the capacitor should be about 2.5 μF.

If we change the value of the capacitor but do not change the resistor, the electrical time-constant will no longer be equal to the physical time-constant. In fact, with a 2.5-μF capacitor and a 156-Ω resistor, the electrical time-constant would be one 10-millionth of the physical time-constant ($156 \times 2.5 \times 10^{-6} = 0.396 = 0.4$ ms). This implies that events occurring in 1 s in the physical system will take 0.1 μs in the electrical system. The effect of compressed time is often advantageous as it can serve as a "time saver." Several different studies can be made on the analog simulation in the time required to perform a single test on the physical system. However, in this case the compression of time is somewhat excessive. The time-constant can be extended by increasing the resistor. If we retain a 2.5-μF capacitor and replace the 156-Ω resistor with one of 1,560,000 (1.56 MΩ), the time-constant would then be 4 s. This is easy enough to work with. Physical events taking place in 1 s would now be observed in 1 ms in the simulation.*

A diagram of one practical analog simulation is shown in Fig. 5.18a.

* The procedure outlined here might be compared with that presented in Sec. 4.9.

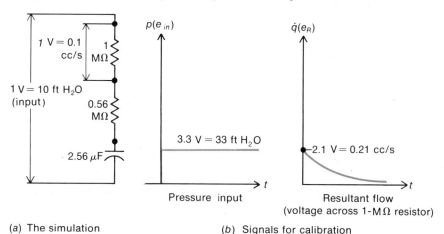

(a) The simulation   (b) Signals for calibration

FIG. 5.18 *Analog simulation of nitrogen-liberation system using practical components.*

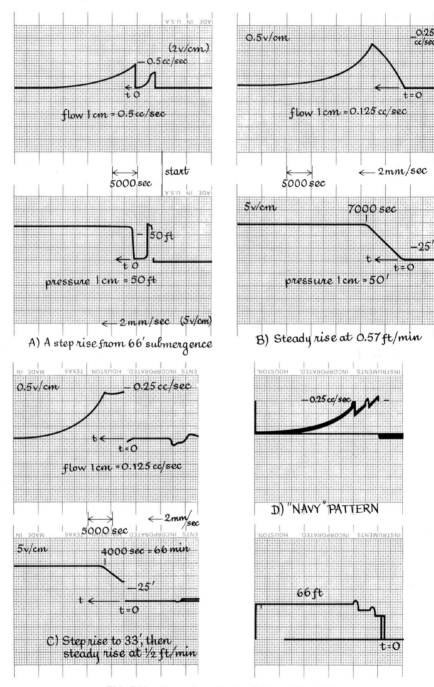

FIG. 5.19 Curves from the simulation of Fig. 5.18.

The scale factors were obtained by using a 3.3-V step to represent 33 ft H_2O (Fig. 5.18b). One volt across the input thus represents 10 ft H_2O. The nitrogen flow is determined by measuring the voltage across a 1-MΩ resistor (chosen for convenience). A 1-V drop across the resistor thus represents a flow of 0.1 cm^3/s.

The circuit can now be used to determine the effect of various ascent profiles. Figure 5.19 shows the various flow patterns that could be expected for different emergence patterns. A step rise to the surface from a 66-ft submergence is not acceptable, as the maximum flow exceeds the stated design limit of 764 cm^3/h or 0.21 cm^3/s (Fig. 5.19a). A steady rise from zero to 66 ft is acceptable if the rise rate does not exceed 0.57 ft/s. This ascent pattern (Fig. 5.19b) would require 116 min to surface. A rapid rise to 33 ft and a steady rise rate of 0.5 ft/min thereafter provide a safe rise pattern. Emergence would then take about 66 min (Fig. 5.19c). In the last set of curves (Fig. 5.19d), the ascent program suggested by the Navy was used. As pointed out earlier, this program gives a somewhat higher "safe-flow average" than the one assumed.

For a simple problem such as this, it probably would not be worthwhile to build an analog to study the system. The mathematical procedure is not complex and leads to quick solutions (although a solution using the navy ascent program would be somewhat tedious). However, in more complex situations, when mathematical solutions are virtually impossible, the time required to develop the analog is well spent. The analog provides great versatility, and solutions to a wide variety of problem situations are easily obtained.

5.6 USE OF STEP–RESPONSE DATA IN IDENTIFICATION

In the preceding section it was shown that, if a first-order representation for the step-response of a system is acceptable, general conclusions can be drawn about the system performance and properties. When one is interested only in the performance of a system, the details of system construction and function are not important. The system can be considered a black box. The contents are of secondary importance; of interest only is that the input and output are properly related. These ideas can be presented diagrammatically as shown in Fig. 5.20. In Fig. 5.20a the nitrogen-decompression system is represented by a box labeled simply "first-order system" (provided the available data indicate such a representation is permissible). The implications of Fig. 5.20a are that the overt characteristics of the nitrogen-decompression system can be studied by any system that gives an acceptable first-order response. The box can therefore be

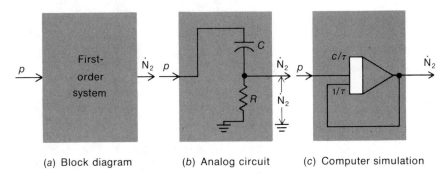

(a) Block diagram   (b) Analog circuit   (c) Computer simulation

FIG. 5.20   *Alternative representation of first-order system.*

composed of a simple RC simulation as shown by Fig. 5.20*b* or by an analog-computer simulation as shown in Fig. 5.20*c*.* Either simulation would provide a system substitute which could be used to study the variation of nitrogen liberation during different decompression stimuli.

When one is interested in the internal construction of a system, concern is not primarily with performance characteristics of the first-order system; rather one's interest is in the system properties implied by the first-order response. If attention is focused on the properties indicated and their correlation with the physiological system, often much insight into the operation of the system can be gained. We can once again use the nitrogen-liberation system to illustrate this concept.

From our elementary examination of the step-response data given in Fig. 5.11 it was concluded that the system had storage and resistive properties. This poses some leading questions. One might ask, for example, what mechanism(s) or process(es) is included in the resistive property and what constitutes the storage property. The latter is straightforward; it encompasses all mechanisms by which nitrogen is contained or held within the physiological system. A number of possibilities exist: The nitrogen could be stored as a gas; it could enter into chemical combination; it could dissolve in the various body tissues and fluids. A quick examination of the physiological system and some library research establish that the most reasonable explanation of nitrogen storage is by dissolution into various body tissues and fluids. (The lungs cannot hold the volumes of gas involved; nitrogen does not combine chemically in physiological reactions.) As a rough breakdown, approximately 60 percent of the body is water, about 20 percent is fatty tissues, and the rest is protein and solid material. The solubility of nitrogen in the latter

* See Prob. 5.4 and Appendix A, steps 20 to 35.

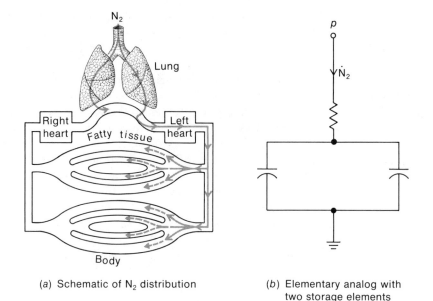

(a) Schematic of N_2 distribution

(b) Elementary analog with two storage elements

FIG. 5.21 *Evolution of models of the nitrogen-liberation system to include information about system components.*

constituent is rather small. Nitrogen is therefore primarily stored in the body fluids and the fatty tissue of the body.

A schematic representation of the system is thus provided by the model shown in Fig. 5.21a. Figure 5.21b shows a modification of the simple RC circuit previously used (see Fig. 5.17, for example) to represent the nitrogen-liberation system. The single storage element is now represented by two storage elements in parallel. We can, in fact, obtain numerical values for these elements. It has been experimentally determined that nitrogen is about five times more soluble in fat than it is in water. For equal weights of fat and water the amount of nitrogen in the fatty tissue would be five times that in the water. Since there is three times as much water in the body as there is fat (60%/20%), the fatty tissue holds $\frac{5}{3}$ as much nitrogen as does the body water. Therefore $\frac{3}{8}$ of the total nitrogen is stored in the water of the body and $\frac{5}{8}$ in the fatty tissue.*

---

* Algebraically: $0.20 \times$ body weight $\times$ solubility in fat = nitrogen in fat
$0.60 \times$ body weight $\times$ solubility in water = nitrogen in water
Therefore

$$\frac{1}{3} \frac{S_f}{S_w} = \frac{N_{2f}}{N_{2w}} \quad \text{or} \quad \frac{5}{3} N_{2w} = N_{2f}$$

$$N_{2w} + N_{2f} = \text{total } N_2 = N_{2w} + \frac{5}{3} N_{2w} \quad \text{or} \quad N_{2w} = \frac{3}{8} \text{ total } N_2$$

Since the total storage for the system was shown to be 25.4 cm³/ft H₂O (see discussion in Sec. 5.5), the storage elements in Fig. 5.21*b* can be taken to be 9.5 cm³/ft H₂O (to represent body water) and 15.9 cm³/ft H₂O (to represent fatty tissue).

The resistive property implied by the earlier analysis must now also be related to the physiological system. This property is taken to mean any mechanism that opposes the delivery of nitrogen to the storage elements. A glance at Fig. 5.21*a* indicates that the resistive property would thus include the flow resistance in the lung, the resistance to flow across the lung membranes into the bloodstream (diffusion resistance), the rate of blood delivered to the various tissues (well-perfused areas receive nitrogen quickly), diffusion from the circulatory system to the storage element, and diffusion through the storage elements. This is not to imply that these are the only resistive effects present in the system; these do, however, constitute most of the obvious contributing effects. If, after a careful analysis, it is found that these do not adequately account for the

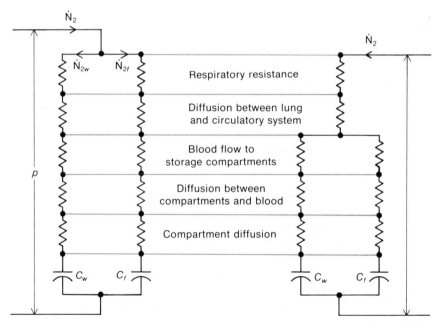

(a) Lumped resistances associated with storage compartments

(b) Representation when common resistive effects are combined

FIG. 5.22 *Resistive-storage representation of two-compartment system.*

total measured effect we would then have to investigate further to determine what mechanisms and effects were not included.

The resistive mechanisms mentioned above are almost entirely distributed effects associated with the distributed storage mechanisms. To
obtain some diagrammatic representation of the system we might consider them as lumped effects associated with the lumped storage properties. Such a representation is shown in Fig. 5.22. In Fig. 5.22a the resistive elements associated with each of the storage elements are shown to
form two parallel branches. In physiology and biophysical work such
parallel branches are often designated as "compartments"; an analysis
involving these branches constitutes separation of the system in component compartments. The use of lumped properties to form conceptual
compartments is thus well established in physiological studies.

Since some of the resistive effects are common to both the water-
storage and fatty-storage compartments, it is possible to combine these
effects into a single element. This is shown diagrammatically in Fig.
5.22b. The respiratory resistance and lung diffusion affect both compartments. These can be combined to form a single branch; the separate
compartments now form two parallel branches and include effects associated only with the individual compartments.

By examining the implications of a first-order approximation we
have essentially developed a more accurate representation of the system
and simultaneously shown a first-order representation to be inadequate.*
The parallel branches of Fig. 5.22 indicate that at least two exponential
functions are needed to describe the step-response of the system. In fact,
we can use the circuit of Fig. 5.22a to provide some estimate of what
these exponential functions should be. If we lump all the resistors in each
branch into single resistors, as shown in Fig. 5.23, we see that, for an
applied step-pressure (P), the nitrogen liberated from each compartment
can be expressed as

$$(5.52) \qquad N_{2w} = PC_w(1 - \epsilon^{-t/\tau_w}) \qquad \tau_w = R_w C_w \qquad t > 0$$

$$(5.53) \qquad N_{2f} = PC_f(1 - \epsilon^{-t/\tau_f}) \qquad \tau_f = R_f C_f \qquad t > 0$$

These equations follow directly from the discussion at the beginning of
Sec. 5.5. The situations can be compared since here there are two RC circuits driven with the same step-pressure.

* This observation is not surprising. Each approximation will lead to deeper studies which
will lead to a more refined approximation, etc. One cannot ever expect a perfect representation, only guidance toward better understanding. If the step-response data of Fig.
5.1b were a perfect first-order response, it would not lead to the conclusion that this is a
single-compartment system. It would say only that all compartments of the system have
the same time-constant. If, for example, the resistive-storage branches shown in Fig. 5.22a
were identical, the circuit of 5.22a would then reduce to that shown in Fig. 5.17.

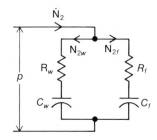

FIG. 5.23 *Simplified representation of Fig. 5.22a.*

The total nitrogen liberated is thus given by

(5.54) $$N_2 = P(C_w + C_f) - P(C_w\epsilon^{-t/\tau_w} + C_f\epsilon^{-t/\tau_f})$$

Equation (5.54) should now describe the liberation curve given by Fig. 5.1*b*. We can, in fact, use this curve in conjunction with Eq. (5.54) to determine the value of the various constants. If both storage elements and both time-constants are to be evaluated, there are four unknowns to determine. We must therefore use four corresponding values of t and N_2 in Eq. (5.54) to form four simultaneous equations. The solutions of these equations provide the desired values for storage elements and time-constants. The procedure is easier said than done, but we outline this suggested approach: For the step-pressure of $P = 33$ ft H_2O, at large t $(t \to \infty)$, N_2 is 840 cm³ (the numerical values are taken from the curve in Fig. 5.1*b*). Equation (5.54) then becomes

(5.55) $$840 = 33(C_w + C_f) - 33(\text{zero} + \text{zero})$$

or

(5.56) $$C_w + C_f = \frac{840}{33} = 25.4 \text{ cm}^3/\text{ft } H_2O \qquad \text{(a not unexpected result)}$$

At $t = 1$ h, the liberated nitrogen is 580 cm³. Therefore Eq. (5.54) is

(5.57) $$580 = 840 - 33(C_w\epsilon^{-1/\tau_w} + C_f\epsilon^{-1/\tau_f})$$

or

(5.58) $$C_w\epsilon^{-1/\tau_w} + C_f\epsilon^{-1/\tau_f} = \frac{260}{33} = 7.87$$

By repeating the procedure at t equal to 2 and 3 h we obtain

(5.59) $$C_w\epsilon^{-2/\tau_w} + C_f\epsilon^{-2/\tau_f} = 3.18$$

(5.60) $$C_w\epsilon^{-3/\tau_w} + C_f\epsilon^{-3/\tau_f} = 1.42$$

Equations (5.56) and (5.58) to (5.60) are four required simultaneous equations; a simultaneous solution is obviously rather tedious. If we were confident in the accuracy of the data and felt that an algebraic solution were necessary, we could use a digital computer to good advantage here. In this case, however, the expense and effort are not warranted. We can obtain some estimate of the numerical values through a trial-and-error procedure or by a graphical technique to be described in the following paragraphs. To complete the above problem, the author used a trial-and-error procedure to obtain the time-constants of the two compartments. It was found that the fatty compartment had a time-constant of 1.22 h and the water compartment had one of 0.24 h. The corresponding values for storage elements were 17.5 cm^3/ft H$_2$O for the C_f term and 8 cm^3/ft H$_2$O for the C_w term. If these values are used in Eq. (5.54), we then have

$$(5.61) \qquad N_2 = 840 - (264\epsilon^{-t/0.24} + 576\epsilon^{-t/1.22})$$

A plot of Eq. (5.61) is shown in Fig. 5.24. For comparison, the original nitrogen-liberation curve (Fig. 5.1b) is also included. The agreement between the curves is far superior to that obtained by the first-order approximation. The excellent agreement between the two curves might lead one to assume that if this system has other storage compartments they are very small or they have the same time-constants as the fatty and water compartments. As always, further experimentation and analysis would be needed to justify such claims.

The question now arises whether it would have been possible to use the step-response data directly to postulate that the nitrogen-liberation system was more appropriately represented as a second-order system than as a first-order system. Although the answer is probably "yes," one should be careful before hastily answering in the affirmative. The first-order approximation was just that: an approximation. It led to the

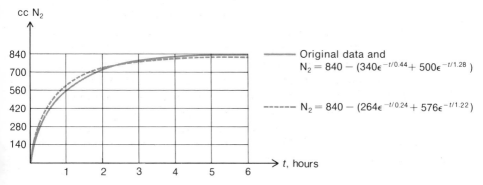

FIG. 5.24 *Comparison of derived and actual nitrogen-liberation curves.*

recognition and description of resistive and storage elements. A physiological interpretation of the meaning of these properties then led to the possibility of a two-compartment representation. This may often be the case, but one should not eliminate the possibility that the physiological interpretation may lead to the conclusion that a single-compartment representation is more realistic. Then the conclusion must be that the system is first-order but nonlinear; that is, its construction is of such a nature that within the range of the stimulating signal linear approximations are invalid. Thus to use arbitrarily a second-order representation because a first-order representation does not appear to be accurate enough may make mathematical sense (a curve fit may be possible) but it may not make physiological sense. Or, to rephrase the last comment, to state that a system is second-order simply because the step-response data indicate that a second-order mathematical representation is possible may not be physiologically justified.

If a two-compartment representation is justified, the step-response data can then be used to evaluate the system constants. One procedure was outlined in the preceding paragraphs. However, the method described was somewhat involved and tedious. A somewhat simpler (though somewhat less accurate) technique involves the use of the semilog representation of the step-response data (see Fig. 5.12) and the two-compartment representation given by Eq. (5.54). It will be recalled, for first-order systems, it was shown that a semilogarithmic plot of the approach to equilibrium resulted in a straight line (Sec. 5.4). Although a second-order system does not approach equilibrium exponentially, its step-response can often be considered exponential at the final stages of the response. Under some circumstances, therefore, the semilogarithmic plot of a second-order system will show an approximate straight line at the end portions of the response. (This is not to be interpreted as a general statement. We shall see later that some second-order systems may have transient responses that are oscillatory. These will not show the straight-line characteristic when plotted semilogarithmically. Such second-order systems will be considered in Chap. 9.) By using the tail end of a semilogarithmic plot of a step-response, one of the exponentials of the second-order (or higher-order) system can be evaluated. This exponential is then removed from the response and the procedure repeated to evaluate the second exponential, etc. This method is often described as "peeling off exponentials."

We can illustrate the procedure just outlined by using Eq. (5.54) and the plot of Fig. 5.12b. When Eq. (5.54) is rewritten to describe the approach to equilibrium we have

$$(5.62) \qquad\qquad 840 - N_2 = A_1\epsilon^{-t/\tau_1} + A_2\epsilon^{-t/\tau_2}$$

where the final value of $N_2 = P(C_w + C_f)$ has been replaced by its numerical equivalent 840 cm³ and the coefficients PC_w and PC_f have been replaced by arbitrary constants A_1 and A_2. The identifying subscripts w and f have been dropped and general subscripts 1 and 2 are used in their stead. The reasons for these changes will be understood in the course of the discussion. When Eq. (5.62) is written in logarithmic form, we have

$$(5.63) \qquad \log (840 - N_2) = \log (A_1 \epsilon^{-t/\tau_1} + A_2 \epsilon^{-t/\tau_2})$$

We now seek values of A_1, A_2, τ_1, and τ_2 so that, if the log $(840 - N_2)$ is plotted vs. time, the result will be the curve shown in Fig. 5.12b.

We cannot yet take advantage of the logarithmic properties of the exponential, as Eq. (5.63) is expressed in terms of the logarithm of the sum of exponentials. However, we note from the original plot (Fig. 5.12a, for example) that after 6 h the response has essentially reached its final value. This implies that after 6 h both exponential functions are practically zero. We know that after approximately four time-constants an exponential decays to about 2 percent of its original value. We can therefore assume that neither τ_1 nor τ_2 exceeds 1.5 h ($6\!\!/\!4$). We now make the arbitrary assumption that τ_1 and τ_2 differ by at least a factor of 2. This is to say that, if τ_1 is less than 1.5 h, τ_2 is less than 0.75 h,* or vice versa. If, however, one of the time-constants is 0.75 h, then after 3 h (four time-constants) the exponential term associated with this time-constant will have decayed to negligible values. This implies that after 3 h Eq. (5.63) can be written in terms of a single exponential term. Specifically, we would have

$$(5.64) \qquad \log (840 - N_2) \approx \log (A_1 \epsilon^{-t/\tau_1} + 0) \qquad \text{for } t > 3 \text{ h}$$

or

$$(5.65) \qquad \underbrace{\log (840 - N_2)}_{y} \approx \underbrace{\log A_1}_{b} - \underbrace{\left(\frac{1}{\tau_1} \log \epsilon\right)}_{m} \underbrace{t}_{x}$$

Equation (5.65) indicates that if the system can be represented as a two-compartment system with different time-constants ($\tau_1 > 2\tau_2$) then after

---

* This assumption is arbitrary from a theoretical point of view. It is, however, somewhat less arbitrary from a practical point of view. If the factor is less than 2 we must depend upon the extreme end of the response curve for numerical values. In this range the response is so small that instrument errors become significant. If we find that by using this assumption a "good fit" is obtained and the time-constants differ by a factor greater than 2, we might then accept this as a valid assumption. If this is not the case, we would have to use the alternative procedure to obtain numerical values or we might have to assume nonlinearity or a higher-order system.

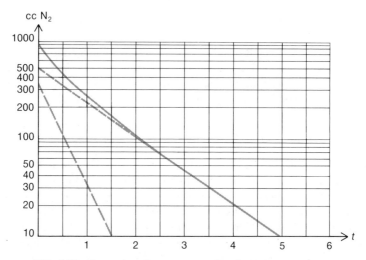

FIG. 5.25   *Separation of step-response data into two compartments.*

3 h the plot of log $(840 - N_2)$ versus t will be an approximate straight line with an intercept log A_1 and a slope given by log $\epsilon/\tau_1 = 0.435/\tau_1$. By using the semilog plot we can therefore evaluate A_1 and τ_1.

As a specific illustration, the semilog plot of Fig. 5.12b is repeated in Fig. 5.25. As can be seen, a straight line through the tail of the response appears to be a good approximation. The slope of the line defines the time-constant of the slower compartment (the exponential with the higher time-constant). From the plot, the slope of the line is seen to be

$$(5.66) \qquad \frac{\log 500 - \log 10}{5} = \frac{1.7}{5} = 0.34$$

The time-constant of the slower compartment is therefore

$$(5.67) \qquad \tau_1 = \frac{0.435}{0.34} = 1.28 \text{ h}$$

The straight-line y intercept defines the compartmental coefficient A_1. This value can be read directly from the graph, and its value is seen to be 500.

One of the exponential terms in Eq. (5.62) has now been evaluated. To determine the other term, we rearrange Eq. (5.62) so that the un-known exponential term stands alone. The semilog plot of the result should yield a straight line with log A_2 as intercept and a slope related to τ_2. To demonstrate,

$$(5.68) \qquad [(840 - N_2) - 500\epsilon^{-t/1.28}] = A_2\epsilon^{-t/\tau_2}$$

$$(5.69) \qquad \log[(840 - N_2) - 500\epsilon^{-t/1.28}] = \log A_2 - \frac{0.435}{\tau_2} t$$

The numerical difference between the original curve and the exponential curve of the slow compartment can be read directly from the semilog plot in Fig. 5.25. At $t = 0$ this difference is seen to be $840 - 500$ or 340; at $t = \frac{1}{2}$ h the difference is $450 - 335$ or 115, etc. A plot of the logarithm of these values vs. time is shown by the dashed line in Fig. 5.25. The slope of this line is given by

$$(5.70) \qquad \frac{\log 340 - \log 10}{1.56} = \frac{1.532}{1.56} = 0.984$$

The time-constant of this compartment is thus

$$(5.71) \qquad \tau_2 = \frac{0.435}{0.984} = 0.44$$

The compartmental coefficient is 340.

Both compartments have now been described quantitatively. In effect, we have derived an equation which can be used to describe the total liberation of nitrogen. If the numerical values obtained are used in Eq. (5.54), we have

$$(5.72) \qquad N_2 = 840 - (340\epsilon^{-t/0.44} + 500\epsilon^{-t/1.28})$$

This result may be compared with previously derived results given by Eq. (5.61). As might be expected, the numbers do not exactly agree. Yet it is interesting to note that if Eq. (5.72) is plotted (see Fig. 5.24) it also very closely follows the original data.*

We see, then, that there may be a number of equations that provide a good fit to the original data. Therefore one cannot draw any highly accurate quantitative conclusions from a single set of data. From this analysis we might conclude that the system has two predominant storage mechanisms. From other investigations we know the physiological system contains significant percentages of fatty tissue and water, that nitrogen is soluble in both, and that nitrogen is more soluble in fat than in water. We might therefore conclude that the two predominant storage compart-

---

* It duplicates the original data so well that differences between the theoretical curve and actual curve are not perceptible. It is somewhat surprising that the graphical results give a better fit than the analytic results do. This is probably due to the slide-rule approximations used in the previous analysis. In general, one would expect the analytic results to be closer than the graphical results, particularly at the points where the curve and the equation were matched.

ments are the water content and fatty tissue of the system. From other relevant physiological data (proportions of fat and water in the body and high solubility of nitrogen in fat) we would accept that the exponential with the larger coefficient represents the fatty compartment and the other represents the water compartment. We could then conclude that the fatty compartment has the longer time-constant.

If we use the numerical values indicated by Eqs. (5.61) and (5.72) we may then write (for the data on this particular individual)

$$(5.73) \qquad\qquad 264 \leq PC_w \leq 340$$

or

$$(5.74) \qquad\qquad 8 \leq C_w \leq 10 \qquad \text{for } P = 33 \text{ ft } H_2O$$

where C_w represents the nitrogen storage in the water compartment in cubic centimeters per foot of water pressure. Similarly,

$$(5.75) \qquad\qquad 500 \leq PC_f \leq 576$$

$$(5.76) \qquad\qquad 15.2 \leq C_f \leq 17.4$$

The time-constant for the fatty compartment can be written

$$(5.77) \qquad\qquad 1.22 \leq \tau_f \leq 1.28$$

Since the time-constant reflects both the resistance and storage in the system ($\tau_f = R_f C_f$) we can then use Eqs. (5.76) and (5.77) to define resistive effects associated with the fatty compartments:

$$(5.78) \qquad \frac{1.22}{17.4} < R_f < \frac{1.28}{15.2} \qquad 0.07 < R_f < 0.084$$

A range of values for the resistance associated with the water compartment can similarly be evaluated:

$$(5.79) \qquad\qquad 0.24 < \tau_w < 0.44$$

Therefore

$$(5.80) \qquad \frac{0.24}{10} < R_w < \frac{0.44}{8} \qquad 0.024 < R_w < 0.055$$

From Eqs. (5.78) and (5.80) we can conclude that the resistive effect associated with the fatty compartment is certainly greater than that of the water compartment. We may now ask why one of the resistances is greater than the other. In the model representation proposed in Fig. 5.22, two resistive mechanisms were associated with each compartment: blood perfusion and diffusion. The questions now arise: Are these two mechanisms sufficient to account for the differences in resistance?

Which of these mechanisms is more significant? Evidently these questions can be answered only by further research (both library and laboratory) and further analysis; this will lead to more refined models, deeper understanding, more questions, and so on.

5.7 EXAMINATION OF FIRST–ORDER STEP–RESPONSE OF A GENERALIZED SYSTEM

In the foregoing sections we examined the relationship between step-response data and system properties for two specific systems. Although the discussions involved specific systems, most of the concepts developed were fairly general and would be applicable to any system. We shall review once again the ideas presented but this time with the use of generalized through- and across-variables. The results obtained in this fashion will not be associated with any one system but will be applicable to all systems having an approximate first-order step-response.

We start by assuming a system is driven by an across-variable x_a (such as pressure) with a step-input and that the through-variable y_t (such as flow) shows an exponential response* (see Fig. 5.26). The equation for the through-variable response can then be written:

$$(5.81) \qquad y_t = Y_t(1 - \epsilon^{-t/\tau})$$

where τ is the time-constant of the system. Y_t (the final or steady-state response of the through-variable) and X_a (the constant across-variable input) define a system property. In terms of specific variables, if Y_t is volume and X_a is pressure, this ratio obviously defines compliance. If Y_t is flow and X_a is pressure, the ratio is conductance. In general, if this

---

* To aid visualization, a pair of specific variables, such as p and q (or $\dot{q}$), can be used to replace the general through- and across-variables.

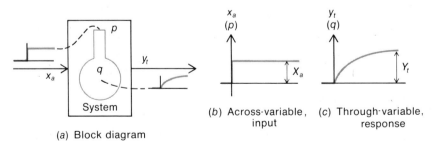

(a) Block diagram   (b) Across-variable, input   (c) Through-variable, response

FIG. 5.26 *First-order step-response of generalized system. [Resistive-compliant system in (a) is included to illustrate how generalized variables relate to an actual system.]*

property is designated by s_p, then

(5.82)
$$\frac{Y_t}{X_a} = s_p$$

Equation (5.81) then becomes

(5.83)
$$y_t = s_p X_a (1 - \epsilon^{-t/\tau})$$

We can now construct a first-order differential equation. If we differentiate Eq. (5.83) we get

(5.84)
$$\dot{y}_t = \frac{s_p X_a}{\tau} \epsilon^{-t/\tau}$$

or

(5.85)
$$\tau \dot{y}_t = s_p X_a \epsilon^{-t/\tau}$$

The sum of Eqs. (5.83) and (5.85) yields the standard first-order differential equation:

(5.86)
$$\tau \dot{y}_t + y_t = s_p X_a$$

This cannot be considered proof that a system with an exponential step-response must be described by a first-order differential equation. However, we can accept such a response would imply that the system can be described by a first-order differential equation.

If Eq. (5.86) is rewritten, we have

(5.87)
$$\frac{\tau}{s_p} \dot{y}_t + \frac{1}{s_p} y_t = X_a$$

In this form, we note the general equation has a solution for any input of the across-variable. If we remove the restriction that X_a is a step-function and allow $x_a(t)$ to be any type of input, then so long as the system properties remain linear, we can determine the through-response. The implications here are that once the system is described on the basis of a step-response (or indeed for any response) it should then be possible to predict the response to any other stimulus.

It should also be noted that the term τ/s_p in Eq. (5.87) must represent a system property that is different from that of s_p. If s_p is a storage property, τ/s_p must represent a resistive property; if s_p is a resistive property, τ/s_p must represent a storage property. We can justify this assertion by examining the units of these terms. τ is known to have the units of time only (T^1). The time exponent of τ/s_p and that of s_p must therefore differ by unity. If the time exponent of s_p is even, it represents a storage property; τ/s_p must then represent a resistive property, as its time exponent must be odd. Conversely, if s_p is a resistive property, it must have an odd time exponent and τ/s_p must then be a storage property.

The identification of s_p as a storage property and τ/s_p as a resistive property can be directly related to the identification of properties in the systems previously discussed. In later sections we shall examine systems where the property s_p will be identified as a resistive property. The term τ/s_p will then define the storage property of the system.

It should be obvious that, if we had started this discussion by assuming the input was a through-variable instead of an across-variable, all x's and y's in the previous discussion would have been interchanged. The general statements about the interpretation of the system constants remain the same to the extent that s_p will define one property; τ and s_p will define the other property. In this case, however, if we retain the definition of s_p as the ratio Y_t/X_a, the product of τ and s_p (τs_p) must be used to define the second system property. This can be verified by performing the same sort of analysis shown by Eqs. (5.81) to (5.87) and will be shown by the example of the next section.

5.8 STEP-RESPONSE DATA IN DYE-DILUTION STUDIES

In order to explore further the implications and use of step-response data, we examine some of the aspects of the dye-dilution technique which is often used in physiological testing. In its simplest form, dye-dilution is the use of a dye to determine fluid volumes (see Prob. 5.5). A known weight of dye in a known volume of fluid provides a color-density standard reference. By introducing a known weight of dye into an unknown volume, one can determine the unknown volume by comparing the resultant color density against the standard. In practice the color density is now generally measured photoelectrically so that the "color standard" is generally a voltage rather than a color wavelength.

The determination of blood volume serves as a good illustration of the use of dye-dilution* for volume measurements in a physiological system. A known weight of dye is introduced into the bloodstream (see Fig. 5.27a). After a few minutes, to allow for complete mixing, a blood sample is withdrawn. The weight of dye in the diluted sample provides a measure of the total blood volume. (Weight of dye introduced ÷ weight of dye per unit volume of sample = volume of blood.)

---

* Although the discussion here is restricted to the use of dyes as indicators, any identifiable material or effect can be used as an indicator. The dilution technique might therefore be more broadly described as the indicator-dilution technique. This broader classification includes the use of dye, radioactive materials, thermal effects, gases, etc., as indicators for dilution studies. A rather complete discussion of indicator-dilution techniques can be found in Ref. 8.

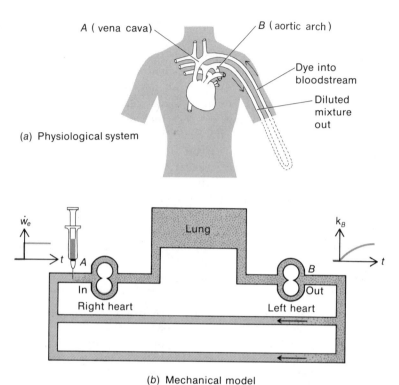

(a) Physiological system

(b) Mechanical model

FIG. 5.27 *Dye-dilution study of circulation.*

The need for a dye that is retained by the bloodstream or has an affinity for blood should be evident. If some of the dye can diffuse out (or be otherwise lost) from the bloodstream, the measurements will be in error. The ideal dyes combine directly with the blood constituents and so remain in the bloodstream for a significant period of time. Probably the most widely used dye of this type is evans blue. This dye combines with the plasma albumin and is eliminated very slowly. Another material that is very effective for blood-volume studies is radioactive chromium (Cr^{51}). This gamma emitter combines directly with the red cells. In practice, a small amount of blood is withdrawn and combined with Cr^{51}. The mixture is checked with a counter to establish a reference and then injected into the bloodstream. After the indicator has had time to be completely dispersed, a sample is withdrawn and checked with a counter. A knowledge of the counts in the reference sample and of the counts in the diluted sample provides a measure of the blood volume.

In physiological testing, the dye-dilution technique is used both for "static"-volume measurements (such as blood volume, plasma volume, etc.) and for fluid-transport determinations (such as cardiac output, renal clearance, etc.). The former application should be easily understood from the brief foregoing discussion. As an illustration of the latter application we examine how dye-dilution techniques may be used for in vivo estimates of cardiac output. To explain this application, reference is made to the gross model of the circulatory system in Fig. 5.27b. Here the heart is shown as a simple mechanical pump; entrance to the right heart from the vena cava is located at point A; exit from the left heart before the aortic arch is located at B. (Corresponding locations in the physiological system are shown in Fig. 5.27a.) To obtain a determination of cardiac output, a steady rate of dye is injected at A ($\dot{w}_e$); from the dye-concentration curve measured at B (k_B versus t in Fig. 5.28b) the cardiac output may be evaluated.

To see how this measurement is made, we note if a steady rate of dye is injected at A the dye content in the heart chambers will increase and the concentration curve at B will rise slowly.* When the dye concentration in the heart chambers becomes equal to the dye concentration of the entering blood, the concentration curve at B will reach a constant level. From the plateau concentration and the rate of dye input, an estimate of blood flow may be obtained. In equation form this may be expressed as

$$(5.88) \quad \frac{\text{Weight of dye injected/unit time}}{\text{Volume of blood entering/unit time}} = \frac{\dot{w}_e}{\dot{q}_B}$$
$$= \text{concentration at } A(k_A) = \text{concentration at } B(k_B)$$

* It should be noted that the concentration curve at B does not rise immediately after the dye is injected. There is a perceptible time lag between the start of injection and the appearance of the dye, because of the time required for the blood to transport the dye from point A to point B. Such time lags, or time delays, are not uncommon during the functioning of a system, and in general such time delays must be considered. In this case, however, the delay does not significantly affect the result and we can neglect the lag effect.

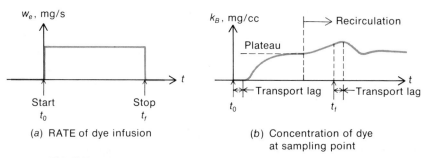

(a) RATE of dye infusion

(b) Concentration of dye at sampling point

FIG. 5.28 *Input-output curves for constant-injection dye-dilution studies.*

or

$$(5.89) \qquad \dot{q}_B = \frac{\dot{w}_e}{k_B} = \frac{\text{mg/s}}{\text{mg/cm}^3} = \frac{\text{cm}^3}{\text{s}}$$

Since $\dot{w}_e$ (the rate of dye injection) is known and k_B (the dye concentration at B) is measured, the average blood flow ($\dot{q}$) can be determined.

It should be evident from Fig. 5.27b that the concentration equilibrium at B is not continuously maintained. If constant-rate dye injection is continued until dye-ladened blood is returned (fed back) to the injection site, the dye concentration of the blood at the sampling site will then also begin to rise, as shown in Fig. 5.28b. The value of k_B used in Eq. (5.89) should therefore be the plateau value before recirculation of dye.

To perform the physiological test as outlined, a catheter is threaded through one of the large veins of the arm (the median cubital vein, for example) into the right heart. Dye is delivered to the heart through this catheter. Another catheter is located in the base of the aortic arch by threading it through one of the large arteries (as the brachial artery). A small steady flow of blood is withdrawn from the heart through this catheter. The dye concentration is continuously measured by passing the blood through an instrument (cuvette oximeter) to obtain a written record of concentration vs. time.

From Fig. 5.28 it can be seen that the initial portions of the input and output data represent an output response to an input step. Although the steady-state portion of the concentration curve provides the desired information about flow, we should still examine some of the ramifications of the step-response data to see what additional information can be obtained. Before we do so, however, we should define the system actually being *measured*. Information about the circulatory system is the prime purpose of the investigation, but the physical properties of the circulatory system (tubular resistance, elastance of conduction tubes, etc.) are not of concern here. If this were the purpose of the study, pressure and flow would be the measured variables. In this case the system under investigation is in reality a dye-distribution system. The delivery of dye and its distribution constitute the measured variables. The mechanisms by which the dye is distributed (or delayed or stored) constitute the system properties. These properties must then be interpreted in terms of the physiological system to provide information about the circulatory system. These comments will be more clearly understood after the following discussion.

With respect to the dye-distribution system, the input-output variables are evidently the rate of dye injection and the dye concentration, respectively. Since the concentration k_B is measured with respect to a reference, it can be designated as an across-variable; the dye injection can thus be considered a through-variable input. From Fig. 5.28 we see that a step-input results in a characteristic step-response. Although Fig. 5.28*b* is not the response of a first-order system (the zero slope at the beginning of the response curve is not characteristic of a first-order system), we can still develop some preliminary ideas if we make the assumption that a first-order representation is a good initial approximation. We can then write (compare with Sec. 5.7)

(5.90)
$$\begin{cases} \dot{w}_e = \dot{W}_{em}\, u(t) \\ y_t = Y_t\, u(t) \end{cases}$$

(5.91)
$$\begin{cases} k_B = K_{BM}\, (1 - \epsilon^{-t/\tau})\, u(t) \\ x_a = X_a\, (1 - \epsilon^{-t/\tau})\, u(t) \end{cases}$$

The ratio of the steady-state values defines a system property of the dye-distribution system.

(5.92)
$$\begin{cases} \dfrac{\dot{W}_{em}}{K_{BM}} = s_p = \underbrace{\dfrac{mg/s}{mg/cm^3}}_{\text{Units}} = \underbrace{(MT^{-1})\,(L^3M^{-1})}_{\text{Basic units}} = L^3T^{-1} \\[4mm] \dfrac{Y_t}{X_a} = s_p \end{cases}$$

From the earlier discussion [see Eq. (5.89)] we know that s_p is flow; this is also evident from the units of s_p. In accordance with our previous definitions, we must identify s_p as a system property associated with the resistive mechanism of the system. In particular, since Y_t/X_a is the slope of the through vs. across characteristics, s_p is identified as conductance. We might well question the significance of associating flow with conductance (or reciprocal flow with resistance). The blood flow is evidently related to the resistive mechanism of the dye-distribution system; the higher the flow, the more quickly is dye distributed, the more rapidly is an equilibrium concentration attained (the shorter the time-constant), and the lower is the concentration plateau.

The differential equation of the system is obtained from Eqs. (5.91) and (5.92). By differentiating the first of these and adding the result to Eq. (5.91) we have

(5.93)
$$\tau \dot{k}_B + k_B = K_{BM}\, u(t)$$

When the relationship shown by Eq. (5.92) is used in Eq. (5.93) we have

(5.94)
$$\tau s_p \dot{k}_B + s_p k = \dot{W}_{em} u(t)$$

where now τs_p defines the storage property of the dye-distribution system. The units of this storage property are those of volume, and it evidently

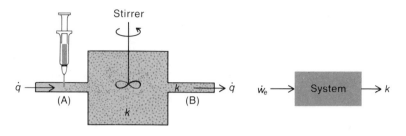

(a) Mechanical model   (b) Block diagram
FIG. 5.29   *Single-reservoir model of a dye-dilution system.*

relates to the dye-storage capability of the blood volume between the injection and sampling sites. In other words, the storage property of the dye-distribution system is related to the circulatory blood volume contained between the input and output points of the dye system.

The relationship between the properties of the dye-distribution system and the circulatory system could also have been established from an elementary "mixing chamber model."* Figure 5.29 shows a lumped model of the system where only the primary effects are included. The vascular bed of the total heart and lung system is represented as a single reservoir of volume V_t. The flow through the bed ($\dot{q}$) is continuously maintained by some means (not shown). A stirrer to ensure complete mixing within the chamber is shown. Dye (the input) is delivered to the system at point A; the concentration at B (the output) is to be determined. In order to set up the appropriate relationship governing the response of the system, we use a sampling procedure similar to the method used in Sec. 5.2. The concentration in the chamber (and hence the concentration at B) at any nth-sample instant is given by the weight of dye in the chamber divided by the chamber volume. Thus

$$(5.95) \qquad k_n = \frac{(W_t)_n}{V_t}$$

where k_n is the concentration of the nth instant, $(W_t)_n$ is the weight of dye in the chamber at the nth instant, and V_t is the volume of the chamber. The weight of the dye in the chamber at the nth instant is the weight

* In the previous analysis input-output data were used to establish system properties. In this analysis a system with known properties is assumed in order to see what response can be expected when a known input is applied. This is essentially the same sort of procedure used in Sec. 5.2. The purpose is to gain insight. It is important to note that this representation neglects feedback; i.e., it does not include any mechanism to provide for recirculation. The model can therefore be used only to interpret the physiological system prior to the advent of recirculation.

just prior to sampling, increased by the weight of dye entering and reduced by the weight leaving, or

$$(5.96) \qquad (W_t)_n = (W_t)_{n-1} + (W_e)_{n-1} - (W_L)_{n-1}$$

where the $n-1$ subscript indicates the sample period before the nth, W_e is the weight of dye entering, and W_L is the weight leaving. From Eqs. (5.95) and (5.96) we have

$$(5.97) \qquad k_n = \frac{(W_t)_{n-1}}{V_t} + \frac{(W_e)_{n-1} - (W_L)_{n-1}}{V_t}$$

Since $(W_t)_{n-1}/V_t$ is the concentration at the $(n-1)$st-sample instant we can write

$$(5.98) \qquad V_t(k_n - k_{n-1}) = (W_e)_{n-1} - (W_L)_{n-1}$$

In the interval between the nth and $(n-1)$st samples, the flow ($\dot{q}$) carries all the injected dye (at A) into the chamber and causes some of the dye to be removed. If Δt represents the time interval between two adjacent samples, the volume of the fluid removed is $\dot{q}\,\Delta t$. The weight of dye removed $[(W_L)_{n-1}]$ is thus $k_{n-1}\dot{q}\,\Delta t$. If the injection rate can be assumed to be constant during the Δt time interval, then the weight of delivered dye may be given by $(\dot{w}_e)_{n-1}\,\Delta t$. Equation (5.98) may now be written

$$(5.99) \qquad V_t(k_n - k_{n-1}) + k_{n-1}\dot{q}\,\Delta t = (\dot{w}_e)_{n-1}\,\Delta t$$

Equation (5.99) is a first-order difference equation which can be directly programmed on a digital computer [see footnote to Eq. (5.16)]. We can convert this to a first-order linear differential equation by dividing through by Δt and allowing Δt to become small. The nth and the $(n-1)$st intervals then merge to an instant of time, and we have

$$(5.100) \qquad V_t \frac{dk}{dt} + \dot{q}k = \dot{w}_e$$

If the injection rate is a step-function $[\dot{W}_{em}\,u\,(t)]$ and dk/dt is written as $\dot{k}$, we can compare Eqs. (5.100) and (5.94), term by term. As expected, the steady flow $\dot{q}$ corresponds to the resistive property s_p, and the chamber volume corresponds to the term τs_p. The time-constant is $V_t/\dot{q}$.

We can now see that flow and volume are related to the properties of the dye-distribution system. We can furthermore see that, if the physiological system can be represented by a single mixing chamber with constant flow, flow and volume can be obtained from the step-response data of the dye-distribution system. The flow would be steady or average flow through the vascular bed, and the volume would be the total vascular volume between the injection and sampling points.

Before we draw any very definite conclusions from the foregoing analyses we must realize that these analyses represent only a beginning; they indicate not what must be expected but rather what might be expected. If vascular volumes and flow are determined in the manner described and provide correct results (as compared with other determinations), we might then state that the single mixing chamber with constant flow closely represents the physiological system. Should there be definite discrepancies, we should not immediately consider the analyses or the model to be wrong. A more instructive and fruitful approach would be to consider the model inadequate and then to determine why it is inadequate and what modifications are needed to improve it. In practice, dye-dilution tests do not accurately indicate vascular volumes (although flow measurements seem fairly good). In an attempt to determine why theory and practice do not agree for ventricular-volume measurements of the heart, it was shown that there could be inadequate mixing in the ventricle.* This is a somewhat surprising observation as it would be expected that the rather forceful action of the ventricle would ensure complete mixing. This could indicate that the ventricle does not create the turbulence that would be expected during the pumping action and implies that the ventricle does not eject its total contents during systole (active pumping).

It is also of interest to note that, in an attempt to explain the differences between theory and practice, it was shown analytically that one possible source of error was the assumption of steady average flow.† Vascular volumes will be incorrectly evaluated by the procedure outlined because in the actual physiological system the flow is pulsatile. With this new knowledge, it would be possible to refine the postulated model so that more accurate results could be obtained. More importantly this could open up the possibility of a new line of investigation: If pulsatile flow affects the evaluation of vascular volumes by dye-dilution, then if vascular volumes were known (by other measurements) it should be possible to use the dye-dilution technique to determine not only the steady flow but also the pulsatile flow. This change in viewpoint can be schematically shown in block-diagram form (Fig. 5.30). In the original analysis a known dye-injection rate and the resultant dye-concentration curve were used to obtain information about the dye-distribution system (Fig. 5.30a). In the proposed investigation (Fig. 5.30b) the dye-distribution system is assumed to be known (or at least understood); the known injection rate is considered to be part of the system. The dye-concentra-

---

* H. J. C. Swan and W. Beck, Ventricular Nonmixing as a Source of Error in Estimation of Ventricular Volume by the Indicator-dilution Technique, *Circulation Res.*, **8**: (1960).

† See Ref. 10.

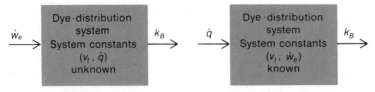

(a) Analysis to define system   (b) Analysis to define input

FIG. 5.30 *Alternative viewpoints for dye-distribution investigations.*

tion response (k_B) is now used to define the (spontaneous) flow input signal $\dot{q}$. From a theoretical point of view, the procedure outlined seems feasible; no claims can be made, however, until actual laboratory tests have been performed.

5.9 PULSE–RESPONSE OF A FIRST–ORDER SYSTEM (DYE–DILUTION)

In Sec. 5.3 we discussed the practical significance of the step-function and described its mathematical properties. We shall review briefly some of the ideas presented:

1. The step-function is assumed to be the description of a signal that suddenly changes from one operating level to another and remains constant at this new level for infinite time.

2. The signal is mathematically represented by $Au(t)$, where $u(t)$ is zero for $t \leq 0$ and unity for $t > 0$. The constant term A is the magnitude of the step.

Strictly speaking and on the basis of the mathematical description of a step, the input signal shown in Fig. 5.28 cannot be properly considered a step-function; it does not remain constant ad infinitum. Yet in the discussion of the dye-distribution system, the dye-injection rate was considered to be a step-function [Eq. (5.90)]. The justification of this procedure lies in the fact that the response was observed only for the period during which the input was constant $(t_0 < t < t_f)$. If we had been interested in the response for a longer period (say $t > t_f$), we could not have used the step-function to describe the input; instead, we would have had to describe the stimulus as a "pulse-function." The pulse-function is the description of a signal that suddenly rises, remains constant for a specified time, and then returns to its original value. Since pulse-type signals are now generally used for dye-dilution studies, we shall employ this system to illustrate the use and significance of the pulse-function.

The earliest dye-dilution studies were made on the basis of constant infusion rates (as described in Sec. 5.8). For these tests, dye was injected at a constant rate (i.e., a step-function), and the responses were measured during the constant-rate-infusion period. At a later date the slug injection or the pulse injection was proposed. There is some controversy as to who initiated the concept of the slug injection,* and there was some controversy as to which technique provided more information. It will be seen that both techniques, constant-rate or sudden slug injection, yield essentially the same information. The information is displayed differently as the response must, of course, be somewhat different. But the information content of one response must be essentially the same as any other response for linear systems. The input-output characteristics must define the system; for the same system, therefore, any input-output pair must provide the necessary information to describe the system. If one stimulus-response pair gave information different from another pair, one would have to conclude that two different systems had been investigated or that the system was driven beyond the limits of assumed linearity.

The slug infusion (or bolus injection, as it is sometimes called) implies that dye is suddenly and rapidly delivered for a short time. A typical time recording of dye delivery would look like the curve in Fig. 5.31a. The rate of dye injection rapidly rises to a maximum value; injection of dye ceases abruptly (when the hypodermic piston bottoms). In order to obtain some estimate of what response may be expected from this type of stimulus, we assume that the model representation in Fig. 5.29 can serve as an approximate representation for a single-chambered vascular bed (such as the heart ventricle). This model is not completely valid but, as has been repeatedly pointed out, we use the model not as an end in

* See the article, History and Developmental Aspects of the Indicator-dilution Technic, by I. J. Fox in Ref. 8.

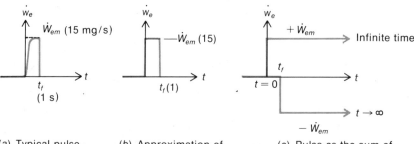

(a) Typical pulse-injection curve

(b) Approximation of injection curve

(c) Pulse as the sum of two-step functions

FIG. 5.31 *Pulse delivery of dye injection.*

itself but as a basis of reference. The model provides information about trends; differences between the model and the actual system can then be more clearly perceived.

If we accept Fig. 5.29 as a first-order approximation of a dye-distribution system, we can also accept that the system is mathematically represented by Eq. (5.100), where the driving function (the injection rate $\dot{w}_e$) is now given by the slug-injection curve in Fig. 5.31a. If a mathematical description of the injection curve were available, a solution for k_B might then be feasible (though probably tedious). Rather than use an exact mathematical representation we use the simplified pulse approximation shown in Fig. 5.31b. With this approximate stimulus waveform we shall see that analysis is simple and straightforward and the results are relatively easy to interpret.

At first glance it might appear that analysis with the pulse-function approximation (instead of the exact stimulus) does not afford any great simplification. However, since a pulse-function can be synthesized by using two step-functions, a solution is readily available. The synthesis procedure is probably most easily explained through the use of a diagram. Figure 5.31c shows two step-functions. The "positive going" step-function has an amplitude of $+ \dot{W}_{em}$ and starts at $t = 0$. As mentioned earlier, it is assumed that this step-function lasts for infinite time. The "negative going" step-function has an amplitude of $- \dot{W}_{em}$ and starts at $t = t_f$. This step-function also lasts for infinite time.* It should be obvious that when these two step-functions are graphically added the result is the pulse-function shown in Fig. 5.31b. The method by which a pulse-function can be synthesized is now self-evident: A positive step-function is added to a delayed negative step-function of the same amplitude. The delay between the two functions is equal to the time duration of the pulse. (See Prob. 5.2.)

A diagrammatic representation of the procedure is shown in Fig. 5.32. In order to describe the procedure mathematically we must develop a representation for a delayed step. This does not present any great difficulty as the notation previously developed for a step-function can easily be used to indicate a delayed step. It will be recalled that the unit step is generally represented by u (x), where the argument x indicates where the step starts. Thus when x is negative, the value of the step remains zero; when x is positive, the value of the step is unity. Therefore if a step is to begin 1 s after the start of an observation period, the parenthetic term should be positive when t is greater than 1; it should be negative when $t < 1$. This is easily arranged by using $t - 1$ as the term included

---

* These step-functions do not necessarily relate to actual signals. For this discussion the negative step-function is only a conceptual convenience; it does not imply that dye is removed from the system.

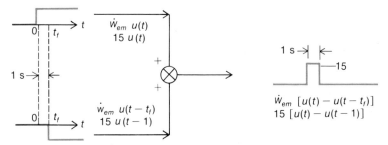

FIG. 5.32   *Synthesis of a pulse-function.*

in the parenthesis of the unit step. Thus

$$\text{(5.101)} \qquad \begin{aligned} u\,(t-1) &= 0 \qquad \text{for } t < 1 \\ u\,(t-1) &= 1 \qquad \text{for } t > 1 \end{aligned}$$

The mathematical representation for a pulse-injection rate is now self-evident:

(5.102)   Pulse of magnitude $\dot{W}_{em}$ and duration $t_f = \dot{W}_{em}[u(t) - u(t - t_f)]$

The bracketed term in Eq. (5.102) might be described as a unit-pulse-function. For values of t less than zero, the bracketed term is zero; for $0 < t < t_f$ the term is unity; for $t > t_f$ the term again reduces to zero.

The solution for the concentration response to a pulse-injection stimulus is now almost trivial. We need only find the solution of a first-order system to a step (which we already have) and subtract the response that would be generated by a delayed step. The procedure is diagrammatically shown in Fig. 5.33*b*.* The mathematical expression for the solution is determined by using Eq. (5.91) to represent the response for the "positive going" step and adding a time-delayed solution for the "negative going" step. Thus

$$\text{(5.103)} \quad k_B = \frac{\dot{W}_{em}}{\dot{q}}\,(1 - \epsilon^{-t/\tau})u(t) - \frac{\dot{W}_{em}}{\dot{q}}\,(1 - \epsilon^{-(t-t_f)/\tau})\,u(t - t_f)$$

---

* In Fig. 5.33*a* a pulse is synthesized and is directly applied to the block representation of the dye-distribution system. In Fig. 5.33*b* each of the components of the pulse is separately applied to blocks representing the system. The separate responses are then added to find the total response. This procedure is often called superposition; it is based upon a principle which states that for linear systems the total response to many individual stimuli simultaneously applied is the same as the sum of the responses that would be obtained if each stimulus acted alone. The imposed condition of linearity is to ensure that the system properties remain the same for all applied stimuli. In a computer simulation the arrangement shown in Fig. 5.33*a* would be favored, as only one block is required. Figure 5.33*b* thus represents a conceptual simulation rather than an actual computer simulation.

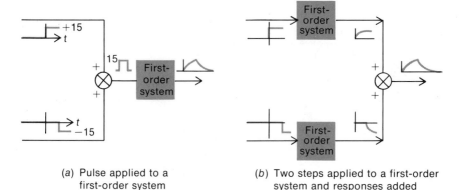

(a) Pulse applied to a
first-order system

(b) Two steps applied to a first-order
system and responses added

FIG. 5.33   *Alternative viewpoints to find response of dye-distribution system to a pulse injection.*

where K_{BM} of Eq. (5.91) has been replaced by its equivalent value $\dot{W}_{em}/s_p$ or $\dot{W}_{em}/\dot{q}$ and τ is defined by $V_t/\dot{q}$ [see the discussion from Eqs. (5.90) to (5.100)]. As Eq. (5.103) indicates, the response for $t < 0$ is zero [both $u(t)$ and $u(t - t_f)$ are zero]; for the time interval $0 < t < t_f$, the response is given by the first term only [$u(t) = 1$ but $u(t - t_f)$ is still zero]; for any time greater than t_f the response is the sum of both terms [both $u(t)$ and $u(t - t_f)$ are unity]. In the second term of Eq. (5.103) all functions of time are expressed as $t - t_f$ rather than t alone. The need for this new time base is evident because all response terms associated with the delayed stimulus must undergo a time shift. The use of $t - t_f$ as an exponent rather than t simply shifts the starting point of the exponential from $t = 0$ to $t = t_f$.

To show the correspondence between the mathematical solution [Eq. (5.103)] and the graphical interpretation, the various component responses and the total response are plotted in Fig. 5.34. To obtain these curves, numerical values were assumed for the various system constants and stimulus. The central volume (V_t) was assumed to be 100 cm³, and the flow rate was taken as 70 cm³/s or about 4,200 cm³/min (the ventricular chamber has been estimated to have an average volume of about 100 cm³ and cardiac output is approximately 4 l/min). For these system values the time-constant is $V_t/\dot{q} = 10/7$. The injection rate ($\dot{W}_{em}$) was taken to be 15 mg/s, and 1 s was used for the pulse duration. With these values, Eq. (5.103) becomes

$$(5.104) \quad k_B = {}^{15}\!/_{70}(1 - \epsilon^{-0.7t})\, u(t) - {}^{15}\!/_{70}(1 - \epsilon^{-0.7(t-1)})u(t - 1)$$

The first term of Eq. (5.104) is shown in Fig. 5.34a, the second term in Fig. 5.34b, and the total response in Fig. 5.34c (the latter figure also includes the pulse-injection stimulus). From the total-response curve we

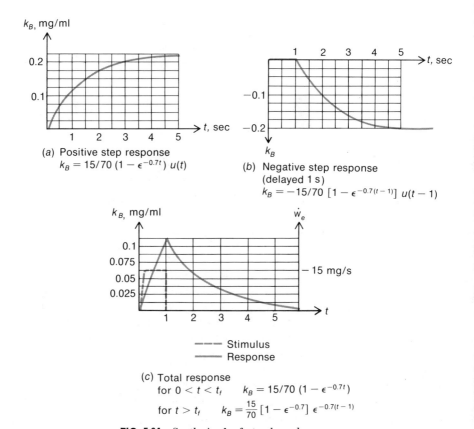

(a) Positive step response
$$k_B = 15/70\,(1 - \epsilon^{-0.7t})\,u(t)$$

(b) Negative step response (delayed 1 s)
$$k_B = -15/70\,[1 - \epsilon^{-0.7(t-1)}]\,u(t-1)$$

-------- Stimulus
——— Response

(c) Total response
for $0 < t < t_f$ $k_B = 15/70\,(1 - \epsilon^{-0.7t})$

for $t > t_f$ $k_B = \dfrac{15}{70}\,[1 - \epsilon^{-0.7}]\,\epsilon^{-0.7(t-1)}$

FIG. 5.34   *Synthesis of a first-order pulse-response.*

note that the first portion of the response is essentially an incompleted first-order step-response (the curve does not reach its steady-state level; it lasts only until the pulse input reduces to zero). This portion of the response is obviously governed by the system constants (the time-constant) and the amplitude of the driver. The tail of the response (after 1 s) must also be a first-order response, and it too must be governed by the system constants. This is intuitively obvious yet it may not be clear from the graphical construction, because the tail of the response is determined by adding two curves. To show analytically that the response after 1 s has the same properties as the response before 1 s we need consider Eq. (5.104) only for t greater than 1. For such values of time, Eq. (5.104) becomes

$$(5.105) \qquad k_B = {}^{15}\!/_{70}(1 - \epsilon^{-0.7t}) - {}^{15}\!/_{70}(1 - \epsilon^{-0.7(t-1)})$$
$$= {}^{15}\!/_{70}(\epsilon^{-0.7(t-1)} - \epsilon^{-0.7t})$$

By algebraic manipulation this expression can be altered so that

$$(5.106) \quad k_B = \tfrac{15}{70}(1 - \epsilon^{-0.7})\epsilon^{-0.7(t-1)} = 0.108 \ \epsilon^{-0.7(t-1)}u(t - 1)*$$

From this equation it can be seen that the tail of the response decays with the same time-constant as the initial portion of the response. The final portion of the response starts from an amplitude of 0.108 mg/cm^3. This amplitude is obviously governed by the magnitude and duration of the pulse and, of course, the system properties. The final value attained at the end of the pulse is often called the initial condition of the subsequent response.

Having seen how the system affects a pulse stimulus, we can now easily understand how the pulse-response can be used to define the properties of an assumed first-order system. The time-constant can be determined from the leading or trailing (initial or final) portions of the curve,† or the system constants can be determined by using a pair of points on the curves. In any case, two items of information must be taken from the curve in order to define the two assumed system constants (the assumption was that the system could be represented by a first-order approximation).

In order to show how the concepts developed in the preceding paragraph can be applied, we examine the dye-dilution system shown in Fig. 5.35. Dye is injected into the superior vena cava; venous blood carries the dye into the right ventricle. The dye concentration is measured in the pulmonary artery (i.e., just after blood leaves the right ventricle; see Fig. 5.35a). The pulse-input infusion is shown in Fig. 5.35b and the concentration response in Fig. 5.35c. The pulse-input curve of Fig. 5.35b is hypothetical; in most cases this information is not recorded. The time at which infusion starts is generally noted, and it is assumed that injection is very rapid so that the pulse width is very small. Although in many cases it is important to know the total-input curve, here useful results can still be obtained so long as we can safely assume that the pulse has a sharp cutoff. For a hypodermic infusion this would be a reasonable assumption as the injection stops abruptly when the plunger reaches bottom.

A comparison of an actual dye-dilution response curve (Fig. 5.35c) and the theoretically derived curve (Fig. 5.34c) shows that there are marked differences. The actual curve, for example, does not return to a

* The unit function $u(t - 1)$ is included to emphasize that this solution is valid only if t is greater than 1.

† In practice, the final response must ordinarily be used. This is particularly true when the stimulus has a large pulse height and a short duration. Under these circumstances the initial portion of the response curve has a steep rise, and it is difficult to make accurate measurements from the data.

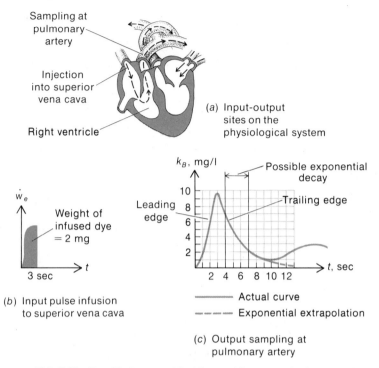

FIG. 5.35   *Dye-dilution test with right ventricle as central volume.*

zero base line, whereas the theoretical curve exponentially decays to zero. This difference can easily be explained by the fact that in the theoretical analysis the input flow was always dye-free blood; recirculation was not considered. In the actual system, recirculation must, of course, take place.* The final nonzero value for concentration and the slight oscillation prior to the steady-state value are thus due to the recirculation in the closed system. The theoretically predicted exponential response, if it exists in the actual system, would therefore be found in the portion of the response curve prior to the advent of recirculation. This would imply that we should examine either the leading edge of the response curve or the beginning of the trailing edge (see Fig. 5.35c). We intuitively realize, how-

---

* The recirculation portion of the concentration curve will not appear if an appropriate indicator is used. If, for example, the indicator diffuses or metabolizes rapidly, the recirculated blood will not be "tagged." The dilution curve will then decay to zero with no oscillation. Thermal-dilution curves show these characteristics. The indicator used is a warm or cold saline solution. A thermistor records the "temperature dilution." However recirculated blood is always at the same temperature so that the dilution curve does not show recirculation characteristics.

ever, that the leading edge will be exponential only if the input was truly a pulse-function, i.e., the input showed an "instantaneous" rise. It would be unrealistic to assume that the input was truly a pulse-function (to achieve instantaneously a constant injection rate is not practically feasible under ordinary circumstances) and, with no recorded information about the input waveshape, it would be safer to ignore this leading edge of the response. On the other hand, if the input infusion cut off sharply, we could expect an exponential decay in the trailing edge of the response. As pointed out earlier, a sharp cutoff of injection rate is practically realizable. If an exponential response can be expected anywhere, therefore, it would most likely be found in the trailing edge before recirculation.

To examine the trailing portion of the response curve for exponential properties, any of the techniques outlined earlier can be used (see Sec. 5.4). The most accurate procedure is to plot this portion of the curve in semilogarithmic coordinates. If the resultant plot is a straight line, the trailing portion of the dilution curve can be said to decay exponentially. The slope of the line would be the time-constant of the system. A semilogarithmic plot of the trailing edge of a dye-concentration curve is shown in Fig. 5.36. The curve is a well-defined straight line in the time interval $3 < t < 8$. After 8 s the curve deviates from a straight line, indicating that recirculation has begun. However, the almost perfect result shown in Fig. 5.36 cannot always be expected. In most cases the plotted points show more scatter, and the straight-line portion does not have the extended time duration shown here. However, in normal subjects some straight-line section can generally be seen.

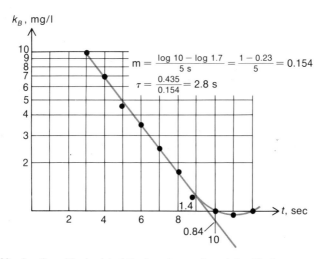

FIG. 5.36 *Semilogarithmic plot of the decaying portion of dye-dilution response curve.*

From the plot in Fig. 5.36 we can easily determine the time-constant of the system. The slope of the line is

$$(5.107) \qquad m = \frac{\log 10 - \log 1.7}{8 - 3} = \frac{1 - 0.23}{5} = 0.154$$

The time-constant is therefore

$$(5.108) \qquad \tau = \frac{0.435}{0.154} = 2.8 \text{ s}$$

It was established earlier that the time-constant is defined by the ratio of the chamber volume (V_t) and the flow rate $(\dot{q})$. Equation (5.108) thus provides one relationship between these two system constants. We can determine both these constants if another relationship between them is obtained or if one of the constants is separately evaluated.

If the input injection had truly been a pulse-function, it would have been a relatively easy matter to determine the flow rate $(\dot{q})$. From Eq. (5.103) we note that just *prior* to the end of the pulse (at $t \approx t_f$) the concentration reaches a maximum value given by

$$(5.109) \qquad k_B(\text{max}) = \frac{\dot{W}_{em}}{\dot{q}} (1 - \epsilon^{-t_f/\tau})$$

If the input had been a pulse of known duration, t_f and $\dot{W}_{em}$ would have been available. (W_e, the weight of the delivered dye, is known. If t_f were known and the injection was a pulse, the delivery rate would have been a constant at $\dot{W}_{em} = W_e/t_f$.) Since the maximum concentration $k_B(\text{max})$ and the time-constant (τ) are available from the concentration curve, $\dot{q}$ could have been determined, as it would then have been the only unknown in Eq. (5.109). This, in a small way, indicates why it may often be desirable to record input as well as output waveforms. In this case there is enough information available to permit an evaluation of the flow rate even though the input was not recorded; however, this is not always the case, and in general it is good policy to record input signals whenever possible. This is particularly true in the early phases of an investigation when the relationship between variables is not yet clearly understood.

The procedure outlined in the preceding paragraph is based upon the premise that the input waveform was completely known. For this system, however, it is possible to evaluate the system constants even if the input waveform is not completely known; all that is required is that the input be abruptly terminated (sharp cutoff) and that the weight of the delivered injectate be known. To see this more clearly we have but to note that if the input has abrupt termination the concentration curve can be expected to decay exponentially. Thus, even if we do not know the time of injection cutoff, we can still evaluate the time-constant of the system (from the slope of the semilog plot of the concentration curve). This provides one relation between V_t and $\dot{q}$. We can obtain a second relation between V_t and $\dot{q}$ even if the exact stimulus waveform is not known so long as we know the weight of the injectate. To illustrate this point, we look for a description of the system that involves the weight of injectate. An examination of Eq. (5.100) (the equation governing the system operation) indicates that it can be converted to the desired form by integration:

$$(5.110) \qquad \int_0^t \left(V_t \frac{dk}{dt} \right) dt + \int_0^t \dot{q} K \, dt = \int_0^t \dot{w}_e \, dt$$

Since the chamber volume (V_t) and the flow rate $(\dot{q})$ are assumed to be constant, Eq. (5.110) can be more simply written:

$$(5.111) \qquad V_t \int_0^t dk + \dot{q} \int_0^t k \, dt = \int_0^t dw_e$$

The time limits of integration are from the beginning of the pulse $(t = 0)$ to some time after the pulse has been completed $(t = t_c$, where t_c is evidently any time after the concentration curve reaches a peak; this condition is required to ensure that all the injectate has been delivered). Then Eq. (5.111) becomes

$$(5.112) \quad V_t[k(t = t_c) - k(t = 0)] + \dot{q} \int_0^{t_c} k \, dt = w_e(t = t_c) - w_e(t = 0)$$

or

$$(5.113) \quad V_t K_{t_c} + \dot{q} \int_0^{t_c} k \, dt = \text{weight of injectate*} \qquad k(t = 0) = 0$$

* This same result could have been obtained in a more basic fashion by using the finite-difference representation of the system as given by Eq. (5.99). If this notation is used, then for the first time interval $n = 1$, and we have

$$V_t(K_1 - K_0) + K_0 \dot{q} \, \Delta t = \dot{w}_{e0} \, \Delta t = \text{weight delivered during first interval}$$

For the second interval we have

$$V_t(K_2 - K_1) + K_1 \dot{q} \, \Delta t = \dot{w}_{e1} \, \Delta t = \text{weight delivered during second interval}$$

For the nth interval

$$V_t(K_n - K_{n-1}) + K_n \dot{q} \, \Delta t = \dot{w}_{en} \, \Delta t = \text{weight delivered during } n \text{th interval}$$

To obtain the total weight delivered we add the weight delivered during all intervals; i.e., we add all the equations. The result is then

$$V_t(K_n - K_0) + \dot{q}(K_0 + K_1 + K_2 + \cdots + K_n) \, \Delta t = W_e$$

If K_0 is zero, the first term becomes simply $V_t K_n$, where K_n is the concentration of the nth instant. In the second term the summation $\sum_{n=0}^{n} K_n \, \Delta t$ is the area under the concentration curve between the zeroth instant and the nth instant. The results are the same as those shown in Eq. (5.113).

Equation (5.113) provides the second relation between V_t and $\dot{q}$, as mentioned earlier. These two terms are the only ones in Eq. (5.113) that cannot be directly evaluated.

As a specific example to illustrate the outlined procedure, let us arbitrarily choose t_c to be 3 s (just at the finish of the pulse infusion). From the response curve (Fig. 5.35c) we find that at 3 s the concentration is 10 mg/l; K_{t_c} is thus 10. The second term in Eq. (5.113) includes the integral of the concentration curve in the interval $\left(\text{that is, } \int_0^{t_c} k \, dt\right)$. This is interpreted as the area under the curve from 0 to 3 s. This numerical value can be obtained by counting squares, if need be, or by use of a planimeter. In any case this value can be determined and is evaluated to be 14.2 mg-s/l. Since the pulse has been completed by 3 s, all the dye has been delivered and W_e is simply 2 mg. Equation (5.113) can now be written

$$(5.114) \qquad 10V_t + 14.2\dot{q} = 2$$

Equations (5.114) and (5.108) can now be used to obtain a simultaneous solution for both V_t and $\dot{q}$. The results of such a solution are $\dot{q} = 2{,}840 \text{ cm}^3/\text{min}$ and $V_t = 133 \text{ cm}^3$. This particular patient has therefore a ventricular delivery rate (or a cardiac output) of 2,840 cm³/min and an average ventricular volume of 133 cm³. The cardiac output for a normal adult is about 5,000 cm³/min; a delivery rate of 2,840 cm³/min for a young child is thus not an unreasonable figure. To interpret the figure for the chamber volume, we must remember that this represents an average value for the ventricular volume. If the ventricle ejected its total contents during systole, the heart delivery per stroke would be taken as 266 cm³. At a normal heart rate of about 70 beats/min the delivery rate would be 70 × 266 or 18,620 cm³/min—far in excess of the delivery rate previously determined. The conclusion is, of course, the ventricle does not eject its entire contents during systole.

To obtain some estimate of the volume remaining in the ventricle* we could assume that, for a delivery of 2,840 cm³/min and a heart rate of 70 beats/min, the ventricle ejects about 40 cm³/stroke. If the minimum ventricular volume is taken as V_R, the maximum volume (for normal operation) is $V_R + 40$ and the average volume is $V_R + 20$. If we accept the figure for average ventricular volume as 133 cm³, then the volume retained in the ventricle is roughly 110 cm³. It appears, then, that the ventricle ejects only about 30 percent of the maximally stored contents.

* The volume retained by the ventricle is sometimes considered to be composed of two separate components: the systolic reserve volume and the residual volume. The latter component is always present; the former depends upon whether the subject is at rest or performing physical exercise. See Ref. 11.

This is a very crude estimate based upon a single study. Further studies (literature and laboratory) would be needed before any definite quantitative results could be given.

In the previous analysis, the upper limit of integration (t_c) was arbitrarily chosen to be 3 s. We could have used any value for t_c so long as it was greater than the pulse duration and so long as the first-order representation [as required by Eq. (5.113)] remains valid. Thus if t_c were chosen to be 6 s we would have

$$(5.115) \qquad V_t K_6 + \dot{q} \int_0^6 k \, dt = 2$$

where now K_6 is 3.5 and $\int_0^6 k \, dt$ ($= 32.2$) is the area under the concentration curve from $0 \le t \le 6$ (see Fig. 5.35c). Equations (5.115) and (5.108) when solved simultaneously should yield the same solutions for V_t and $\dot{q}$ as were determined previously. This procedure remains valid so long as Eq. (5.113) remains valid; Eq. (5.113) remains valid so long as the concentration curve decays exponentially.

It was seen in an earlier discussion that the actual concentration curve decays exponentially for a limited range. The only reason it did not continue to decay exponentially to zero was that there was dye recirculation in the closed system. If a completely diffusible dye (or indicator) were used, the effects of recirculation would have been minimized and possibly eliminated. Under these circumstances we would have expected the concentration curve to decay continuously to zero in an exponential fashion. We can therefore construct a hypothetical response curve which excludes recirculation by simply extending the exponential response of an actual dye-concentration curve (see Fig. 5.35c). Equation (5.113) would then be valid for all values of time greater than 3 s (V_t and $\dot{q}$ remain unaffected by the proposed extension of the exponential response). This presents a direct method for evaluating $\dot{q}$ without using a simultaneous solution. Since Eq. (5.113) is now valid for all time greater than 3 s, it is valid as time grows large (i.e., as $t_c \to \infty$). As t_c becomes larger, K_{t_c} becomes very small (as $t_c \to \infty$, $K_{t_c} \to 0$) and Eq. (5.113) can now be written

$$(5.116) \qquad \text{Zero} + \dot{q} \int_0^\infty k \, dt = 2$$

The integral is now interpreted to be the total area under the hypothetical concentration curve which was exponentially extrapolated to eliminate recirculation. Since this integral can be evaluated, $\dot{q}$ is the only unknown in Eq. (5.116) and it can therefore be determined. This provides a simpler method to evaluate the flow as it permits $\dot{q}$ to be directly evaluated without the need of simultaneous equations.

To carry out the procedure suggested in the preceding paragraph we must first extend the actual concentration curve in an appropriate exponential fashion. This is easily done by simply extending the straight-line portion of the semilogarithmic plot beyond the curved section due to recirculation (see Fig. 5.36). From this extended line we can obtain corresponding values for k and t (e.g., at $t = 9$ s, $k = 1.4$; $t = 10$, $k = 0.84$, etc). When plotted in linear coordinates, these points will generate a curve that will decay exponentially with a time constant of 2.8 s (see Fig. 5.35c). We need now to evaluate the area under this artificially constructed curve. As a practical matter, since an exponential curve decays to practically zero in four time-constants (11 s in this case), we could probably evaluate the area between zero and 17 s and neglect any area beyond 17 s with little loss in accuracy.* If we carry out the suggested procedure we find that the integral (the area) has a value of about 41.8 mg/l/s. The flow rate is now determined to be 2/41.8 l/s or 2,870 cm^3/min, which agrees very closely with the value previously obtained. The chamber volume V_t can now be determined by substitution in Eq. (5.108).

The procedure outlined above is the one commonly used in dye-dilution studies.

5.10 CONCEPT OF THE IMPULSE–FUNCTION

The impulse-function, like the step-function and the pulse-function, is a mathematical representation of a signal excitation encountered in research investigations. The very descriptive nature of the name "impulse-function" indicates qualitatively the types of signals that may be associated with this function. Such signals are evidently characterized by high-amplitude, short-duration waveforms. A flicker of light applied to the eye, a sharp blow to the patellar tendon, a click applied to the ear are all illustrative of stimuli that would be intuitively classified as impulsive excitations.

Impulse stimuli are often employed to disturb systems so that the

---

* This procedure would be employed if mechanical methods (planimeter, box count, etc.) were used to evaluate the area. A combination of mechanical measurement and mathematical integration could be used. One might, for example, "count boxes" between 0 and 6 s (area = 32.2) and then integrate the mathematical expression describing the curve for the area beyond 6 s. As the curve is known to be an exponential with a 2.8-s time-constant, this area would be

$$\int_0^\infty K_6 \epsilon^{-t/2.8}\, dt = 3.5 \int_0^\infty \epsilon^{-t/2.8}\, dt = -2.8[3.5\epsilon^{-t/2.8}]_0^\infty = (2.8)(3.5)(1 - 0)$$
$$= (2.8)(3.5) = 9.8$$

By this procedure the total area would then be 42 mg/l/s.

return to equilibrium can be observed. In terms of energy, the impulse delivers energy to storage elements of the system; at the end of the impulse the stored energy is released; the released energy causes a measurable response. If the response to the impulsive stimulus is analyzed, an estimate of the gross operating characteristics of the system may be obtained. The common procedure of tapping an instrument is an example of an impulse used to excite a system so that its responses may be observed. The tap is the impulsive driver; the energy delivered is stored in the inertia and elastic members of the instrument. At the end of the tap, the stored energy (probably a stretched spring in the case of an instrument) is released. The released energy causes motion of the instrument mechanisms which may be observed at the instrument "readout." A responsive instrument usually oscillates a bit before it settles down at equilibrium; a sluggish or slow-responding instrument usually returns to equilibrium in an exponential fashion with no oscillation. (We shall examine some of the characteristics of instrument response in later sections.)

To place the impulse-function in proper perspective we review briefly the properties of the pulse-function. This function represents stimuli that rapidly change from one signal level to another and then rapidly return to the original operating level at a later time (see Fig. 5.37a). The term "rapidly" is used in a relative sense; if the response is measured in seconds, a stimulus that changes to a new level in milliseconds, maintains this level for seconds, and then returns to its initial value in milliseconds could be considered a pulse-function with small error (see Fig. 5.37a). (A stimulus is not considered a pulse-function if the stimulus rise time or decay time is of the same order of magnitude as the stimulus duration.) The impulse-function is related to the pulse-function as it also represents signals that rise and fall rapidly. The basic difference between the two is that of time duration. If the duration of a stimulus is of the same order of magnitude as the time scale of the response, the stimulus can be appropriately represented by a pulse. If the duration of the stimulus is an order of magnitude less than the time scale of the response, then the stimulus can be considered an impulse excitation. Thus, if an excitation signal rises and falls in microseconds and the response is measured in milliseconds, it would probably be valid to consider the excitation an impulse-function (see Fig. 5.37b). Although this duration criterion is somewhat arbitrary and qualitative, it serves to place the impulse-function in proper perspective. At the end of this section a more quantitative criterion will be suggested.

The total response of a system to an impulsive stimulus consists of two parts, as does the pulse-response (see Fig. 5.37b). The first portion of

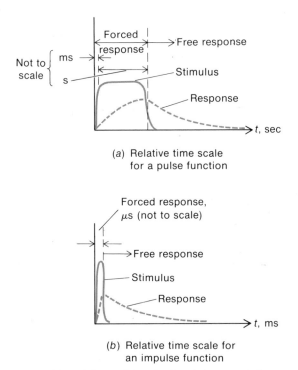

(a) Relative time scale
for a pulse function

(b) Relative time scale for
an impulse function

FIG. 5.37 *Pulse- and impulse-function representations.*

the response occurs while the impulse is delivered; the second occurs after the impulse has terminated. As the latter response takes place during a period when the system is not driven by an external stimulus, this response is often called the "free response"; the response during the excitation is called the "forced response." Since an impulsive stimulus is of short duration, the forced response must also be of short duration.

We digress briefly to note that, since the free response takes place when the system is essentially unstimulated, the shape of this portion of the curve must depend primarily on the system properties. It would appear, then, that the system can be completely defined from the free response. Past experience indicates that this is not the case. We have seen in previous examples that some information about the relationship between system properties can be obtained from the free response (e.g., the time-constant) but the individual system properties cannot be completely described from this information alone. To define the system properties completely one needs either some additional specific information about the system properties or information supplied by the forced response. Since the forced response to an impulse excitation is very short, it might seem that it would be more difficult to define a system with such an input excitation. As a matter of fact, however, if the area under the impulse is known, a system can be more easily defined when this excitation is used than when a pulse is used.

In order to develop a technique that can be used to determine the system properties from an impulse-response, we start with the procedure described in Sec. 5.9. Although this method was developed for a quasi-pulse excitation, it is still valid even though an impulse excitation is applied. To review this technique we once again examine the dye-dilution system shown in Fig. 5.35a. Instead of a pulse stimulation, however, we shall use a much more rapid infusion, such as shown in Fig. 5.38a. Here 2 mg of cardio-green dye was administered in $\frac{1}{2}$ s (instead

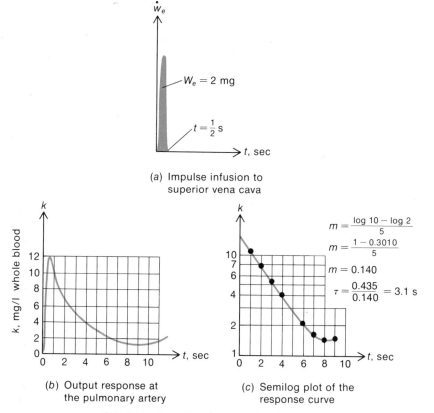

(a) Impulse infusion to
superior vena cava

(b) Output response at
the pulmonary artery

(c) Semilog plot of the
response curve

$$m = \frac{\log 10 - \log 2}{5}$$

$$m = \frac{1 - 0.3010}{5}$$

$$m = 0.140$$

$$\tau = \frac{0.435}{0.140} = 3.1 \text{ s}$$

FIG. 5.38 *Dye-dilution test with impulse excitation.*

of 3 s as in Sec. 5.9). The resultant dye-concentration curve at the pulmonary artery is shown in Fig. 5.38b. When recirculation is ignored, a semilogarithmic plot of the tail of the response has straight-line characteristics (see Fig. 5.38c). This system thus has a first-order response with a time-constant of 3.1 s, and Eq. (5.100) can be used to describe the system. Since the exact waveform of the input is not known but the weight of injectate is known, we use the integral form of Eq. (5.100) as given in Eq. (5.113).*

$$(5.113) \qquad V_t K_{t_c} + \dot{q} \int_0^{t_c} k \, dt = W_e = 2 \text{ mg}$$

In order to use Eq. (5.113) we must determine appropriate coefficients for the system constants V_t and $\dot{q}$. These can be obtained from the response curve for any value of t after the termination of the impulse (and of course before recirculation). If we arbitrarily choose t to be 4 s, then, from the curve in Fig. 5.38b, K_{t_c} is 4.4 mg/l. The area under the curve in the interval $0 < t < 4$ $\left(\text{that is, } \int_0^4 k \, dt \right)$ may be roughly evaluated by counting squares. By this procedure the integral is determined to be 25 mg/l/s. Equation (5.113) can now be written

$$(5.117) \qquad 4.4V_t + 25\dot{q} = 2$$

If we use the fact that the time-constant is defined by the ratio $V_t/\dot{q}$ we can replace $\dot{q}$ in Eq. (5.117) by $V_t/3.1$. Then V_t is the only unknown, and its value can be determined ($V_t = 160$ cm³; $\dot{q} = 3{,}100$ cm³/min).

The above review reemphasizes the fact that, if a stimulus gives rise to a free response, the system properties can be evaluated if the area under the stimulus curve is known. The review also serves to point out that if the stimulus is of short duration (i.e., impulsive) the area under the *forced-response* curve is of little importance. The latter point was not specifically delineated in the analysis, but it is clear that the area used to evaluate $\int k \, dt$ was predominantly under the free-response curve. The area under the forced-response curve contributed very little to the evaluation of the integral. (The forced-response curve rose very sharply; the area under this curve was therefore of small numerical value.) This fact has interesting implications. To use Eq. (5.113) we had to evaluate $\int_0^{t_c} k \, dt$, where t_c could be any time after the termination of the impulse. We arbitrarily chose t_c to be 4 s. Any value down to ½ s (the end of the forced response) would have served as well. If we chose t_c as ½, i.e., the

---

* The weight of the injectate delivered is always given by the area under the stimulus curve (i.e., the integral of the injection rate over the duration of the stimulus).

time at which the impulse is just completed, then the integral would be equivalent to the area under the forced-response curve. Since this area is quite "small," we might, as a first estimate, neglect the second term in Eq. (5.113). If we do so, then V_t can be immediately evaluated. At $\frac{1}{2}$ s, K_{t_e} is 12 mg/l and V_t is then $\frac{2}{12}$ l or 167 cm³. The excellent agreement between this value for V_t and the value previously determined is testament to the fact that the area under the forced response could be neglected with small error. It should be intuitively obvious now that, if an impulse were of shorter duration, closer results would be expected, and if the impulse were of sufficiently short duration, then for all practical purposes V_t would be exactly equivalent to W_e/K_0 (where K_0 is the initial value of the response).

Probably the most striking aspect of this development is that (in retrospect) the relationship between the central volume (V_t), the injectate weight (W_e), and the initial concentration (K_0) could have been intuitively anticipated. It seems entirely reasonable (now) that if the infusion is rapid enough practically all the dye will be delivered to the central volume before any can be washed away. (See Prob. 5.7.) The initial concentration in the central volume must therefore obviously be the weight of dye delivered divided by the central volume. We can now see that, if an impulse infusion can be used, we do not need to evaluate areas under the response curve to determine the system properties. The central volume can be determined immediately from the ratio of the injectate weight to the initial concentration. The time-constant is determined from the free-response curve. From these two values the flow rate can be evaluated.

Although the observation in the preceding paragraph refers to a specific first-order system, we can intuitively conclude that it must be equally applicable to first-order systems in general. The general statement of this concept is: When an impulsive driver is applied to a first-order system it establishes the initial conditions of the subsequent free response. The ratio of the impulsive area to the initial conditions is related to one of the properties of the system. This statement can be written in equation form:

$$(5.118) \qquad s_p = \frac{\int_{0-}^{0+} (\text{IMP})_y \, dt}{X_{a0}} = \frac{\text{area under impulse}}{\text{initial value of response}}$$

The symbol $(\text{IMP})_y$ is used to indicate an impulsive through-variable driver (an impulsive delivery of dye, for example). The limits of integration shown ($0^- < t < 0^+$) imply that the peak of the impulse occurs at $t = 0$; the impulse starts just before $t = 0$ and ends just after $t = 0$. A more popular notation for a unit impulse (and the one that will henceforth be used) is the symbol $\delta(t)$. This symbol automatically indicates that an impulsive driver with unit area was applied at the time indicated by the argument t. Thus $10\ \delta(2)$ would mean an impulse with an area of 10 units was applied at $t = 2$. If this notation is used, an impulsive through-variable driver could be written as $A_t\ \delta(t)$, where A_t would be the area under the impulsive driver. Equation (5.118) would now be written

$$(5.119) \qquad s_p = \frac{\int_{0-}^{0+} A_t\ \delta(t)\ dt}{X_{a0}} = \frac{A_t}{X_{a0}}$$

Evidently $\int_{0-}^{0+} \delta(t)\ dt$ is interpreted as unity. Although this notation may appear to be clumsy at this point, it will be seen later that it has meaningful and useful applications. From the foregoing discussion, it should be obvious that if the impulse driver is an across-variable (e.g., a sudden pressure) then a system property may be determined by

$$(5.120) \qquad s_p = \frac{\int_{0-}^{0+} A_a\ \delta(t)\ dt}{Y_{t0}} = \frac{A_a}{Y_{t0}}$$

where A_a is the area under the across-variable impulse and Y_{t0} is the initial value of the through-variable response.

In order to use Eqs. (5.119) and (5.120) one must first establish that a stimulus can be considered impulsive. This judgment must be based upon a consideration of the excitation and the response. Although the rule-of-thumb criterion suggested on page 227 often provides a practical basis for judgment, there are times when a more quantitative criterion may be desired. We can establish such a criterion by examining Eq. (5.113) once more. From this equation we can see that an excitation may be considered impulsive if the second term, $\dot{q} \int_{0}^{t_c} k\ dt$, can be neglected when compared with $V_t K_{t_e}$, where t_c is the duration of the impulse. Since $\dot{q}$ is equivalent to V_t/τ, this second term can be written $V_t \left(\int_{0}^{t_c} k\ dt \right) \Big/ \tau$. The integral term in the latter expression is approximately equivalent to $(K_{t_e})(t_c)/2$ (the area under the forced response is considered to be approximately triangular; see Fig. 5.38b). Equation (5.113) can now be written

$$(5.121) \qquad V_t K_{t_e} + V_t K_{t_e} \frac{t_c}{2\tau} = W_e$$

or

$$V_t K_{t_e} \left(1 + \frac{t_c}{2\tau} \right) = W_e$$

From the last equation we can easily see how the stimulus duration

(t_c) should be related to the response if the stimulus is to be considered impulsive. In particular, if the duration of the stimulus is one-tenth the system time-constant, $t_c/2\tau$ is 0.05 and the value of V_t will be about 5 percent high if the stimulus is considered impulsive. For most practical purposes a 5 percent error would not be excessive; we can therefore conclude that, if the stimulus duration is less than one-tenth the system time-constant, the stimulus may be considered impulsive. In a given situation, then, if an applied stimulus evokes a free response, the free response is first used to evaluate the system time-constant. If the stimulus duration is less than 10 percent of this value, the stimulus may be considered impulsive and a property of the system may be evaluated by Eq. (5.119) or (5.120). If a system has an oscillatory free response or the decay is shown to be the sum of two or more exponentials, the system is evidently of higher order than first. A stimulus would then be considered impulsive if its duration is less than 10 percent of the shortest time-constant or less than 10 percent of the time between the adjacent peaks of an oscillatory response (i.e., the period). We shall examine the effect of an impulsive driver on second-order systems more thoroughly in a later chapter.

We close this section with a few final comments regarding the significance of the impulse-function. At this point we regard the impulse-function simply as a description of a stimulus that exists for a brief instant. Although such a stimulus may sometimes be described by other functions (e.g., a pulse-function), data analysis is somewhat easier if the stimulus can be considered impulsive. This is especially true if only the area of the excitation waveform is known and the system is of higher order than first. In the next chapter we shall examine some of the mathematical implications of the impulse concept. It will then be seen that not only does the impulse-function serve to describe a physically realizable signal but it can also be used to establish some important mathematical concepts.

Problems

5.1 In the footnote associated with Eq. (5.16) it was indicated that the criterion for a stable solution by finite-difference methods depends upon the term $(G/C)\,\Delta t$: For $(G/C)\,\Delta t > 2$ the solution is unstable.

(a) If G/C is 2.5, compare the true solution $[p_c(t) = 2(1 - \epsilon^{-2.5t})$; Eq. (5.29)] with those which would be obtained by using $\Delta t = 0.75$, 0.8, and 0.9.

(b) If G/C is large (say 25), Δt must be chosen "small" if the solution is to remain stable. This means, to reach any given time, many intervals must be used. Thus, if Δt is 0.075 and the value of p_c when t is 0.3 s is required, five successive evaluations of p_c must be made (that is, $0.075 \times 4 = 0.3$). However, with G/C large (shorter time-constant) the system approaches the final value more rapidly than when G/C is small. Show that, when G/C is 25, p_c reaches its final level in about 0.2 s. What is p_c at 0.3 s?

5.2   The mathematical description of the step-function shown in Fig. P5.2a is given by $p(t) = u(t)$. This notation indicates automatically that the signal $p(t)$ is zero if $t < 0$ and $p(t)$ is unity for $t \geq 0$. In some experimental situations it is desirable to excite a system with a step at the beginning of the experiment (Fig. P5.2a) and then to apply a second step at a later time (Fig. P5.2b).

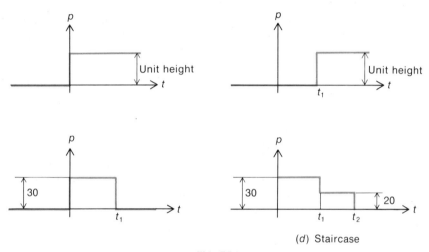

(d) Staircase

FIG. P5.2

(a) *Draw* the waveform that would represent the total-excitation signal for $-\infty < t < \infty$. Propose a physiological situation in which such an excitation signal might be used.

(b) To represent the delayed step shown in Fig. P5.2b, a modified and more general interpretation of the unit-step notation is used: $u(+\text{ or } 0)$ is now considered to be unity and $u(-)$ to be zero. This notation implies the unit function is unity when the parenthetic terms give rise to a positive number or zero and the unit function is zero when the parenthetic term is negative. Thus $u(t - 5)$ is interpreted as zero when $t < 5$ [then $t - 5$ gives rise to a negative number and $u(-)$ is zero]; it is interpreted as unity when $t \geq 5$ [then $t - 5$ is either zero or positive and $u(0,+) = 1$]. Write the equation for $p(t)$ of the waveform shown in Fig. P5.2b.

(c) Write a single expression to describe an excitation consisting of a step applied at zero and a second step applied at t.

(d) Figure P5.2c is a graphical representation of a pulse excitation. Show that this excitation may be considered the summation of a step applied at $t = 0$ and another applied at $t = t_1$. Write a mathematical expression for the pulse-function.

(e) Write a mathematical expression to represent the excitation waveform shown in Fig. P5.2d.

5.3 A man descends rapidly to 33 ft below sea level and remains there for 3 h. If it is *assumed* that nitrogen is absorbed in the same manner as it would be liberated, the time-constant for absorption would be the same as for liberation.

(a) How much N_2 is held in storage after 3 h? ("Normal" compressed air is delivered to the subject.)

(b) After 3 h the man rapidly descends another 33 ft and remains at 66 ft for 1 h. Draw a plot of the descent curve. Assuming the time-constant for absorption is still the same as that for liberation, draw a plot of the nitrogen absorbed as a function of time.

(c) After 1 h at 66 ft (total of 4 h submergence) the man is to be returned to sea level. (His system has not absorbed the amount of nitrogen it could have absorbed at 66 ft H_2O.) What pressure profile would you recommend for the man on his return to sea level?

5.4 Figure 5.20c shows one configuration for the representation of a first-order system. However, Fig. 5.20c is a simplified version of the computer circuit shown in Fig. P5.4a. To show that the latter circuit closely represents a first-order system, assume the operational amplifier draws practically no current and the

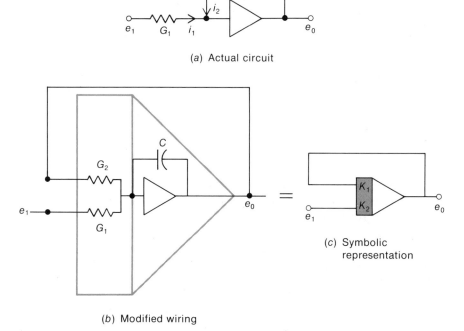

(a) Actual circuit

(b) Modified wiring

(c) Symbolic representation

FIG. P5.4

input voltage to the amplifier is neglectably small (see Appendix A, steps 21 to 26). A nodal equation at the input node of the amplifier will then be a first-order differential equation.

(a) Write the differential equation represented by Fig. P5.4a.

(b) Determine the values that should be used for G_1, G_2, and C if the representation in Fig. P5.4a is to solve Eq. (5.50). What will be the values of e_1 and e_0?

(c) If the circuit wiring is modified slightly, Fig. P5.4a can be drawn as shown in Fig. P5.4b. Show that the portion enclosed by the phantom lines in Fig. P5.4b acts essentially as an integrator. (See Appendix A, step 35.)

(d) The final simplified version of the first-order representation is shown in Fig. P5.4c. The signals entering the integrator are added and then integrated. Write an equation for e_0 using this configuration, and keep the result in mind when reading the early sections of Chap. 6.

5.5 Inulin is a dye often used in physiological studies to determine plasma volume and the plasma-clearance efficiency of the kidney. In a particular test, inulin is introduced at a rate of $\dot{W}_i$ g/min and 3 h later the inulin injections are abruptly terminated. A plot of inulin injection vs. time is shown in Fig. P5.5a.

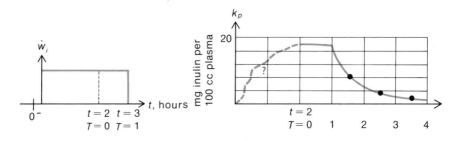

(a) Input-inulin injection   (b) Output-plasma concentration

FIG. P5.5

The concentration of the inulin in the plasma is continuously sampled and is plotted in Fig. P5.5b.* (The inulin concentration is reduced by the continuous dye-removal action of the kidney.)

(a) Write an equation to describe the input waveform for $0^- < t < \infty$.

(b) Suppose the observed response data were not considered until $t = 2$ h (that is, $T = 0$; the time $T = 0$ marks the beginning of the experimental period). Write an expression describing the input waveform $0 < T < \infty$. What are the quiescent values for the input and output waveforms?

(c) During the period $1 < T < \infty$ the urine is collected and the total

* Such curves are fairly standard. See, for example, Refs. 6 and 7.

quantity of inulin measured is 0.5 g. What was the volume of the plasma in the physiological system?

(d) Determine an appropriate expression to be used to approximate the output waveform (K_p as a function of t) as a first-order step-response for the period $T \geq 1$. (Use a semilog plot to obtain two values for the time-constant to indicate the limits between which the true data must lie.)

(e) At what rate is inulin removed from the plasma by the kidney when $T = 1$? Is this removal rate maintained throughout the clearance period? Plot the clearance rate vs. time. From such information how could the rate of plasma processing by the kidney be estimated?

(f) From the data available, the system appears to have storage and resistive properties. Describe these properties qualitatively in terms of the physiological characteristics of the system.

(g) Determine the numerical limits of the storage and resistive properties based upon the figures determined in (c).

5.6 To obtain blood samples from the heart during a dye-dilution test, catheters are generally threaded through an artery into the right heart. Blood is withdrawn through the catheter by a pump and passed through a cuvette oximeter. The oximeter measures the color density of the blood and sends the information to a recorder (see Fig. P5.6a). The information recorded contains the response of the physiological system and of the instrument system. To assess the effect of the instrument system, it may be tested separately as shown in Fig. P5.6b.* The catheter is connected to a mechanical valve which can switch the catheter fluid supply from one tank to another. In an actual test the catheter is first connected to the tank with undyed blood and then rapidly switched to the tank with dyed blood. For practical purposes the switching can be considered instantaneous. The recorded response curve is shown in Fig. P5.6c.

(a) Write a first-order differential equation to describe the catheter system. Describe all variables and evaluate any numerical values that can be determined.

(b) The overall operation of the dye-dilution recording system may be described by the block-diagram configuration in Fig. P5.6d. Dye is injected ($\dot{W}_e$); it passes through the system and delivers a concentration signal K_B. The signal enters the instrumentation system, which then delivers a recorded signal K_R. If it is assumed that the physiological system and the instrument system can both be described by first-order differential equations, what differential equation would describe the overall operation, where $\dot{W}_e$ is the input and K_R is the output? (If $y = dx/dt$, then dy/dt is d^2x/dt^2.)

(c) In an actual test the input injection has the waveshape shown in Fig. P5.6e; the output waveshape (after recirculation has been eliminated) is also shown in this figure. Show that the cardiac output can still be evaluated by

$$\dot{q} = W_e \Big/ \int_0^\infty K_R \, dt \ (= W_e/\text{area under concentration curve}) \text{ even if the instru-}$$

---

* One such experimental arrangement is described by W. Mitzner, W. Blesser, and E. Glassman, On line Correction of Dye Dilution Curves by a Compensation System, *Digest, 7th Intern. Conf. Med. Bioeng.*, 1967. See also Ref. 8 and I. J. Fox and E. H. Wood, Blood Flow Measurements, in Ref. 9.

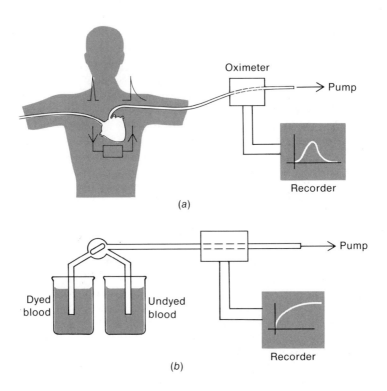

(a)

(b)

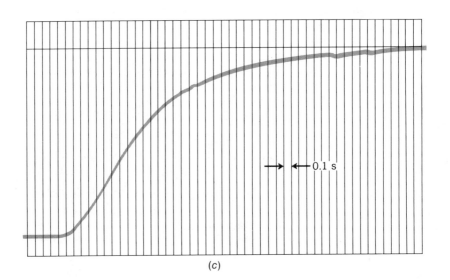

(c)

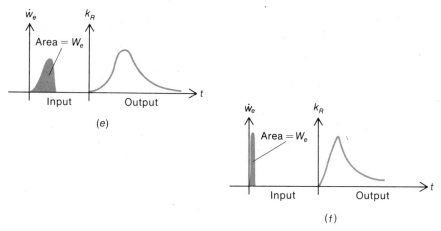

FIG. P5.6

ment time-constant is not zero. Can the response curve still be used to evaluate the physiological time-constant? How can the central volume (V_c) of the physiological system be determined?

(d) Suppose for a particular test the input injection is closely impulsive, as shown in Fig. P5.6f, and the response appears as shown in the same figure (corrected for recirculation). The cardiac output now cannot be determined by $\dot{q} = W_e \Big/ \int_0^\infty K_R \, dt$. Why not? What additional information must now be taken from the response curve to evaluate $\dot{q}$?

5.7 In Sec. 5.9 it was assumed a dye was injected into a system which had a flow of 70 cm³/s and a central volume of 100 cm³ (Fig. 5.34). The dye was introduced as a pulse injection at a rate of 15 mg/s for 1 s (Fig. 5.31). A diagrammatic sketch of this situation is shown in Fig. P5.7. The equation describing the input curve was given as

$$(5.102) \qquad \dot{w}_e = \dot{W}_{em}[u(t) - u(t - t_f)] = 15[u(t) - u(t - 1)]$$

The equation to describe the output was given as

$$(5.103) \quad k_B = \frac{\dot{W}_{em}}{\dot{q}} (1 - \epsilon^{-t/\tau})u(t) - \frac{\dot{W}_{em}}{\dot{q}} (1 - \epsilon^{-(t-t_f)/\tau})u(t - t_f)$$

$$= \frac{15}{70} (1 - \epsilon^{-0.7t})u(t) - \frac{15}{70} (1 - \epsilon^{-0.7(t-1)})u(t - 1)$$

If events just after the pulse is completed ($t = t_f = 1$) are considered, the equation for the concentration curve can be written

$$(5.106) \qquad\qquad k_B = 0.108\epsilon^{-0.7(t-1)} \qquad \text{mg/cm}^3$$

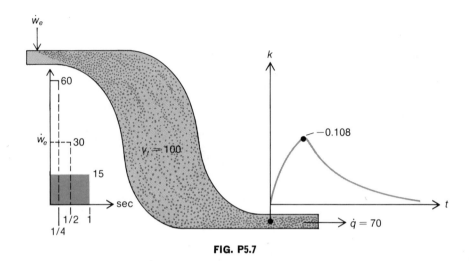

FIG. P5.7

(a) At $t = t_f = 1$, what is the concentration of the dye in the tank? How much dye (weight) is in the tank at $t = t_f = 1$?

(b) Since dye was delivered at 15 mg/s for 1 s, 15 mg of dye was delivered to the system. Yet, from part (a), the tank does not contain 15 mg of dye after the pulse injection has been completed. Why the discrepancy?

(c) Suppose dye is injected at a rate of 30 mg/s but the injection lasts only ½ s (see dashed input curve in Fig. P5.7; t_f is now ½ s). Fifteen milligrams of dye is still delivered to the system, but the response curve will now be different. Determine the response curve (k_B) for the period $t > t_f = \frac{1}{2}$. What is the concentration k_B at $t = t_f = \frac{1}{2}$? How much dye (weight) is in the tank when $t = t_f$? Why is this figure different from the figure calculated in part (a) and the weight of dye delivered?

(d) Suppose dye is injected even more rapidly but for less time (say, at 60 mg/s for ¼ s) so that the weight of dye delivered remains the same. How does this affect the results of part (c)?

(e) In general, suppose the injection rate is very high but the duration of injection is short. The input may be considered impulsive with a specific weight of dye delivered. Find the equation for the concentration curve after the impulse has been completed. What is the concentration curve at this time? How much weight of dye is in the tank? Explain.

REFERENCES

1. Armstrong, H. G.: "Principles and Practice of Aviation Medicine," The Williams & Wilkins Company, Baltimore, 1952.
2. Butler, R., and E. Kerr: "An Introduction to Numerical Methods," Pitman Publishing Corporation, New York, 1961.

3. Submarine Medicine Practice, *U.S. Navy Bur. Med. Surg.*, *Navmed P5054 P41*.
4. Groxes, G., and W. Munk: A Decompression Gauge for Divers, *Scripps Inst. Oceanog. Rept.* 53–64, Dec. 10, 1964.
5. Bradner, H., and R. Stuart MacKay: Biophysical Limitations on Deep Diving, *Bull. Math. Biophys.*, **25**:(1963).
6. Pitts, R. F.: "Physiology of the Kidney and Body Fluids," The Year Book Medical Publishers, Inc., Chicago, 1963.
7. Guyton, A.: "Textbook of Medical Physiology," 3d ed., W. B. Saunders Company, Philadelphia, 1966.
8. Wood, Earl H. (ed.): Symposium on the Use of Indicator-dilution Techniques in the Study of the Circulation, *Am. Heart Assoc. Monograph* 4, 1962.
9. Glasser, O. (ed.): "Medical Physics," The Year Book Medical Publishers, Inc., Chicago, 1960.
10. Gerd, J. A., M. D. Cropp, and A. C. Burton: Theoretical Considerations and Model Experiments on the Validity of Indicator Dilution Methods for Measurements of Variable Flow, *Circulation Res.*, **18** (1): (January, 1966).
11. Rushmer, R. F.: "Cardiovascular Dynamics," 2d ed., W. B. Saunders Company, Philadelphia, 1961.

CHAPTER SIX

THE TRANSFER FUNCTION

6.1 INTRODUCTORY COMMENTS

Lest we lose sight of the difference between reality and supposition, it is important to remember that we have been dealing with systems in which the system properties are presumed constant. These have been called "linear systems."

There is a danger that such terminology may lead to the conclusion that linear systems are a separate class of systems different from another class called nonlinear systems. This is an artificial separation not in keeping with reality. Every system is nonlinear if the system excitation is of sufficient magnitude so that changes in system properties can be easily detected. On the other hand, every system is linear if the system excitation is sufficiently small so that either system properties remain constant or the changes in these properties are too small to be detected. Thus there are not two separate system categories; there are only two viewpoints about the properties of a system.

The linear viewpoint is used when the properties of a system remain relatively constant during an investigation; the nonlinear viewpoint must be used when the constancy of system properties cannot be assumed.

One possible way to evolve a system representation is always to start with the assumption of linearity. This does not imply that it is believed the system will behave in a linear manner nor does it imply a denial of the evidence of one's senses or data by calling an obviously nonlinear output the response of a linear system. The philosophy of the linear approach is simply to enter the problem area with some semblance of organization. Linearized analysis provides this organized approach, and to that extent its utility is unquestioned. The danger of a linearized approach is that one begins to treat it as a law of nature instead of only a device of man. This, however, is not the fault of the linearized approach but is due to the shortcomings of the user who has lost sight of the limitations of his tool. If the investigator maintains contact with reality and relegates linearized thinking to its proper place, it will not come as any shock to him to find that the resistive or storage property that he evaluated in a system is not constant or that a lumped representation is too approximate as the system under investigation is intrinsically distributed.

These remarks may appear redundant and unnecessary, as they have been made before. However, it cannot be emphasized too strongly that the linear approach represents only a way to start a problem, not necessarily *the* way to solve one. It is particularly appropriate to keep in mind the utility and limitations of linearized thinking at this point, as we now shall develop some additional concepts dealing with systems and signals. These new concepts will be applicable only to situations where linearized approximations are acceptable. Although these new concepts may help in understanding a system and they may sometimes indicate how to obtain a nonlinear simulation of a system, it must be remembered they are not a panacea and cannot be considered a solution to all problems encountered in research and development.

Before any new material is introduced, we shall review the foregoing chapters briefly, so that we may take stock of our present position. This will make clear why there is a need for further development. Chapters 1 to 4 were devoted to the introduction of the system concept and system properties. The properties were defined, and examples were given to illustrate the concepts of storage and resistance separately.

In Chap. 5 we examined some simple systems with combined properties. At that point we established that certain combinations of system properties respond in a predictable and describable fashion when excited with some specific types of stimuli. We established, for example, that if a first-order system is excited by a step-function of the form $Au(t)$ it will have a response of the form $A(1 - \epsilon^{-t/\tau})$ during the excitation and of the form $A\epsilon^{-t/\tau}$ if the excitation is reduced to zero [see Eq. (5.106)]. Finally, although it was not specifically stated, we established that the response of a first-order system will be $A\epsilon^{-t/\tau}$ when the excitation is an impulse of the form $A \delta(t)$.*

Although correspondence between stimulus, system, and response was developed by applying a known stimulus to a known system (viewpoint shown by Fig. 5.2b), it was shown later that the input-output information thus obtained (i.e., the stimulus-response data) can also be used to identify an unknown system (viewpoint shown by Fig. 5.2a). From such an identification, the system can then be described mathematically and the response of the system to other stimuli can be predicted. This procedure was employed in the investigation of the nitrogen-liberation system (Secs. 5.4 and 5.5). Prior to the discussion of this system, it was established that a step applied to a first-order system yields a response $A(1 - \epsilon^{-t/\tau})$. It was then shown that the step-response of the nitrogen-liberation system was approximately represented by $A(1 - \epsilon^{-t/\tau})$. This established that the nitrogen-liberation system could be considered a first-order system governed by a first-order differential equation. Once the mathematical law that governed the system was established, its response to any arbitrary excitation could then be predicted.

The procedure outlined in the preceding paragraph is applicable to any system investigation, but it is somewhat impractical and of limited utility, as it depends upon the ability to recognize a variety of stimulus, system, and response combinations. If a standard (or known) excitation yields a recognizable output that permits an identification, one can then offer some description of the system. If not, all that can be said is that the system is not familiar. Thus, the only way such a procedure can be effective is to have available an extensive listing of signal and system combinations. This would entail testing a wide variety of known systems with a set of stimuli. To be practical, the stimuli should be a set of standard excitations (e.g., impulses, steps, etc.). The resultant responses of such tests would then have to be recorded and tabulated. This is not

---

* This relationship was not actually developed in any formal manner. Instead it was more or less implied in the discussion of the pulse-function and the impulse-function (Secs. 5.9 and 5.10). In Sec. 5.9 it was shown that after a pulse excitation the free response of a first-order system is of the form $A\epsilon^{-t/\tau}$; in Sec. 5.10 it was shown that the impulse-response of a first-order system consists entirely of a free response and that this free response has the same exponential decay as a pulse excitation. It might be intuitively concluded, therefore, that the impulse-response of a first-order system is given by $A\epsilon^{-t/\tau}$. Although this is not a rigorous proof, it can be accepted as a heuristic confirmation that this relationship exists. We shall establish this relationship more carefully in a later section.

an impossible task* but it is clumsy and must necessarily be incomplete. (It would be a hopeless task to attempt to tabulate all possible systems and property combinations.) Furthermore, it suffers one obvious drawback: It assumes a standard stimulus can be applied to all systems. There are many instances when input-output signal information for a system is available but the input is not in standard form nor can a standard stimulus be applied because of the nature of the system. (In vivo pressure-flow information is an illustration.) To define such a system one must depend entirely on an interpretation of the input-output data even though the input may not be in a suitable form. As a practical matter, then, it would be most desirable if a procedure could be developed whereby a system may be defined from signal data even though the signal stimulation cannot be varied at will. In this chapter we examine systems and signals with this viewpoint in mind.

To develop a technique whereby stimulus-response data may be used to define a system, we divide the discussion into two parts. The first is devoted to the evolution of a different form of system representation. This form complements the differential-equation and circuit representations, and it has an added advantage as it provides a more direct link between signal information and system configuration. The second part of the discussion is devoted to an analytic description of signal information which allows signals to be used directly for a quantitative definition of a system. When we have completed these appointed tasks, we should be able to develop a quantitative description of the block T in Fig. 6.1 from the given signal information. In particular, we hope to be able to describe the excitation signal $6\epsilon^{-5t}$ and the response signal

---

* Short tables of transient-response curves (i.e., impulse- and step-responses) for various systems can be found in the literature. See, for example, Ref. 1, pp. 1–24, 25; also Ref. 2, p. 48, table 2-1.

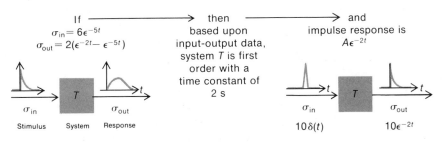

If ⟶ then ⟶ and
$\sigma_{in} = 6\epsilon^{-5t}$
$\sigma_{out} = 2(\epsilon^{-2t} - \epsilon^{-5t})$
based upon input-output data, system T is first order with a time constant of 2 s
impulse response is $A\epsilon^{-2t}$

σ_{in} T σ_{out}

Stimulus   System   Response

$10\,\delta(t)$ σ_{in} T σ_{out} $10\epsilon^{-2t}$

(a) Response to one input   (b) Response to another input

FIG. 6.1 *Illustrative example of the use of input-output-signal information.*

$2(\epsilon^{-2t} - \epsilon^{-5t})$ so that they may be used to establish that the block T may be represented by a simple RC circuit with a time-constant of $\frac{1}{2}$ s ($RC = \frac{1}{2}$) and that this system will have a response $10\epsilon^{-2t}$ when the excitation is impulsive of the form $10\delta(t)$. It will be seen that such a system description is possible; when the block is so described it is generally called a transfer function. The significance of this term is discussed in the next and later sections.

6.2 SYSTEMS AS OPERATORS

Thus far we have interpreted the block diagram in Fig. 6.1 from a primarily mechanical or physical point of view. By this interpretation we visualize the block as an enclosure containing the system; the block and the system are thus imagined to be physically equivalent. A stimulus is applied to the system (block), an energy interchange takes place within the system, and the system delivers a response signal that indicates some internal functions or operations have taken place.

There is an alternative interpretation of the block diagram, however, which is particularly useful from a mathematical point of view. In this interpretation the block is considered a "signal modifier." A signal enters the block, is modified by the block, and exits from the block in altered form. This viewpoint implies that the block functions as a transmission or transfer agent which transfers the signal from input to output. In so doing, the block "operates" upon the signal to alter its characteristics. In this context, the term "operates" is used in the sense of "acts upon." Because of this "transfer-operate" feature, the block is often described as a transfer function.

By the interpretation put forth in the preceding paragraph, we now visualize the block as a collection of operators. These act upon the input signal and convert it from one form to another. The output is then considered a modified version of the input. The way in which a particular input is converted to a particular output is a reflection of the kind of operator functions performed by the block. As the block has also been associated with the system and its related properties, the system properties may be expressed as an operator function. In short, a system may be considered a particular collection of properties that operate on stimulus waveforms to convert them to appropriate response waveforms. The operations required to convert a particular stimulus to a particular response is a reflection of the properties associated with the system. We can now assess the twofold task before us: On the one hand, we must learn how to represent systems as signal modifiers or operators; on the other hand, we must learn how signals are related to the "operation functions." It will be seen that both concepts may be incorporated into a mathematical expression which may then be used to describe the "transfer function" of the block.

6.3 SOME OPERATIONAL NOTATIONS

To gain some experience with the notation and concepts associated with operational representation, we examine first some very simple relationships between input and output signals. Some examples of simply related signal-pairs are shown in Fig. 6.2. In Fig. 6.2a it is relatively easy to see that the signal-pair are probably related by a multiplicative constant. The output signal (σ_{out}) at any instant is a duplication of the input

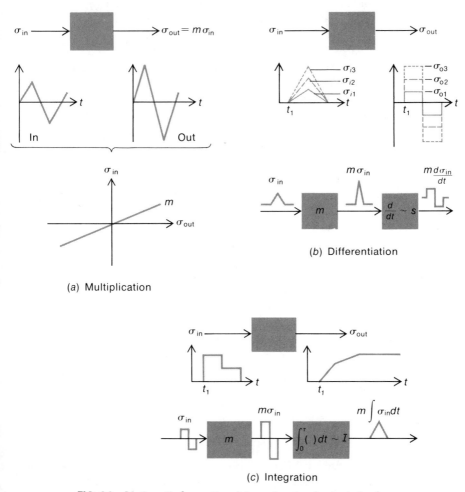

(a) Multiplication

(b) Differentiation

(c) Integration

FIG. 6.2 *Mathematical operations between input and output signals.*

except for a magnitude change. One can easily verify this assertion by plotting σ_{in} against σ_{out} (with time removed), as shown in Fig. 6.2a. A straight-line plot indicates that the signals are indeed proportional. The slope of the line represents the constant of proportionality. In this case, therefore, the block is designated as a multiplicative operation. If it can be established that a block represents a multiplicative constant, then the block represents a system with only one system property. The specific property may be *either* storage *or* resistance, depending upon the character of the measured variables. Thus, if in Fig. 6.2a σ_{in} is pressure (p) and σ_{out} is flow ($\dot{q}$), the system property is resistive (that is, $\dot{q} = mp$; m is evidently a proportionality constant with units of conductance). On the other hand, if σ_{in} is pressure (p) and σ_{out} is volume (q), then the system property is storage. From previous experience, it should be relatively easy to establish the validity of these statements.

In some cases the mathematical relationship between input and output may still be relatively simple though not as simple as pure multiplication. In Fig. 6.2b, for example, we see immediately that the block does not act as a multiplicative constant. At first glance it may be difficult to understand exactly what the block does do. A little careful thought and observation, however, indicate that, as the slope of the input signal increases, the amplitude of the output signal increases. It would appear, then, that the block acts or operates as a "differentiator"; the block delivers an output signal (σ_{out}) that is proportional to the derivative of the input signal (σ_{in}). If the output signal is exactly equal to the derivative of the input, the constant of proportionality is unity and the block acts as a pure differentiator. If the output is the derivative of the input to within a multiplicative constant, then the block acts as a multiplier as well as a differentiator. The single block might then be represented by two blocks to indicate the two mathematical operations involved. One block represents the multiplicative operation as indicated by the constant multiplier m; the other represents the derivative operator as indicated by the d/dt notation. The d/dt notation proposed by Leibniz to designate the operation of differentiation is still often used in most introductory texts. However, other notations have also gained popularity, and differential operators are now often designated as D, λ, p, s, and a variety of others. Thus $d\sigma_{in}/dt$, $D\sigma_{in}$, $\lambda\sigma_{in}$, etc., may all be equivalent representations, depending upon the text. At this point in the development we shall use the letter s to represent the differential operator; then, $s\sigma_{in}$ is interpreted as $d\sigma_{in}/dt$. (It will be seen later that s has added significance.) The block labeled s in Fig. 6.2b is therefore interpreted as a system that performs a differential operation. The signal delivered from the block is the derivative of the signal delivered to the block.

Figure 6.2c shows a situation similar to the one just described. In this case, however, the block acts as an integrator in the sense that the output signal at any instant is equivalent to (or proportional to) the area under the input curve. If the output curve is identical to the integral of the input curve, the constant of proportionality is unity. If the output and the integral of the input are equivalent to each other to within a multiplicative constant, the block may be described as a multiplier and an integrator and may be represented by two blocks, as shown in Fig. 6.2c. The block labeled m is the multiplicative operator, and the one labeled I is the integral operator.*

If it can be established that a system may be represented by an integral or differential block in tandem (or in series) with a multiplicative block, the system has only one property. Once again, however, it must be pointed out that no specific statements can be made about the type of system property until the character of the measured variables has been checked. If, for example, σ_{in} in Fig. 6.2b is volume (q) and σ_{out} is pressure (p), the system is purely resistive [that is, $p = m(dq/dt) = msq$ or $p = m\dot{q}$; therefore m is a resistive coefficient]. In a similar fashion, if, in Fig. 6.2c, σ_{in} is flow ($\dot{q}$) and σ_{out} is pressure (p), then the block diagram represents a storage property [that is, $p = m \int_0^t \dot{p} \, dt = mI\dot{q} = mq$; therefore m is a storage coefficient]. These results may not appear to be in agreement with our previous definitions of resistance and storage. In the first case, the ratio of the variables (p/q) seems to define a storage property, yet the system is resistive. In the second case, the ratio of the variables ($p/\dot{q}$) seems to define a resistive property, yet the system is one of pure storage.

This discrepancy may be explained by noting that the definitions for system properties were based upon the ratios of static or steady measurements. In Fig. 6.2b and c both signals are not static (or time-invariant). We cannot form the proper ratio and therefore cannot define the system properties directly. This would seem reasonable until we note that in Fig. 6.2a the signal ratio defined a system property even though the signals were not static. Apparently there is an inconsistency, and we must revise our definition for a system property so that these new situations may be included. The revision is simple: We remove the condition that the signals must be constant and instead now state that if the ratio

---

* As a temporary notation we shall use the symbol I instead of $\int_0^t (\ ) \, dt$ to represent the integral operation. We interpret I simply as an abbreviation of $\int_0^t (\ ) \, dt$, and so $\int_0^t x \, dt$ and Ix have identical meanings. The reasons for this temporary notation will become apparent later.

of two variables is time-independent then this ratio may be used to define a system property. The signals in Fig. 6.2a can form a time-independent ratio; this ratio can define a system property. The units of the ratio govern the type of property. Since the input and output signals of Fig. 6.2b and c do not form a time-independent ratio, these signals cannot, in their present form, be used to define system properties. However, by appropriate integration or differentiation we can obtain signals that form time-independent ratios. These signals can then be used to define system properties. The type of system property defined is, of course, governed by the units of the ratio. This revised definition is useful for many cases, although not for all, as it is not always possible to convert data to an appropriate form. In such cases one must resort to a deeper analysis to identify the system properties. We shall call attention to these situations when they arise.

6.4 SIMPLE OPERATIONS COMPARED WITH THE OPERATION OF A FIRST–ORDER SYSTEM

When a system contains a single property, the input and output signals are simply related to each other. In particular, if the system is purely resistive or purely storage, the output signal is related to the input signal by a multiplicative constant and/or integration or differentiation. The relationship between the signals is not so clear-cut, however, when a system has both resistive and storage properties. In first-order systems input and output are not related by the product of simple operations alone (e.g., multiplication and differentiation); the relationship also includes a summation of operations.

To illustrate this point, we once again examine the response of a simple first-order system. For variety, we shall use an instrument for illustration rather than a physiological system. Figure 6.3 shows the response of a simple mercury-in-glass thermometer (such as a clinical thermometer) when it is immersed in a bath of hot water. The excitation in this case is a step temperature change; the response is the temperature reading. As can be seen, the response closely resembles the step-response of a first-order system. From previous experience we know that this system must have properties that can be represented by a storage and a resistive element (see Prob. 6.1). We can readily appreciate that input and output are not proportionately related by multiplication, integration, or differentiation by using two approaches: (1) We can show such individual operations do not yield results that correspond to the actual result; (2) we can determine the operations that govern the relationship between the input and output.

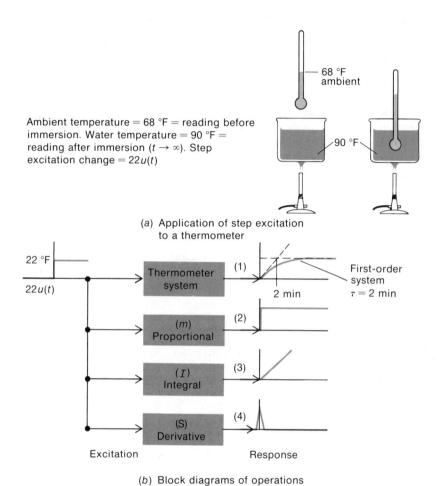

Ambient temperature = 68 °F = reading before immersion. Water temperature = 90 °F = reading after immersion ($t \to \infty$). Step excitation change = $22u(t)$

(a) Application of step excitation to a thermometer

(b) Block diagrams of operations on a step function

FIG. 6.3 *Effect of simple operations compared with that of a first-order system.*

The first procedure is straightforward; we need only multiply, integrate, and differentiate the input and compare the resultant curves with the output obtained from the physical system. The results are shown in Fig. 6.3*b*. The first curve in Fig. 6.3*b* is the actual step-response of the thermometer system. The second curve is the response that would be expected from a simple multiplicative operation. The output is simply a magnified (or attenuated) replica of the input. The third curve is the effect of an integral operator. The step-function is converted to a "ramp-function" which continuously rises at constant slope [output = I $22u(t) = \int_0^t 22u(t)\ dt = 22tu(t)$; for $t > 0$, $u(t) = 1$ and output is $22t$]. The last curve would be the response if the system were a differential operator. The output would rise instantaneously and would drop back to zero immediately thereafter, giving rise to an impulsive output.* None of these curves is a duplication of the response curve of the actual system. Therefore the system does not act as a simple multiplier, integrator, or differentiator.

To use the second approach, we construct the differential equation that governs the system. For this situation the relationship between stimulus and response is given by

$$(6.1) \qquad\qquad \tau \frac{d\sigma_{\text{out}}}{dt} + \sigma_{\text{out}} = \sigma_{\text{in}}$$

where σ_{in} is the temperature-excitation signal applied to the instrument, σ_{out} is the change of thermometer reading at any instant, and τ is the system time-constant (which is related to the thermal capacity and thermal resistance of the instrument; see Prob. 6.1.) The relationship stated by Eq. (6.1) may be developed by the procedure outlined in Sec. 5.5. If the response is assumed to be a first-order step-response, then from the plot in Fig. 6.3 we approximate σ_{out} to be

$$(6.2) \qquad\qquad \sigma_{\text{out}} = 22°\text{F}\ (1 - \epsilon^{-t/\tau})$$

By differentiation

$$(6.3) \qquad\qquad \tau \frac{d\sigma_{\text{out}}}{dt} = 22\epsilon^{-t/\tau}$$

When Eq. (6.2) is added to Eq. (6.3) the result is Eq. (6.1). As can be seen, the output and input are related by a summation of operational functions. The derivative of the output (with an appropriate coefficient) added to the output is equivalent to the input. Evidently, therefore, we

---

* The derivative of a curve is the slope at every point. The derivative of a step therefore is zero immediately before and after the step-rise (the slope of the curve is zero at these times). During the step-rise the slope is very high; the derivative at this time must therefore also be very high. The derivative of a step thus rises sharply from zero and then drops immediately back to zero. Evidently the derivative of a step is an impulse; the area under the impulse (i.e., the integral of the impulse) is the magnitude of the step. The foregoing statements may be concisely stated in mathematical form as follows:

$$s[22u(t)] = \frac{d}{dt}\,22[u(t)] = \frac{d(22)}{dt}\,u(t) + 22\,\frac{du(t)}{dt} = 0 + 22\,\delta(t)$$

where $\delta(t)$ is the symbol for the impulse-function (see Sec. 5.10). These ideas will be encountered again and discussed in a later section.

could not expect to derive the output from the input by multiplication, integration, or differentiation alone.

It is of interest to note, however, that for limited time intervals under some circumstances it may be assumed that a first-order system does in actuality act as an integrator. If we examine the response of the thermometer system just discussed, for example, we note that, in the interval $0 < t < \tau/2$, the response rises almost linearly (see Fig. 6.3b, curve 1). We might, in fact, describe this portion of the response as a straight line with slope $22°F/\tau$ (see footnote on page 177). In this range the output could thus be described by

$$(6.4) \qquad \sigma_{\text{out}} = \left(\frac{22}{\tau}\right) t$$

Since the input is a constant and the output (for the range considered) is a ramp, the system could be considered an integrator with a multiplicative constant $1/\tau$ (see Fig. 6.3b, curve 3).

The integrator properties of a first-order system can be established directly from the differential equation. If Eq. (6.1) is rewritten,

$$(6.5) \qquad \tau \frac{d\sigma_{\text{out}}}{dt} = \sigma_{\text{in}} - \sigma_{\text{out}}$$

If the observation time is very much less than the time-constant $(t < \tau)$ then σ_{out} never becomes very large, and it may be neglected compared with σ_{in}. Therefore

$$(6.6) \qquad \tau \frac{d\sigma_{\text{out}}}{dt} \approx \sigma_{\text{in}}$$

or

$$(6.7) \qquad \sigma_{\text{out}} \approx \frac{\int_0^t \sigma_{\text{in}} \, dt}{\tau}$$

If σ_{in} is a step of $22°F$ then Eq. (6.7) reduces to Eq. (6.4). We might extend this concept and state that if the time-constant of a first-order system is large compared with the time scale of the measured response, the system may act as an integrator. Electrical integrators can thus be constructed by using large RC values in a first-order network (see Prob. 6.2).

6.5 THE FIRST–ORDER OPERATOR

From the discussion in the preceding section we can now readily see that, when a system contains two properties, the input and output signals are

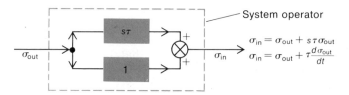

FIG. 6.4 *Operational diagram of a first-order system.*

not related by the product of operational functions alone (e.g., multiplication and differentiation) but are also related by the summation of operational functions. In this particular system (a first-order system) the signals are related so that the output (σ_{out}) added to its derivative [multiplied by an appropriate factor, that is, $\tau(d\sigma_{out}/dt)$] is equivalent to the input [Eq. (6.1)]. We can translate this statement into block-diagram form as shown in Fig. 6.4 and could, in fact, consider the phantom block surrounding the single operational blocks to be a description of a first-order operator. The differential equation that governs the operation of the first-order system can thus be used to define a block diagram that relates output and input. Unfortunately, however, the diagram does not present the information in a convenient form. We should like to see how to operate on the input to get the output; the differential equation shows how we must operate on the output to get the input.

To convert the differential equation (and the block diagram of Fig. 6.4) into a form that displays the information in a more convenient fashion, we begin by examining the first-order system when the time-constant is "large." In Sec. 6.4 it was seen that under such circumstances the first-order system can act as an integrator. The block diagram of the system can then be represented as shown in Fig. 6.5. In this diagram input and output are oriented in the desired manner. The input enters a block (an approximation of the system); the block operates on the input to deliver the appropriate output.

The representation in Fig. 6.5 is acceptable as long as the time of observation is small compared with the time-constant (that is, $t_0 \ll \tau$); as the time of observation increases, the approximation becomes progressively poorer. To illustrate: If a first-order system [e.g., Eq. (6.1) or

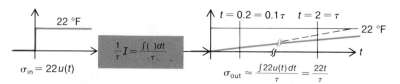

FIG. 6.5 *Representation of a first-order system when $0 < t \ll \tau$.*

(6.5)] has a time-constant of 2 min and the input is a step of magnitude 22, then after 0.2 min ($t = 0.1\tau$) the output calculated from the exact solution [Eq. (6.2)] would be $22(1 - \epsilon^{-0.1})$ or 2.09. The approximate solution as determined from Eq. (6.7) (or Fig. 6.5) would be 2.2. The correspondence between the two solutions is good (5 percent). On the other hand, if t is 2 min ($t = \tau$), the exact solution is 13.9 and the approximate solution is 22. The disparity between exact and approximate solutions is significant (over 35 percent error).

In order to retain the format presented by Fig. 6.5 (namely, the input is converted to the output by some "operator") and yet be accurate over an extended period, the diagram must be modified. To determine what modifications should be included, we compare the actual and approximate differential equations of the system as given by Eqs. (6.5) and (6.6). The validity of Eq. (6.6) is based upon the assumption that, in the range $0 < t < \tau$, σ_{out} remains approximately constant at zero. Thus σ_{out} may be omitted from Eq. (6.5) and the result is (6.6). For the particular situation presented in the preceding paragraph, this approximation is very good for the interval $0 < t < 0.2$ min. We could probably use this same approximation for a second time period $0.2 < t < 0.4$, and we would probably still get very good results. However, the accuracy obtained during the second time interval could not be expected to be as good as that obtained during the first interval.

The reason is self-evident: During the first time interval σ_{out} starts at zero and grows to 2.09; during the second interval σ_{out} starts at 2.09 and grows to 3.988 [as obtained by solving Eq. (6.2); the graphical results are shown by the solid curve C in Fig. 6.6]. It is obvious that σ_{out} does not remain zero (or constant) for either interval and that an error must be generated by such an assumption. However, the assumption that σ_{out} is zero is closer to the truth for the first interval than it is for the second.

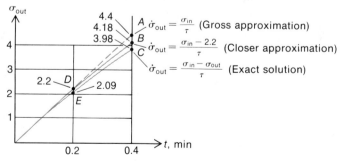

FIG. 6.6 *Comparison of exact and approximate solutions for step-response of a first-order system with a time-constant (τ) of 2 min.*

We would expect, therefore, that error generated by this approximation would be less during the first period than during the second period. We can easily demonstrate that this is indeed so by a graphical comparison between the approximate solution and the exact solution for the two time periods (see Fig. 6.6). When σ_{out} in Eq. (6.5) is assumed to be zero, the error at the end of the first period is given by the vertical line DE; the error at the end of the second period is given by the line AC. Clearly the error grows continuously.

From the previous discussion it is now easy to see how we may reduce the error developed by the first-order representation of Fig. 6.6. Instead of assuming that σ_{out} in Eq. (6.5) is always constant at zero, we should change the assumed constant for each period. The new constant could be the maximum value attained during the previous period. We can use the example of the previous discussion to illustrate the procedure. It was seen that when the approximation given by Eq. (6.6) was used [that is, σ_{out} in Eq. (6.5) was zero] σ_{out} attained a value of 2.2 when t was 0.2 min (an error of about 5 percent). If we continue to use Eq. (6.6) for a second time interval we incur an error of approximately 10 percent. To reduce this error we use a new approximation for Eq. (6.5) and now assume that σ_{out} for the period $0.2 < t < 0.4$ is not constant at zero but constant at $2.2°$. Under these circumstances, the approximation for Eq. (6.5) becomes

(6.8) $\qquad \tau \dfrac{d\sigma_{out}}{dt} \approx \sigma_{in} - 2.2 \qquad\qquad 0.2 < t < 0.4$

or

(6.9)* $\qquad \sigma_{out}\Big]_{0.2}^{t} \approx \dfrac{\displaystyle\int_{0.2}^{t} (\sigma_{in} - 2.2)\, dt}{\tau} \qquad 0.2 < t < 0.4$

By using this new approximation for the period $0.2 < t < 0.4$ we find that σ_{out} increased approximately 1.98 during this interval. Thus σ_{out} at 0.4 min is evaluated to be $2.2 + 1.98$ or 4.18. The exact value of σ_{out} at 0.4 min is 3.98 [as calculated from Eq. (6.2)]. The new approximation thus yields a result that is 5 percent in error (as compared with the 10

* The expression for σ_{out} during the period $0.2 < t < 0.4$ could be written

$$\sigma_{out}(0.2 < t < 0.4) = \sigma_{out}(0.2) + \dfrac{\displaystyle\int_{0.2}^{t} (\sigma_{in} - 2.2)\, dt}{\tau}$$

or

$$\sigma_{out}(0.2 < t < 0.4) = 2.2 + \dfrac{\displaystyle\int_{0.2}^{t} (22 - 2.2)\, dt}{\tau} = 9.9t\,\Big]_{0.2}^{t}$$

$$= 0.22 + 9.9t$$

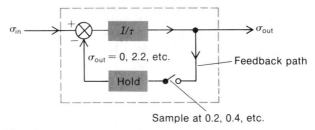

FIG. 6.7 *Block diagram to approximate the step-response of a first-order system.*

percent error incurred when σ_{out} was assumed zero). In the graphical comparison of Fig. 6.6a the error at $t = 0.4$ min is given by line AB when σ_{out} is assumed to be zero for the interval $0 < t < 0.4$. The error is given by line BC when we assume σ_{out} is zero for $0 < t < 0.2$ and 2.2 for $0.2 < t < 0.4$.

We have established that we can retain the representation proposed by Fig. 6.5 without error buildup if we modify the diagram appropriately. The procedure is simply to deduct a suitable constant from the input at the proper time. A schematic diagram of one possible arrangement is shown in Fig. 6.7. The dashed block represents the overall operational function performed by the first-order system; the internal blocks and symbols represent the individual operations performed within the first-order block when the system operates on the input to yield an output (Figs. 6.7 and 6.4 should be compared). To explain the block diagram of Fig. 6.7 we note that σ_{in} is delivered to one side of an adder. The output of the adder ($\sigma_{in} - \sigma_{out}$) thus represents an approximation of the right side of Eq. (6.5). This output is then delivered to the operator block I/τ; the output of the block is the desired system output.

To obtain the proper constants for σ_{out} a "feedback path" with a "sample-and-hold" circuit is included in the diagram. The reason for this arrangement should be evident: The sample switch samples the output at discrete instants of time (for the previous illustrative example these instants would be $t = 0, 0.2, 0.4$, etc., min). The "hold" circuit registers and retains (memorizes) the last value obtained by the sampling switch. To illustrate the operation, we again use the previous example: If σ_{in} is 22 then, at $t = 0$, σ_{out} is zero and the output of the I/τ block is $(22/2)t = 11t$. At $t = 0.2$ the sample switch takes a reading and the hold circuit memorizes the value $(22/2)(0.2)$ or 2.2. The signal delivered to the I/τ block now becomes $22-2.2$ or 19.8. The signal delivered from the block is $2.2 + (19.8/2)T = 2.2 + 9.9T$ (where T is the time after 0.2 min, that is, $T = t - 0.2$, $0.2 < t < 0.4$). At $t = 0.4$ min the output attains a value of 4.18. The sampler transmits this value to the hold circuit so that the adder now delivers $22 - 4.18 = 17.92$ to the integration block, and so the cycle continues. The process outlined here may be compared with that presented by the computer block diagram proposed in the footnote on page 167 [see (Eq. 5.16)]. It is easily seen that the processes depicted are essentially the same. The sampled-data mechanism (the sample-and-hold circuit) serves to supply the digital information required by the diagram presented earlier. The diagram in Fig. 6.7 is, in fact, sometimes called a sampled-data computer system or a sampled-data control system.

The diagram in Fig. 6.7 shows schematically that the first-order system may be considered a composite of various operator functions. When these operators are appropriately assembled, the result is an overall operator function which converts applied stimuli to the response expected of a first-order system. This does not yet serve any useful purpose except to establish conceptually that it may be possible to describe a first-order system by a mathematical operator that properly converts an input to an output. To apply this schematic representation of an operator function, we first improve the approximation so that the system can deliver more accurate results. The procedure is fairly straightforward; the sampling rate is increased by decreasing the sampling interval. For the previous example, a sample would be taken, say, every 0.1 min rather than every 0.2 min. It should be obvious that this procedure tends to decrease the error between the response generated by the system in Fig. 6.7 and the exact response as given by Eq. (6.2). The system will evidently generate an exact response when the sampling rate approaches infinity. Under these circumstances the sample-and-hold circuits continuously deliver the output signal, and the diagram of Fig. 6.7 reduces to the single-loop feedback system shown in Fig. 6.8 (see Prob. 5.4). Feedback systems will be discussed at greater length in Chap. 10. Here we simply note that a first-order system may be represented by a single-loop feedback system. This does not mean that a first-order system must be a feedback system. It does, however, imply that a first-order system (and indeed any system) may be represented by or may actually be constructed as a feedback system.

We can use the conceptual implications of Fig. 6.8 in conjunction

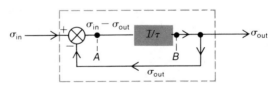

FIG. 6.8 *Reduction of the sampled-data system to a single-loop feedback system.*

with Eq. (6.5) to develop a mathematical notation to represent a first-order operator. Equation (6.5) may be rewritten as

$$(6.10) \qquad \sigma_{\text{out}} = \frac{1}{\tau} \int (\sigma_{\text{in}} - \sigma_{\text{out}}) \, dt = \frac{I}{\tau} (\sigma_{\text{in}} - \sigma_{\text{out}})$$

This equation may be converted to a recursion relationship by simple resubstitution of σ_{out}. The left side of Eq. (6.10) is used as a substitution for the σ_{out} which appears on the right side. (Physically this may be compared to the operation of the models in Figs. 6.7 and 6.8 when the output is fed back to modify the input.) After one such substitution, Eq. (6.10) becomes

$$(6.11) \qquad \sigma_{\text{out}} = \frac{I}{\tau} \left[\sigma_{\text{in}} - \frac{I}{\tau} (\sigma_{\text{in}} - \sigma_{\text{out}}) \right] = \frac{I}{\tau} \sigma_{\text{in}} \left(\frac{I}{\tau}\right)^2 \sigma_{\text{in}} + \left(\frac{I}{\tau}\right)^2 \sigma_{\text{out}}$$

where $(I/\tau)^{2}*$ is interpreted as two successive integrals of σ_{in}; that is,

$$(6.12) \qquad \left(\frac{I}{\tau}\right)^2 \sigma_{\text{in}} = \frac{\int_0^t \int_0^t \sigma_{\text{in}} \, (dt)^2}{\tau^2}$$

After a second substitution the relationship becomes

$$(6.13) \qquad \sigma_{\text{out}} = \frac{I}{\tau} [\sigma_{\text{in}} - \text{Eq. } (6.11)]$$

$$\sigma_{\text{out}} = \frac{I}{\tau} \left[\sigma_{\text{in}} - \frac{I}{\tau} \sigma_{\text{in}} + \left(\frac{I}{\tau}\right)^2 \sigma_{\text{in}} - \left(\frac{I}{\tau}\right)^2 \sigma_{\text{out}} \right]$$

$$\sigma_{\text{out}} = \left[\frac{I}{\tau} - \left(\frac{I}{\tau}\right)^2 + \left(\frac{I}{\tau}\right)^3 \right] \sigma_{\text{in}} - \left(\frac{I}{\tau}\right)^3 \sigma_{\text{out}}$$

* The use of $(I/\tau)^2$ to represent two successive integrations is quite unambiguous. It may be compared to the use of $D^2(\ )$ or $s^2(\ )$ to represent $d^2(\ )/(dt)^2$ (i.e., two successive differentiations of a quantity). The shorthand notation is derived by assuming operator symbols may be treated as algebraic quantities and the fundamental laws of algebra may be applied to these symbols. A justification of this procedure for differential operators is presented in many of the older mathematics texts (see, for example, H. W. Reddick and F. H. Miller, "Advanced Mathematics for Engineers," 2d ed., p. 40, John Wiley & Sons, Inc., New York, 1947). The author does not, however, know of a formal presentation where in integral operators may also be handled in this fashion (although in Reddick and Miller, *op. cit.*, chap. XI, integral operators, designated as Q, are introduced and are manipulated as algebraic quantities). We justify our procedure here by noting that algebraic simplification of an operator function is accepted only to the extent that it provides an abbreviated representation of a longer operational form. The shorthand notation for the operator can always be expanded again to produce the original operator function if we wish to perform the actual mathematical operations.

The final results which we obtain (and those developed later) can be obtained in a more rigorous fashion by Laplace-transform techniques. Although such a development is more formal and exacting, it is also more abstract. The approach offered here lacks the rigor and elegance of the formal development but it maintains a closer contact with the physical-system concept.

The final value of σ_{out} at any instant would be the result after n substitutions as n becomes infinite. We would then have

$$(6.14) \quad \sigma_{\text{out}} = \left[\frac{I}{\tau} - \left(\frac{I}{\tau}\right)^2 + \left(\frac{I}{\tau}\right)^3 - \cdots \left(\frac{I}{\tau}\right)^n\right] \sigma_{\text{in}} - \left(\frac{I}{\tau}\right)^n \sigma_{\text{out}}$$

It can be shown that the last term of this expression vanishes as $n \to \infty$,[*] so that Eq. (6.14) can now be written

$$(6.15) \quad \sigma_{\text{out}} = \lim_{n \to \infty} \left[\frac{I}{\tau} - \left(\frac{I}{\tau}\right)^2 + \left(\frac{I}{\tau}\right)^3 - \cdots \left(\frac{I}{\tau}\right)^n\right] \sigma_{\text{in}}$$

The bracketed expression in Eq. (6.15) may be compared to an infinite geometric series with I/τ as the common ratio. If the operator $(I/\tau)^n$ causes a function to vanish when n becomes infinite, then I/τ behaves as an algebraic fraction (see last footnote). The series in Eq. (6.15) may thus be compared to a geometric series with a common ratio less than unity. This series can be expressed in more compact form by using the equation for the sum of an infinite geometric series with a common ratio less than unity. Equation (6.15) then becomes

$$(6.16) \quad \sigma_{\text{out}} = \left[\frac{I/\tau}{1 + I/\tau}\right] \sigma_{\text{in}}$$

Here the only significance of the more compact notation of Eq. (6.16) is that it may be expanded into the more meaningful notation of Eq. (6.15). However, Eq. (6.15) was developed as a result of the algebraic manipulations indicated by the block diagram of Fig. 6.8. Equation (6.16) must therefore represent a shorthand notation of the mathematical manipulations indicated by Fig. 6.8. If we compare the implications of the block diagram in Fig. 6.8 and the statement of Eq. (6.16), we see that the bracketed term in Eq. (6.16) represents the operations performed by the

---

[*] The validity of this contention can be illustrated by some simple examples. If σ_{out} were t^a, where a is some finite number, then $\int_0^t t^a \, dt = t^{a+1}/(a+1)$. After n integrations this becomes $t^{a+n}/(a+n)!$. The limit of this expression as $n \to \infty$ is zero. Therefore the limit of $(I/\tau)^n t^a$ is zero as n becomes infinite [$\lim_{n \to \infty} (I/\tau)^n \sigma_{\text{out}} = \lim_{n \to \infty} (I/\tau) t^a = 0$]. In a similar manner, if σ_{out} were ϵ^{-at} (a finite), after a first integration between the limits 0 and t we have $(1 - \epsilon^{-at})/a$. After "n infinite integrations" the limit of the resultant expression is zero. This cannot be considered a general proof that $\lim_{n \to \infty} (I/\tau)^n \sigma_{\text{out}} = 0$. However, this and other considerations (i.e., a knowledge of the "correct" final result) permit acceptance that the limit of the expression is indeed zero.

Since $(I/\tau)^n$ operating on a term causes the term to disappear as $n \to \infty$, it appears the operator I/τ acts as a fractional number (i.e., a number less than unity). Any fraction raised to an infinite power is zero; if $1/A$ is a fraction and ϕ is any finite function, then $\lim_{n \to \infty} (1/A)^n \phi = 0$. Since $\lim_{n \to \infty} (I/\tau)^n = 0$, we see I/τ may be thought of as an algebraic fraction. We use this concept in the development of a final operator function.

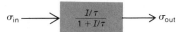

FIG. 6.9   *Transfer-function representation of a first-order system.*

phantom block in Fig. 6.8. In effect, then, the feedback representation of a first-order system included within the phantom block may be replaced by a single block, as shown in Fig. 6.9. In short, therefore, the bracketed term in Eq. (6.16) (and the labeled block of Fig. 6.9) may now be identified as one form of a first-order operator function or a first-order transfer function. This transfer function operates on the stimulating signal to deliver a first-order response.

We can easily illustrate that the representation of Fig. 6.9 gives correct results by using a stimulus that yields a known response. It is known, for example, that a first-order system will deliver a decaying exponential when stimulated by an impulsive excitation. If an impulsive excitation is applied to the system of Fig. 6.9 then σ_{in} is given by $A\delta(t)$, and the output is written

$$\text{(6.17)} \qquad \sigma_{out} = \frac{I/\tau}{1 + I/\tau} A\,\delta(t)$$

If we expand the transfer function into its equivalent series and let each term operate on the input we then have

$$\text{(6.18)} \qquad \sigma_{out} = \frac{I}{\tau} A\,\delta(t) - \left(\frac{I}{\tau}\right)^2 A\,\delta(t) + \cdots \left(\frac{I}{\tau}\right)^n A\,\delta(t)$$

The first term on the right side of Eq. (6.18) is the integral of an impulse-function with area A. This integral is the area under the impulse curve (see footnotes in Sec. 6.4 and Sec. 5.10). This first term therefore is A/τ. The second term on the right side of Eq. (6.18) is the second integral of the impulse; this is evidently the integral of A/τ or $(A/\tau^2)t$. By performing the successive integrations as indicated, Eq. (6.18) can be written

$$\text{(6.19)} \qquad \sigma_{out} = \frac{A}{\tau} - \frac{A}{\tau^2}t + \frac{A}{\tau^3}\frac{t^2}{2!} - \frac{A}{\tau^4}\frac{t^3}{3!} + \cdots$$

or

$$\sigma_{out} = \frac{A}{\tau}\left[1 - \frac{t}{\tau} + \frac{1}{2!}\left(\frac{t}{\tau}\right)^2 - \frac{1}{3!}\left(\frac{t}{\tau}\right)^3 + \cdots\right]$$

The bracketed term in the last expression is the series expansion for $\epsilon^{-t/\tau}$ so that the final result may be written

$$\text{(6.20)} \qquad \sigma_{out} = \frac{A}{\tau}\,\epsilon^{-t/\tau}$$

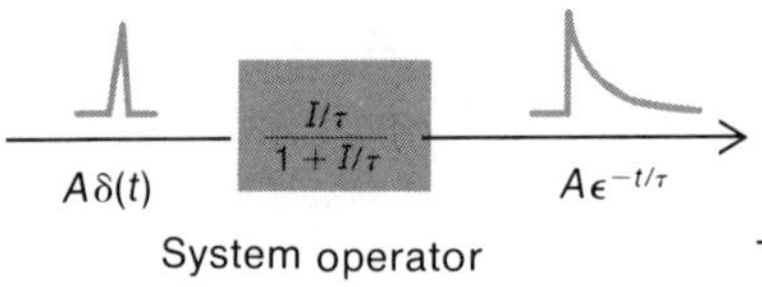

System operator   Transfer function   Impulse response

$$\left[\frac{I/\tau}{1+I/\tau}\right] A\delta(t) = A\left[\epsilon^{-t/\tau}\right]$$

(a) Block diagram   (b) Transform pair

FIG. 6.10   *Impulse-response of first-order system.*

The result given by Eq. (6.20) is the correct response of a first-order system stimulated by an impulsive excitation.

The first-order operator (or transfer function) developed in this section has far greater significance than has been implied. It will be seen later that this first-order transfer function forms the foundation for the development of transfer functions of any linearized system and leads to the description of system transfer functions through an examination of stimulus-response data. Here, however, the first-order transfer function is just one form of an analytic description of a first-order system. We use this description to indicate the properties of a block (as a constant G might indicate the multiplicative properties of a block) and to indicate how the stimulus and response of a first-order system are related.

In Fig. 6.10, therefore, the first-order operator automatically implies that the system will deliver an exponential response when an impulsive excitation is applied. The equation accompanying the block diagram describes how the transfer function operates on the impulsive input to produce the output. The transfer function $[(I/\tau)/1 + I/\tau]$ and the impulse-response are generally called a transform-pair. In most texts such transform-pairs are tabulated in terms of differential operators (s) rather than integral operators (I). In order to convert the integral-operator function to an expression in s we must examine how integral and differential operators are related. This relationship is developed in the next section.

6.6 OPERATOR FUNCTIONS IN TERMS OF s

Operator functions such as the one developed in the preceding section represent another form of a system model. Such functions are mathematical models of a system in the same sense as a differential equation is a mathematical model of a system. Both analytically relate the output of a system to the input, and both include constant terms (coefficients or multiplicative factors) that are related to system properties. The basic difference between these mathematical representations is that the differential equation expresses the *input* as a function of output terms, whereas the operational model derived in Sec. 6.5 expresses the *output* as a function of operations on the input. To show this more forcefully, we compare a block-diagram representation of the integral operational model (Fig. 6.11a) and an operator model derived directly from the differential equation (Fig. 6.11b). The latter model is essentially the same as that shown in Fig. 6.4 except that the blocks are combined into a single block.*

Although the integral operational model derived in Sec. 6.5 was developed in a rather special manner, it was basically derived from the differential equation. It would seem reasonable to expect that the operational model derived from the differential equation (as a function of s; see Fig. 6.11b) and the operational model derived by the procedure outlined in Sec. 6.5 should be related. We should, in fact, intuitively expect that these operational models would be reciprocally related. We can support this contention by using the block diagrams of Fig. 6.11. Since the first block operates on the input to give the output and the second operates on the output to give the input, we should expect that, if these blocks were connected in tandem, the second block would deliver the same signal that was applied to the first block. (This is shown by phantom lines indicating the course of an applied signal σ_A.) The effect of each of the blocks is thus canceled when they are connected in tandem. We can describe this cancellation mathematically by expressing the signals delivered from each block in terms of the signal applied to the block. Thus, if σ_A is applied to the first block, the signal delivered from this block is

$$(6.21) \qquad (\sigma_{\text{out}})_1 = \frac{I/\tau}{I/\tau + 1} \sigma_A$$

* The justification for this procedure can be developed directly from the differential equation when the derivative is written as a function of s.

$$\tau \frac{d\sigma_{\text{out}}}{dt} + \sigma_{\text{out}} = \tau s \, \sigma_{\text{out}} + \sigma_{\text{out}} = (\tau s + 1)\sigma_{\text{out}} = \sigma_{\text{in}}$$

The parenthetic term is an operational function acting on σ_{out}. Evidently operational functions can be expressed in terms of s and I.

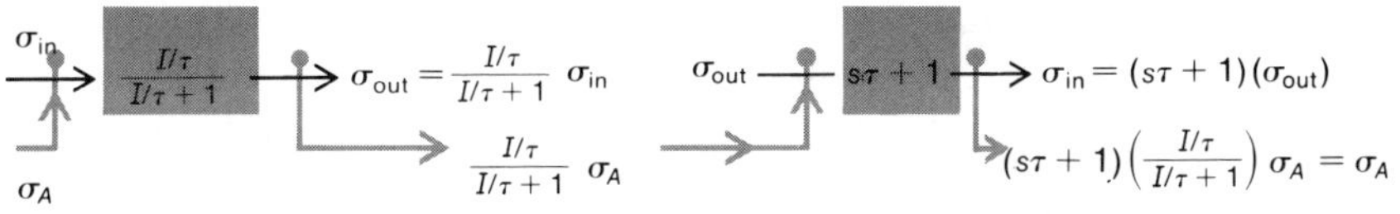

(a) Integral representation   (b) Differential representation

FIG. 6.11   *Models using differential and integral operators.*

where $(\sigma_{out})_1$ represents the output signal delivered from the first block. If this signal is applied to the second block, the output of the second block $[(\sigma_{out})_2]$ is

(6.22) $$(\sigma_{out})_2 = (s\tau + 1)(\sigma_{out})_1$$

By substitution of Eq. (6.21) into Eq. (6.22), we have

(6.23) $$(\sigma_{out})_2 = (s\tau + 1)\frac{I/\tau}{I/\tau + 1}\sigma_A$$

Since we intuitively require that $(\sigma_{out})_2$ be the same as σ_A, we must conclude that the effect of the dual operation is equivalent to multiplication by unity and therefore that $(I/\tau)/(I/\tau + 1)$ and $\tau s + 1$ may be interpreted as reciprocal functions. In equation form, we can thus write

(6.24) $$\frac{1}{s\tau + 1} \simeq \frac{I/\tau}{1 + I/\tau}$$

(where the symbol $\simeq$ is read "is equivalent to"). If we consider the operators as algebraic quantities and work out the implications of this equivalency we conclude that s (differentiation) and I (integration) can also be considered reciprocal functions.

The equivalency of $1/s$ and I does not seem unreasonable, since integration and differentiation are inverse operations. We can, in fact, establish that $1/s$ and I may be rationally considered equivalent operations in a more direct manner if we examine the significance of integration followed by differentiation (or vice versa.) When initial conditions are zero (or operation around a known quiescent point is implied) the integral of a differential is the same as the derivative of an integral. In equation form this can be written

(6.25) $$\int \frac{d\sigma}{dt}\,dt = \sigma = \frac{d}{dt}\left(\int \sigma\,dt\right)$$

This states mathematically that for zero initial conditions integration and differentiation are defined as reverse operations; an integral operation "cancels" a differentiation, and a differential operation "cancels" an integration. If we use I to represent the integral operation, the integral of σ is $I\sigma$. Differentiation of the integral should then be equivalent to the original function σ. In equation form this is written

(6.26) $$sI\sigma = \sigma$$

Equation (6.26) indicates that as a result of the double operation (i.e., integration followed by differentiation) the original function remains unaltered. Thus s operating on $I\sigma$ (or I operating on $s\sigma$) seems to be equivalent to multiplication by unity. If we accept this interpretation, then we might interpret the dual operator sI (or Is) as unity. If we use algebraic notation and algebraic laws to simplify this dual operator, we can then interpret I as $1/s$. With this interpretation we can now describe the integral operator of a first-order-system operator (Fig. 6.9) by an equivalent operator in terms of s [as implied by (Eq. 6.24)].

Once we accept that operator functions can always be written in terms of s, we can then see that it is fairly simple to define a system directly from the differential equation. To do so, we note that the differential equation given by Eq. (6.1) can be written

$$(6.27) \qquad \tau s \sigma_{out} + \sigma_{out} = \sigma_{in} \qquad \text{or} \qquad \sigma_{in} = (\tau s + 1)\sigma_{out}$$

Equation (6.27) is a simple algebraic equality. If we perform the same operation on both sides of the equation, we do not destroy this equality. We can therefore integrate both sides, differentiate them, or apply an operation of the form $1/(\tau s + 1)$. If we use the latter function to operate on both sides of Eq. (6.27), we have

$$(6.28) \qquad \frac{1}{\tau s + 1}(\tau s + 1)\sigma_{out} = \frac{1}{\tau s + 1}\sigma_{in}$$

The inverse operations cancel each other, and we have

$$(6.29) \qquad \sigma_{out} = \frac{1}{\tau s + 1}\sigma_{in}$$

$1/(\tau s + 1)$ is the desired first-order transfer function that operates on the input to produce the output. It is evidently the reciprocal of the operator function that acts on the output signal to generate the input signal as indicated by the differential equation.

Rather than use the terms "operator functions" and "transfer functions" interchangeably we shall restrict the use of the latter term to the operator functions that act on the *input* to yield the *output*. Thus the expressions in both blocks in Fig. 6.11 would be considered operator functions but only the operator function of Fig. 6.11a would be considered a transfer function.

The concepts presented in the preceding paragraph form the basis for evaluating system transfer functions from any linear differential equation. The procedure may be described as a transform technique related to the Laplace-transform method for solving differential equations. Although the foregoing development is not rigorous, the conclusions can be justified by more elegant derivations which can be found in any advanced mathematics text dealing with or including chapters on operational methods and the Laplace transform.* In such texts the "kernel function" generally used to transform an algebraic function to its Laplace-transform equivalent is ϵ^{-st}. The transformed quantity is thus generally expressed in terms of s. Most transform tables are therefore presented in terms of the s parameters. Such tables are directly applicable to our work as the transforms developed here are identical to those given in standard tables except for the addition of initial conditions. The exact Laplace transform of an algebraic quantity includes the effect of initial conditions. Since we assume operation around a known quiescent point, a formal mathematical expression of initial conditions is not necessary and the standard transforms reduce to the transforms used here.

6.7 AN INTRODUCTORY ILLUSTRATION OF THE USE OF THE TRANSFER FUNCTION

Before we examine how the transfer function may be determined directly from signal information, we shall develop some facility with the transform concept by studying some simple systems. The purpose of these studies is twofold: On the one hand, they may be considered "practice sessions" to establish familiarity with operator expressions and how they may be manipulated; on the other hand, they will serve to develop interpretive facility of the operator functions so that they may be translated into meaningful results. We use radioactive systems as illustrative examples, as these may be introduced at a relatively simple level where physical reasoning and intuition can be used to establish and interpret the results obtained. The interpretations and techniques developed are general, however, and applicable to all systems.

We begin by using the mathematical expression that describes the law of radioactive decay. In equation form, this law is written

$$(6.30) \qquad\qquad -\frac{dN}{dt} = KN$$

In words the law states that the rate of nuclear decay $(-dN/dt)$ is proportional to the amount of radioactive material present (or number N of radioactive molecules available). To the author, this sequence of development seems to put the emphasis in the wrong place. Stating the "law" in this fashion suggests that the law is derived analytically and nature obligingly obeys it. In reality, the law as stated by Eq. (6.30) is derived not by mathematical reasoning alone but from an approximate mathematical representation of the decay curve of radioactive materials which presumably transform directly to the stable state with no inter-

*See, for example, Ref. 3 or 4.

mediary steps. Furthermore, this approximate representation is acceptable only if there is a statistically large number of disintegrating molecules compared with the number of stable molecules. If the latter condition is not fulfilled, Eq. (6.30) is not a valid description of radioactive decay.* Therefore, we shall not use Eq. (6.30) directly but instead shall examine physical data to see how the "law" given by Eq. (6.30) may have been determined. Of course, this is a tongue-in-cheek procedure as there is now available some 80 years of experience in radioactive studies and the analytic results are already known. Nevertheless, this procedure will serve to show the empirical nature of the law given by Eq. (6.30) and to emphasize that, although a transform may be obtained from a mathematical expression, this expression may well have been derived from physical data. Ultimately, therefore, transforms may be determined directly from physical or graphical data.

We begin with the assumption that it is known that radioactive materials disintegrate into another state by particle and wave emissions. These emissions may be detected by the luminous effects they produce. The amount of particles that disintegrate may thus be determined by counting light flashes with an appropriate instrument (e.g., a scintillation counter). To perform a specific test, radioactive gold (Au^{198}) is enclosed in a shielded chamber with a shutter. A counter mechanism is located behind this shutter so that decay particles can be counted when the shutter is removed. (See Fig. 6.12. One of the statistical implications of this procedure is immediately evident; the exposed face of the gold sample is assumed to be representative of any other face, i.e., the distribution of radioactive particles is uniform throughout the sample.) Each count indicates one radioactive particle of gold has decayed to another (assumed stable) state. In this case the decay product is mercury (Hg^{198}). If we record the number of disintegrations and plot them against the time required to make the count, we obtain the curve shown in Fig. 6.12b. From past experience we assume that this curve may be represented by an expression of the form $1 - \epsilon^{-t/\tau}$. An appropriate logarithmic plot establishes that this is indeed the case, and we find the disintegra-

---

* If the number of disintegrating particles is too small, the decay curve is very erratic and discontinuous and a curve fit is not practical. A good discussion of this point (with illustrative curves) is given in D. S. Riggs, "The Mathematical Approach to Physiological Problems," p. 130, The Williams & Wilkins Company, Baltimore, 1963.

Interested readers can find a number of excellent texts on the subject of radioactive-decay schemes and the theoretical aspects of atomic fission. A very comprehensive treatment of the various aspects of atomic disintegration is given by R. D. Evans in Ref. 5. Although this book is over 10 years old, it is still reasonably up to date. A rather interesting and simple (and very short) explanation of radioactive phenomena is given by Chadwick (one of the early leaders in the field) in Ref. 6. A more up-to-date text which provides more detail than Chadwick's short primer but not the detail given by Evans is Ref. 7.

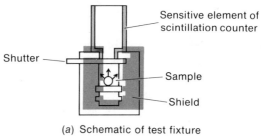

(a) Schematic of test fixture

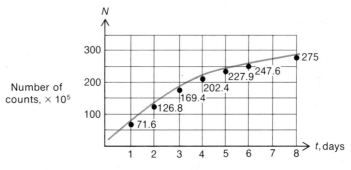

(b) Output characteristics

FIG. 6.12 *Decay of radioactive gold Au^{198}. (Curve constructed from numerical values given by E. H. Quimby and S. Feitelberg, "Radioactive Isotopes in Medicine and Biology," Lea & Febiger, Philadelphia, 1963.)*

tion may be given by

$$(6.31) \qquad N_D = N_{p0} (1 - \epsilon^{-t/\tau}) = (314 \times 10^5) (1 - \epsilon^{-t/3.9})$$

where N_D is the number of detected disintegrations, τ is the time-constant (about 4 days), and N_{p0} (314×10^5) is the maximum number of particles that can disintegrate. We see that N_{p0} must also represent the number of radioactive particles (nuclides) present at the beginning of the test. Since the number of unstable particles that are left at any instant is given by the difference between the original number of nuclides and the number that have decayed, we have

$$(6.32) \qquad N_L = N_{p0} - N_D$$

where N_L represents the number of nuclides still remaining in the unstable state. If we use Eq. (6.31) to replace N_D in Eq. (6.32), the result is

$$(6.33) \qquad N_L = N_{p0}\epsilon^{-t/\tau}$$

We can now see how the "law" stated by Eq. (6.30) may be derived. If we wished to determine the rate at which the remaining nuclides disintegrate, we would differentiate Eq. (6.33) and obtain

(6.34) $$\frac{dN_L}{dt} = -\frac{1}{\tau} N_{p0}\epsilon^{-t/\tau}$$

or

$$-\tau \frac{dN_L}{dt} = N_{p0}\epsilon^{-t/\tau}$$

If we now use Eq. (6.33) to replace $N_{p0}\epsilon^{-t/\tau}$ in Eq. (6.34), the result will be Eq. (6.30), where K is evidently the reciprocal time-constant. As the rate of nuclide disappearance is the same magnitude as the rate of nuclide disintegration [i.e., by Eq. (6.32), $-dN_L/dt = dN_D/dt$] we see that the rate of nuclide disintegration (dN_D/dt) is proportional to the number of nuclides still remaining in the unstable state (N_L). The statistical nature of this process is evident, since Eq. (6.30) does not involve the fate of single particles but of populations of particles. In effect, the equation states that the greater the number of nuclides, the more probable it is that there will be many disintegrations at any instant.

It is appropriate here to note the connection between the time-constant τ and the nuclide "half-life" which is commonly used as an identifying property of a radioactive material. The half-life is defined as the time required for half the remaining nuclides to decay to a stable state. In mathematical terms the half-life (T_H) is the time at which N_L in Eq. (6.33) has become $N_{p0}/2$. By substitution into Eq. (6.33) we have

(6.35) $$\frac{N_{p0}}{2} = N_{p0}\epsilon^{-T_H/\tau}$$

Therefore

(6.36) $$\epsilon^{-T_H/\tau} = \frac{1}{2} \qquad \text{or} \qquad T_H = \tau \ln 2 = 0.693\tau$$

The half-life is thus about 0.7 time-constant. For Au^{198} this is about 2.7 days.

With this very brief introduction we are now prepared to examine the transform aspects of the simple decay system described in the foregoing paragraphs. From the experimental data and the description of the mechanical operating procedure we have sufficient information to formulate a system transfer function. Prior to the beginning of the experimental procedure $(t \leq 0)$, the number of counts registered by the counting mechanism is zero. When the shutter is suddenly raised to expose the radioactive sample, the instrument starts its counting operation. We can artificially describe this situation by assuming that opening the shutter is comparable to the sudden introduction of a nuclide supply into the shielded chamber of Fig. 6.12a. The input to the chamber system might therefore be described as

(6.37) $$N_p = N_{p0}u(t)$$

If we consider the output of the chamber system to be the instrument reading (this assumes a perfect instrument), Eq. (6.31) may be used to represent the output signal. We are now familiar with this pair of equations (in fact, are probably tired of them; however, they lead to better things) and can immediately write the system differential equation as

(6.38) $$\tau \dot{N}_D + N_D = N_p$$

or

$$(\tau s + 1)N_D = N_p$$

or

$$\frac{N_D}{N_p} = \frac{1/\tau}{s + 1/\tau}$$

The last form of Eq. (6.38) may be considered the transfer function of the decay system of radiogold Au^{198} to its stable form Hg^{198}. The decay system may be represented by a block-diagram description as shown in Fig. 6.13a. N_p represents the nuclide supply function; N_D represents the number of nuclides that have degenerated to the stable form. We can use this diagram to indicate how the number of nuclides left may be determined. If we deduct the nuclides destroyed (N_D) from those available (N_p) we have the number that remain in the unstable state [see Eq. (6.32)]. This operation is indicated by the phantom lines in Fig. 6.13a. When we examine how a system transfer function may be simulated with physical elements we shall then see that by such added connections it is relatively easy to measure a variety of signal responses, such as N_L, which may be too complex to determine analytically. In this case, however, N_L can be described very simply: Since $N_L = N_p - N_D$ and N_D is given by

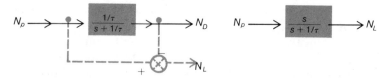

(a) Output is the number
of nuclides destroyed

(b) Output is the number
of nuclides remaining

FIG. 6.13 *Block-diagram description of Au^{198} decay.*

Eq. (6.38), N_L (in terms of s) may be described as

(6.39)
$$N_L = N_p - \frac{N_p}{\tau s + 1} = N_p \frac{s\tau}{\tau s + 1}$$

or

$$\frac{N_L}{N_p} = \frac{s}{s + 1/\tau}$$

The nuclide-supply and nuclide-remaining system thus has a transfer function given by Eq. (6.39), and it may be represented by the block diagram of Fig. 6.13b (although that shown in Fig. 6.13a is also adequate).

When N_p is specified, we can use the block diagram of Fig. 6.13b [i.e., Eq. (6.39)] to obtain an expression for the nuclides remaining at any instant. Thus, if N_p is $N_{p0}u(t)$, then N_L may be written as $N_{p0}u(t)$ $[s/(s + 1/\tau)]$. Unfortunately, we cannot as yet interpret this s function as a function of time (and cannot therefore use the transfer function to its fullest advantage). If we were required to obtain a solution for N_L when N_p is specified, we could use Eq. (6.39) to obtain the differential equation that relates these variables. By so doing we could then obtain an equation that is amenable to solution by standard techniques. To construct the differential equation from the transfer function, Eq. (6.39) is written in expanded form, and we have

(6.40)
$$\left(\frac{s + 1}{\tau}\right) N_L = sN_p$$

or

$$\frac{dN_L}{dt} + \frac{1}{\tau} N_L = \frac{dN_p}{dt}$$

If N_p is given we can now use the differential form of Eq. (6.40) to determine a solution for N_L. As a trivial check we note that, if N_p is a step-function $[N_{p0}u(t)]$, we already know N_L is given by $N_{p0}\epsilon^{-t/\tau}$ [see Eq. (6.33)]. If we use Eq. (6.40), then, when N_p is given as $N_{p0}u(t)$, dN_p/dt is defined by $N_{p0}\,\delta(t)$ (see the discussion in Sec. 6.4) and Eq. (6.40) becomes

(6.41)
$$\dot{N}_L + \frac{1}{\tau} N_L = N_{p0}\,\delta(t)$$

In this form N_L is essentially the response of a first-order system excited by an impulsive excitation (see the end of Sec. 6.1; a rapid review of Sec. 6.5 would also be appropriate here), and the solution is given by

(6.42)
$$N_L = N_{p0}\epsilon^{-t/\tau}$$

This, of course, agrees with the solution for N_L obtained earlier.

We also note that we could have converted Eq. (6.41) to the equivalent s notation. If we did so, we would have

$$(6.43) \qquad N_L = \frac{1}{s + 1/\tau} N_{p0} \, \delta(t)$$

This is essentially the same function described in Fig. 6.10 except I is now replaced by $1/s$, A is N_{p0}, and N_L is directly written as $N_{p0}\epsilon^{-t/\tau}$.

One might well wonder why the s form of Eq. (6.39) differs from that of (6.43), as the latter was derived from the former by converting to a differential equation and then reconverting. There was, however, one intermediary step. After Eq. (6.39) was converted to a differential equation, N_p was defined as a step-function $N_{p0}u(t)$. This term was then differentiated to give an impulse. After this differentiation, the resultant expression was reconverted to the operational form. The new transformed representation contains an s function operating on an impulse-function. In this form a solution for N_L may be obtained simply by inspection. We would now expect that Eq. (6.39) could probably have been used directly to obtain a solution for N_L if the input-step $[N_{p0}u(t)]$ could be converted into an equivalent impulse-function representation. For this excitation the conversion is almost self-evident. The derivative of a step is an impulse. Mathematically, this may be expressed as

$$(6.44) \qquad \frac{du(t)}{dt} = \delta(t)$$

or

$$su(t) = \delta(t)$$

or

$$u(t) = \frac{\delta(t)}{s}$$

(The last form indicates that the step is defined as the integral of an impulse, which is also obvious from an intuitive point of view.) We can now see that if we use this form for $u(t)$ then $N_{p0}u(t)$ may be written $(N_{p0}/s) \, \delta(t)$. If this is used in Eq. (6.39) it reduces to the form shown by Eq. (6.43) and an immediate solution is possible. The pattern begins to unfold, and we can now begin to see the intimate connection between the operator functions, the signal functions, and the system function. Before we formally connect these ideas, however, we examine a few more illustrative situations.

6.8 BLOCK DIAGRAMS OF COUPLED SYSTEMS

To gain further experience with the transfer-function concept we examine a situation that gives rise to a block diagram composed of coupled blocks.

In such representations the output of one block serves as the input to another; the second block, however, does not affect the operation of the first. A familiar and real system arrangement which may give rise to a coupled-block configuration is an instrumented physiological system. An excitation to the physiological system causes the system to respond; this may be represented by one block. The response may then be detected by an instrument system; this is represented by another block. If it can be assumed that the physiological response is not altered* by the instrument system, we can then assume that the physiological response serves as the excitation to the instrument. The resultant block diagram then consists of two blocks in series (similar to the way the first block is connected to the second in Fig. 6.11).

Another example of a real system that gives rise to a coupled-block-diagram representation is the radioactive material that decays to a stable state in two or more stages. Such materials decay to intermediate products which are also radioactive. These radioactive products further decay to another stage which may or may not be radioactive. This process continues until a final stable state is reached. Since each stage of radiation continues independently of any of the following stages, no loading effects are encountered, and the total decay scheme may be represented by a series of blocks connected in tandem. Each block represents one stage of decay.

In view of the experience and background developed in Sec. 6.7 we shall use a radioactive system to study the evolution and implications of a coupled-block-diagram representation. As a specific example we shall examine the decay of radioactive tellurium (Te^{131}) to its final stable state, xenon (Xe^{131}). This material does not decay directly to its final state. Instead, the radioactive particles (nuclides) of tellurium decay first to an intermediate nuclide, radioactive iodine; this nuclide then decays to the final stable form of xenon. Such a two-stage decay sequence is often called a mother-daughter (or parent-daughter) decay scheme.

To perform the suggested study we start by using the procedure employed in Sec. 6.7. Instead of immediately proposing any mathematical representation, we examine experimental data. To obtain these data we use an experimental arrangement similar to the one described earlier (see Fig. 6.12) except that we now use radioactive tellurium instead of radiogold and record in counts per minute rather than actual counts. Two comments about the experimental arrangement are appropriate here: The first deals with the "purity" of the radioactive sample and the second with the counting procedure. For the study that involved the use of

---

* I.e., the instrument does not cause a "loading effect."

radiogold, radioactive purity was of no concern. Radiogold decays directly to the stable state; the only nuclides that could be present were those of Au^{198}. In this suggested study, however, two radioactive nuclides may be (and probably are) initially present. The parent (tellurium) nuclide is present; its decay product I^{131} (the daughter nuclide) may also be in evidence. When the shutter is opened, therefore, the counting instrument may be exposed to an *initial* supply of two kinds of decaying nuclides. Although this is realistic, it unfortunately complicates the preliminary discussion and does not materially add to an understanding of the decay process. For simplicity, then, we assume that until the shutter is opened the I^{131} produced is either mechanically or chemically removed from the tellurium (possibly by resin extraction). We can therefore assume that when the shutter is opened the nuclides in the sample are purely radioactive tellurium. We shall see later how this assumption may be modified to include the initial presence of I^{131} nuclides.

Counts per minute rather than actual counts are used in this discussion primarily to indicate that such measurements may also be used in actual studies; counts per minute are, in fact, sometimes a more practical measurement because the numbers are not so unwieldy as those obtained by total count. This is particularly true when surveying a highly active source (high disintegration rate) for a long period of time. Historically, total counts (made by eye) were the first measurement scale used. Later it was found that nuclide disintegration caused ionization which could be detected as a current; the higher the disintegration rate, the higher the current. Rate counting then became popular, as current could be related to rate, and so continuous readings were possible. In recent years, because of accuracy requirements, background noise, and advances in scintillation devices and photomultipliers, total counting methods are again used to a considerable degree. Evidently rate counting and absolute counts are related. The integral of the rate is the same as the number of counts. Obviously, then, it does not matter how the data are acquired (count or rate count) so long as it is understood *what* data have been acquired and how they may be converted to a more appropriate form.

We are now ready to perform the suggested experiment and to proceed with the system study. When the shutter is raised (Fig. 6.12) and the count rate recorded, the decay curve and its semilogarithmic plot appear as shown in Fig. 6.14. In both curves the ordinate is the number of recorded counts per second. Readings are taken until the count rate is too small to be significant (in this case after about 20 days). We now make the following observations: If $\dot{N}_c$ represents the number of counts per second, then, based upon the evidence of the data, $\dot{N}_c$ cannot

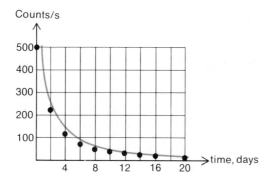

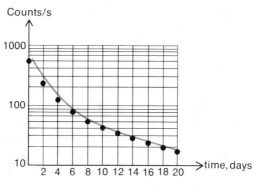

(b) Semilogarithmic plot

FIG. 6.14 *Decay of radioactive tellurium (Te^{131}) to stable xenon (hypothetical curve).*

be described as the response of a linear first-order system. (Nor, therefore, can N_c, the total count, be so described.) From the data we must conclude either that Te^{131} decays to Xe^{131} in a nonlinear fashion or that it decays as a higher-order system or that the Te^{131} nuclide disintegration and the counts are not in a one-to-one correspondence.*

With no other information, we would be at an impasse and could not make any further statements about the nature of the system. We

---

* If we use these data we cannot derive the general law of radioactive decay directly as given by Eq. (6.30). This equation is based upon a linear first-order decay. The data obtained in this case do not show the first-order decay. The conclusion is therefore that either Eq. (6.30) is not a general law or we must reinterpret the significance of the equation. If we revise our viewpoint and state that Eq. (6.30) describes only the disintegration process through a single stage of decay [from one nuclide to another particle (stable or unstable)] and not necessarily the scintillation counts associated with radioactive disintegration, this equation can be used to explain the decay curve of Te^{131} to Xe^{131}. We shall subsequently examine this assertion from an analytic point of view.

would have to depend upon library or laboratory research to decide how to proceed further. In this case two bits of information are readily available: (1) About 80 years of experience have revealed that radioactive decay is not affected by external conditions (temperature, pressure, product supply, formation, etc.). The decay constants are thus indeed constants; system time-constants are therefore also constant. In short, radioactive systems may be closely approximated by linear representations. (2) It is known (through hindsight, to be sure) that Te^{131} decays in at least two stages. (There may be other intermediate stages but they would be so short-lived as to be essentially undetectable.)

With this information we might now propose that Te^{131} decay may be represented by a linear system of at least second-order. Each order might represent one stage of nuclide decay. This does not explain the shape of the count curve and its relation to radioactive disintegration, but it allows us to propose a possible connection. If the single-stage decay of a single nuclide is accompanied by a single count and this single decay of the nuclide results in a stable particle ($Au^{198} \rightarrow Hg^{198}$), then the disintegration process is represented directly by the count curve. If, however, the single-stage decay of a nuclide results in another nuclide ($Te^{131} \rightarrow I^{131}$), then one count is produced by the first disintegration; at some time later another count is produced by the decay of the second-stage (daughter) nuclide. In this case, however, the count curve cannot represent the disintegration curve directly. The decay process is a statistical one; the decay of the daughter (counts due to the daughter disintegration) depends upon the number of daughter nuclides present. This in turn depends upon the number of parent nuclides that have decayed (counts due to parent disintegration). Both processes progress at the same time, and the count curve should reflect this information. It is not immediately obvious how this information is included in the count curve although it should be obvious that the disintegration process may not be directly reflected by it. To determine how to extract this information (and to examine the feasibility of the proposed description of the decay and count process) we must obtain an analytic representation of the mother-daughter decay scheme. To do so we use a block-diagram description of the decay process.

We begin by assuming that nuclide disintegration proceeds on a statistical basis; i.e., the rate of disintegration is proportional to the number of nuclides present. This implies that Eq. (6.30) may be used to describe the disintegration of Te^{131} to I^{131} and that the block diagram of Fig. 6.13 may be used to represent this stage of the tellurium-nuclide decay. This first stage of the tellurium decay process is shown by the cross-hatched block in Fig. 6.15. The input to the system is the number

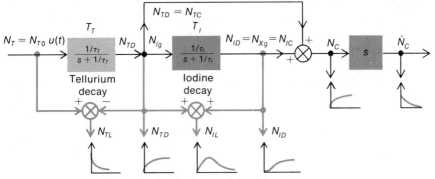

FIG. 6.15   *Block diagram of Te^{131} decay sequence.*

of tellurium nuclides (N_T) supplied after the shutter is opened (it is assumed the iodine nuclides produced prior to the opening of the shutter are removed). Since no additional tellurium nuclides are introduced after the sample is exposed, N_T is a step-function of $N_{T0}u(t)$, where N_{T0} is the number of tellurium nuclides present when the experiment is started ($t = 0$). The output of the tellurium block (N_{TD}) represents the number of disintegrated tellurium particles at any instant. As each disintegrated tellurium particle gives rise to an iodine nuclide, we have $N_{TD} = N_{Ig}$, where N_{Ig} represents the number of iodine nuclides generated. Furthermore, since the decay of each Te^{131} nuclide gives rise to one count we have $N_{TD} = N_{TC}$, where N_{TC} represents the count produced by the decay of tellurium particles. These signals are shown in the block diagram of Fig. 6.15.

If we assume now that the decay of I^{131} to Xe^{131} depends only on the supply of I^{131} and that the statistical nature of the decay process is not changed [i.e., Eq. (6.30) is still applicable], the decay of iodine may also be represented by the first-order transfer function of Fig. 6.13. We can therefore represent the second stage of tellurium decay by the same kind of function as used to represent the first stage. However, the decay constant (or time-constant) may be different for each stage. We account for this possible (probable) difference by assigning a time-constant τ_T to the tellurium block and a different time-constant τ_I to the iodine block (see Fig. 6.15). The supply for the iodine block is, of course, derived from the output of the tellurium block. The output of the second stage is the number of iodine nuclides that have decayed (N_{ID}). This number must be equal to the number of xenon particles generated (N_{Xg}) and must ultimately be equal to the number of tellurium particles initially present (N_{T0}). Since the decay of each iodine nuclide must give rise to a count,

$N_{IC} = N_{ID}$, where N_{IC} represents the number of counts due to iodine decay. The total counts (N_C) produced by the decay of tellurium to xenon is evidently given by $N_{TC} + N_{IC} = N_C$. The total counts can therefore be represented by the output of a device that is driven by the response signals from the tellurium and iodine blocks. This diagrammatic representation is shown in Fig. 6.15. To obtain count rate $(\dot{N}_C)$ we differentiate the count signal (N_C) as shown.

We have now set up a block-diagram description of the tellurium decay system which displays the operational processes of the system (as theorized). It is now a simple matter to obtain a transform description of any of the signals shown*; we have only to follow the directions indicated by the diagram. To define the number of tellurium nuclides that have decayed, for example, we note from the diagram

$$(6.45) \qquad \frac{N_{TD}}{N_T} = \frac{1/\tau_T}{s + 1/\tau_T}$$

or

$$N_{TD} = \frac{1/\tau_T}{s + 1/\tau_T} N_T$$

If we substitute for the excitation $[N_T = N_{T0}u(t) = (N_{T0}/s)\,\delta(t)]$ we have

$$(6.46) \qquad N_{TD} = \frac{N_{T0}/\tau_T}{s(s + 1/\tau_T)}\,\delta(t)$$

This simply derived expression represents the number of tellurium particles that have decayed at any instant. We cannot yet convert this transformed expression into a meaningful signal; this is the subject matter of the next section. We therefore circumvent the problem here and use two devious approaches which will demonstrate further how signals systems and operators are interrelated. Both approaches are based upon the knowledge of how to interpret $[A/(s + a)]\,\delta(t)$. This term was described as part of a transform-pair (see Fig. 6.10) and was again encountered in connection with Eqs. (6.42) and (6.43). By this time we should be willing to accept that this s function can be interpreted as $A\epsilon^{-at}$. To use this concept in Eq. (6.46) we rewrite this equation as

$$(6.47) \qquad sN_{TD} = \frac{N_{T0}/\tau_T}{s + 1/\tau_T}\,\delta(t)$$

---

* We shall see in a later chapter that such block diagrams may be easily simulated with an analog computer so that the various signals may be measured directly. This, as always, assumes some numbers are known. Often, in order to analyze a system or theorize about its operation, it is more effective to describe the system in a mathematical sense (if it can be done).

Since sN_{TD} is interpreted as $\dot{N}_{TD}$ or dN_{TD}/dt we now have

$$(6.48) \qquad \frac{dN_{TD}}{dt} = \frac{N_{T0}}{\tau_T} \epsilon^{-t/\tau_T} \qquad \left[\text{since } \frac{\delta(t)}{s + 1/\tau_T} = \epsilon^{-t/\tau_T} \right]$$

To determine N_{TD} we can integrate the decay rate $(\dot{N}_{TD})$ from the beginning of the experiment $(t = 0)$ to any time thereafter $(t = t)$. Then

$$(6.49) \qquad N_{TD} = \frac{N_{T0}}{\tau_T} \int_0^t \epsilon^{-t/\tau_T} \, dt = N_{T0}(1 - \epsilon^{-t/\tau_T})$$

The result given by Eq. (6.49) is reasonable. It states that at the beginning of the experiment no tellurium particles have been destroyed $(N_{TD} = 0)$; after infinite time $(t \rightarrow \infty)$ all the tellurium nuclides that existed at the beginning of the experiment have now decayed to some other form $(N_{TD} = N_{T0})$. The combination of Eqs. (6.46) and (6.49) might be considered the definition of another transform-pair comparable to that shown in Fig. 6.10.

We can arrive at the same conclusion by using a different approach. Earlier in this section it was shown that in a decay process the nuclides of a fixed supply that are left at any instant are given by

$$(6.42) \qquad N_L = N_{P0}\epsilon^{-t/\tau}$$

or

$$(6.43) \qquad N_L = \frac{N_{P0}}{s + 1/\tau} \, \delta(t)$$

If we write that the nuclide supply (N_T) reduced by the number destroyed (N_{TD}) leaves the number remaining (N_{TL}), we have

$$(6.50) \qquad N_T - N_{TD} = N_{TL}$$

If the supply is a step $N_{T0}u(t)$ then, for $t > 0$, N_T is N_{T0} and N_{TL} is $N_{T0}\epsilon^{-t/\tau_T}$; therefore

$$(6.51) \qquad N_{T0} - N_{TD} = N_{T0}\epsilon^{-t/\tau_T}$$

or

$$N_{TD} = N_{T0}(1 - \epsilon^{-t/\tau_T})$$

which is in agreement with Eq. (6.49). If the s equivalent is used in Eq. (6.50) we have

$$(6.52) \qquad \frac{N_{T0}}{s} \, \delta(t) - N_{TD} = \frac{N_{T0}}{s + 1/\tau_T} \, \delta(t)$$

or

$$(6.53) \qquad N_{TD} = \frac{N_{T0}}{s} \, \delta(t) - \frac{N_{T0}}{s + 1/\tau_T} \, \delta(t)$$

Equation (6.53) is then simply an s representation of N_{TD}, just as Eq. (6.46) is an s representation of N_{TD}. If Eq. (6.53) is expressed in combined form, the result will be Eq. (6.46). (This is easily verified by combining the two fractions.) From the other point of view, it must now be evident that Eq. (6.53) is Eq. (6.46) expanded into a summation of fractions. This process is often described as expansion into partial fractions and will be mentioned again later. Of importance now is the fact that s functions of the form shown by Eq. (6.46) may be expanded into a form shown by Eq. (6.53). It is, in general, relatively easy to interpret the expanded form. To illustrate very briefly and summarize the procedure:

1. If

$$N_{TD} = \frac{N_{T0}/\tau_T}{s(s + 1/\tau_T)}\,\delta(t)$$

2. then by partial-fraction expansion this may be expressed as

$$N_{TD} = \frac{N_{T0}}{s}\,\delta(t) - \frac{N_{T0}}{s + 1/\tau_T}\,\delta(t)$$

3. This may be interpreted as

$$N_{TD} = N_{T0}u(t) - N_{T0}\epsilon^{-t/\tau_T}u(t) = N_{T0}(1 - \epsilon^{-t/\tau_T})u(t)$$

To show the utility of this concept we now examine another signal in the block diagram of Fig. 6.15. The easiest one to handle at this stage is the presence of iodine nuclides (i.e., the number of iodine nuclides left, N_{IL}). Radioactive-iodine particles are produced continuously by the decay of tellurium particles. Thus the number of iodine particles generated (N_{Ig}) is equal to the number of tellurium particles destroyed (N_{TD}). At the same time that iodine nuclides are produced, some degenerate to the stable form Xe^{131}. If we represent the nuclides of iodine that are destroyed by N_{ID} we can translate the foregoing statements into a concise algebraic expression:

(6.54) $$N_{IL} = N_{TD} - N_{ID}$$

This expression is represented by the operation of the second summer shown in Fig. 6.15. If (for simplicity of notation) we call the transfer function of the tellurium block $T_T(s)$ and that of the iodine block $T_I(s)$, then from the diagram we have

(6.55) $$\frac{N_{TD}}{N_T} = T_T(s) \quad \text{or} \quad N_{TD} = T_T(s)N_T = N_{Ig}$$

(6.56) $$\frac{N_{ID}}{N_{Ig}} = T_I(s) \quad \text{or} \quad N_{ID} = T_I(s)N_{Ig} = T_I(s)[T_T(s)N_T]$$

By appropriate substitutions into Eq. (6.54), we have

(6.57)
$$N_{IL} = N_T T_T(s) - N_T T_T(s) T_I(s)$$
$$= N_T T_T(s)[1 - T_I(s)]$$

If we again assume that the tellurium supply is a step-function $(N_{T0}/s)\,\delta(t)$ and we replace the transfer functions $T(s)$ by the appropriate s functions, we get

(6.58)
$$N_{IL} = \frac{N_{T0}}{s} \frac{1/\tau_T}{s + 1/\tau_T} \left(1 - \frac{1/\tau_I}{s + 1/\tau_I}\right) \delta(t)$$

or

$$N_{IL} = \frac{N_{T0}}{\cancel{s}} \frac{1/\tau_T}{s + 1/\tau_T} \frac{\cancel{s}}{s + 1/\tau_I} \delta(t)$$

or

$$N_{IL} = \frac{N_{T0}}{\tau_T} \frac{1}{(s + 1/\tau_T)(s + 1/\tau_I)} \delta(t)$$

The last form of Eq. (6.58) is essentially a solution for the number of iodine particles present at any instant. We can arrive at a final solution in terms of time by either obtaining the equivalent differential equation or expanding the final form of Eq. (6.58) into partial fractions. By the first approach the factors $s + 1/\tau_T$ and $s + 1/\tau_I$ are combined by multiplication into a single expression $s^2 + (1/\tau_T + 1/\tau_I)s + 1/\tau_T\tau_I$.

(6.59)
$$\left[s^2 + \left(\frac{1}{\tau_T} + \frac{1}{\tau_I}\right)s + \frac{1}{\tau_I\tau_T}\right] N_{IL} = \frac{N_{T0}}{\tau_T} \delta(t)$$

If s^2 and s are interpreted as d^2/dt^2 and d/dt, respectively, we have

(6.60)
$$\ddot{N}_{IL} + \left(\frac{1}{\tau_T} + \frac{1}{\tau_I}\right) \dot{N}_{IL} + \frac{1}{\tau_I\tau_T} N_{IL} = \frac{N_{T0}}{\tau_T} \delta(t)$$

which is a linear second-order differential equation with impulsive excitation which may be solved by standard techniques.

By the second approach, we must first expand into partial fractions, and so Eq. (6.58) may be written in the form

(6.61)
$$N_{IL} = \frac{N_{T0}}{\tau_T} \left(\frac{A}{s + 1/\tau_T} + \frac{B}{s + 1/\tau_I}\right) \delta(t)$$

where A and B are to be evaluated. If we can express Eq. (6.58) as shown by Eq. (6.61) we can then immediately interpret each fraction as an equivalent time function. There are a variety of ways by which the

numerators of partial fractions may be evaluated.* One rather straightforward way (though not the shortest) is to combine the fractions between the parentheses of Eq. (6.61). The numerator of the result is
$s(A + B) + (A/\tau_I + B/\tau_T)$. By Eq. (6.58) this numerator must also be
unity. This equality is possible if $A + B$ is zero and $A/\tau_I + B/\tau_T$ is
unity. This pair of relationships constitutes a pair of simultaneous
equations which when solved give

$$(6.62) \qquad A = -B = \frac{\tau_T \tau_I}{\tau_T - T_I}$$

Substitutions of these values into Eq. (6.61) then gives

$$(6.63) \qquad N_{IL} = N_{T0} \frac{\tau_I}{\tau_T - \tau_I} \left(\frac{1}{s + 1/\tau_T} - \frac{1}{s + 1/\tau_I} \right) \delta(t)$$

which can now be interpreted as

$$(6.64) \qquad N_{IL} = N_{T0} \frac{\tau_I}{\tau_T - \tau_I} \left(\epsilon^{-t/\tau_T} - \epsilon^{-t/\tau_I} \right)$$

We could easily verify that these results are "correct" by simply substituting appropriate values derived from Eq. (6.64) into Eq. (6.60). Of
greater importance is that the results appear reasonable; they lead to
conclusions that make physical sense.

According to Eq. (6.64) there are no iodine nuclides when t is zero.
This is in agreement with our initial premise that the sample is pure
tellurium when the experiment is started. The equation states further
that after a significant time ($t \gg \tau_I, \tau_T$; that is, $t \to \infty$) all the iodine
nuclides will have decayed to xenon. This, too, is consistent with what
one might expect physically. A qualitative sketch of the N_{IL} curve is
shown in Fig. 6.15. It must be noted, however, that we cannot as yet
obtain the actual curve analytically, nor can we practically obtain the
curve experimentally. In the latter instance we would have continuously
to remove and assay all I^{131} produced from Te^{131}. (Although such a procedure may be feasible, there are many practical mechanical and instrumentation difficulties.) In the former instance we cannot construct the
curve without the time-constants, and these are not yet available. If it
were important to know the shape of this curve (as it might be to a
pharmaceutical house that manufactures isotopic materials) it would be
cheaper and easier to use an analytic determination of the curve if it can

---

* Partial-fraction expansion can be found in most texts dealing with transform analysis
and related subjects. One good description of the technique appears in Ref. 8, pp. 240 and
280. A short description of this procedure is included in Appendix C but it is suggested that
interested readers refer to other texts if they wish a detailed explanation of the topic.
Of importance here is the fact that such a procedure is possible.

be established that (1) the analytic description developed describes system operation and (2) the necessary time-constants can be evaluated. One way to satisfy both these requirements is to develop an analytic determination of the count curve. It can then be seen how well the theoretical prediction fits the experimental data and simultaneously how the system constants are related to the curve.

The method by which a mathematical representation of the various signals shown in Fig. 6.15 is obtained should now be evident. To describe the count curve, we perform the required operations as shown in the block diagram of Fig. 6.15. Thus $\dot{N}_C = sN_C$ (as is obvious) but

$$N_C = N_{IC} + N_{TC}$$

Since $N_{TC} = N_{TD}$ and $N_{IC} = N_{ID}$, we have

(6.65)
$$\dot{N}_C = s(N_{TD} + N_{ID})$$

The parenthetic terms in Eq. (6.65) were described earlier by Eqs. (6.55) and (6.56) (and are evident from the block diagram). If we substitute for these terms, we get

(6.66)
$$\dot{N}_C = s[N_T T_T(s)] + N_T T_T(s) T_I(s)$$
$$= s[N_T T_T(s)][1 + T_I(s)]$$

N_T has previously been described as a step $(N_{T0}/s)\,\delta(t)$, and the various transfer functions may be obtained directly from the diagram. By appropriate substitution Eq. (6.66) becomes

(6.67)
$$\dot{N}_C = N_{T0} \frac{1/\tau_T}{s + 1/\tau_T} \left(1 + \frac{1/\tau_I}{s + 1/\tau_I}\right) \delta(t)$$

or

$$\dot{N}_C = N_{T0} \frac{(1/\tau_T)(s + 2/\tau_I)}{(s + 1/\tau_T)(s + 1/\tau_I)} \delta(t)$$

We now see that in order to effect a solution for $\dot{N}_C$ we must expand the last form of Eq. (6.67) into partial fractions. The result of this operation is

(6.68)
$$\dot{N}_C = \frac{N_{T0}}{\tau_T - \tau_I} \left(\frac{2 - \tau_I/\tau_T}{s + 1/\tau_T} - \frac{1}{s + 1/\tau_I}\right) \delta(t)$$

We can finally interpret these results to be

(6.69)
$$\dot{N}_C = \frac{N_{T0}}{\tau_T - \tau_I} \left[\left(2 - \frac{\tau_I}{\tau_T}\right) \epsilon^{-t/\tau_T} - \epsilon^{-t/\tau_I}\right]$$

If our model development has been correct, Eq. (6.69) should represent the count curve in Fig. 6.14. One way to determine whether there is correspondence between the theoretical equation and the actual data is to determine whether Eq. (6.69) may be adjusted to fit the experimental curve. This may be done by using points from the experimental curve to evaluate the equation constants. We can then use the constructed equation to check the experimental data. We note from the curve, for example, that $\dot{N}_C = 500$ counts/s when $t = 0$. When these values are used in Eq. (6.69), we have

$$(6.70) \qquad\qquad 500 = \frac{N_{T0}}{\tau_T}$$

By using two more pairs of point values, we would obtain two additional equations; with these three equations we could solve for the three unknown values N_{T0}, τ_T, τ_T. It would then be possible to reconstruct Eq. (6.69) with known constants. If we carried out this suggested procedure we would find that N_{T0} may be taken as 900, τ_T as 1.8 days (with a corresponding half-life of 1.25 days), and τ_I as 11.5 days ($T_H = 8.08$ days). Equation (6.69) could then be rewritten as

$$(6.71) \qquad\qquad \dot{N}_C = 408\epsilon^{-0.56t} + 92\epsilon^{-0.086t}$$

When Eq. (6.71) is plotted, the generated curve agrees with the experimental data. In an actual situation, this correspondence would give confidence that this model and the theoretical analysis are applicable to the physical system. [In this case it shows only that our algebra was correct and we obtained the answers we should have obtained. In Fig. 6.14 the "experimental" curve was labeled hypothetical. This curve was in essence constructed from Eq. (6.69). The correspondence between the curve and the equation cannot therefore come as a surprise. However, the purpose of the exercise was not to prove or disprove a theory which has evolved and been checked again and again for over three-quarters of a century; it was simply an exercise to develop facility and understanding of the transfer-function concept. To that extent it is hoped it served its purpose. It would be unjustified to read any more into the discussion.]

If the equation derived for the counting rate represents the physical data closely, we can use the equation to study how the count curve is constructed and how system information can be obtained from the physical curve. We note from Eq. (6.71), for example, that the count curve is composed of the sum of two exponential functions. The time-constants of these exponentials are the time-constants of the mother and daughter nuclides. (This may appear to be a trivial observation but it should be noted that the time-constant of a physical curve is not always simply

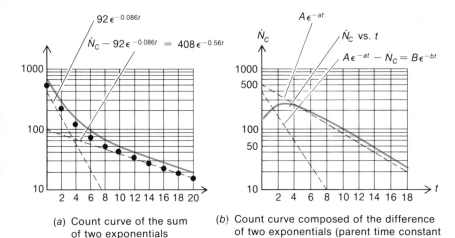

(a) Count curve of the sum of two exponentials

(b) Count curve composed of the difference of two exponentials (parent time constant is much larger than daughter)

FIG. 6.16 *Evaluation of time-constants from semilog representation of count curve.*

related to the constants of a physical system. We shall see later that in higher-order systems time-constants are related to resistive and storage combinations which are not obvious by inspection.) The fact that curve time-constants and nuclide time-constants are the same provides an easy method for evaluating the nuclide time-constants (and their half-lives); the procedure is simply to peel off exponentials from the experimental curve in the manner previously described (Fig. 6.16a). The time-constants of the component exponential functions are the nuclide time-constants.

A number of other observations regarding the construction of the count curve may be made from Eq. (6.69). For example, we may note that the shape of the count curve will vary depending on the relative values of the time-constants. If, for example, the parent time-constant is larger than that of the daughter, Eq. (6.67) predicts a count curve that will first rise and then fall. [See Fig. 6.16b. This can be readily established by differentiating Eq. (6.67) and finding the slope at $t = 0$. When $\tau_T > \tau_I$, this slope is positive.] This prediction can be physically demonstrated by the count curve obtained from the mother-daughter reaction of barium to lanthanum to cerium.* To obtain the nuclide time-constants in this case, we still use the peel-off procedure but we now deduct the count curve from the first exponential approximation instead of deducting the exponential from the count curve (see Fig. 6.16b). As the parent time-constant

---

* $Ba^{140} \rightarrow La^{140} \rightarrow Ce^{140}$ (stable). The parent (barium) has a half-life of 12.8 days and the daughter (lanthanum) has a half-life of 40 hr or about 1.7 days. The actual decay curve may be found in Ref. 9, p. 310.

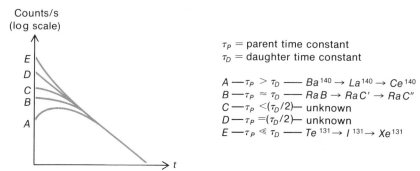

Counts/s
(log scale)

τ_P = parent time constant
τ_D = daughter time constant

$A \;-\tau_P > \tau_D \;----\; Ba^{140} \rightarrow La^{140} \rightarrow Ce^{140}$
$B \;-\tau_P \approx \tau_D \;----\; RaB \rightarrow RaC' \rightarrow RaC''$
$C \;-\tau_P <(\tau_D/2)\!-\;$ unknown
$D \;-\tau_P =(\tau_D/2)\!-\;$ unknown
$E \;-\tau_P \ll \tau_D \;----\; Te^{131} \rightarrow I^{131} \rightarrow Xe^{131}$

FIG. 6.17 *Shape of count-per-second curve for various relative values of time-constants.*

approaches and exceeds the daughter time-constant, Eq. (6.69) predicts that the count curve will start progressively higher, as shown in Fig. 6.17. When the mother and daughter nuclides have almost equal time-constants the count curve will become practically horizontal at $t = 0$ (curve B; physical verification is the decay of RaB to RaC to RaC' or RaC'', etc. The parent has a half-life of 27 min, the daughter of 20 min). When the time-constant of the mother nuclide is less than that of the daughter, the count curve starts and continues to fall with a negative slope, as shown by curves C, D, and E in Fig. 6.17. A physical example of E is the tellurium-iodine-xenon sequence previously discussed.

The author has not been able to find physical examples to justify curves C and D. This may have an interesting explanation. When the parent nuclide has a time-constant one-half that of the daughter nuclide, Eq. (6.69) reduces to $(2N_{T0}/\tau_I)\epsilon^{-t/\tau_I}$. Equation (6.69) thus implies that when the daughter has twice the half-life of the parent the count curve will decay as a single exponential (on a straight line when plotted semilogarithmically). However, in an actual test such straight-line characteristics of a decay curve would probably be attributed to a single-stage disintegration. If Eq. (6.69) may be trusted, it is possible that some count curves that have been considered the decay curve of a single-stage disintegration are in reality the decay of a mother-daughter sequence where the daughter half-life is approximately twice that of the mother. In such cases the count curve could not be used to detect the presence of both nuclides; radiochemical techniques would probably be required. The author has not yet had the opportunity to investigate this situation further. Should a search show that this situation exists, this would serve as a good illustration of how theory may be used to direct further research.

It should be pointed out that Eq. (6.69) cannot be used with blind faith to obtain numerical values. Some of the results can be meaningless. In the example chosen, N_{T0} was evaluated to be 900. One might ask: 900 what? This term might be interpreted as the number of potential counts available from the parent nuclides at the time of shutter opening. There are at least two difficulties (among others) with this interpretation. The first is that one can never be certain every exploding nuclide has been counted. The counting instrument is never 100 percent efficient; it therefore does not count everything it intercepts. Furthermore, the instrument is not generally exposed to the effects of all nuclide disintegrations; it intercepts the effects of only a fraction of the total nuclide supply. All that can be said, therefore, is that N_{T0} and $\dot{N}_C$ both represent a fixed though unknown fraction of the potential counts and count rate.

It may be recalled that we derived Eq. (6.69) on the premise that only parent nuclides were present at the beginning of the experiment. This was an artificiality. In most cases one does not (or cannot) arrange to have only the parent nuclide present at the beginning of a test. Therefore no great significance can be attributed to the numerical values associated with N_{T0} in Eq. (6.69). We can however, show that the ambiguity surrounding the nuclide supply does not affect the evaluation of the nuclide time-constant (provided there is a statistically large nuclide sampling). This is reasonable from an intuitive point of view, since the half-life is essentially a probability figure. For a given material the half-life implies that within a fixed time one-half the available nuclides will probably have disintegrated; in a similar period one-half the remaining nuclides will have decayed, etc. So long as the amount of nuclides remains large, the half-life decay period can be expected to remain consistent.

To show that the half-life can be determined even though there is some question about the source and quantity of nuclide supply, we can use the same block diagram previously developed, except that we must now redefine the excitation terms. A sketch of the new block diagram is shown in Fig. 6.18. Instead of N_{T0} as an excitation, we now use fN_{TT}, where N_{TT} represents the total parent-nuclide supply and f is some fraction (to indicate that only a portion of the disintegrating particles may be counted). Furthermore, we now include an excitation to represent the possibility of the presence of daughter nuclides at the inception of the test. This excitation is shown as fN_{IT}, where N_{IT} represents the total iodine-nuclide supply at time zero. The initial daughter-nuclide supply is added to the daughter nuclides generated by tellurium disintegration. The total daughter supply $[N_{I_g} + fN_{IT}u(t) = N_I]$ is then delivered to the daughter-decay block. We can now obtain a solution for $\dot{N}_C$ by perform-

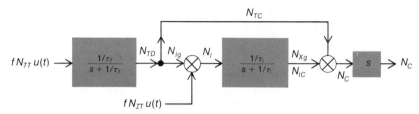

FIG. 6.18 *Mother-daughter decay scheme when parent and daughter nuclides are present at beginning of test.*

ing the required operations as indicated by the diagram. By inspection we can easily see that the solution for $\dot{N}_C$ is essentially the same as previously obtained except that we must now add a term to account for the effect of the daughter-nuclide supply. This term is given by

$$(6.72) \quad \dot{N}_C \text{ [due to } fN_{IT}u(t)] = \frac{fN_{IT}}{s}\,\delta(t)\,\frac{1/\tau_I}{s+1/\tau_I}\,(s) = \frac{fN_{IT}}{\tau_I}\frac{\delta(t)}{s+1/\tau_I}$$

Equation (6.72) can now be added to Eq. (6.68) (which represents the effect of the parent supply) to give the effect of both nuclide supplies on the total count:

$$(6.73)^* \quad \dot{N}_C = \frac{fN_{TT}}{\tau_T - \tau_I}\left(\frac{2 - \tau_I/\tau_T}{s+1/\tau_T} - \frac{1}{s+1/\tau_I}\right)\delta(t) + \frac{fN_{IT}}{\tau_I}\frac{\delta(t)}{s+1/\tau_I}$$

By collecting terms and converting to a time representation, we have

$$(6.74) \quad \dot{N}_C = \left[\frac{fN_{TT}}{\tau_T - \tau_I}\left(2 - \frac{\tau_T}{\tau_I}\right)\right]\epsilon^{-t/\tau_T} + \left[\frac{fN_{IT}}{\tau_I} - \frac{fN_{TT}}{\tau_T - \tau_I}\right]\epsilon^{-t/\tau_I}$$

The relationship representing the count-rate curve is now given by Eq. (6.74). A comparison of this equation with the one previously obtained [Eq. (6.69)] indicates that the form of the equation has not changed. The count curve is still represented by the sum of two exponentials; the time-constants of the exponentials still represent the decay constants of mother and daughter nuclides, and these constants would still be obtained by a "peel-off" procedure. However, the constant coefficients of the exponential terms (shown in brackets) have changed. Although these constant terms can be evaluated, the numerical values obtained have little significance unless the fraction f can be determined. Unless this term can be evaluated there is no way to use the numerical

---

* Equation (6.73) illustrates the use of (not the proof of) the principle of superposition, which states that in linear systems the effect of a number of stimuli is the sum of the effects of each individual stimulus acting alone.

values to obtain either N_{TT} or N_{IT}. This may appear to be a negative result but it is just as important to know that something cannot be done as it is to know that it can be done. Such knowledge can prevent the futile waste of valuable time.

6.9 SIGNALS TO AND FROM OPERATOR FUNCTIONS

The discussions of the last two sections have already covered the material of this section to some degree. We have seen how some response signals may be expressed in terms of the s-transform notation and also how time representations of these signals may be obtained from the transform representation. (This conversion from the s representation to the time representation is generally expressed as "transforming from the s domain to the time domain.") The sequence of the last form of Eq. (6.58) and Eqs. (6.61) and (6.63), for example, illustrates the basic principle behind the transition from the s to the time domain. The composite form of the s function [the combined form of Eq. (6.58)] is separated into simpler components by a partial-fraction expansion [Eq. (6.61)]; the simple components are then interpreted individually [Eq. (6.64)]. In reality, the mechanical process of partial-fraction expansion (Appendix C) serves to minimize the memorization needed to use transform techniques. Though each s transform has an equivalent time function (which constitute a transform-pair; see Fig. 6.10), and while it is possible to memorize an extensive list of transform-pairs, by using the partial-fraction technique complex transforms are decomposed into simpler recognizable forms so that only relatively few transform-pairs need be committed to memory. A short list of transform-pairs is shown in Table 6.1. Of the eight pairs shown the first four are the only ones that need be memorized (and probably have already been); the rest can be obtained by partial-fraction expansion.

Facility with the expansion technique is not obtained by reading explanations but rather by practicing the procedure. It is suggested, therefore, that the reader use partial fractions to derive the last four transform-pairs shown in Table 6.1. Explanations of special cases will be given as the need arises during the discussion of various illustrative problems. (For those who prefer working from tables, a slightly more comprehensive list of transforms is included in Appendix D. Even more extensive lists may be found in recent editions of the "Handbook of Chemistry and Physics"* and in most engineering handbooks under the heading of Laplace transforms.)

To some extent we "put the cart before the horse" when we discuss transformation from the s domain (expressions in s) to the time domain

* 48th ed., Chemical Rubber Publishing Company, Cleveland, Ohio, 1966.

TABLE 6.1

	Transform-Pairs	
Graphical Representation	$f(t)$	$F(s)$
1	$A\delta(t)$	$A\delta(t)$
2	$Au(t)$	$\dfrac{A}{s}\,\delta(t)$
3	$\left\{\begin{array}{c}Au_R(t)\\ \text{or}\\ Atu(t)\end{array}\right.$	$\dfrac{A}{s^2}\,\delta(t)$
4	$A\epsilon^{-at}$	$\dfrac{A}{s+a}\,\delta(t)$

	Transform-Pairs	
Graphical Representation	$f(t)$	$F(s)$
5	$A(1-\epsilon^{-at})$	$A\,\dfrac{a}{s(s+a)}\,\delta(t)$
6	$\dfrac{A}{a-b}(\epsilon^{-bt}-\epsilon^{-at})$	$\dfrac{A}{(s+a)(s+b)}\,\delta(t)$
7	$\dfrac{1}{ab}+\dfrac{1}{a-b}\left(\dfrac{\epsilon^{-at}}{a}-\dfrac{\epsilon^{-bt}}{b}\right)$	$\dfrac{1}{s(s+a)(s+b)}\,\delta(t)$
8	$\dfrac{t}{a}-\dfrac{1}{a^2}(1-\epsilon^{-at})$	$\dfrac{1}{s^2(s+a)}\,\delta(t)$

(expressions in t) before we discuss transformation of time functions into s functions. As a practical matter, one does not generally have signals in terms of s without first having had some signal either expressed mathematically or shown graphically as a function of time. The response of a known system to a known stimulus is expressed in s only if the stimulus and response signals may both be expressed as transformed functions. The reason the transformation process was presented in the reverse order is that s transforms may be converted to time functions by using a relatively straightforward procedure. There is practically no guesswork associated with this process. Once a signal transform is available, only one time function can be obtained, and it is obtained in a routine manner. Thus, when a signal transform $\delta(t)/(s + a)(s + b)$ is to be converted into the time domain, we can (mechanically) expand this function into partial fractions and can then immediately express the result as $1/(a - b)(\epsilon^{-bt} - \epsilon^{-at})$ (Table 6.1, sixth curve). If physical signals are expressed in terms of appropriate time functions, the reverse procedure is also straightforward. To use the above example: If a signal is expressed by

$$(6.75) \qquad f(t) = \frac{1}{a - b} (\epsilon^{-bt} - \epsilon^{-at})$$

the transformed function would be

$$(6.76) \qquad F(s) = \frac{1}{a - b} \left[\frac{\delta(t)}{s + b} - \frac{\delta(t)}{s + a} \right] = \frac{\delta(t)}{(s + b)(s + a)}$$

We see therefore that time functions can sometimes be easily converted to s functions simply by memorizing a few transform-pairs. In many cases, however, time functions are not expressed in a convenient mathematical form or are not expressed mathematically at all. To handle such situations, the s transform must be developed for each case separately. This may involve some guesswork, interpretation, and approximations. (The procedure to convert from time functions to s functions is not as straightforward as transition from the s to time domain.) We examine the transformation of aperiodic (nonperiodic or nonrepetitious) waveforms in this section. In Chap. 8 we shall consider the representation of periodic or repetitive waveforms (exemplified by the square wave or sinusoidal functions).

In a laboratory or experimental situation, signals are generally obtained in some recorded form. These signals are represented as a graphical function of time. To express these time recordings in terms of s, the signals are first expressed mathematically as a function of time and then these time functions are converted to an equivalent function of s.

Once more, therefore, the problem of curve fitting arises. As mentioned earlier, curve fitting is beyond the scope of a text of this nature. Instead of using any formal technique, therefore, we shall depend upon intuitive and approximate approaches.

The first point to be made regarding curve fitting for our particular application is that it would be helpful if the time functions used were easily transformable to the s domain. If, for example, all waveforms could be expressed as some combination of exponentials and/or step-functions, it would then be fairly simple to convert such representations to equivalent s functions. We have already seen how some signal waveforms may be approximated by the summation of exponential functions. The signal is redrawn on semilogarithmic coordinates and an attempt is made to peel off exponentials. The sum of these exponentials should represent the given waveform. Unfortunately, this very useful technique is not applicable in all cases. Some waveforms are obviously not related to exponential functions. The initial portion of the heat pulse shown in Fig. 6.19, for example, is clearly not exponential. To represent such waveforms other techniques must be used.

Theoretically, it is possible to express any waveform by an appropriate summation of step-functions. In effect, this assumes that any curve may be approximated by a "staircase-function" such as that shown in Fig. 6.19a. The greater the number of steps, the more accurate is the representation. For such a representation the transform of the waveform is simply the s transform of a series of steps. In Fig. 6.19a, for example,

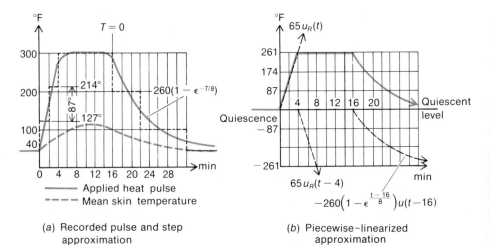

(a) Recorded pulse and step approximation

(b) Piecewise-linearized approximation

FIG. 6.19 *Transient heat pulse used to test pilot response to heat exposure (adapted from Bioastronautics Data Book, NASA Publ. SP 3006, p. 119).*

the temperature function with respect to quiescence [that is, $f(0) = 0$] is represented by

$$(6.77) \quad f(t) = 87u(t) + 87u(t - 2) + 87u(t - 4) - 100u(t - 16)$$
$$- 100u(t - 22) - 60u(t - 32)$$

where each step is superimposed on the previous step at the indicated time. It is now a simple, though tedious, matter to write the corresponding transform of the given waveform. The result is the sum of the step transforms:

$$(6.78) \quad F(s) = \frac{87\delta(t)}{s} + \frac{87\delta(t - 2)}{s} + \frac{87\delta(t - 4)}{s} + \frac{100\delta(t - 16)}{s}$$
$$- \frac{100\delta(t - 22)}{s} - \frac{60\delta(t - 32)}{s}$$

Although it is always possible to approximate a waveform by a sequence of steps, the result may be too unwieldy to be of practical use. To simplify the representation yet retain the advantage of simple conversion to the s domain, the use of steps is combined with other easily transformable functions. As an alternative representation of the waveform in Fig. 6.19 we might therefore use steps and an exponential function. The waveform could then be written

$$(6.79) \quad f(t) = 87u(t) + 87u(t - 2) + 87u(t - 4)$$
$$- 260(1 - \epsilon^{-(t-16)/8})u(t - 16)$$

The last (exponential) term in Eq. (6.79) has replaced the last three step-functions in Eq. (6.78). The unit function [$u(t - 16)$] and the time scale of the exponential are arranged so that the exponential begins after $t = 16$. A brief examination of Fig. 6.19a indicates why this delay is required. The transform of the heat pulse is now written as

$$(6.80) \quad F(s) = \frac{87\delta(t)}{s} + \frac{87\delta(t - 2)}{s} + \frac{87\delta(t - 4)}{s} - 260\frac{\frac{1}{8}\delta(t - 16)}{s(s + \frac{1}{8})}$$

where the exponential is transformed according to the fifth transform-pair shown in Table 6.1 [or could be derived as shown by Eqs. (6.75) and (6.76)].

In many cases the use of steps is practically precluded by the shape of the waveform. The heat pulse of Fig. 6.19 is a case in point. From appearances alone, we would expect that a better fit could be obtained with ramp-functions than with steps. Since ramps are the integrals of steps, they are also easily transformable. By using ramps, therefore, we retain the advantage of ease of transformation and simultaneously provide increased versatility (curve fitting by ramps and steps is evidently a piecewise-linearized representation of the time signal). If we use this concept for the initial portion of the signal in Fig. 6.19 we could represent the curve by

$$(6.81) \quad f(t) = 65u_R(t) - 65u_R(t - 4) - 260(1 - \epsilon^{-(t-16)/8})u(t - 16)$$

where $u_R(t)$ represents a unit-ramp function and the coefficient (65°F/min) represents the slope of the ramp. This ramp can be written directly as a function of time:

$$(6.82) \quad 65u_R(t) = 65tu(t)$$

$$(6.83) \quad -65u_R(t - 4) = -65(t - 4)u(t - 4)$$

Equation (6.82) describes a function that starts at $t = 0$ and rises with a slope of 65°/min, and Eq. (6.83) describes one that starts at $t = 4$ and drops with a slope of 65°/min. These functions are shown as dashed lines in Fig. 6.19b. We can easily see that for the period $0 < t < 4$ only the first ramp exists [i.e., Eq. (6.82)]; this approximates the given waveform for this interval. At any time after 4 min both ramp-functions exist. We can intuitively see that the ramps cancel each other to produce a curve of fixed amplitude.* This approximates the constant temperature shown by the given waveform. The addition of the exponential at $t = 16$ then brings the temperature down to approximately the shape of the given waveform. The transform of the signal [Eq. (6.81)] can now be directly written:

$$(6.84) \quad F(s) = \frac{65}{s^2} \delta(t) - \frac{65}{s^2} \delta(t - 4) - 260 \frac{\frac{1}{8}\delta(t - 16)}{s(s + \frac{1}{8})}$$

6.10 ILLUSTRATIVE EXAMPLE USING TRANSFORMED SIGNALS

The procedure for curve fitting proposed in the foregoing paragraphs is useful from an analytic point of view, particularly when the analysis involves the response of a known system to a known waveform. To illustrate the utility in this connection we examine the response of a mercury-in-glass thermometer (similar to the one mentioned in Sec. 6.4) to a temperature pulse with the waveform given in Fig. 6.19. A block diagram of the problem is shown in Fig. 6.20. The instrument is assumed

---

* If intuition fails,

$$65tu(t) - 65(t - 4)u(t - 4) = 65t - 65(t - 4)$$

when $t > 4$. By expansion, this becomes $65t - 65t + 260$. The result is thus the constant maximum temperature of the pulse.

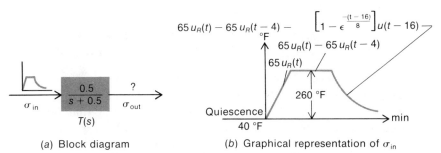

(a) Block diagram   (b) Graphical representation of σ_{in}

FIG. 6.20   *System response to an arbitrary waveform.*

to be a first-order system* with a time-constant of 2 min. From the block diagram we can immediately write

$$(6.85) \qquad \sigma_{out} = \frac{0.5}{s + 0.5}\, \sigma_{in}$$

If we use Eq. (6.84) as the transformed representation of the applied signal, then the response becomes

$$(6.86) \qquad \sigma_{out} = \frac{0.5}{s + 0.5}\left[\frac{65}{s^2}\,\delta(t) - \frac{65}{s^2}\,\delta(t - 4) - \frac{32.5\delta(t - 16)}{s(s + 0.125)}\right]$$

* In Sec. 6.4 we established that the thermometer is governed by a first-order differential equation [Eq. (6.1)]:

$$(6.1) \qquad \tau\,\frac{d\sigma_{out}}{dt} + \sigma_{out} = \sigma_{in}$$

Rewriting in terms of s we have

$$(\tau s + 1)\sigma_{out} = \sigma_{in}$$

The transfer function σ_{out}/σ_{in} is thus

$$T(s) = \frac{\sigma_{out}}{\sigma_{in}} = \frac{1}{\tau s + 1} = \frac{1/\tau}{s + 1/\tau}$$

For a time-constant of 2 min this becomes

$$T(s) = \frac{0.5}{s + 0.5}$$

The 2-min time-constant is somewhat unrealistic for medical instrumentation. In most cases more responsive instruments must be used for thermal measurements. Even the clinical thermometer, which is rather slow, has a time-constant of about $\frac{1}{2}$ min (a steady reading is expected after about 2 min or four time-constants). The long time-constant was chosen here to show the effect it may have on the fidelity of the instrument during dynamic response. This will become clear during the course of the discussion. It should be borne in mind, however, that this is an illustrative problem and the numerical results may be somewhat unrealistic.

or

$$\sigma_{\text{out}} = \frac{32.5\delta(t)}{s^2(s + 0.5)} - \frac{32.5\delta(t - 4)}{s^2(s + 0.5)} - \frac{16.25\delta(t - 16)}{s(s + 0.5)(s + 0.125)}$$

(NOTE: The quiescent level of the system is at 40°F. The amplitude of the temperature pulse is thus 260° above quiescence, as shown in Fig. 6.20.) We can now use Table 6.1 to determine the corresponding time functions for each of the transforms shown in Eq. (6.86). As a brief exercise, however, the first function of Eq. (6.86) will be expanded into simpler recognizable components. Since the denominator of the function is third-order, there will be three fractional components:

(6.87) $$\frac{32.5}{s^2(s + 0.5)} = \frac{A}{s^2} + \frac{B}{s} + \frac{C}{s + 0.5}$$

If we multiply through by s^2 and permit s to approach zero, we find A is evaluated to be 65; if we multiply through by $s + 0.5$ and permit s to approach -0.5, C is evaluated to be 130. Probably the easiest way to determine B is to convert the s terms to the time domain. (See also Appendix C.) Then

(6.88) $$\frac{32.5}{s^2(s + 0.5)} \delta(t) = (65t + B + 130\epsilon^{-t/2})u(t)$$

Since there can be no temperature change immediately after excitation, Eq. (6.88) must reduce to zero when t is 0^+. Then B must be -130, and Eq. (6.88) becomes $65t - 130(1 - \epsilon^{-t/2})$. Now that we have evaluated the first term of the total response [Eq. (6.86)] we can write the second term by inspection. It is identical to the first ramp-response except that it is negative and delayed 4 min. The equation for this second term is given by

(6.89) $$f(t) = -[65(t - 4) - 130(1 - \epsilon^{-(t-4)/2})]u(t - 4)$$

and plots of both these functions are shown by curves A and B in Fig. 6.21. The sum of the first two ramp-response terms is shown as the gray dashed curve in the interval $0 < t < 16$.

The last term of Eq. (6.86) may be evaluated directly from the seventh transform-pair in Table 6.1. The result for this particular case is

(6.90) $$\frac{16.25\delta(t - 16)}{s(s + 0.5)(s + 0.125)}$$

$$= -\left(260 + \frac{260}{3}\epsilon^{-(t-16)/2} - \frac{1,040}{3}\epsilon^{-(t-16)/8}\right)u(t - 16)$$

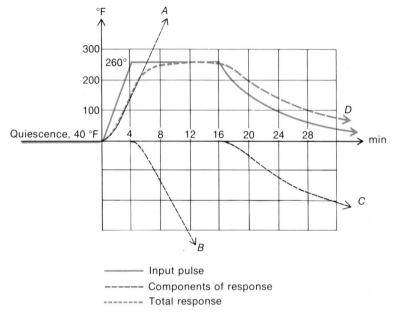

- ———————— Input pulse
- – – – – – Components of response
- ·········· Total response

FIG. 6.21 *Comparison of applied temperature pulse and thermometer response.*

The last response component is negative and is therefore plotted below the time axis, as shown by curve *C* in Fig. 6.21. The total response of the thermometer is now the sum of the separate response curves (*A*, *B*, and *C* in Fig. 6.21); the final result is shown as the gray dashed curve *D*.

We can readily see that by the procedure outlined the (temperature) response to almost any waveform can be predicted. This capability can be advantageous in analytical work and may be particularly useful for evaluating the usefulness of an instrument (compared with its price) for measurements where a rough estimate of the signal input is available. In the case just examined, the thermometer is all but useless for most of the measurements to be made. To show how the thermometer response compares with the actual applied signal, the true temperature pulse is included in Fig. 6.21 (shown as a solid curve). It is clear that the thermometer does a very poor job of measuring the actual temperature. It is almost 100 percent in error at the 4-min mark and does not give reasonable readings until 12 min after the start of the pulse. Furthermore, as soon as the pulse begins to decay, the thermometer again reads in error. In fact, the thermometer reads correctly only after the excitation is constant for a "long time." Whenever the excitation changes, the ther-

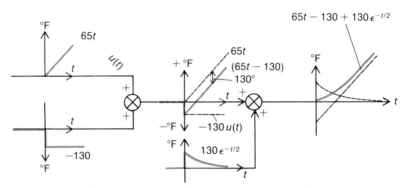

FIG. 6.22 *Thermometer response to ramp excitation.*

mometer reading is in error. This error is due to the dynamic lag of the instrument which is in turn related to the time-constant of the instrument; the longer the time-constant, the greater is the dynamic lag, and hence the greater the dynamic error.

To put this on a somewhat more quantitative basis we examine the ramp-response of the instrument. This is given by Eq. (6.88), which represents the instrument response in the interval $0 < t < 4$ when the temperature excitation is a ramp of slope $65°/$min. From Eq. (6.88) we can see this response is composed of three components. The first is a ramp with the same slope as the excitation. This seems reasonable; if the temperature rises at a steady rate, the thermometer reading also should rise at the same rate. This term may be recognized as the forced response (see Sec. 5.10). The second term is a negative step. This term is of particular interest as it reflects the steady-state error that would be expected for long-term operation. Although the instrument rises at the same rate as the stimulus, it never records the correct temperature as it always lags behind the excitation temperature. This can be clearly understood when we add the first two components of the response as shown schematically in Fig. 6.22. The sum of the ramp and the negative step is shown as the output signal of the first summer. Since $65t$ represents the excitation signal, ultimately the temperature reading will be $130°$ below the actual temperature (an unbelievable yet possible result; however, recall the comment made in a previous footnote). If we examine the source of this $130°$ term, we note it was the value of B (or C) in Eq. (6.87). If we trace back further we find that $130°F$ is essentially derived from the products of the excitation slope ($65°/$min) and the time-constant (2 min). In effect, then, the longer the time-constant, the greater is the dynamic lag and the greater the error. We can now begin to appreciate the effect instrument time-constants may have on the fidelity of measurements. An instrument with a long time-constant may be perfectly accurate (though slow) for final measurements of step-functions but completely impractical for dynamic measurements.

The last term of the ramp-response [Eq. (6.88)] is a decaying exponential and may be considered the free response. It is due to the system properties only and will exist for almost any excitation that the system experiences. This system exponential reflects the system lag; it holds the system back (in a mathematical sense) so that the response starts gradually, falls behind, and stays behind. The effect of the exponential is shown by the final-response curve in Fig. 6.22.

It was implied earlier that aperiodic signals may be conveniently transformed when the signal is approximated by straight-line segments. This does not imply that this same procedure cannot be used for periodic signals. Linearized representations of periodic signals are, in fact, sometimes more descriptive and accurate than any other representation might be. The signal shown in Fig. 6.23 may be used to illustrate this point. In Chap. 8 we shall see how such waveforms may be decomposed into a sum of periodic waveforms of different frequencies (i.e., sinusoidal signals). We intuitively realize now, however, that such waveforms may be more accurately represented by straight-line segments in the manner employed to represent the signal of Fig. 6.19. The easiest way to handle this is to decompose the given waveform into an average signal (a steady component; electrical engineers refer to this as a dc component) and a variable signal (ac component). The first signal may be represented as a step and the second a triangular waveform which varies about zero. The signal function might thus be described as

$$(6.91) \qquad\qquad \sigma(t) = Mu(t) + f(t)$$

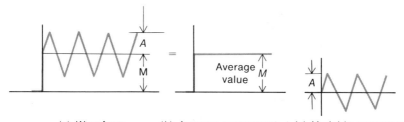

(a) Waveform   =   (b) Average component + (c) Variable component

FIG. 6.23 *Triangular wave—example of a periodic waveform composed of straight-line elements.*

where

(6.92) $f(t)$ = triangular waveform with amplitude $\pm A$

The triangular waveform can now be represented by a series of ramp-functions as shown in Fig. 6.24. The slope of the first ramp is A/t_i; every succeeding ramp has twice this slope ($2A/t_i$; the reason for this becomes obvious when one examines the magnitude of the various ramps at $2t$, $4t_i$, etc; see Fig. 6.24). The variable signal may thus be written

(6.93) $f(t) = mu_R(t) - 2mu_R(t - t_i) + 2mu_R(t - 3t_i) + \cdots$

where m is A/t_i. The transform of the given signal [Eq. (6.91)] is now

(6.94) $\sigma(s) = \dfrac{M}{s}\,\delta(t) + \dfrac{m}{s^2}\,\delta(t) - \dfrac{2m}{s^2}\,\delta(t - t_i) + \dfrac{2m}{s^2}\,\delta(t - 3t_i) - \cdots$

or

$$\sigma(s) = \frac{1}{s}\left(M - \frac{m}{s}\right)\delta(t) - \frac{-2m}{s^2}\sum_{n=1,3,5}^{\infty}(-1)\,\frac{N+1}{2}\,\delta(t - nt_i)$$

where the last term represents the summation of an infinite number of impulsive functions.

We can now use the representation given by Eq. (6.94) to determine the response of a known system to the triangular waveform. As a simple illustration we assume once more that the waveform is a temperature signal applied to a thermometer (the system representation is shown by the block diagram in Fig. 6.25a). To take advantage of our previous analysis we assume the thermometer has a single time-constant (similar to the one employed earlier); to be more realistic, however, we assume

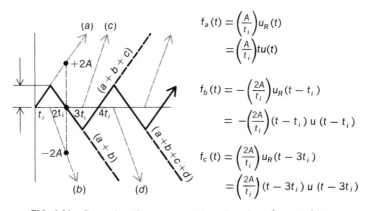

$$f_a(t) = \left(\frac{A}{t_i}\right)u_R(t)$$
$$= \left(\frac{A}{t_i}\right)tu(t)$$

$$f_b(t) = -\left(\frac{2A}{t_i}\right)u_R(t - t_i)$$
$$= -\left(\frac{2A}{t_i}\right)(t - t_i)\,u\,(t - t_i)$$

$$f_c(t) = \left(\frac{2A}{t_i}\right)u_R(t - 3t_i)$$
$$= \left(\frac{2A}{t_i}\right)(t - 3t_i)\,u\,(t - 3t_i)$$

FIG. 6.24   *Ramp-function representation of a triangular waveform.*

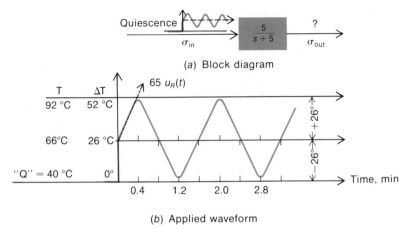

(a) Block diagram

(b) Applied waveform

FIG. 6.25 *System response to triangular waveform.*

the time-constant to be 0.2 min instead of 2 min. The instrument is at an ambient temperature of 40°C and then is suddenly inserted into a small chamber (a calorimeter for small animals, for example) where the temperature is cycled as shown in Fig. 6.25b. We seek to determine the thermometer response. Symbolically this response can be written

$$(6.95) \qquad \sigma_{\text{out}} = \frac{5}{s+5} \sigma_{\text{in}}$$

where σ_{in} is given by Eq. (6.94). For the waveform shown, M is evidently 26°C, m is 65°C/min, and t_i is 0.4 min. The response to the step component $[(26/s)\,\delta(t)]$ can be written by inspection:

$$(6.96) \quad \sigma_{\text{out}}(\text{due to step or average value}) = \frac{26 \times 5}{s(s+5)}\,\delta(t) = 26(1 - \epsilon^{-5t})$$

This component of the response is shown as curve A in Fig. 6.26. The ramp-response components can also be written by inspection. The response to the first ramp component of the excitation is given by

$$(6.97) \quad \sigma_{\text{out}}(\text{due to first ramp}) = \frac{65}{s^2}\frac{5}{s+5}\,\delta(t) = [65t - 13(1 - \epsilon^{-5t})]u(t)$$

[This result may be compared with that given by Eq. (6.88).] This ramp-response component is shown by curve B in Fig. 6.26. The response due to the second ramp component is

$$(6.98) \quad \sigma_{\text{out}}(\text{due to second ramp}) = \frac{130}{s^2}\frac{5}{s+5}\,\delta(t - 0.4)$$
$$= [130(t - 0.4) - 26(1 - \epsilon^{-5(t-0.4)})]u(t - 0.4)$$

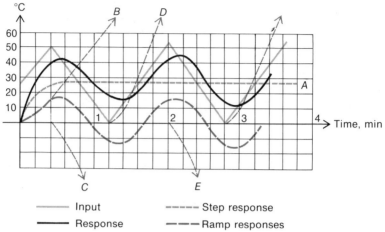

FIG. 6.26 *Thermometer response to triangular excitation.*

This response is shown by curve C in Fig. 6.26. As a general expression for all succeeding ramps we can write

$$(6.99) \qquad \sigma_{out} = 130(t - 0.4n) - 26(1 - \epsilon^{-5(t-0.4n)})u(t - 0.4n)$$

where n is an odd number ($n = 1, 3, 5, \ldots$). Curve D in Fig. 6.26 shows the component ramp-response when n is 1 and curve E shows the response when n is 3. To find the total response the response components are added. The procedure is shown diagrammatically by the block diagram of Fig. 6.27. The response curve is the solid curve in Fig. 6.26. For comparison the excitation signal is shown as a solid gray curve.

At the beginning, the thermometer response is very poor. After the step-function transient has died away (about 0.8 min, i.e., four time-constants) the thermometer response has an average value equal to the average value of the excitation; the response is also a periodic (cyclic) waveform, as the excitation is, and it has the same repetition rate (period for the waveform to repeat itself). The correspondence between the response and the excitation is, however, rather superficial; if we did not know the input was a triangular waveform we *certainly* would not be able to guess that the response was due to a triangular periodic excitation. The waveform of the response has no sharp peaks as the excitation has; the maximum and minimum values are all blunt round curves. Although the response (in this case) rises and falls along ramplike segments with the same slope as the excitation, these segments are not collinear with the excitation. Instead, they "lag" (fall behind) the excitation to form a parallel tracing. The blunted extrema and the response lag cause the thermometer to be in error; the error is related to the product of the signal slope and the instrument time-constant [see the discussion associated with Eq. (6.88) and Fig. 6.22].

For this situation, where the signal is periodic, we can express the slope in terms of the amplitude (A) and the time to peak (t_i); thus $m = A/t_i$.* The error is therefore related to the product $A\tau/t_i$. In this form we can see the error is proportional to the signal amplitude and the ratio of the time-constant and the signal period. Periodic signals are more popularly described by their frequency (ω). If we use this terminology we might say the error is proportional to the product of signal amplitude, time-constant, and signal frequency. The discussion of periodic signals is injected here to show that we already have the capability to handle such signals. However, we shall discuss these signals in a more detailed fashion in Chap. 8 to develop other more versatile concepts and techniques to facilitate analysis of systems when subjected to such stimuli.

To conclude this section, some mention should be made of the formal analytic technique used to convert signals in the time domain to the

---

* See Fig. 6.24 and the associated discussion. The "peaking time" is essentially one-fourth of the repetition time, or the time it takes the wave to repeat itself. From Fig. 6.25b the repetition time is seen to be 1.6 min or $4t_i$. The repetition time is generally described as the period of the waveform. The period is thus the time per unit cycle. The reciprocal function (i.e., cycles per unit time) is often called the repetition rate of the signal or the signal frequency (ω). We shall meet these concepts again in Chap. 8.

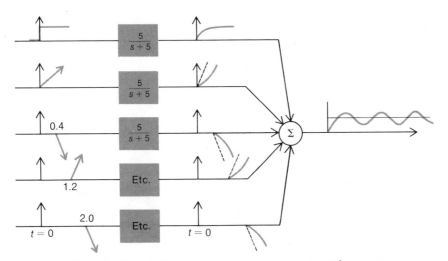

FIG. 6.27 *Superposition of response components to give total response.*

s-domain representation. A formal statement of the transformation is given by

$$(6.100) \qquad \mathcal{L}f(t) = \int_0^\infty \epsilon^{-st} f(t) \, dt$$

The left side of this equation is the symbolic notation indicating the time-function $f(t)$ is to be transformed into the s domain. It is read: "The Laplace transform of the function of t is . . . ." If, for example, $f(t)$ were a decaying exponential of the form ϵ^{-at} we would write

$$(6.101) \qquad \mathcal{L}\epsilon^{-at} = \int_0^\infty \epsilon^{-st} \epsilon^{-at} \, dt$$

If we perform the operations indicated on the right side of Eq. (6.101)* we would find that $\mathcal{L}\epsilon^{-at}$ is given by $1/(s + a)$. This function agrees with the transform developed by the less formal technique presented in the earlier sections of this chapter. The basic difference between the results is the use of the impulse-function. For the procedure developed earlier, the s function is interpreted as an operator which acts upon the impulse-function to give the desired time signal. In the formal technique the impulse-function does not appear in the transform, and the s function is not considered an operator but simply a transformed equivalent of a time function. We shall have occasion to use the formal definition given by Eq. (6.100) in a later section.

6.11 TRANSFER FUNCTIONS FROM INPUT–OUTPUT–SIGNAL INFORMATION

Thus far in this chapter we have established the following for limited ranges of excitations:

1. A system may be represented by a transfer function. This transfer function is composed of s operators and constant terms (which depend upon the system properties).

2. The transfer function represents a mathematical model of a system in the sense that it analytically relates input and output. To this extent it is the same as the differential equation that describes the system operation.

3. The transfer function and the differential equation can be derived from each other via the s-operator notation. These model representations differ, however, in that the transfer function operates on the input to yield the output; the differential equation describes the input as a function of the output.

$$* \int_{t=0}^{t=\infty} \epsilon^{-st} \epsilon^{-at} \, dt = \int_{t=0}^{t=\infty} \epsilon^{-(s+a)t} \, dt = \left[\frac{\epsilon^{-(s+a)t}}{-(s+a)} \right]_{t=0}^{t=\infty} = \frac{1}{s+a}.$$

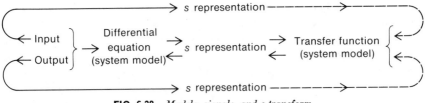

FIG. 6.28   *Models, signals, and s transform.*

4. Signals may be transformed to and from operator functions operating on an impulse-function.

In the development thus far one cannot fail to note the implied intimate relationship between the differential equations, systems, and signals. The common denominator seems to be the s-operator function. This is clearly demonstrated when a simple diagram such as that in Fig. 6.28 is used to assess our present position. As can be seen, both signals and mathematical models may be converted into or from an s transform.

The diagram in Fig. 6.28 also illustrates how signal information has been used to define the system transfer function. Stimulus-response data were used first to define a differential equation. This equation was then converted to operational form by using the s notation where appropriate. The transfer function was then derived from the transform of the differential equation. It would therefore seem reasonable that, since transfer functions are expressions in s and signals can be expressed directly in s notation, the transfer function can be obtained directly from signal information without obtaining the differential equation as an intermediary step. The next step in the development of signals and systems relationships is therefore to examine if and how transfer functions may be defined directly from signal information (as indicated by the dashed lines in Fig. 6.28). We shall see that not only can the transfer function be evaluated directly from the input-output signals but that it can often be determined more easily than the differential equation. To some extent, some of the concepts to be examined in this section have already been introduced and used in earlier sections. The discussion in this section does not therefore involve new material but is a recapitulation and reorganization of ideas already expressed.

The procedure by which systems are defined from signals can be "discovered" by simple verbal statements that describe how a system converts a stimulus to a response. A stimulus is applied to a system; this stimulus may be described by an operator function acting on an impulse [a step is described as $1/s$ operating on $\delta(t)$; see Fig. 6.29]. The system responds and delivers an output signal. This output signal is also described by an operator function acting on an impulse. [If the response is a rising experimental of the form $1 - \epsilon^{-at}$, it would be defined by $a/s(s + a)$ operating on $\delta(t)$.] The operator function used to describe the response is presumed to be related to the stimulus operator function $(1/s)$ and a modifying operator function associated with the system $[a/(s + a)]$. The latter operation $[a/(s + a)]$, which is needed to convert the input to the output, is described as the system transfer function. By this interpretation we see that the impulse-function serves somewhat as a reference signal. An operator acting on the impulse defines the input and the output. The modification required to transform the input operator into the output operator is attributed to the system. We can easily convert these verbal statements into equivalent mathematical expressions by assigning appropriate symbols to the various functions involved. If $T_{in}(s)$ represents the input-signal operator, then the input signal is given by

(6.102) $$\sigma_{in} = T_{in}(s)\ \delta(t)$$

If $T_{out}(s)$ represents the output-signal operator, then

(6.103) $$\sigma_{out} = T_{out}(s)\ \delta(t)$$

If we now describe the system transfer function as $T(s)$, then from the block diagram relating input, output, and system (e.g., Fig. 6.29) we have

(6.104) $$\sigma_{out} = T_s(s)\sigma_{in}$$

From Eqs. (6.102) to (6.104) we can write

(6.105) $$T_s(s) = \frac{\sigma_{out}(s)}{\sigma_{in}(s)} = \frac{T_{out}(s)\ \delta(t)}{T_{in}(s)\ \delta(t)} = \frac{T_{out}}{T_{in}}(s)$$

Equation (6.105) clearly indicates that the system transfer function is defined by the ratio of the output-signal operator to the input-signal operator. Both operators are functions that act on the impulse to produce

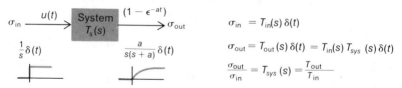

FIG. 6.29 *System transfer function from signal information.*

the input and output signals, respectively. This equation finally provides the stimulus-system-response relationship that has been alluded to in many of the earlier discussions.

As a simple illustration of the formal procedure just developed, we examine the specific example of Fig. 6.1 to justify the contention that the signals shown define a first-order system with a time constant of $\frac{1}{2}$ s. The procedure in outline is as follows:

1. $\sigma_{in} = 6\epsilon^{-5t} = \dfrac{6}{s + 5}\, \delta(t)$ (Table 6.1, fourth pair)

2. $\sigma_{out} = 2(\epsilon^{-2t} - \epsilon^{-5t}) = 2\left(\dfrac{1}{s + 2} - \dfrac{1}{s + 5}\right)\delta(t)$

$$= \dfrac{6}{(s + 2)(s + 5)}\, \delta(t)$$

3. $T_s = \dfrac{6/[(s + 2)(s + 5)]}{6/(s + 5)} = \dfrac{1}{s + 2}$

If this system is excited by an impulsive signal $10\,\delta(t)$ the output will be $[10/(s + 2)]\,\delta(t)$ or $10\epsilon^{-2t}$, as indicated by Fig. 6.1b. The system is evidently first-order with a time-constant of $\frac{1}{2}$ s. In terms of model simulations, this system could be represented by a circuit with an RC product of $\frac{1}{2}$ s, as mentioned at the end of Sec. 6.1. We shall use this concept frequently in the problems and examples in the following chapters.

The mechanics of defining system transfer functions from aperiodic signals based upon $\delta(t)$ is now fairly routine. The graphical signal is represented by a mathematical expression; the mathematical expressions are described by operator functions acting on $\delta(t)$; the transfer function is the ratio of the operators. The means the investigator uses to convert signal information to a mathematical representation depends to a large extent upon the faith the researcher has in his data and instrumentation. If he feels his data are highly reproducible and accurate, he may enlist the aid of a specialist and a computer to obtain an "exact" mathematical expression by elaborate curve-fitting techniques. In the early phases of an investigation, however, it is unlikely that such steps would be warranted and, in general, the investigator would depend upon approximations and/or pattern recognition to obtain some estimate of the representative mathematical equations (or possibly the overall system transfer function).

Problems

6.1 A diagrammatic sketch of a mercury-in-glass thermometer bulb is shown in Fig. P6.1. When the thermometer at ambient temperature T_a is immersed in a bath of temperature T_B, $(T_B > T_a)$, heat flows into the thermometer.

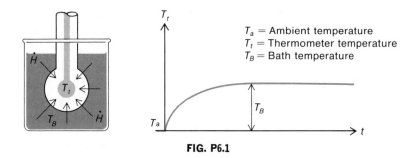

T_a = Ambient temperature
T_t = Thermometer temperature
T_B = Bath temperature

FIG. P6.1

(a) What is the waveform of the input or stimulus? Describe the input with a mathematical equation.

(b) The thermometer reading rises exponentially from the ambient reading to the final reading T_B. What differential equation describes the thermometer temperature T_t?

(c) Outline the procedure by which you would determine the storage and resistive properties of the system. Where are these system properties located; i.e., what parts of the thermometer provide the resistive and storage effects?

(d) What changes should be made in the thermometer design to decrease the time response (time-constant) of the instrument?

6.2 A series RC circuit may serve as an integrator if no current is drawn from the circuit. Thus, if i_c is zero in the circuit in Fig. P6.2, it can be shown that for a short time the voltage across the capacitor may serve as the integral of the input

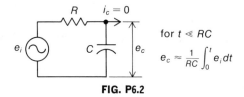

for $t \ll RC$

$$e_c \approx \frac{1}{RC} \int_0^t e_i \, dt$$

FIG. P6.2

voltage. The longer the system time-constant, the longer is the period during which the circuit can act as an integrator. By reasoning similar to that employed in Sec. 6.4, show e_c is closely equal to the integral of e_i when $t \ll RC$.

6.3 If a step is applied to a first-order system, the response may be represented as $A \, \delta(t)/s(s + a)$. This could be interpreted as $A(1/s)[\delta(t)/(s + a)]$, where $1/s$ can be considered an integration between zero and t and $\delta(t)/(s + a)$ may be considered ϵ^{-at}. Show that by performing the indicated operation of integration the correct result for the output response can be obtained. Will the impulse-

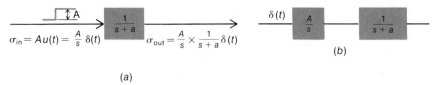

(a)

(b)

FIG. P6.3

response of the system shown in Fig. P6.3b be the same as the step-response of the system in Fig. P6.3a?

6.4 In Chap. 5 we examined some nitrogen-liberation responses which could be expected for various pressure (ascent) profiles. Some of these responses were determined from a model simulation; others were simply stated. These responses can now be calculated by defining the system transfer function and the signal excitations.

(a) What is the transfer function of the nitrogen-liberation system when flow is considered the output and pressure during ascent is considered the input? [This may be determined from the differential equation that describes the system; see Eq. (5.48).]

(b) Obtain an equation for the flow ($\dot{N}_2$) when the diver is subjected to a step-function-ascent profile from 66 ft of water to sea level.

(c) The Navy recommends the staircase-ascent pattern shown in Fig. 5.16. Derive an equation to describe the flow of liberated nitrogen.

(d) If the diver is returned to sea level so that the pressure decreases are linear with time [see Eq. (5.49) and Fig. 5.19], what should be the slope of the pressure curve to ensure that the nitrogen flow does not exceed the proposed "safe" limit of 764 cm³/h before the man surfaces? What is the minimum ascent time possible for a ramp-ascent profile?

6.5 A sample contains two radioactive materials, each of which has a single-stage decay scheme independent of the other. When such a sample is tested with a counter, the counter will register the decay of both materials. The count curve will thus consist of two time-constants which reflect the half-lives of both materials. A block-diagram representation of this situation is shown in Fig. P6.5a.

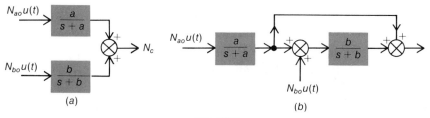

(a)

(b)

FIG. P6.5

(a) Determine the equation that would describe the output count curve.

(b) How are the time-constants of the count curve related to the half-lives of the radioactive substances?

(c) If a sample contains two radioactive materials that bear a mother-daughter relationship to each other (i.e., one material decays into a second material which then decays into a stable product), the system may be represented by the sequential block diagram in Fig. P6.5b. Determine the equation that describes the output count curve for this situation. How are the time-constants of the count curve related to the half-lives of the mother-daughter materials?

(d) Is there any way to use the count curves alone to discriminate or identify which decay scheme is under observation?

6.6 In Prob. 5.6 it was noted the instrument system (catheter and cuvette oximeter) which measures the color density of the blood for dye-dilution studies does not respond as a pure multiplier. The step-response of the instrument assembly seems to indicate it may be approximately represented as a first-order system.

(a) From the data given in Prob. 5.6, determine the transfer function of the instrument system. What is the time-constant of the system? In what way do the catheter diameter, catheter length, and pumping rate affect the time-constant?

(b) The first-order characteristics of the instrument system must affect the signal information derived from the physiological system. Evidently, then, the final recorded data reflect not only the characteristics of the physiological system but also the instrument system. The schematic diagram indicating how the signal information is processed by both physiological system and instrument system is shown in Fig. P5.6a. If we can consider the physiological system to be first-order [see Eq (5.100)], what is the overall transfer function $K_R/\dot{W}_e$ for the block diagram in Fig. P5.6d?

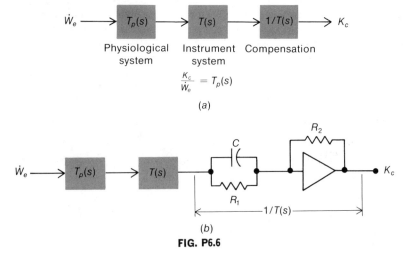

FIG. P6.6

(c) For an impulsive excitation $[\dot{W}_e = \dot{W}_{em}\delta(t)]$ what is the response K_B and K_R?

(d) When the instrument time-constant is one-half the physiological time-constant, does K_R closely represent K_B? What if the instrument has twice the time-constant of the physiological system?

(e) It is sometimes possible to compensate for the effect of an instrument (at least for a limited range of operating conditions) by adding a block to the system which tends to cancel the transfer function of the instrument system. A block diagram of the modified system is shown in Fig. P6.6a; an analog-computer network to realize the required compensation is shown in Fig. P6.6b. If the instrument system has a transfer function $0.2/(s + 0.2)$, find a set of values for R_1, R_2, and C so that the network will cancel the effect of the instrument.

6.7 A thermistor is mounted on a subject's arm and an infrared source beams heat to the area monitored by the thermistor (Fig. P6.7a). When the source is switched on to a steady low intensity, the thermistor responds as a first-order system with a time-constant of 0.1 s.

(a) If the source is switched on and off at a repetition rate of 1 pulse/s with a pulse width of ½ s (see Fig. P6.7c), determine the expected thermistor

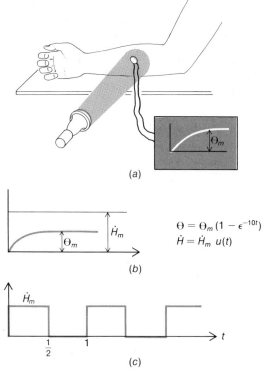

(a)

$$\Theta = \Theta_m (1 - \epsilon^{-10t})$$
$$\dot{H} = \dot{H}_m\, u(t)$$

(b)

(c)

FIG. P6.7

response. What would be the maximum reading of the thermistor? What is the average reading?

(b) Suppose the pulse width and repetition rate are reduced to $\frac{1}{4}$ and $\frac{1}{2}$ s, respectively. What is the maximum reading of the thermistor now? What is the average reading? How is the maximum error of the thermistor reading related to the repetition rate or frequency of excitation?

REFERENCES

1. Truxal, J. G. (ed.): "Control Engineers' Handbook," McGraw-Hill Book Company, New York, 1958.
2. Harman, W. W., and D. W. Lytle: "Electrical and Mechanical Networks," McGraw-Hill Book Company, New York, 1962.
3. Aseltine, J. A.: "Transform Method in Linear System Analysis," McGraw-Hill Book Company, New York, 1958.
4. Craig, E. J.: "Laplace and Fourier Transforms," Holt, Rinehart and Winston, Inc., New York, 1964.
5. Evans, R. D.: "The Atomic Nucleus," McGraw-Hill Book Company, New York, 1955.
6. Chadwick, J.: "Radioactivity and Radioactive Substances," Sir Isaac Pitman & Sons, Ltd., London, 1947.
7. Quimby, E. H., and S. Freitelberg: "Radioactive Isotopes in Medicine and Biology," Lea & Febiger, Philadelphia, 1963.
8. Lynch, W. A., and J. G. Truxal: "Signals and Systems in Electrical Engineering," McGraw-Hill Book Company, New York, 1962.
9. Overman, R. T., and H. H. Clark: "Radioisotope Techniques," McGraw-Hill Book Company, New York, 1960.
10. Bioastronautics Data Book, *NASA Publ. SP* 3006.

CHAPTER SEVEN

THE IMPEDANCE CONCEPT

7.1 INTRODUCTORY COMMENTS

Prior to the development of the transfer-function concept we represented systems by a circuit analog or an appropriate differential equation. These two representations were seen to be interchangeable, each derivable from the other. In some cases the analog was derived first (Sec. 5.2), and the differential equation was then derived from the model; sometimes the differential equation was derived first (Sec. 5.5), and then the model configuration was proposed. The choice of which procedure to be used depended upon the information available and the result desired. If we knew the construction of the system to some extent, we postulated a model first. If this knowledge was not available, we had to depend upon input-output data (and hence the differential equation) to describe the system. In any case, having once obtained one model representation, we could get the other. Neither representation provided more information than the other; they were merely different ways of presenting the same information.

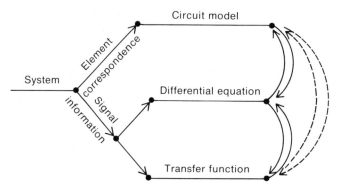

FIG. 7.1   *Transitions between system and model representations.*

The transfer function represents another method to describe a system. Like the two mentioned above, it does not present more information about a system; it simply displays the information in a different fashion. It is therefore not always better or worse than the other methods but only better or worse for some purposes. The transfer function is similar to the differential equation in the sense that it may also be derived from signal information. It differs from the differential equation, however, in that it operates upon the input to define the output. It is often more convenient, therefore, as it can be used to indicate how a system modifies input information to yield output information.

A diagrammatic summary of the ideas presented in the foregoing paragraphs is shown in Fig. 7.1. As indicated in the diagram, if element correspondence is used, the circuit analog is the most directly derived representation of the system under study. If the input-output-signal data are used, then the differential equation or the transfer function is the most directly derived representation. Some elementary illustrations of all three techniques have been given in earlier chapters. It should be noted, however, that these examples were specifically chosen, and they should not be used as criteria to judge the relative complexity of each of the techniques. The illustrations in which the differential equation was derived from signal data, for example, were deceptively easy. They are not representative of the complexities associated with the translation of signal information into an appropriate differential equation. In general, this procedure is not simple or obvious. The transfer function, on the other hand, is somewhat easier to derive than has been indicated. Once a mathematical expression has been developed to represent the signal information, it is no longer necessary to guess how the input and output signals may be related. The time functions that describe these signals can be transformed into appropriate functions of s (by synthesis, as in Fig. 6.19, or by use of tables); from this information the system transfer function can be derived directly.

The versatility of the transfer-function concept has not yet been fully demonstrated but we shall see that this viewpoint provides a powerful tool for investigating and understanding systems. However, it should be kept in mind that both the differential equation and the transfer function are based upon the assumption that the system may be described by a linear lumped representation. This implies that within the range of interest it will be possible to express the signals in terms of time functions that can be converted to a power series. If this approximation is not acceptable, the transfer-function concept is not applicable and it is necessary to search for a nonlinear relationship between the signals.

The interchangeability of model representations (as indicated by the heavy solid lines in Fig. 7.1) has been demonstrated by illustrative examples in previous sections. It has been shown, for example, that the differential equation may be derived from a model simulation (Sec. 5.2) and that an analog simulation may be proposed on the basis of a differential s equation (Sec. 5.5). We have also seen that the differential-equation and transfer-function models are interchangeable representations of a system (Secs. 6.6 and 6.7). Since the circuit model and the transfer function are interchangeable with the differential equation, the circuit model and the transfer function are interchangeable with each other. At our present stage of development, we must depend upon the differential equation to effect this transformation. It seems intuitively obvious, however, that it should be possible to derive a transfer function from a circuit model (and vice versa) without developing the differential equation as an intermediary step. This direct interchangeability between the circuit model and the transfer function (as indicated by the dashed lines in Fig. 7.1) is feasible and straightforward. It depends upon the description of circuit elements as impedances and, in general, upon the impedance concept. The concept of circuit elements as impedances is developed in the next section; the utility of this viewpoint will be demonstrated in later sections.

7.2 CIRCUITS INTO TRANSFER FUNCTIONS: THE IMPEDANCE CONCEPT

To show how networks and transfer functions may be related, we can use one of two approaches: We can either show how networks may be derived from transfer functions or show how transfer functions may be derived from networks. The first is generally thought of as network synthesis

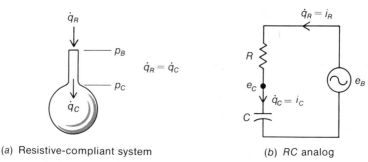

(a) Resistive-compliant system   (b) RC analog

FIG. 7.2   *Kirchhoff's current law.*

and the second as network analysis. We shall use the second procedure here as it is simpler and serves adequately to demonstrate the concepts to be developed.*

Networks may be analyzed by using either Kirchhoff's current law or his voltage law. The first of these states that the sum of the current entering a junction must equal the sum of the currents leaving the junction. (See Appendix A, step 1.) The second states that the sum of the voltage drops around a closed loop must be zero. In general, the former technique is simpler and conceptually clearer, and it is used extensively to analyze networks.

We have in effect already used Kirchhoff's current law to examine a system. In Sec. 5.2, we examined the simple resistive-compliant system of Fig. 7.2a which was represented by the analog circuit shown in Fig. 7.2b. We obtained the equation for the chamber pressure by stating the flow through the resistive element must be equivalent to the flow into the compliant element. In the electrical analog of Fig. 7.2b, this is equivalent to the statement: The current to node e_C is equal to the current from the node e_C. In equation form, this is written

$$(7.1) \qquad\qquad \dot{q}_R = \dot{q}_C$$

---

* Network synthesis has been studied in great detail, and there is a wealth of information on the subject. One text which may be appropriate for readers with backgrounds equivalent to that established here is Ref. 1. Another is Ref. 2.

We cannot go into the details of network synthesis, nor would it be desirable to do so, as such studies should be the subject of a separate text. Here we are interested in network synthesis only to the extent that we understand the implications of such a procedure and that we can refer to proper sources should the need for synthesis arise. We therefore examine the correspondence between circuits and transfer function only so that we may understand how these concepts are related. Readers interested in the mechanics and details of network synthesis should refer to the abundant literature available on the subject.

We use electrical networks and electrical variables in this section as a matter of convenience. The electrical terms may be analogous to the physical quantities of other physical systems.

or

(7.1a)
$$i_R = i_C$$

If we now express these currents in terms of circuit elements and node voltages, we can develop equations relating any pair of variables.

To illustrate the final comment of the last paragraph: If we express the current flow in the resistor as

(7.2)
$$i_R = (e_B - e_C)G$$

and the current flow into the capacitor as

(7.3)
$$i_C = C \frac{de_C}{dt} \quad \text{[see Eq. (4.14)]}$$

then from Eqs. (7.1a), (7.2), and (7.3) we obtain an equation relating the node voltage e_C to the supply voltage e_B:

(7.4)
$$(e_B - e_C)G = C \frac{de_C}{dt}$$

or

(7.5)
$$Ge_B = C \frac{de_C}{dt} + Ge_C$$

Equation (7.5) is the same first-order differential equation as Eq. (5.7). The transfer function relating e_C (the output) to e_B (the input) is derived simply by converting Eq. (7.5) to the s domain:

(7.6)
$$\frac{e_C}{e_B} = \frac{G_G}{sC + G} = \frac{1}{RCs + 1} = \frac{1}{s\tau + 1}$$

In a comparable manner, we can find the transfer function relating the current flow [i_R or i_C (as these are equal)] to the applied voltage. If we use Eq. (7.3) to replace $C(de_C/dt)$ in Eq. (7.5) we have

(7.7)
$$Ge_B = i_C + Ge_C$$

If we then change Eq. (7.3) to integral form [see Eq. (4.13)] so that e_C can be written $e_C = (1/C)\int i\,dt$ (initial conditions omitted), we can then replace e_C in Eq. (7.7) to get

(7.8)
$$Ge_B = i_C + \frac{G}{C} \int i_C \, dt$$

The transfer function relating i_C as the output to e_B as the input is then

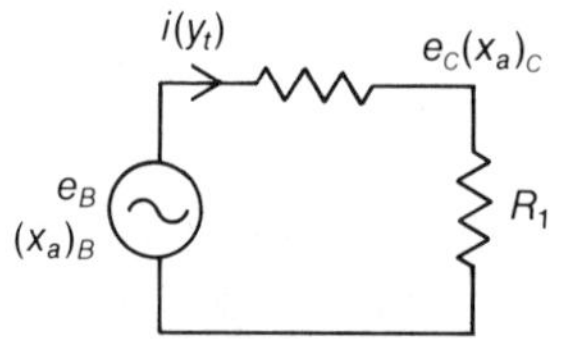

Total resistance $= R + R_1 = \dfrac{e_B}{i} = \dfrac{x_a}{y_t}$

Total conductance $= \dfrac{1}{R + R_1} = \dfrac{i}{e_B} = \dfrac{y_t}{x_a}$

FIG. 7.3   *Current-voltage relations in a series-resistive circuit.*

written

$$(7.9) \qquad \frac{i_C}{e_B} = \frac{G}{1 + (G/C)(1/s)} = \frac{1}{R + 1/sC}$$

Equation (7.9) is of particular interest in this discussion. The left side of this equation has already been encountered under different circumstances. In Chap. 3 this ratio i/e was described as the conductive property of a system, and the units were given as reciprocal ohms (or mhos). Although in this chapter the ratio i/e in Eq. (7.9) is described as a transfer function (with no other physical significance), by previous association the temptation to assign conductive units to this operator is very strong. In fact, the denominator of the transfer function is composed of R and $1/sC$. The first of these has the units of ohms. If R and $1/sC$ are additive, we must assume that $1/sC$ also has the units of ohms. The transfer function would then have the units of conductance. This viewpoint becomes more strongly reinforced if the capacitor of Fig. 7.2 is replaced by a resistor, as shown in Fig. 7.3. We would not now hesitate to define the conductance as $1/(R + R_1)$, nor would we hesitate to write

$$(7.10) \qquad \frac{i}{e_B} = \frac{1}{R + R_1}$$

The pattern now begins to unfold. The correspondence between Eqs. (7.10) and (7.9) is too pointed to be ignored, and on the basis of this comparison one might interpret $1/sC$ as the "resistance" of a capacitor and the transfer function formed by the ratio of an across- to a through-variable as a "conductance." This intuitive viewpoint is very useful except that the nomenclature must be changed to avoid confusion. Resistance implies frictional or lossy properties (as energy loss). To speak of the "resistance" of a *pure* capacitor (or storage element in general) is contradictory. By our definition, a pure-storage element has only storage capability; there is no lossiness associated with it. To indicate that $1/sC$ has similar implications as resistance (both have units of ohms) but that it does not indicate resistance as we know it, we describe $1/sC$

as the impedance of a capacitor.* Sometimes this is called the parametric impedance of a capacitor, to distinguish it from the sinusoidal impedance (a related concept that will be examined in the next chapter). In either case, impedance implies "hindrance of flow" under certain conditions but not irretrievable loss of energy, as might be associated with a resistive element. In keeping with this new terminology, we now refer to $R + 1/sC$ as the "total impedance" of a series RC circuit; similarly, the transfer function e/i can be designated as the impedance of a network.

With this intuitive interpretation of the significance of $1/sC$, we now examine the basic current-voltage relationship for a capacitor to show how the impedance for this storage element may be established directly from its defining equation. The defining equation for a capacitor is Eq. (7.3) $[i = C(de/dt)]$. In operational form this is written as $i = Cse$. If e is taken as the stimulus (input) to the capacitor and i as the response (output), then the transfer function of the capacitor is e/i or $1/sC$. The ratio of voltage to current (the across- to a through-variable) is now also described as impedance. The impedance of the capacitor may therefore be written

(7.11)   Impedance of capacitor $= Z = \dfrac{1}{sC}$

where Z is the standard symbol for impedance.

We can now easily extrapolate these ideas to describe the impedance of an inductivelike storage element. The current-voltage relationship for such an element is

$$(7.12) \qquad e = L\frac{di}{dt}$$

By transforming ($e = sLi$) and forming the ratio e/i, the impedance of the inductor is

$$(7.12a) \qquad \frac{e}{i} = Z_L = sL$$

We shall have occasion to use these relations for storage-element impedance in later sections.

It is appropriate also to introduce here the concept of admittance.

* This has been called the reactance of a capacitor. Impedance is then taken to mean the combined lossy and storage effects of a "nonpure"-storage element. Thus the resistance of a *lossy* storage element (such as a flexible hose) and the *reactance* of this element (i.e., its effect as a storage element) make up the impedance of the element. Strictly speaking, then, the impedance of a nonlossy element ($R = 0$) is its reactance. This seems to add more confusion, and for our purposes we shall not use the term reactance but shall refer only to the impedance of an element. In most cases the meaning will be clear. Ambiguous situations will be separately considered.

Admittance is reciprocal impedance in the same sense as conductance is reciprocal resistance. As ohms are the units of impedance, the units of admittance are reciprocal ohms or mhos. The standard symbol for admittance is Y. Since admittance and impedance are reciprocally related, we have $Y = 1/Z$. The admittance of a capacitor is thus $Y_C = sC$ and that of an inductor is $Y_L = 1/sL$.

With this introduction it now becomes very simple to derive directly the transfer functions related to a network. The transfer function relating i_C and e_B in Fig. 7.2b is, of course, $i_C/e_B = Y$ since $Y = 1/Z$ and $Z = R + 1/sC$.

$$(7.13) \qquad \frac{i_C}{e_B} = \frac{1}{Z} = \frac{1}{R + 1/sC}$$

Similarly, since $e_C = i_C Z_C$, we can replace i_C in Eq. (7.13) and obtain

$$(7.13a) \qquad \frac{e_C}{e_B} = \frac{Z_C}{Z}$$

(This is the more general form of the voltage-divider relationship derived in Appendix A, step 18. See also Prob. 7.2.) Since Z_C is $1/sC$, we have

$$(7.13b) \qquad \frac{e_C}{e_B} = \frac{1/sC}{R + 1/sC} = \frac{1}{s\tau + 1}$$

which is the relationship given by Eq. (7.6). We see that, by using model equations and the impedance concept, it is possible to derive transfer functions directly from a network without first determining the differential equation to relate the network variables.

7.3 THE TRANSFER FUNCTION FROM IMPEDANCE: PREDICTION OF PERFORMANCE

To show how the newly developed concept of impedance may be used to advantage, we examine the system in Fig. 7.4a. A flexible compliant tube is connected to an expansible, single-port vessel. Of interest is the flow induced by a pressure applied at the entrance of the tube. An analog simulation of the system is shown in Fig. 7.4b. The two resistors with an intermediate capacitor (R_b, C_b, R_s) represent the flexible conducting tube (see Fig. 4.20); the remaining capacitor represents the compliance of the expansible single-port vessel (see Fig. 4.3).

From the statement of the problem, we require the transfer function of the system where the flow is the output and the applied pressure is the

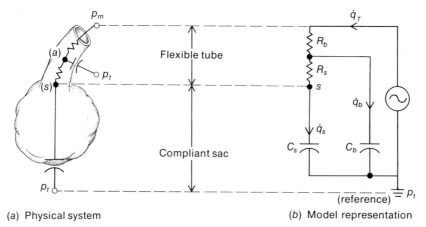

(a) Physical system   (b) Model representation

FIG. 7.4 *An analog representation of a flexible tube feeding a single-port compliant vessel.*

input (see Fig. 7.5*a*). In mathematical notation, we might state that we require $T(s)$, where

(7.14) $$T(s) = \frac{\dot{q}}{p_m}$$

Since $\dot{q}/p_m$ is analogous to the ratio of current to voltage (based on the circuit analog in Fig. 7.4*b*), the transfer function is defined by the admit-

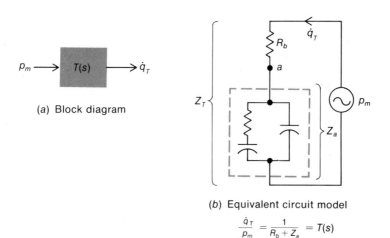

(a) Block diagram

(b) Equivalent circuit model

$$\frac{\dot{q}_T}{p_m} = \frac{1}{R_b + Z_a} = T(s)$$

FIG. 7.5 *Alternative representations for system in Fig. 7.4.*

tance of the analog. We might, in fact, immediately write

(7.15) $\qquad T(s) = \text{total circuit admittance} = Y_T = \dfrac{\dot{q}_T}{p_m}$

or

$$T(s) = \frac{1}{\text{total circuit impedance}} = \frac{1}{Z_T} = \frac{\dot{q}_T}{p_m}$$

Because R_b is in series with a combination of other circuit elements (see Fig. 7.5b) it is easier to use the second form of Eq. (7.15) to evaluate the system transfer function. From Fig. 7.5b we can easily see that if Z_a represents the combined impedance effect of R_s and C_s in parallel with C_b, the total circuit impedance can be written

(7.16) $\qquad Z_T = R_b + Z_a{}^*$

Because of the parallel arrangement of elements, it is easier to evaluate Z_a by using admittances. In Fig. 7.6a the impedance Z_a is redrawn to

---

* It should be intuitively evident that *impedances* in *series* are added to find the total impedance effect. To prove this relationship, one can use Kirchhoff's current law to establish that the same current must flow through all impedances in series. The rest is algebraic manipulation. See Prob. 7.1. A similar comment can be made about admittances. If it is not intuitively evident that *admittances* in *parallel* are added to determine total admittance, this relationship can be established by appropriate application of Kirchhoff's laws (Prob. 7.1).

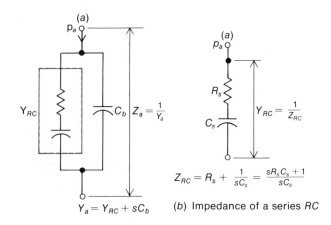

(a) Admittance of Z_a

(b) Impedance of a series RC

FIG. 7.6 *Impedance and admittance of circuit combinations.*

show the combination of parallel elements. From the diagram we see that

$$(7.17) \qquad Z_a = \frac{1}{Y_a} = \frac{1}{Y_{RC} + sC_b}$$

where Y_{RC} is the admittance of the series combination of R_s and C_s (see Fig. 7.6b). This value is determined from the series impedance as shown in Fig. 7.6b. Therefore Z_a can be written

$$(7.18) \qquad Z_a = \frac{1}{sC_s/(sR_sC_s + 1) + sC_b} = \frac{sR_sC_s + 1}{s[sR_sC_sC_b + (C_s + C_b)]}$$

and Z_T is finally given by

$$(7.19) \qquad Z_T = R_b + \frac{sR_sC_s + 1}{s[sR_sC_sC_b + (C_s + C_b)]}$$

After much algebraic manipulation* we can finally write the system transfer function as

$$(7.20) \qquad T(s) = \frac{s(s + 1/\tau_{bs})/R_b}{s^2 + s(1/\tau_{bs} + 1/\tau_b) + 1/\tau_b\tau_s} = \frac{\dot{q}_T}{p_m}$$

where the various time-constants are as defined in the footnote. This may appear to be a complex procedure but after a little practice the mechanics are routine; when applicable, the impedance evaluation provides a relatively straightforward technique for obtaining a system transfer function.

Once the transfer function relating flow and pressure is obtained, other transfer functions can be easily determined. The total volume delivered to the system, for example, is defined by $q_T = (1/s)\dot{q}_T$ (i.e., if

* To outline the algebra:

$$R_b + \frac{sR_sC_s + 1}{s[sR_sC_sC_b + (C_s + C_b)]} = \frac{s[sR_sC_sR_bC_b + (C_s + C_b)R_b] + sR_sC_s + 1}{s[sR_sC_sC_b + (C_s + C_b)]}$$

$$Z_T = \frac{[s^2 + s(C_s + C_b)R_b/R_sC_sR_bC_b + sR_sC_s/R_sC_sR_bC_b + 1/R_sC_sC_bR_b]R_sC_sR_bC_b}{s[s + (C_s + C_b)/R_sC_sC_b]R_sC_sC_b}$$

$$Z_T = \frac{\{s^2 + s/[R_s(C_sC_b/C_s + C_b)] + s/R_bC_b + 1/R_sC_sC_bR_b\}R_b}{s\{(s + 1)/[R_sC_sC_b/(C_s + C_b)]\}}$$

$$Z_T = \frac{[s^2 + s(1/\tau_{bs} + 1/\tau_b) + 1/\tau_s\tau_b]R_b}{s(s + 1/\tau_{bs})}$$

where

$$\tau_{bs} = R_s \frac{C_sC_b}{C_s + C_b}$$

$C_sC_b/(C_s + C_b)$ is the equivalent of C_s and C_b in series

$$\tau_b = R_bC_b$$

$$\tau_s = R_sC_s$$

initial conditions are neglected, $q = \int \dot{q}\, dt$). The transfer function relating volume to pressure is thus

$$(7.21) \qquad \frac{q_T}{p_m} = \frac{(1/s)\dot{q}}{p_m} = \frac{1}{s}\, T(s) = \frac{(s + 1/\tau_{bs})/R_b}{s^2 + s(1/\tau_{bs} + 1/\tau_b) + 1/\tau_b\tau_s}$$

In a similar manner, if we wish to describe only the flow into the single-port vessel (into C_s) we can do so by noting that the flow through the $R_s C_s$ combination is proportional to the pressure drop across this series impedance. In equation form this is written

$$(7.22) \qquad \dot{q}_s = \frac{p_a}{Z_{RC}}$$

where $\dot{q}_s$ is the flow to C_s, p_a is the pressure drop across the $R_s C_s$ combination, and Z_{RC} is the impedance of this combination (see Fig. 7.6b). To define p_a we note this pressure drop is due to the total flow into the combination of $R_s C_s$ in parallel with C_b. From Fig. 7.6a, this relationship may be written

$$(7.23) \qquad p_a = \dot{q}_T Z_a$$

where p_a, $\dot{q}$, and Z_a are as described in Fig. 7.6a. Since $\dot{q}_T$ has already been defined in terms of the excitation p_m [see Eq. (7.15)] Eq. (7.22) can be written

$$(7.24) \qquad p_a = \frac{Z_a}{Z_T}\, p_m = \frac{Z_a}{Z_a + R_b}\, p_m{}^*$$

and $\dot{q}_s$ is now given by

$$(7.25) \qquad \dot{q}_s = \left(\frac{Z_a}{Z_a + R_b}\right)\frac{p_m}{Z_{RC}}$$

We can express Eq. (7.25) in terms of element components by expanding the various impedance terms. Thus, if we use

$$Z_a = \frac{1}{Y_a} = \frac{1}{(Y_{RC} + sC_b)}{}^{\dagger}$$

then

$$(7.26) \qquad \dot{q}_s = \frac{p_m}{(1 + R_b Y_a)Z_{RC}} = \frac{p_m}{Z_{RC} + R_b Z_{RC}(Y_{RC} + sC_b)}$$

and

$$(7.27) \qquad \dot{q}_s = \frac{p_m}{Z_{RC} + R_b + R_b Z_{RC} s C_b}$$

* See Prob. 7.2.
† From Eq. (7.17) or Fig. 7.6a.

Further expansion is possible if Z_{RC} is replaced by $R_s + 1/sC_s$. When we make this substitution and collect terms, we finally get

$$(7.28) \qquad \frac{\dot{q}_s}{p_m} = \frac{sC_s/\tau_b\tau_s}{s^2 + s(1/\tau_{bs} + 1/\tau_b) + 1/\tau_b\tau_s}$$

It is of interest to note that $\dot{q}_s/p_m$, $\dot{q}_T/p_m$, and q_T/p_m all contain the same function of s in the denominator. This is not an unusual result, and it might have been expected intuitively. The variables of $\dot{q}_s$, $\dot{q}_T$, and q_T are intimately related; any changes in one must necessarily cause corresponding changes in the other. Since the responses of these interrelated variables are governed by the denominators of their transfer functions,* it would seem reasonable to expect these denominators to be similar or identical. In general, the through- and across-variables in a network are defined by transfer functions with the same denominator.

With the transfer function relating flow and pressure now available, we can predict flow for a proposed pressure stimulus. If, for example, we postulate the application of a step-input pressure of amplitude P_m, then p_m is $P_m u(t)$ or $(P_m/s)\, \delta(t)$; the flow into the system then becomes

$$(7.29) \qquad \dot{q}_T = \frac{P_m}{R_b} \left[\frac{s + 1/\tau_{bs}}{s^2 + s(1/\tau_{bs} + 1/\tau_b) + 1/\tau_b\tau_s} \right] \delta(t)$$

Although it would now be possible to find a general solution for $\dot{q}_T$, the arithmetic is rather tedious and the result is complex. For illustrative purposes, we shall obtain a solution of Eq. (7.29) when specific values are assigned to the various system elements. This will serve to review the mechanics of partial fraction expansion and to illustrate how the component elements affect measured data. To relate this to physiological systems, we shall assume the network of Fig. 7.4b is a model of the respiratory system. The correspondence between the physiological elements and the circuit elements may be established by reference to Fig. 7.7.

The respiratory system is made up of conducting air passages which are ultimately terminated by the alveolar sacs. (The latter are microscopic, elastic baglike tissues. Gas interchange between the lungs and blood vessels takes place at the alveolar level.) The upper air passages are for the most part rather rigid with little extensibility. They may be represented by the resistor labeled R_t in Fig. 7.7c. The flexible airways have some extensibility and are represented by resistors and an intermediate capacitor (R_s, C_b, R_s in Fig. 7.7c). The alveoli are by far the most flexible components of the system. These elastic, single-port elements are collectively represented by a capacitor C_s. When the various elements in

---

* The time responses are determined by the terms $\delta(t)/(s + a)$; the denominator factors thus govern the time response of the system variables.

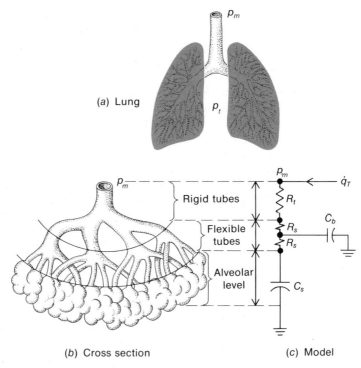

(a) Lung

p_m

p_t

Rigid tubes

Flexible tubes

Alveolar level

p_m

R_t

R_s

R_s

C_b

C_s

$\dot{q}_T$

(b) Cross section   (c) Model

FIG. 7.7 *Development of a lung model.*

Fig. 7.7c are combined and redrawn, the result is the circuit of Fig. 7.4b (where $R_t + R_s = R_b$).

It is easy to describe the qualitative relationship between the electrical and physiological components, but it is not so easy to decide what numerical values should be used in the circuit simulation. The physiological system is not neatly separated into elements as proposed by the network, and measurements on the physiological system do not necessarily reflect the magnitude of the separate elements. A choice of numbers for model components may often involve guesswork to start and experimentation to verify the numbers assigned. As we are interested in this particular problem only as an illustrative example, we shall arbitrarily assume a set of values for the various circuit elements.* Thus C_b will be taken as 0.02 l/cm H_2O; C_s will be 0.18 l/cm H_2O; the resistance of the upper respiratory tract (R_b) will be taken as 1.6 cm $H_2O/l/s$, and that of the lower tract (R_s) will be taken as 0.08 cm $H_2O/l/s$.

* To preserve continuity, the explanation of how these numbers were chosen will be deferred to a supplementary discussion at the end of this chapter.

With these numerical values, we are now prepared to use Eq. (7.29) to evaluate the waveform for the air flow $\dot{q}_T$ to (or from) the respiratory system when the pressure across the system undergoes a step change (P_m) from one constant level to another. When the various equation constants are determined,* Eq. (7.29) becomes

$$(7.30) \qquad \dot{q}_T = \frac{P_m}{R_b}\left(\frac{s + 70}{s^2 + 109s + 273}\right)$$

If we had a table which included the operator function given by Eq. (7.30), we could immediately express $\dot{q}_T$ in the time domain. Without such a table, it is necessary to convert the s into a recognizable form so that it can be interpreted as a function of time. We do not have any experience with the transform in Eq. (7.30) as we have not encountered a transfer function with a quadratic expression in the denominator. Our experience has been only with denominators expressed as the product of monomial factors. However, since quadratics are the expanded form of two monomial factors, we can factor the denominator and convert Eq. (7.30) into a familiar form. We can then write Eq. (7.30) as

$$(7.31) \quad \dot{q}_T = \frac{P_m}{R_b}\frac{s + 70}{(s + 106.4)(s + 2.6)} = \frac{P_m}{R_b}\frac{s + 1/\tau_{bs}}{(s + 1/\tau_1)(s + 1/\tau_2)}$$

The solution for $\dot{q}_T$ may now be obtained by partial-fraction expansion in the manner described in the previous chapter:

$$(7.32) \quad \dot{q}_T = \frac{P_m}{R_b}\left(\frac{0.35}{s + 106.4} + \frac{0.65}{s + 2.6}\right) = \frac{P_m}{R_b}\left(\frac{A}{s + 1/\tau_1} + \frac{B}{s + 1/\tau_2}\right)$$

or

$$\dot{q}_T = \frac{P_m}{R_b}\left(0.35\epsilon^{-106.4t} + 0.65\epsilon^{-2.6t}\right) = \frac{P_m}{R_b}\left(A\epsilon^{-t/\tau_1} + B\epsilon^{-t/\tau_2}\right)$$

* From a footnote on an earlier page we have

$$\tau_{bs} = \frac{R_s C_s C_b}{C_s + C_b}$$

By substitution of the appropriate numerical values as "derived" above, we have

$$\tau_{bs} = \frac{(0.8)(0.18)(0.02)}{0.18 + 0.02} = \frac{0.00288}{0.2} = 0.0144 \text{ s}$$

In a similar manner,

$$\tau_s = R_s C_s = (0.8)(0.18) = 0.144$$

$$\frac{1}{\tau_s} = 70$$

$$\tau_b = R_b C_b = (1.6)(0.02) = 0.032$$

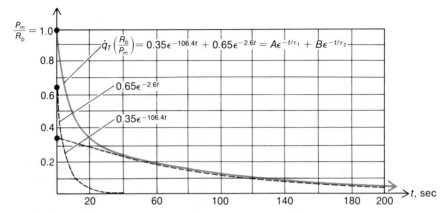

FIG. 7.8 *Flow response of the system in Fig. 7.4 when stimulated by a step-pressure.*

Equation (7.32) constitutes a solution for the flow response of the system shown in Fig. 7.4b when a step excitation pressure is applied at p_m. A plot of this response curve is shown in Fig. 7.8. (The plot is normalized with respect to the coefficient P_m/R_b.) This curve represents the flow waveform that would be detected at the mouth if interthoracic pressure were suddenly changed in a stepwise fashion.

Theoretical studies such as the one just described not only help to predict system performance; they can be used also to indicate how practical data may be utilized to evaluate system properties. As a simple illustration of this point, if we accept the model of Fig. 7.4b as a representation of the lung (see Fig. 7.7), then Eq. (7.32) indicates that the resistance R_b can be determined directly from the flow response to a step-excitation pressure. In particular, Eq. (7.32) indicates that immediately after the application of the step pressure (i.e., at $t \approx 0$) the flow is given by $\dot{q}_T = P_m/R_b$. This implies that, if the flow response to an applied step is recorded from the physiological system, the resistance R_b (which might be interpreted as the resistance of the upper respiratory tract) can be evaluated as

$$(7.33) \qquad R_b = \frac{P_m}{\dot{q}_T \text{ (at } t = 0)}$$

It might appear that the conclusion given by Eq. (7.33) is not of much use as it is restricted to the response of a step excitation. It will be shown that this result is fairly general and is applicable to any suddenly applied pressure stimulus. This may appear somewhat surprising yet it is not unreasonable, because a system subjected to an abrupt pressure rise does not know whether the pressure will drop or remain constant thereafter. We would expect, therefore, that an abrupt pressure change of a given magnitude would always elicit the same initial response from a system regardless of how the pressure signal subsequently appeared.

Another important point established by the theoretical considerations in the early part of this section is the complex relationship that may exist between system properties and the response-curve time-constants. It should be evident from the earlier discussions that it is relatively easy to define the time-constants of the response curve when the system elements are defined; however, it is not so simple to use the curve time-constants to evaluate the system elements. We can demonstrate the latter comment by examining the illustrative example of Figs. 7.7 and 7.4b.

When the various circuit elements for Fig. 7.4b were defined, it was easy to show that the flow curve should decay with the time-constants $\tau_1 = 1/106.4$ and $\tau_2 = 1/2.6$ [see Eqs. (7.31) and (7.32)]. Let us now reverse the situation and propose that the flow curve is available and the time-constants are found to be $\tau_1 = 1/106.4$ and $\tau_2 = 1/2.6$ (evaluated by peeling off exponentials). How would we relate this information to the system elements? To answer this question, we can go backward and note that the curve time-constants (τ_1 and τ_2) are defined by the factors of the denominator of the transfer function [see Eqs. (7.31) and (7.32)]. The denominator of the transfer function, however, is defined by the system elements [see Eqs. (7.29) and (7.30)]. The factored form of the transfer function must therefore be used to relate the system elements to the *curve* time-constants. In short, to relate the *curve* time-constants to the system properties, the second form of Eq. (7.31) should be used where τ_1 and τ_2 are now defined by

$$(7.34)^* \qquad \frac{1}{\tau_1} = \left[\left(\frac{1}{\tau_{bs}} + \frac{1}{\tau_b} \right) + \sqrt{ \left(\frac{1}{\tau_{bs}} + \frac{1}{\tau_b} \right)^2 - \frac{4}{\tau_b \tau_s} } \right] \div 2$$

$$(7.35)^* \qquad \frac{1}{\tau_2} = \left[\left(\frac{1}{\tau_{bs}} + \frac{1}{\tau_b} \right) - \sqrt{ \left(\frac{1}{\tau_{bs}} + \frac{1}{\tau_b} \right)^2 - \frac{4}{\tau_b \tau_s} } \right] \div 2$$

To find how the time-constants of the *curve* are related to the properties of the system we must express these constants in terms of system properties. To do so, we replace the constants τ_{bs}, τ_b, and τ_s in Eqs. (7.34) and (7.35) by equivalent expressions in R_s, R_b, C_s, and C_b (see the footnote on page 323 for the definitions of these constants). The result after substitution and rearrangement is given by Eq. (7.36). (As the

---

* These are obtained by using the quadratic formula $(-b \pm \sqrt{b^2 - 4ac})/2a$ to factor the denominator of Eq. (7.29).

algebra is somewhat tedious it has been omitted and only the final result shown.)

$$(7.36) \quad \frac{1}{\tau_1}, \frac{1}{\tau_2}$$

$$= \frac{\left(\dfrac{1 + C_s/C_b}{R_s C_s} + \dfrac{1}{R_b C_b}\right) \left\{1 \pm \sqrt{1 - \dfrac{4 R_s C_s R_b C_b}{[R_b(C_s + C_b) + R_s C_s]^2}}\right\}}{2}$$

Although a rather unwieldy expression, this serves to emphasize that the time-constants of the *curve* are defined by the system properties; the relationships between these time-constants and the system elements are, however, not obvious or simple and they cannot be obtained by inspection. This point was mentioned in Sec. 6.8 where it was stated that "the time-constant of a physical curve is not always simply related to the constants of a physical system." This is a good illustration of the possible complex effect system properties may have on recorded curves.

To show how we can use the ideas just presented in a more general situation, we shall again use the respiratory system for an illustrative example and again assume this system may be represented by the model of Fig. 7.4b. In this instance, however, we shall not assume a step excitation and known circuit elements. Instead, we shall assume the pressure excitation during a respiratory test appears as shown in Fig. 7.9a and the resultant flow response curve is given by Fig. 7.9b.* The circuit elements have not been defined; we expect to use the recorded data to evaluate these elements. From the information in these curves, it should be possible to set up four equations to determine the four unknown system constants (R_s, R_b, C_s, and C_b) associated with the circuit of Fig. 7.4b.

To obtain the required relationships, we can use the experience gained in the previous discussion and can expect that the four needed equations will be determined from the initial flow, two time-constants of the flow curve, and one other relationship. The latter relationship was

---

* Such data might be obtained during a respiratory test in which the patient exerts maximal expiratory effort after maximal inspiration. Unfortunately, we cannot be completely realistic in this case as it is difficult to find appropriate data in the literature. Pressure vs. time curves are not generally recorded during tests of forced respiration (or they are not often reported in the literature). The author has devoted some effort to a literature search and has found pressure-volume and pressure-flow and flow-volume curves (expirograms) reported frequently, but very few researchers report simultaneous recordings of the pressure-time curve and the flow-time curve. From one paper in which such curves appear (Ref. 4, Fig. 7A) it is possible to see that the pressure curve during expiration starts with a sharp rise and decays gradually thereafter. Unfortunately, the reported pressure curves were not labeled, and so it was not possible to relate pressure and flow curves in a quantitative manner.

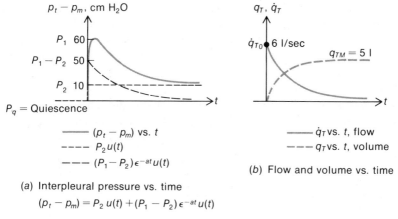

(a) Interpleural pressure vs. time

$$(p_t - p_m) = P_2\, u(t) + (P_1 - P_2)\, \epsilon^{-at}\, u(t)$$

(b) Flow and volume vs. time

FIG. 7.9 *Pressure and flow curves during maximal expiratory effort.*

not specifically mentioned in the previous discussion but it was mentioned earlier in this section and should be obvious by inspection. The volume of gas exhaled during expiration is related to the change of interpleural pressure by the system compliance. During the expiration maneuver the interpleural pressure undergoes a change of P_2 units. (See Fig. 7.9; interpleural pressure, at the start, is given by the quiescent pressure P_q. At the end of the test, interpleural pressure is P_2 units above quiescence. The total change in interpleural pressure is thus P_2.) Because of this pressure difference the internal gas volume has been decreased. This volume is, of course, the total volume of air exhaled and can be determined from the recorded-volume vs. time curve (q_{TM} in Fig. 7.9b). Since P_2 and q_{TM} are both static measurements, their ratio defines the total compliance of the system, and we may write

(7.37)
$$C_s + C_b = \frac{q_{TM}}{P_2}$$

We now have one of the required equations.

To obtain another equation which relates the system constants, we examine the pressure-flow relationship at the onset of the pressure stimulus. From an earlier discussion we would expect that the initial pressure and the contingent flow may somehow define the upper-airway resistance. If we work out the details, we would find that, although the applied pressure is not a step, this airway resistance may again be defined by Eq. (7.33) ($R_b = P_m/\dot{q}_{T0}$). We can prove this in a general way by using what is called the initial-value theorem.* However, we shall use a more intuitive approach based upon the resolution of a signal into components. A signal with an abrupt rise (or fall) can always be represented as the sum of a step and some other time function. The pressure curve of Fig. 7.9a, for example, may be represented as shown in Fig. 7.10. The actual signal is assumed to be composed of a step-function $P_1u(t)$ and another signal (assumed to be a negative signal with exponential characteristics). If we examine the response due to each of the components (Fig. 7.10b) we see the step gives rise to an initial response; the other signal starts

---

* This is one of the general properties of the Laplace transform. A proof of this theorem is given in most texts dealing with transform analysis (e.g., J. A. Aseltine, "Transform Method in Linear System Analysis," p. 122, McGraw-Hill Book Company, New York, 1958). The theorem states that if a transform in s is multiplied by s and the limit taken as $s \to \infty$ the result will be the initial value of the time function. If we use Eq. (7.29) as an example,

$$\dot{q}_T = \left(\frac{P_m}{R_b}\right) \frac{s + 1/\tau_{bs}}{s^2 + s(1/\tau_{bs} + 1/\tau_b) + 1/\tau_b\tau_s}$$

[The $\delta(t)$ is omitted when Laplace-transform notation is used.] To find $\dot{q}_T$ when t is zero, we multiply the transform function by s and let $s \to \infty$:

$$\dot{q}_T(0) = \lim_{s \to \infty} s \left[\frac{P_m}{R_b} \frac{s + 1/\tau_{bs}}{s^2 + s(1/\tau_{bs} + 1/\tau_b) + 1/\tau_b\tau_s}\right] = \frac{P_m}{R_b}$$

In mathematical notation this theorem may be written

$$\lim_{s \to \infty} s [\dot{q}_T(s)] = \dot{q}_T(0)$$

It is appropriate here to mention also the final-value theorem. This theorem states that if a transform is multiplied by s and the limit is taken as s becomes zero the result will be the final value of the time function (if such a value exists; oscillatory functions, for example, do not have a final value). In mathematical notation the theorem states

$$\lim_{s \to 0} sF(s) = f(t) \qquad \text{as } t \to \infty$$

As an illustration, if we were interested in $q_T(\infty)$ (the total *volume* delivered) for the above example, we could then write

$$q_T = \frac{1}{s} \left[\frac{P_m}{R_b} \frac{s + 1/\tau_{bs}}{s^2 + s(1/\tau_{bs} + 1/\tau_b) + 1/\tau_b\tau_s}\right] = \text{volume delivered at any instant}$$

From the final-value theorem,

$$q_T(\infty) = \lim_{s \to 0} s \left\{\frac{1}{s} \left[\frac{P_m}{R_b} \frac{s + 1/\tau_{bs}}{s^2 + s(1/\tau_{bs} + 1/\tau_b) + 1/\tau_b\tau_s}\right]\right\} = \frac{P_m}{R_b} \frac{\tau_s\tau_b}{\tau_{bs}}$$

By appropriate substitution this becomes

$$q_T(\infty) = \frac{P_m}{R_b} \frac{R_sC_sR_bC_b}{R_sC_sC_b/C_s + C_b} = P_m(C_s + C_b)$$

This relationship should be compared with Eq. (7.37).

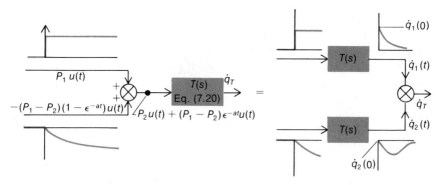

(a) Composite signal applied to network   (b) Components applied to network

FIG. 7.10   *Response of a system to a stimulus with an abrupt change.*

at zero and therefore does not provide any contribution to the initial flow response. The total initial response is thus due to a step acting alone; this was given by Eq. (7.33). For the particular case under investigation we can therefore write

(7.38)
$$R_b = \frac{P_1}{\dot{q}_{T0}}$$

This is the second of the required relationships.

We digress here to examine the implications of Eq. (7.38) from a circuit point of view. If we examine Eq. (7.38) and the circuit of Fig. 7.11b we note that the equation implies the capacitors (and R_s) are of no sig-

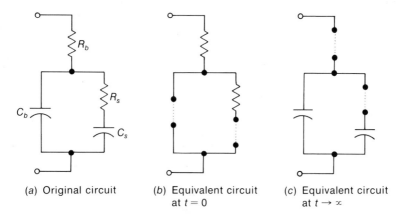

(a) Original circuit   (b) Equivalent circuit at $t = 0$ (c) Equivalent circuit at $t \to \infty$

FIG. 7.11   *Initial and final responses of an RC circuit.*

nificance during the sudden application of the stimulus; the circuit responds as if R_b were the only impedance present. From a network point of view, we might interpret this to mean that at $t = 0$ the circuit responds as if the capacitors were short-circuited (where shorting implies the capacitors are bypassed with zero-resistance conductors). A diagrammatic representation of a circuit that can be used to evaluate the initial response is shown in Fig. 7.11b. This circuit may be considered an equivalent model of the original model for representation at $t = 0$. (It is to be noted that, by shorting C_b, R_s is also shorted and R_b is the only significant opposition to flow.) This concept provides a convenient way to determine the initial response of a system when capacitive and resistive elements are present. The procedure is to consider capacitors as shorted elements and calculate the response based upon the resistors alone. A similar simplification is possible when inductivelike elements are included in the circuit. In this case, however, the inductors are replaced by open circuits and the initial response is calculated on this basis. We shall examine such a situation separately should it arise in a future discussion.

The equivalent circuit that may be used to determine the final response of an RC network is shown in Fig. 7.11c. The correspondence between this circuit and Eq. (7.37) should be self-explanatory. (See also the comments in the supplementary discussion at the end of this chapter.) Similar simplifications are possible when inductive elements are present except inductors are considered short circuits as $t \to \infty$.

To return to the evaluation of the system constants in the circuit of Fig. 7.4b, we must still obtain two more equations before these constants may be determined. Equations (7.37) and (7.38) provide two relationships between the circuit elements; since four circuit elements are shown, two more equations are needed. These two equations are in effect already available. Equation (7.36) shows how time-constants of the response curve are related to the network parameters. Although these relationships were derived on the basis of a step-function stimulus, they are also applicable for most aperiodic excitations that result in a transient response composed of a sum of exponentials. (Under some special circumstances this may not be true; such cases, however, are the exception rather than the rule, and for the quasi-linear systems we have here we can safely ignore these special situations.)

We can justify this contention by using the same procedure which was employed to show that the flow-response to a step consisted of two exponentials [see the discussion following Eq. (7.33)]. We use a slightly more general procedure in this case, however, and instead of proposing that the circuit excitation is a step we propose that the pressure stimulus is given by $p(t)$; the transform of this signal is $P(s)$. The flow into the circuit can now be written

$$(7.39) \quad \dot{q}_T = \frac{P(s)}{Z(s)} = P(s)\,T(s) = \frac{P(s)}{R_b} \left[\frac{s(s + 1/\tau_{bs})}{s^2 + s(1/\tau_b + 1/\tau_{bs}) + 1/\tau_b\tau_s} \right]$$

where $T(s)$ is given by Eq. (7.20). The quadratic denominator is factorable, and by employing the notation used earlier Eq. (7.39) may be rewritten

$$(7.40) \qquad \dot{q}_T = \frac{P(s)}{R_b} \left[\frac{s(s + 1/\tau_{bs})}{(s + 1/\tau_1)(s + 1/\tau_2)} \right]$$

where τ_1 and τ_2 are defined by Eq. (7.36). When $P(s)$ is specified, a total solution for $\dot{q}_T$ may be obtained. However, regardless of what $P(s)$ is, when Eq. (7.40) is expanded into partial fractions, two of these fractions will appear as $A/(s + 1/\tau_1)$ and $B/(s + 1/\tau_2)$. Specifically, Eq. (7.40) will expand into the function

$$(7.41) \qquad \dot{q}_T = \frac{A}{s + 1/\tau_1} + \frac{B}{s + 1/\tau_2} + \text{other terms}$$

where A and B are constants depending upon the system constants (τ_1 and τ_2) and the constants associated with the pressure excitation.* When Eq. (7.41) is converted to the time domain we have

$$(7.42) \qquad \dot{q}_T = A\epsilon^{-t/\tau_1} + B\epsilon^{-t/\tau_2} + \text{other terms}$$

This now shows that regardless of what the excitation is, the transient-flow response should be expected to include two exponential functions that have time-constants related to the system constants as given by Eq. (7.36); these would be the free-response terms (see the discussion of Sec. 5.10). Any other terms included in a description of the response curve would be considered the forced-response terms and would be attributed to the excitation. If the excitation is known, the free-response terms may be evaluated.

The method by which the response curve can be used to obtain equations relating to the system constants should now be clear: A mathematical expression for the response curve is proposed (by curve fitting, peeling exponentials, etc.). This description should contain exponential terms related to the system and excitation. The excitation curve is also described mathematically, and this is then compared with the response

---

* Under special circumstances one or both of these partial fractions may not appear. This will occur when $P(s)$ is defined so as to include the terms $s + 1/\tau_1$ and $s + 1/\tau_2$ in the numerator; these terms will cancel the equivalent ones in the denominator of the transfer function. This is, however, a rather special situation and would be the exception rather than the rule; such situations are of interest primarily from an academic point of view. For the practical situations we are examining here, we can safely ignore this unusual possibility.

curve. The exponential terms that do not appear in the stimulus curve are the free-response terms, and the time-constants of these terms are related to the system properties. By this procedure the additional equations required to determine the system constants of the proposed-model configuration are obtained.

To demonstrate the use of the concepts outlined in the previous discussion we should investigate an actual situation as an illustrative example. As mentioned earlier, however, there do not appear to be available in the literature appropriate data that can be used to investigate the transient behavior of the respiratory system. For the purposes of illustration, therefore, we shall use the hypothetical data shown in Fig. 7.9. The pressure excitation rises rapidly to a peak value (P_1) of 60 cm H_2O and decays exponentially thereafter with a time-constant of $\frac{3}{4}$ s ($a = \frac{4}{3}$) to a final value of 10 cm H_2O (P_2). (In a realistic situation the properties of this recorded curve would be established by analysis.) The flow- (and volume-) response curve is shown qualitatively in Fig. 7.9b and with numerical values in Fig. 7.12. From the previous discussion, we see we can immediately evaluate the upper-airway resistance (R_b) and the total system compliance ($C_s + C_b$): From Eq. (7.38) we have $R_b = \frac{60}{6}$ or 10 cm H_2O; from Eq. (7.37) $C_s + C_b = \frac{5}{10}$. To specify the other system constants we need two more equations. If we analyze the flow curve we find that it is composed of three exponential terms with time-constants 0.2, 0.75, and 6 s; the second of these is identified as the time-constant of the excitation pressure ($a = \frac{4}{3}; \tau = \frac{3}{4}$). The other two must therefore be associated with the system constants and must thus be given by Eq. (7.36). If we replace R_b by 6 and C_s by $0.5 - C_b$ (as derived a few sentences previously), Eq. (7.36) may be written

$$(7.43) \quad \frac{1}{\tau_1} = \frac{1}{0.2} = 5$$

$$= \frac{\left[\dfrac{0.5}{R_s(0.5 - C_b)C_b} + \dfrac{1}{10C_b} \right] \left\{ 1 + \sqrt{1 - \dfrac{40R_sC_b(0.5 - C_b)}{[5 + R_s(0.5 - C_b)]^2}} \right\}}{2}$$

$$(7.44) \quad \frac{1}{\tau_2} = \frac{1}{6}$$

$$= \frac{\left[\dfrac{0.5}{R_s(0.5 - C_b)C_b} + \dfrac{1}{6C_b} \right] \left\{ 1 - \sqrt{1 - \dfrac{40R_sC_b(0.5 - C_b)}{[5 + R_s(0.5 - C_b)]^2}} \right\}}{2}$$

These rather formidable-looking equations constitute a pair of simultaneous equations with two unknowns. A solution is not obvious or easy; the point is, however, that a solution is possible. By some means (a

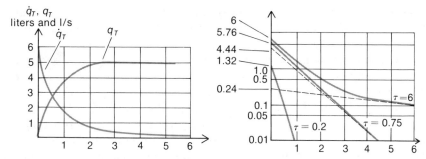

FIG. 7.12 *Flow response to the pressure excitation shown in Fig. 7.9.*

computer, trial and error, or by plotting) we could ultimately determine that C_b is 0.1 l/cm H_2O and R_s is 3 cm H_2O/l/s. With C_b known, we can finally find C_s to be 0.4 l/cm H_2O. The values of the four circuit elements are now known.

The above discussion would have been much more informative if realistic data could have been used. As such data do not appear to be available,* this may suggest the possibility of a worthwhile research investigation that would involve the simultaneous acquisition of pressure-flow data during transient respiratory maneuvers. From such data (and an appropriate examination of them) it might be possible to identify various system constants as implied by the circuit model. The circuit model may, of course, be too crude to serve as an adequate description of the system; it does, however, indicate the reasonable possibility of different regional resistances and compliances and how these system

* Since this writing, another paper was discovered in which the pressure and flow vs. time curves appear (see Ref. 4). Unfortunately, there was no opportunity to use the data of this paper, but it was interesting to note the flow curves shown in the published article resemble the theoretical curves given by Fig. 7.9. The pressure curve in the published paper is quite erratic and has a longer plateau than the curve we have used, but it rises sharply, as proposed by Fig. 7.9a. On the basis of this sharp rise, the R_b term as evaluated in the published recording is about 8 cm H_2O/l/s (compared with the value of 10 cm H_2O/l/s as determined from the data proposed here). It would appear, then, that the curves proposed by Figs. 7.9 and 7.12 are not unrealistic.

It is of further interest to note the resistance R_b is appreciably larger than the generally accepted value for the respiratory resistance (approximately 2 cm H_2O/l/s). This difference might be attributed to two conditions: First, the data here (and in the reference mentioned) are based upon a maximum expiratory effort. Such a respiratory maneuver would tend to compress many of the tubular airways and to increase the respiratory resistance. A second condition which should be considered is the technique used to measure the respiratory resistance. Such measurements are generally made within the range of normal breathing by methods that may reflect only a small portion of the respiratory resistance (see the supplementary discussion at end of this chapter). Before any specific statements can be made regarding the numerical values proposed, further experimentation would be required.

properties may be grossly estimated. This may then serve to guide an investigation of diseases (and drugs) that affect the various sections of the bronchial tree.

The discussion and investigation using the model circuit also bring to mind a number of suggestions and thoughts that might be considered when carrying out a research program as suggested above. The first is the advisability of using data from a forced-expiration test taken from maximal inspiration. For such a wide range of operating signals the resistances and compliances would be expected to vary appreciably during the test. This would affect the flow curve in a complex fashion, and there would be little hope of understanding the relationship between the flow response and the system properties. A more effective procedure might be to start a "moderate" forced-expiration test from a comfortable inspiration level. From these data a first estimate of the various R's and C's could be obtained. A second test from a higher inspiratory level would then indicate how the properties change.

The second suggestion regarding such a research investigation concerns the "choice" of excitation signal. The simpler the excitation signal, the easier the analysis will be. It may therefore be useful to examine the feasibility of a controlled-pressure stimulus. This may be possible if the patient undergoing the test can monitor his own intrapleural pressure (actually intraesophageal pressure is the measurement used; see the discussion associated with Fig. 4.13 in Sec. 4.4). If the pressure measurement is displayed on an oscilloscope (or any other visual display) so that the subject can see the effect of his actions, he might learn to generate a relatively flat pressure curve. This might then be represented as a step-function and the mechanics of analysis would thereby be reduced significantly. We see once more that models and systems concepts can be useful not only for analysis but as a tool to guide a research investigation.

7.4 IMPEDANCE FROM THE TRANSFER FUNCTION: THE IDENTIFICATION PROBLEM

In the preceding section we saw that if a circuit model may be used to represent a physical system, a system transfer function can be derived from the model by evaluating the circuit impedance. The validity and applicability of the transfer function depend, of course, on the adequacy of the circuit representation. If the circuit model is a close representation of the physical system, the transfer function provides a good mathematical description of the system; if not, the transfer function does not accurately predict system response and a revision of the circuit is required.

In this section we examine the reverse problem: We use the signal information to determine a transfer function and consider this a circuit impedance. From this impedance function we can postulate a circuit model and can thereby gain insight into the identification of various system properties. By this approach we are, in essence, synthesizing the circuit model from the transfer function, which was derived from signal information.

The transformation from transfer functions to circuit models is not a new technique; network synthesis has been studied in great detail recently, and there is now a wealth of information on the subject.* In fact, the technique has been developed to such a degree that tables are available to indicate which networks can be used to simulate a given transfer function.† In some cases the procedure is routine and a network synthesis can be programmed on a computer (or a computer-aided synthesis is feasible). As mentioned earlier, to go into the details of network synthesis is beyond the scope of this book; we examine the concept only to indicate how it is related to the investigation of system properties. Readers interested in the theory and mechanics of network synthesis should refer to the abundant literature on the subject. Some appropriate references are listed at the end of this chapter.

To show how the impedance concept may be used for identification, we shall examine the diffusion process as related to physiological systems. The diffusion system was chosen for illustration to avoid repetition (and monotony) and to show the wide applicability of the concepts thus far discussed. Unfortunately, however, diffusion as a subject is not generally covered to the same extent as other physical processes (e.g., heat flow, fluid flow, etc.; this is particularly true for those educated in the engineering sciences). It may be appropriate here, therefore, to include a brief qualitative discussion of diffusion as it relates to physiological processes before we embark on a quantitative examination of a diffusion system. (For those familiar with diffusion concepts, this discussion may be superfluous; these readers may skip the next few pages and go directly to the discussion of Fig. 7.15.)

Diffusion is the process whereby dissolved molecular particles become uniformly distributed throughout a fluid as a result of spontaneous molecular motion. This process is sometimes described as the intermingling of materials due to molecular motility. A homey example of diffusion is the dissolution of a grain of salt when dropped into a container of fresh water. The salt dissociates (into ions) and by diffusion will ultimately become distributed throughout the water (Fig. 7.13). The distribution of the salt molecules is due primarily to the incessant kinetic

* Texts appropriate for readers with backgrounds equivalent to that established here are Refs. 1 and 2.

† For example, see F. E. Terman, "Radio Engineers' Handbook," McGraw-Hill Book Company, New York, 1943.

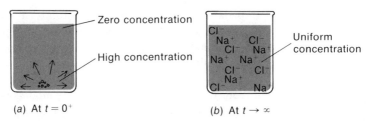

(a) At $t = 0^+$

(b) At $t \to \infty$

FIG. 7.13 *Diffusion in a fluid.*

motion of molecular particles. This motion causes a haphazard but continuous drift of salt particles from an area of high concentration to one of low concentration until there is an equal concentration throughout the fluid. As diffusion is due to molecular motion, the process is obviously temperature-sensitive. Diffusion is thus more rapid at elevated temperatures.

Although the foregoing illustration involves a fluid solution, it should not be inferred that diffusion is restricted to fluid media. Diffusion may take place in solids also. A good example of solid diffusion is found in the sandwiched coins introduced by the U.S. Mint. To save silver, quarters and dimes were coined by sandwiching a copper core between silver surfaces. Within a year one could see discoloration of the silver due to the diffusion of the copper. It is not surprising to learn, therefore, that diffusion of radioactive tracers into bone (or madder root as a coloring agent) serves as a technique to study bone metabolism and teeth formation.

Diffusion is the prime mechanism by which cellular tissue (and cells in general) obtains the necessary ingredients to support the life processes and discard the generated waste products. In most living systems nourishment and the product of metabolism are carried to and from a central source of supply (or elimination) by some fluid carrier. In mammals the blood supply serves as a carrier; in insects one of the carriers is in the form of a gas. These carriers deliver nourishment (or oxygen) to and collect waste from various localities within the physiological system. Final dissemination of nutrients to the cells and waste collection from the cells, however, are through the process of diffusion.*

* If living organisms had to depend *only* on diffusion for their oxygen supply, at least one dimension of the organism would probably not exceed 1 mm. Flat organisms would thus be 1 mm or less in thickness; spherical organisms would be 1 mm or less in diameter. This limitation is due to the low diffusion rate of oxygen and the rate of oxygen consumption needed for metabolism. A more detailed explanation of this natural limitation may be found in Ref. 5, chap. 9. This book contains a number of interesting and well-developed comparisons between various forms of living systems and might be described as an exposition of "living models of living systems."

A highly schematic view of the diffusion-transport system between the vascular (or blood) compartment and a cellular compartment is shown in Fig. 7.14. Endothelial cells are joined at their edges to form the tubular structure of the capillary tubes. The cells are bound together by what is conventionally called an intercellular cement (hyaluronic acid: see Ref. 5, Fig. 5.18). Material to and from the blood within the capillaries diffuses through the endothelial cells and through the intercellular cement. Transport between the cell surface and the capillary surface is through diffusion within the extracellular fluid. Transport between the intracellular and extracellular fluids is by diffusion through the cell wall.

We should at this point discriminate between active and passive transport processes. The mass transport process described in the previous paragraph is considered passive as it depends only on the concentration differences that exist between two points. We shall use the binary approach to describe the active transport process and define processes that do not depend upon concentration gradients to induce flow. This category includes transport by some carrier (as the blood) or ion-exchange transport where chemical reactions cause material modifications to produce or facilitate ion transport. Active transport does not, therefore, depend upon concentration gradient and may in fact oppose concentration gradient. The hypothetical and little understood "sodium pump" mechanism in the cell is an example of an active transport process. For the most part we shall not consider any aspects of the active transport mechanisms but shall be concerned only with the passive processes. We shall describe such processes in general by the all-inclusive term diffusion. Therefore diffusion (at least as used here) will be considered the passive process of mass transport due to concentration gradients. By

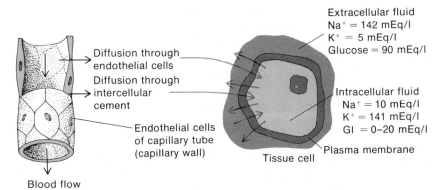

Extracellular fluid
$Na^+ = 142$ mEq/l
$K^+ = 5$ mEq/l
Glucose $= 90$ mEq/l

Diffusion through endothelial cells

Diffusion through intercellular cement

Endothelial cells of capillary tube (capillary wall)

Tissue cell

Intracellular fluid
$Na^+ = 10$ mEq/l
$K^+ = 141$ mEq/l
Gl $= 0–20$ mEq/l

Plasma membrane

Blood flow

FIG. 7.14 *Diffusion coupling between capillaries and cells.*

this definition osmosis would be considered a special case of diffusion where water migrates from areas of high water concentration to areas of low water concentration.*

One further comment may be made regarding the diffusion processes indicated by Fig. 7.14. In order to transport material between the cell interior and the capillary interior, diffusion must take place through a continuous medium (interstitial or extracellular fluid) and through boundary media (such as the endothelial cells and the cell membrane). However, in all cases the diffusion mechanism is essentially the same. The only difference is the point of view. In the first case (the interstitial fluid) the diffusion process is through a distributed system; in the other cases diffusion is through systems in which lumped properties are used to represent distributed properties. The endothelial cells and the cell membrane are in reality distributed systems but because of relative values it becomes convenient to consider them lumped systems. This will be more clearly understood when we consider the quantitative aspects of diffusion.

With this rather lengthy preliminary discussion, we are now in a position to apply impedance and synthesis concepts to problems involving diffusion and to show how these concepts may be used for identification of system properties. The first example will be a very simple one where the circuit will be obvious by inspection. This will serve to introduce the new terminology and concepts. The second example will be somewhat more involved and will more clearly indicate how impedance and synthesis may be used to identify system properties.

We could begin a quantitative examination of the diffusion process by stating Fick's law of diffusion and then showing how this may be applied to experimental observations. Rather than use this sequence, we shall use the reverse approach: We shall presume ignorance and attempt to derive the law from experimental observations. As in the case of nuclear radiation, this must be considered a tongue-in-cheek procedure as we know Fick's law exists (postulated 1855); any techniques or conclusions we evolve must evidently be prejudiced by hindsight. Nevertheless this procedure will lend insight into some of the applications and limitations of quantitative description of the diffusion mechanism. Further, it will serve to remind us that Fick did not have Fick's principle to guide him! He, too, had to depend upon a quantification of experimental data.

We start with the experimental knowledge that the cells of muscle

---

* There are other interpretations of mass-transport processes. Snell et al., for example, lists three mechanisms (see Ref. 6, p. 192). The first is described as transport by passive factors; the second as active transport; the last, transport by chemical conversion. In general, however, these interpretations (and others) do not differ significantly from that stated in the text.

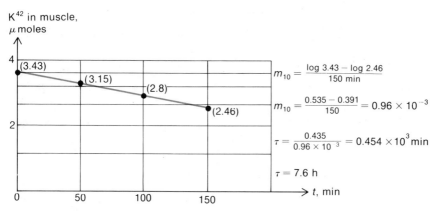

FIG. 7.15   *Efflux at 25°C of potassium from 57.2 mg of frog sartorius muscle incubated for 40 h at 4°C in normal Ringer's solution labeled with K⁴² [adapted from G. N. Ling, Cell Membrane and Cell Permeability, Ann. N. Y. Acad. Sci., **137**:2 (1966)].*

tissue take up potassium ions when immersed in a solution of high potassium-ion (K^+) concentration. Conversely, when such cells are saturated with K^+ they deliver K^+ when immersed in solutions of low K^+ concentration. Ling,[*] among others, has investigated the efflux of K^+ ions from saturated muscle cells. The procedure outlined by Ling is to incubate frog sartorius muscles in Ringer's solution,[†] labeled with isotopic potassium, K^{42}, for a long period of time at constant temperature (see the legend under Fig. 7.15 and the description of the input signal in Fig. 7.16). This permits the muscle to take up K^{42} ions until a fixed internal concentration is attained. The muscle tissue is then washed in a nonlabeled Ringer solution ($K^{42} = 0$) so that some of the K^{42} may diffuse out. A scintillation counter is used to determine the weight of K^{42} in the muscle as washing proceeds. A semilogarithmic plot of the results is shown in Fig. 7.16. The straight-line semilogarithmic "response" suggests a first-order system.

To establish this contention on a more quantitative basis we must translate the data and procedure into mathematical terms. One possible approach is to organize the data in a block-diagram format where input and output signals have been appropriately indicated. This step is in itself often a productive one as it forces a researcher to focus attention

* Ref. 7. That particular issue (entitled Biological Membranes: Recent Progress) contains a wealth of information regarding recent membrane investigations.

† Ringer's solution is a "physiological fluid" that contains the necessary ingredients to maintain cell life for an extended period. It is one of many physiological fluids (see Ref. 10, p. 39).

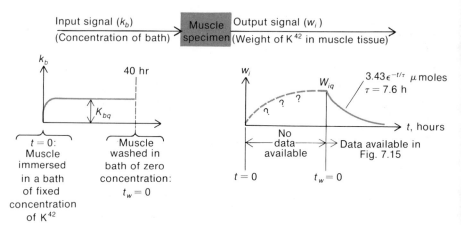

FIG. 7.16   *Block diagram to represent test for study of K^{42} diffusion in frog sartorius muscle.*

on the pertinent variables involved in an investigation. These are not always as obvious as might be imagined. In this case the significant variables are relatively clear-cut: The input signal is the bath concentration used to bathe the muscle specimen; the output signal is the weight of potassium ions retained by the muscle during washing in nonlabeled solution. An appropriately labeled block diagram is shown in Fig. 7.16. The input signal is represented by k_b (concentration of bath) and the output signal by w_i (weight of K^{42} in tissue). The input waveshape is shown as a pulse to indicate the tissue is first immersed in a bath of fixed concentration ($k_b = K_{bq}$) for 40 h and then washed (at $t_w \geq 0$) in a solution of zero K^{42} concentration ($k_b = 0$).

We can only guess about the first portion of the response waveform as no data were obtained during the 40-h incubation period. The curve for the uptake of K^{42} is thus unknown. During the washing period, however, data are available in the form of a semilogarithmic plot (Fig. 7.15). From this information we can describe the response signal as an exponentially decaying signal, as shown by the solid curve in Fig. 7.16. We are now in a position to describe the input-output relationships analytically and thus define the muscle-tissue transfer function. Since the data available are for the period $t > 40$ h (that is, $t_w > 0$) we can consider conditions only during the washing period. There are a number of ways to describe the situation during this period. The simplest would be to assume that prior to the washing period the system is at quiescence. The concentration of K^{42} in the muscle and in the bath is constant at the same value; the weight of K^{42} within the muscle tissue remains constant with time (see Fig. 7.17). At $t_w = 0$ we assume a negative step is

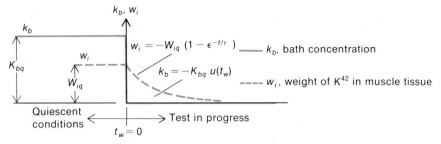

FIG. 7.17 *Input- and output-signal representations of washing muscle in nonlabeled solution.*

applied to the bath; in this manner the bath concentration is reduced to zero. The labeled potassium within the cell then automatically decays according to the recorded data. We can describe these signals analytically as

$$(7.45) \qquad k_b = -K_{bq}u(t_w) \qquad \text{or} \qquad k_b(s) = \frac{-K_{bq}}{s}\,\delta(t)$$

and

$$(7.46) \qquad w_i = -W_{iq}(1 - \epsilon^{-t/\tau})$$

or

$$w_i(s) = \frac{-W_{iq}/\tau}{s(s + 1/\tau)} \qquad \text{(See Table 6.1, curve 5)}$$

where K_{bq} and W_{iq} are the quiescent values of the bath concentration and the weight of labeled potassium in the muscle tissue, respectively. From the operational forms given by Eqs. (7.45) and (7.46) we can now determine the muscle transfer function to be

$$(7.47) \qquad T(s) = \frac{\sigma_{\text{out}}(s)}{\sigma_{\text{in}}(s)} = \frac{w_i(s)}{k_b(s)} = \frac{W_{iq}/K_{bq}\tau}{s + 1/\tau}$$

Equation (7.47) justifies the contention made earlier: The muscle tissue seems to behave as a first-order system.

To gain further insight into the physiological system we might consider the use of a circuit analog. Intuitively (and from previous background) we realize that such an analog would consist of some simple combination of a resistance and a storage element. (The system is first-order.) Some possible choices are shown in Fig. 7.18. To decide which of these circuits would be appropriate we could determine the impedance of each of the circuits and attempt to match them with the transfer function given by Eq. (7.47). If we did this, we would find that any of the

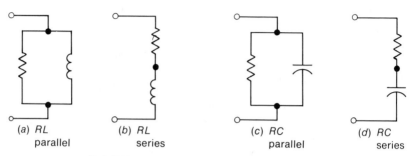

| (a) RL | (b) RL | (c) RC | (d) RC |
| parallel | series | parallel | series |

FIG. 7.18 *First-order resistive-storage combinations.*

circuits in Fig. 7.18 could serve as an analog, depending upon how we identify analogous variables (see Prob. 7.5).

To retain the analog associations developed earlier, we shall use a reverse procedure: We shall first attempt to relate analogous variables in a familiar manner; on this basis we shall then attempt to identify the system transfer function as an impedance or an impedance ratio. The variables with which we presently are concerned are concentration (k_b) and weight of labeled potassium; the potassium flows to or from the muscle, depending upon whether the ion concentration in the muscle is higher or lower than that in the bath. The concentration gradient between the muscle and the bath seems to govern the ion flow (see Fig. 7.13). From the latter description of the relationship between the variables of the diffusion system, we associate concentration with voltage ($k_b \sim e$) and ion *flow* with current ($\dot{w}_i \sim i$). For this choice of analogous variables the ratio $\dot{w}_i/k_b$ is equivalent to the ratio i/e; the latter ratio is identified as a circuit admittance (see also Table 3.1). In mathematical notation we can therefore write

(7.48)
$$\frac{\dot{w}_i}{k_b} = Y$$

or

$$\frac{k_b}{\dot{w}_i} = Z$$

If we examine our original transfer function [Eq. (7.47)], we now see that the analog circuit we seek will have an impedance function equivalent to $1/sT(s)[w_i/k_b = T(s); sw_i/k_b = \dot{w}_i/k_b = sT(s); k_b/\dot{w}_i = 1/sT(s)]$. The circuit impedance function would thus be

(7.49)
$$Z = \frac{s + 1/\tau}{sW_{iq}/K_{bq}\tau} = \frac{K_{bq}\tau}{W_{iq}} + \frac{1}{s(W_{iq}/K_{bq})}$$

In this form we can finally recognize that the circuit we seek is the familiar series RC circuit shown in Fig. 7.19c, where R is given by $K_{bq}\tau/W_{iq}$ and C by W_{iq}/K_{bq}. This result might have been anticipated, based upon past experience. The point to be established here, however, is that the transfer function can be used to identify an impedance function, and from this a model circuit can be proposed. In the next section we shall see that, although the circuit simulation is not obvious by inspection, the transfer function indicates what model might be appropriate.

Now that we have associated the various system constants with electrical elements, we should examine the physical significance of the two components. The capacitive element is easy to interpret from the meaning of the magnitude of the capacitor. Its value is given by the ratio of constant (or quiescent) terms. These are essentially a ratio of steady input and output signals; a check of units indicates that the ratio defines a storage property of the system (as it should; based upon our choice of analogous variables $\dot{w}_i \sim i$ and $k_b \sim e$, capacitors should represent storage elements). As a specific check, if concentration is given in micromoles per cubic centimeter and weight of K^{42} is given in micromoles, then W_{iq}/K_{bq} defines a storage property of the system with units of volume. We would expect that in this case the volume units for the storage property have some significance. It would seem reasonable, in fact, to relate this property to the "internal muscle volume" where K^{42} may enter as a solute. We can, in a way, justify this view because the quiescent bath concentration (K_{bq}) and the quiescent concentration of K^{42} within the muscle must be the same; the latter concentration must be the quiescent weight of K^{42} distributed throughout the volume of the muscle tissue. We may therefore consider K_{bq} to be W_{iq}/V_i, where

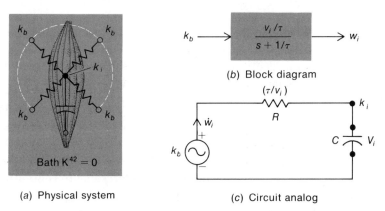

(a) Physical system   (b) Block diagram   (c) Circuit analog

FIG. 7.19 *Representations of muscle during washing operation.*

V_i represents the muscle volume where K^{42} may exist as a solute.* We see now that the storage property of the muscle defined by W_{iq}/K_{bq} is simply V_i, and Eq. (7.47) can now be written

(7.50) $$\frac{w_i(s)}{k_b(s)} + \frac{V_i/\tau}{s + 1/\tau} \quad \text{(See Fig. 7.19}b\text{)}$$

As a quick check of our circuit model, we note that the system resistive property is defined by $K_{bq}\tau/W_{iq}$. This can now be written τ/V_i. In this form the resistive property is defined by the time-constant and a storage property. This is not an unexpected result, since the time-constant of an RC circuit is given by the product of R and C, where C represents the system storage property. The ratio of the time-constant and the storage element should thus define the system resistive property.

To obtain a better "feel" for the physical significance of this resistive property (represented by the electrical resistor) we examine some of the implications of the circuit model that has been developed. The circuit clearly indicates that the flow of solute (labeled potassium) from the bath to the muscle volume is opposed by some resistive mechanism. This resistive property resides beneath the muscle surface area somewhere between the external bath and the "internal muscle volume." Although this "muscle volume" has not been clearly or specifically defined, it is a convenient way to differentiate between the "inside" and the "outside" of the muscle system. If we accept this viewpoint, we can then refer to a concentration of K^{42} within the muscle volume. If we paraphrase this last statement in terms of analogous quantities, we can refer to the voltage of the storage element (or the voltage across the capacitor) and can locate a point in the circuit to represent this internal concentration. The voltage at the node labeled k_i in the circuit of Fig. 7.19c may thus be considered the analog representation of the internal concentration within the muscle system.

The picture of the physical system we have evolved may now be described by the diagram in Fig. 7.19a. The muscle interior is seen as a storage volume or a repository for K^{42}. This is schematically represented by a capacitor. The concentration of K^{42} within the muscle is designated as k_i. The muscle surface is thought of as a "flow barrier" which impedes the transfer of solute from the outside of the muscle to the inside. This resistive property is schematically represented by resistors (which are essentially connected in parallel).

We can now easily obtain a mathematical relationship to describe Fick's law, which deals with the diffusion of solutes due to concentration

---

* Although it is a simple matter to assign an algebraic symbol to represent this volume, it is no simple matter to describe this volume physically (if it can be done at all).

gradients. From the circuit of Fig. 7.19a we note the flow of K^{42} may be given by

$$(7.51) \qquad \qquad \dot{w}_i = G(k_b - k_i)$$

Equation (7.51) states the diffusion of a solute through a surface is proportional to the concentration gradient across the surface and some property of the surface. This equation is a simple version of Fick's (first) law of diffusion, which was first stated in 1885. The constant of proportionality G (the conductance) is a function of the physical dimensions of the surface and the properties of the surface (i.e., the material). To eliminate the effect of area, Fick's law is generally expressed in terms of flow per unit area. If flow is given by $\dot{w}_i$ and the surface area is taken as A, then the flow per unit area is $\dot{w}_i/A$. The term $\dot{w}_i/A$ is sometimes called the solute flux, or simply flux; the units are mass per area (i.e., μmoles/cm^2). If we use the symbol $\mathcal{F}$ to represent flux, then Eq. (7.51) becomes

$$(7.52) \qquad \qquad \mathcal{F} = \frac{G}{A}(k_b - k_i) = \mathcal{P}(k_b - k_i)$$

This is another form of Fick's law in terms of the conductance per unit surface area G/A. The latter term is generally described as the *permeability* of the surface. It has the units of length/time (e.g., cm/s) and is sometimes represented by the symbol $\mathcal{P}$.

The permeability of a surface is not entirely a property of the surface material. It also depends to some extent upon the geometry of the system. If we apply Eq. (7.52) to a cellular structure in particular, we can see how permeability may be related to a system dimension as well as to the system material. In Fig. 7.14 the intracellular and extracellular fluids of a cell are separated by a boundary plasma membrane. A mass flow (flux, $\mathcal{F}$) will be generated through the membrane if the extracellular concentration (k_b) and intracellular concentration (k_i) are not equal and the membrane permeability is not zero. For a fixed concentration difference it is evident that the greater the membrane permeability, the greater the flux. It should be intuitively clear now that the membrane permeability is not a function of only the material constituent of the membrane. It may also be considered to be dependent upon the membrane thickness; for a given membrane, the thicker the material, the less the flux and hence the less the permeability. We can express this concept mathematically by assuming that permeability varies inversely with membrane thickness. This is generally expressed in equation form as

$$(7.53) \qquad \qquad \mathcal{P} = \frac{D}{x}$$

where D, the diffusion coefficient (or diffusivity), is considered to be a function of the membrane composition only and x is the membrane thickness. With this new relationship we can now write Fick's law in a more popular form:

$$(7.54) \qquad \frac{dw_i}{dt} = \frac{DA}{x}(k_b - k_i)$$

or

$$\mathfrak{F} = D\frac{k_b - k_i}{x}$$

The term $(k_b - k_i)/x$ is interpreted as the concentration difference per unit membrane thickness and is often called the concentration gradient across the membrane. This gradient is generally thought of as the change in concentration per change in thickness and is expressed as dk/dx. One therefore often sees Fick's law expressed as

$$(7.55) \qquad \frac{dw_i}{dt} = DA\frac{dk}{dx}$$

To complete the discussion of this section and to gain an appreciation for the numerical values and units involved in diffusion studies, we shall estimate some of the properties of the frog sartorius muscle from the experimental data of Fig. 7.15. We cannot obtain accurate results as there are not sufficient data available about the sample specimen for such purposes. However, by using rough approximations we can obtain an indication of the order of the magnitudes involved. Thus, if we estimate the density of muscle tissue to be approximately that of water (muscle tissue is about 75 percent water and 25 percent protein; the density of protein is closely equivalent to that of water), then the volume of 57.2 mg of muscle tissue is about 0.0572 cm^3. If we assume that the K^{42} ion is distributed throughout this volume, then 0.0572 cm^3 may be taken to be the value of V_i in Eq. (7.50). Since the time-constant of the system was shown to be approximately 0.45×10^3 min or 27×10^3 s (see Fig. 7.15) we can estimate the muscle diffusion resistance to potassium flow to be $(27 \times 10^3)/(57 \times 10^{-3})$, or about 0.5×10^6 s/cm^3 ($\tau = RV_i$; therefore $\tau/V_i = R$). The reciprocal property of conductance is thus about 2×10^{-6} cm^3/s.

To evaluate the permeability of the muscle surface we need the surface area of the muscle. Unfortunately, this information is not available, and we can only make some crude guesses. If we arbitrarily assume that the shape of the muscle sample is spherical, then the surface area that contains a volume of 0.057 cm^3 would be about 0.72 cm^2. The permeability in this case would be approximately $(2 \times 10^{-6})/0.7$ cm/s or about 3×10^{-6} cm/s $(\mathscr{P} = G/A)$. We would suppose that this would be the highest figure that could be expected, as a sphere has the minimum surface area for a given enclosed volume. If we assume the muscle sample is a cylinder about $3\frac{1}{2}$ cm long (the approximate length of frog sartorius muscle), then the surface area that includes a volume of 0.057 cm³ would be about 1.4 cm² and the permeability would then be approximately 1.5×10^{-6} cm/s. An accepted figure for the permeability of skeletal muscle to the passage of a K^+ flux is about 1×10^{-6} cm/s.* In view of the rough approximations employed, the results derived are surprisingly close to those obtained by more thorough and accurate determinations.

The permeability of a material to flux flow is a property of the composition and the length of the diffusion path [Eq. (7.53)]. For a situation such as diffusion into muscle tissue the composition of the diffusion path varies (i.e., the diffusion path is not homogeneous), and the diffusion path is not clearly defined. Therefore we cannot apply Eq. (7.53) directly to muscle tissue unless we make some gross simplifications about the muscle structure. It is known that a strip of muscle tissue is composed of elongated muscle fibers (see Fig. 7.20). Each of these fibers is enclosed in a membrane sheath which may be described as a plasma membrane. The diffusing solute must pass through these membranes to reach the fiber interior. If we lump all the membrane sheaths into one membrane enclosure, we obtain a model as shown in Fig. 7.20c. The membrane thickness is equivalent to the perpendicular paths through fiber membranes that the solute must travel to become distributed throughout the muscle-fiber interiors. If we assume that the specimen used in this discussion is a cylinder about $3\frac{1}{2}$ cm long by 0.15 cm diameter ($\mathscr{P}$ is then about 1.5×10^{-6} cm/s) there will be about 30 fibers across the diameter (each fiber is assumed to be about 50μ). The thickness of a plasma membrane is of the order of 100 Å.† The total thickness of all the membranes along a diameter is 6000 Å; one membrane wall of the model is thus about 3000 Å or 3×10^{-5} cm. If we now assume that the permeability of the muscle is limited primarily by the membrane (i.e., the fiber interiors are highly permeable), the diffusion constant of the membrane $(= \mathscr{P}x)$ is $(1.5 \times 10^{-6})(3 \times 10^{-5})$ cm²/s to the passage of K^{42}; thus D of the membrane for K^{42} is 4.5×10^{-11} cm²/s. It is of interest to note that if the membrane were composed entirely of water the diffusion coefficient would be about 10^{-3} cm²/s. It appears, then, that water is at least 1 million

* Scheer gives the permeability of frog muscle to K^{42} as $\log(1/\mathscr{P}) = 6.0$, where $\mathscr{P}$ is in centimeters per second. From this relation, $\mathscr{P}$ is 1×10^{-6} cm/s. (Ref. 8, p. 245, table 21.)

† Ref. 9; see chap. VI in particular. This book provides excellent background for membrane studies in general.

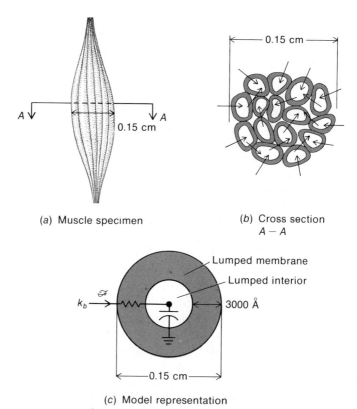

(a) Muscle specimen

(b) Cross section
A − A

(c) Model representation

FIG. 7.20   *Muscle model to calculate membrane diffusion coefficient.*

times more permeable to K^{42} than is the membrane. This might indicate how effective a diffusion barrier the cell membrane can be.

7.5   CIRCUIT MODEL OF HIGHER–ORDER SYSTEMS

The example of the preceding section introduced the relationship between the transfer function and circuit impedance. The simplicity of the network may, however, have deemphasized the potential utility of associating an impedance (and hence a network) with a transfer function. In the case chosen, the transfer function was familiar, and the circuit model that could be used to simulate this function was already known. In most cases the association between model and transfer function is not obvious by inspection. This is particularly true for systems with transfer functions higher than second order and is often true even for second-order systems. In such cases the impedance concept (and network-synthesis techniques) may be used to good advantage. In this section we shall examine a system that gives rise to a second- and third-order transfer function and show how a circuit model may then be developed from an impedance interpretation.

Having gained some experience with diffusion processes, we shall use this system once more to illustrate the relationship between transfer functions, impedances, and network syntheses. In this example, however, we examine sodium rather than potassium-ion diffusion. Ling (Ref. 7) investigated this aspect of diffusion by using a procedure similar to the one previously outlined. A muscle specimen was kept under incubation for 25 h at 2°C in Ringer's solution labeled with isotopic sodium, Na^{22}. The specimen was then washed in sodium-free solution for various time intervals. The amount of sodium retained by the muscle was determined by a radioactive-assay procedure after each washing. The results of the tests were plotted on semilog paper, as shown by the solid curve in Fig. 7.21. The tail of the response curve shows definite straight-line characteristics. However, the curved section at the initial portion indicates that we cannot assume the response may be described by a single exponential function with a constant coefficient. Either the exponent must be considered time-variant, or we must assume the curve may be represented by two or more functions with invariant time-constants. The former premise would be used if we expect that a nonlinear representation is necessary, and the latter if we assume a linear analysis is adequate. As a linear analysis is easier and can be quickly checked out, it is generally employed first, and as the linear analysis illustrates the concept to be demonstrated it will be the one used here.*

* If it appears that the system has been stimulated beyond the range where a linear analysis might be useful, ideas derived from "linear thinking" can still be used to gain insight into the operation of the system. In this case, for example, the response curve finally appears to have exponential (linear) characteristics. It would seem reasonable to assume that this system may actually have first-order properties but that the system time-constant does not remain fixed. This indicates that the system might be represented by a simple RC network as shown below where R and/or C is variable. This might be interpreted to mean that diffusion resistance (or the effective diffusion volume, or both) varies with time, with the applied concentration, or with the degree of solute dispersion. These assertions would each

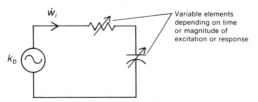

in turn have to be checked either by a literature search or further experimentation. In the latter case, it would probably be wise to reduce the excitation (if possible) to ascertain

Concentration of Na^{22}in cell μmoles/g

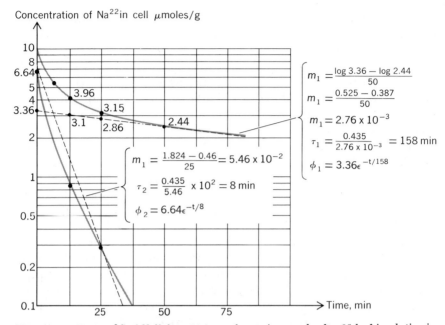

$$m_1 = \frac{\log 3.36 - \log 2.44}{50}$$

$$m_1 = \frac{0.525 - 0.387}{50}$$

$$m_1 = 2.76 \times 10^{-3}$$

$$\tau_1 = \frac{0.435}{2.76 \times 10^{-3}} = 158 \text{ min}$$

$$\phi_1 = 3.36\epsilon^{-t/158}$$

$$m_1 = \frac{1.824 - 0.46}{25} = 5.46 \times 10^{-2}$$

$$\tau_2 = \frac{0.435}{5.46} \times 10^2 = 8 \text{ min}$$

$$\phi_2 = 6.64\epsilon^{-t/8}$$

FIG. 7.21   *Efflux at 0°C of Na^{22} from 72.8 mg of sartorius muscle after 25 h of incubation in Na^{22}-labeled Ringer's solution at 2°C. [Adapted from G. N. Ling, Cell Membrane and Cell Permeability, Ann. N. Y. Acad. Sci., 137:2 (1966).]*

A linear analysis implies that the curve of Fig. 7.21 be described in terms of two or more exponential functions with fixed time-constants. One way to determine these is to peel off exponentials by the method described in Sec. 5.6. A straight line is drawn tangent to the tail of the response curve. The difference between the tangent line and the original curve is then plotted on semilogarithmic coordinates. A straight line is drawn tangent to this new curve and the procedure is repeated. This technique was employed to analyze the solid curve in Fig. 7.21. The semilogarithmic plots of the presumed exponential components are indicated by the dashed curves in Fig. 7.21. One straight-line function seems

whether the initial slope of the response curve may be changed to agree more nearly with the final slope. If this condition can be reached, it may be assumed that a first-order representation for the system is valid, and it could be proposed that system properties had been affected measurably by the applied stimulus. In fact, this procedure could show if the system resistance was stimulus-dependent. If the ratio of initial flow to applied excitation is reasonably constant, it would be assumed the resistive property stayed relatively constant; the storage property of the system must then have varied with the excitation. By evaluating the initial time-constant, some estimate of this variation can be obtained.

entirely justified; the second only roughly approximates the true curve. Although this approximation is admittedly crude, we accept it in this case as it provides sufficient accuracy for our purpose of demonstration. (The reader can refine this approximation as outlined by Prob. 7.6.) It should be added, however, that this simplified approximate procedure often makes sense from a mathematical and physiological point of view. It is pointless to embark upon an elaborate, complicated mathematical treatment until the need for such analysis has been demonstrated and the validity of the analysis justified by examining the physiological implications.

We can now see that, on the basis of the linearized analysis, the Na^{22}-efflux curve may be represented by exponentials with time-constants $\tau_1 = 158$ min and $\tau_2 = 8$ min (Fig. 7.21). As the ratio of these time-constants exceeds $2:1$ by a considerable factor, we may feel confident about the validity of the peel-off process and the numerical values obtained (see the footnote on page 199). We can now represent the efflux curve mathematically by

$$(7.56) \qquad \frac{w_i}{72.8 \times 10^{-3}\,g} = 3.36\epsilon^{-t_w/158} + 6.64\epsilon^{-t_w/8} \qquad \mu moles/g \text{ of muscle}$$

where w_i is the weight of tagged sodium in micromoles that remains in the muscle tissue after washing in nonlabeled solution for a time duration t_w.

From the description of the experimental procedure in the introduction of this section, it is clear that the block-diagram representation in Fig. 7.16 is also applicable to this investigation. The details of the experiment have, of course, been changed. In this case labeled sodium was used instead of labeled potassium; incubation time was 25 h instead of 40 h. The resultant response is now given by the efflux curve in Fig. 7.21. The difference between the sodium and potassium efflux curves (Fig. 7.15) is patently evident. The two systems (the sodium diffusion system and the potassium diffusion system) obviously have different properties. To examine the properties of the sodium system, we first determine the system transfer function; we need, therefore, a mathematical description of the input and output signals. The description of the input signal for this situation is the same as that of the potassium diffusion system and the mathematical relationship given by Eq. (7.45):

$$(7.45) \qquad k_b = K_{bq}u(t_w) = -\frac{K_{bq}}{s}\,\delta(t_w)$$

The output signal in this case is different and is almost that given by Eq. (7.56), but Eq. (7.56) is not complete as it includes initial conditions (that is, it does not describe conditions around a quiescent point).* We can modify Eq. (7.56) to give the desired result if we include the quiescent level of labeled sodium to make w_i zero when t_w is zero. Equation (7.56) then becomes

$$(7.57) \quad \frac{w_i}{0.0728} = (3.36\epsilon^{-t_w/158} + 6.64\epsilon^{-t_w/8} - W_{iq})u(t_w) \quad \mu\text{moles/g}$$

where W_{iq} is the quiescent weight of labeled sodium in the muscle. The numeric value of W_{iq} ($=10$ μmoles) can be obtained from the plot of Fig. 7.21 at $t_w = 0$. (From the way this problem was set up, the steady state is considered a condition where the muscle contained a constant level of Na^{22}. From the diagrams in Figs. 7.16 and 7.17 and the plot of Fig. 7.21 we can see that this would be interpreted as a quiescent weight of labeled sodium of 10 μmoles.) We can easily satisfy ourselves that Eq. (7.57) is of the correct form by permitting t_w to become zero. At this time w_i is zero, indicating initial conditions have been eliminated.

We can now transform the output signal to the s domain and then evaluate the system transfer function. The transformation of Eq. (7.57) is routine: From memory (or by reference to Table 6.1) we can write

$$(7.58) \quad \frac{w_i(s)}{0.0728} = \left(\frac{3.36}{s + \frac{1}{158}} + \frac{6.64}{s + \frac{1}{8}} - \frac{10}{s} \right) \delta(t_w) \quad \mu\text{moles/g}$$

or

$$\frac{w_i(s)}{0.0728} = \frac{-0.855(s + 0.0092)}{s(s + 0.125)(s + 0.00628)} \delta(t_w) \quad \mu\text{moles/g}$$

Now $T(s)$ $[=\sigma_{\text{out}}(s)/\sigma_{\text{in}}(s)]$ can be defined by the ratio of Eqs. (7.45) and (7.58):

$$(7.59) \quad T(s) = \frac{w_i(s)}{k_b(s)} \approx \frac{(0.855)(0.0728)}{K_{bq}} \frac{s + 0.01}{(s + 0.125)(s + 0.0063)}$$

where K_{bq} represents the bath concentration at quiescence. Unfortunately, a numerical value for this term was not recorded. However, if we once more use the approximation that muscles have the same density as water [see the discussion following Eq. (7.55)] then 72.8 mg of muscle has a volume of 0.0728 cm^3. At quiescence the muscle contains 10 μmoles/g of labeled sodium (see Fig. 7.21) or a total of 10 $\times$ 0.0728 μmoles. Therefore K_{bq} can be taken as (10 $\times$ 0.0728)/0.0728 cm^3 or 10 μmoles/cm^3.

---

* Our development requires that the mathematical equations describe *changes* about an operating point—not absolute values. Therefore, the equation required must describe the weight of Na^{22} *above or below* the quiescent level of 10 μmoles. Since at $t_w = 0$ (the beginning of the experiment) there can be no change above or below quiescence (i.e., the system *is* quiescent), the equation should be zero when $t_w = 0$. As Eq. (7.56) now stands, it does not have this property.

Equation (7.59) can now be written

$$(7.60) \quad T(s) = \frac{w_i(s)}{k_b(s)} \approx 62 \times 10^{-4} \frac{s + 0.01}{(s + 0.125)(s + 0.0063)} \quad \text{cm}^3$$

Equation (7.60) is a mathematical description of the system in terms of the transfer function. The constant terms reflect something about the system properties. From a "design" point of view, Eq. (7.60) is adequate as it stands. It can be used to predict response and to provide insight into the kind of excitation necessary to elicit a given response. From an identification point of view, however, the transfer function must be further examined to ascertain how the system properties are related to the transfer-function constants. One way to do this is to attempt to define a network that has the same properties as the system under investigation and then to try to interpret the meaning of the network elements physiologically. The network may be defined by associating a transfer function of the system with a network impedance (or admittance).*

As pointed out previously, a number of possible interpretations may be used (as illustrated by Fig. 7.18). To obtain a familiar kind of representation we first associate a set of system variables with a familiar set of circuit variables and then attempt to define an impedance function. If we use the analogy established in Sec. 7.4 ($k_b \sim e$ and $\dot{w}_i \sim i$; see the discussion following Fig. 7.18) an admittance transfer function would be the ratio $k_b/\dot{w}_i$. For this particular situation we might therefore consider the admittance function to be

$$(7.61) \quad \frac{\dot{w}_i}{k_b} = 62 \times 10^{-4} \frac{s(s + 0.01)}{(s + 0.125)(s + 0.0063)} \frac{\mu\text{moles/s}}{\mu\text{moles/cm}^3} = \text{cm}^3/\text{s}$$

Equation (7.61) is essentially the same as Eq. (7.60) except the ratio is now $(dw_i/dt)/k_b = \dot{w}_i/k_b$ instead of w_i/k_b. The transfer function given by Eq. (7.61) thus contains an s in the numerator not included in Eq. (7.60).

We are now in a position to search for a network with the admittance function given by Eq. (7.61). If we could not utilize a network-synthesis technique and we did not have any experience with this transfer function, we would be at an impasse at this point; it would not be easy to guess which network would satisfy the admittance requirements. We have, however, already encountered this admittance transfer function in a previous discussion. A review of Sec. 7.3 and an examination of Eq. (7.20) will indicate that the circuit in Fig. 7.4b has the desired admittance function. This circuit may therefore be used to simulate the system described by Eq. (7.60). Even for some relatively complex systems, it is

---

* We shall assume impedance and admittance can be used interchangeably in such general discussions unless one or the other is to be specifically mentioned. These particular situations will be described individually so that no ambiguity exists.

sometimes possible to use previous experience to associate a circuit analog with a given system impedance. This is, however, not generally the case; then some network-synthesis technique must be employed. As mentioned earlier, a detailed discussion of synthesis techniques will not be included here. However, later in this section one synthesis technique will be introduced to illustrate how the desired network could have been obtained from the given admittance function. Now, however, we shall take advantage of the established correspondence between a known circuit and a desired transfer function and use Eq. (7.20) in conjunction with Eq. (7.61) to determine numerical values for the circuit elements of Fig. 7.4b.

If we examine Eqs. (7.20) and (7.61) we see that the values of the various circuit components may be obtained by direct comparison. If, for example, we examine the coefficients of the two equations we note that $1/R_b$ may be associated with the constant coefficient of Eq. (7.61). In particular, we can write

$$(7.62a) \qquad \frac{1}{R_b} = 62 \times 10^{-4} \text{ cm}^3/\text{s} = G_b$$

By similar associations we can also write

$$(7.62b) \qquad \frac{1}{\tau_{bs}} = 0.01 = \frac{C_s + C_b}{C_s C_b} \frac{1}{R_s}$$

$$(7.62c) \qquad \frac{1}{\tau_{bs}} + \frac{1}{\tau_b} = 0.125 + 0.0063$$

$$(7.62d) \qquad \frac{1}{\tau_b \tau_s} = 0.125 \times 0.0063$$

By simultaneous solution we can now evaluate C_b to be 0.051 cm³, G_s to be 1.74×10^{-4} cm³/s, and C_s to be 0.027 cm³.* The model circuit with appropriate numerical values included is shown in Fig. 7.22.

To discuss in detail the possible implications of the circuit model in Fig. 7.22 would involve an understanding of the diffusion process beyond the level for which this book is intended. One or two simple items are of interest, however. The first is the existence of two storage repositories. From the circuit model it would appear that Na^{22} may be

* From Eqs. (7.62c) and (7.62b) we have

$$0.01 + \frac{1}{\tau_b} = 0.125 + 0063 = 0.01 + \frac{G_b}{C_b}$$

Since G_b is 62×10^{-4} cm³/s [Eq. (7.62a)] we can now write $0.125 + 0.0063 - 0.01 = (62.4 \times 10^{-4})/C_b$; therefore C_b is 0.051 cm³.

Since R_b and C_b are known, τ_b is known. We can therefore determine $\tau_s = R_s C_s$ from Eq. (7.62d). Since $R_s C_s$ and C_b are now available, they can be used in Eq. (7.62b) and C_s can be evaluated; R_s may then also be evaluated.

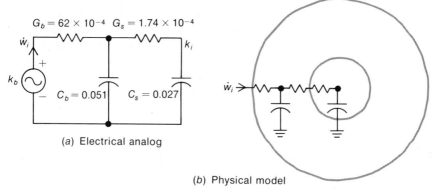

(a) Electrical analog

(b) Physical model

FIG. 7.22 *Representations of Na²² diffusion system.*

stored within two separate volumes of the muscle structure. If we accept
the model of muscle structure in Fig. 7.20c, we might consider one
storage volume to be the muscle cell wall and the other to be the internal
cell volume. If we associate these storage volumes with the capacitors
of the circuit analog, we can correlate the analog and the physical model
by a diagram such as Fig. 7.22b. From this diagram and the numerical
values associated with the circuit components, we note that the storage
repository within the cell (0.027 cm³) seems to be less than the available
storage volume in the wall. We might interpret this to mean the material
within the cell saturates at a relatively low sodium-ion concentration;
this would be in keeping with the physical observation that the Na
concentration within a cell tends to remain low.

The circuit analog indicates also that there are two resistive mechanisms associated with the sodium diffusion system in muscle. The
correlative diagram of Fig. 7.22b shows one possible interpretation of
this resistance distribution. R_s is assumed to be composed of two components: One component resides in the cell wall; the other resides within
the cell cytoplasm. If we can assume the former component represents
roughly one-half the total wall resistance (that is, $R_w/2$, where R_w is the
total wall resistance) we can write $R_s = (R_w/2) + R_c$, where R_c represents the cytoplasm component. The other resistance R_b is shown wholly
within the cell wall and might be considered the other half of the total
wall resistance ($R_w/2 = R_b$). If we accept this distribution of resistances,
the cell-wall resistance can be taken as 322 s/cm³ ($R_w/2 = R_b = 161$;
therefore $R_w = 322$ s/cm³) and the cytoplasm resistance as 5,590 s/cm³.
This crude analysis seems to imply that opposition to sodium-ion flow
may reside almost entirely within the cell cytoplasm and not within the

cell membrane, as generally postulated. This is, in fact, one of Ling's comments[*]; he states the cell membrane may not be a diffusion barrier to Na^+-ion efflux as postulated by Pfeffer. One cannot make any firm comments about the sodium diffusion system based upon this rather simplified analysis. There seems to be need for more work in this area, as the problem of Na^+-ion flux through cell membranes has by no means been resolved. Much more analytic and laboratory work remains to be done before questions involving diffusion to and from cells can be answered.

If there is a recognizable impedance function, network synthesis is unnecessary. In this last illustrative example the impedance function was familiar and it was possible to propose a circuit analog (by memory) that would serve as a system simulation. One cannot expect always to be this "lucky," and the more complex the systems become, the less lucky one is apt to be. In such cases it would be desirable to know that some techniques exist whereby an appropriate model can be derived. As was mentioned earlier, it is not the intent here to study network synthesis; to show what such a technique might involve, however, we shall examine one method of network synthesis: the development of an RC ladder network by continued fraction expansion.

The best way to illustrate the synthesis of the RC ladder network is to show how the impedance function of such a network may be obtained. A two-stage ladder network is shown in Fig. 7.23d. Figure 7.23a, b, and c shows the network in various stages of assembly. In Fig. 7.23, R_1 is shown in series with some impedance Z_{AB} (where Z_{AB} includes the network elements R_2, C_1, and C_2). The impedance of this simple combination is $R_1 + Z_{AB}$, which can be written $R_1 + 1/Y_{AB}$. In Fig. 7.23b, Y_{AB} is seen to be a combination of C_1 in parallel with another network designated Y_{CD}. Thus Y_{AB} can be written $sC_1 + Y_{CD}$; the total impedance can now be written

$$(7.63) \qquad Z_{\text{total}} = R_1 + \frac{1}{sC_1 + Y_{CD}} = R_1 + \frac{1}{sC_1 + 1/Z_{CD}}$$

In Fig. 7.23c Z_{CD} is evaluated to be $R_2 + Z_{ED}$ or $R_2 + 1/sC_2$, so that finally Z_{total} becomes

$$(7.64) \qquad Z_{\text{total}} = R_1 + \cfrac{1}{sC_1 + \cfrac{1}{R_2 + 1/sC_2}}$$

$$= \frac{s^2 R_1 C_1 R_2 C_2 + sC_1 R_1 + sC_2 R_1 + 1}{s^2 C_1 C_2 R_2 + (sC_2 + sC_1)}$$

The first form of Eq. (7.64) is often called a continued fraction. To obtain

[*] See Ref. 7, p. 839.

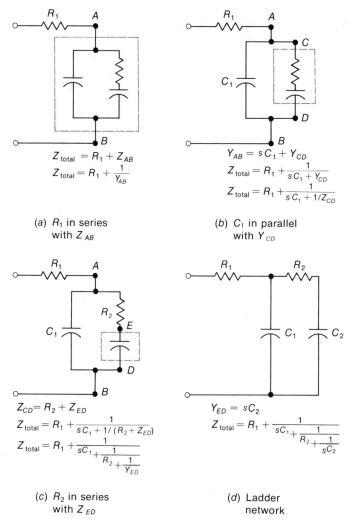

$$Z_\text{total} = R_1 + Z_{AB}$$
$$Z_\text{total} = R_1 + \frac{1}{Y_{AB}}$$

(a) R_1 in series
with Z_{AB}

$$Y_{AB} = sC_1 + Y_{CD}$$
$$Z_\text{total} = R_1 + \frac{1}{sC_1 + Y_{CD}}$$
$$Z_\text{total} = R_1 + \frac{1}{sC_1 + 1/Z_{CD}}$$

(b) C_1 in parallel
with Y_{CD}

$$Z_{CD} = R_2 + Z_{ED}$$
$$Z_\text{total} = R_1 + \frac{1}{sC_1 + 1/(R_2 + Z_{ED})}$$
$$Z_\text{total} = R_1 + \frac{1}{sC_1 + \dfrac{1}{R_2 + \dfrac{1}{Y_{ED}}}}$$

(c) R_2 in series
with Z_{ED}

$$Y_{ED} = sC_2$$
$$Z_\text{total} = R_1 + \frac{1}{sC_1 + \dfrac{1}{R_2 + \dfrac{1}{sC_2}}}$$

(d) Ladder
network

FIG. 7.23 *Impedance of two-stage ladder network.*

the total impedance in closed form, the various fractional expressions must be combined until a single final fraction is obtained [as shown by the second form of Eq. (7.64)].

We can now see how to determine a ladder-network configuration when given a network impedance function. If such a network is realizable,* the required network can be obtained by reversing the process of

---

* There is no guarantee that a given impedance function can be simulated by a ladder network. If a ladder network cannot be realized, some other synthesis technique must be employed.

combining the terms of a continued fraction; i.e., continued fraction expansion is used. The expansion simply involves dividing the numerator by the denominator to form a single term and another fraction. The latter fraction is then inverted and the procedure is repeated. The best way to illustrate this procedure is to perform an actual expansion. As an illustrative example, we use the impedance form of Eq. (7.61). This is written as

$$(7.65) \qquad \frac{k_b}{\dot{w}_i} = Z = \frac{(s + 0.125)(s + 0.0063)}{(62 \times 10^{-4})s(s + 0.01)}$$

or

$$Z = \frac{161s^2 + 21.1s + 0.129}{s^2 + 0.01s}$$

If we divide the denominator into the numerator of Eq. (7.65) we obtain

$$(7.66) \qquad Z = 161 + \frac{19.5s + 0.129}{s^2 + 0.01s}$$

This first form can be interpreted as a resistance of 161 units in series with an impedance given by $(19.5s + 0.129)/(s^2 + 0.01s)$; see Fig. 7.24a.

Equation (7.66) is now rewritten so that the second term appears in terms of an admittance (Y_1):

$$(7.67) \qquad Z = 161 + \frac{1}{(s^2 + 0.01s)/(19.5s + 0.129)}$$

$$= 161 + \frac{1}{(s/19.5) + [0.0034s/(19.5s + 0.129)]}$$

The first term of the admittance function is interpreted as a shunt capacitor as shown in Fig. 7.24b. To continue, we again invert the resultant fraction in the second form of Eq. (7.67) and repeat the process:

$$(7.68) \qquad \frac{19.5s + 0.129}{0.0034s} = 5{,}750 + \frac{1}{0.0264s}$$

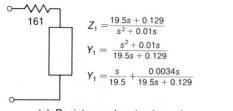

$$Z_1 = \frac{19.5s + 0.129}{s^2 + 0.01s}$$

$$Y_1 = \frac{s^2 + 0.01s}{19.5s + 0.129}$$

$$Y_1 = \frac{s}{19.5} + \frac{0.0034s}{19.5s + 0.129}$$

(a) Resistor and series impedance

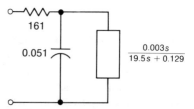

$$\frac{0.003s}{19.5s + 0.129}$$

(b) Resistor, capacitor shunt, and parallel admittance

FIG. 7.24 *Ladder realization from transfer function.*

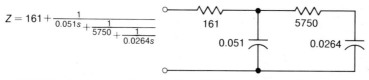

$$Z = 161 + \cfrac{1}{0.051s + \cfrac{1}{5750 + \cfrac{1}{0.0264s}}}$$

FIG. 7.25 *Ladder network for* $Z = (161s^2 + 21.1s + 0.129)/(s^2 + 0.01s)$.

The total expansion of the given impedance function is shown in Fig. 7.25 and with it the synthesized network. This network may be compared with that shown in Fig. 7.22a.

SUPPLEMENTARY DISCUSSION

In Sec. 7.3 it was shown that the circuit model of Fig. 7.4b may be considered a lumped representation of the respiratory system. The qualitative correspondence between the network and the physiological system was described briefly by Fig. 7.7 and the associated discussion. To assign numerical values to the circuit elements related to the physiological system, data derived from physiological measurements must be used and these data interpreted in terms of the circuit elements. In Sec. 7.3 various numbers were assigned to the circuit elements. The relationship between these numbers and physiological measurements and properties is described in this supplementary discussion.

The overall compliance of the physiological system may be taken as 0.2 l/cm H_2O. (See the discussion of Fig. 4.13.) This compliance is a measure of the storage of the system and is determined by the change in stored volume per change in pressure under no-flow conditions (see Sec. 4.2). The storage capability of the circuit analog is governed by the capacitors. Under no-flow conditions, the resistors of the circuit are of no significance (there is no voltage drop; the resistors may therefore be considered short circuits), and the circuit analog reduces to two capacitors in parallel. Therefore $C_b + C_s$ must provide for the storage capability of the analog; we then have $C_b + C_s = 0.2$ l/cm H_2O.

The saclike alveoli are far more extensible than the flexible airways; as an educated guess, we could consider 90 percent of the system compliance due to these elements and only 10 percent due to airway flexibility. With these figures, C_s would then be 0.18 l/cm H_2O and C_b would be 0.02 l/cm H_2O.

The overall resistance of the physiological system has been reported to be about 2 cm H_2O/l/s. It is somewhat difficult, however, to know the true significance of this figure as it can be interpreted in different ways, depending upon the method used for measurement. To illustrate this point, we examine one procedure which has been used to evaluate the respiratory resistance and then note the questions that may be raised about the interpretation of the results.*

---

* This and other methods are described by J. H. Comroe, Jr., R. E. Forster II, A. B. Dubois, W. A. Briscoe, and E. Carlsen, "The Lung—Clinical Physiology and Pulmonary Function Tests," 2d ed., The Year Book Medical Publishers, Inc., Chicago, 1962.

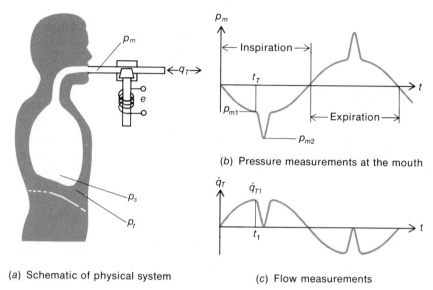

(a) Schematic of physical system

(b) Pressure measurements at the mouth

(c) Flow measurements

FIG. 7.26 *Technique to measure airway resistance.*

The procedure to be described makes use of measurements during interrupted flow. The patient breathes through a tube equipped with a solenoid valve (see Fig. 7.26); the valve can be triggered to close for an instant at any time during the respiratory cycle. The pressure and flow measurements taken during normal and interrupted breathing are used to estimate the resistance of the respiratory tract.

To show how these measured values may be related to resistance evaluations we note that, during a no-flow condition, the pressure throughout the respiratory system should be the same. Thus, if the flow in the respiratory system is zero everywhere, pressure measurements at the mouth will reflect pressures at the alveolar level. This concept is used to determine the pressure drop across the respiratory tract during a breathing cycle. Pressure at the mouth is continuously recorded during normal breathing (see Fig. 7.26b); at some instant (t_1) the solenoid is actuated and breathing is briefly interrupted. The effect on the pressure and flow curves is shown in Fig. 7.26b and c, respectively. Prior to valve closure, the pressure at the mouth is p_{m1} and the flow is $\dot{q}_{T1}$; when the valve closes for an instant, the mouth pressure drops to p_{m2} and the flow drops to zero. If the period of valve closure is a no-flow condition everywhere in the lung, and if the valve is closed long enough for the system to stabilize, the pressure p_{m2} may then be considered the alveolar pressure (p_s in Fig. 7.26a). This pressure may therefore be taken to be closely equal to the alveolar pressure just prior to valve closure. The pressure drop across the respiratory tract just prior to valve closure would thus be $p_{m1} - p_{m2}$; the flow into the lung at this time was $\dot{q}_{T1}$. These measured values are used to define the resistance of the respiratory tract. In short, $R = (p_{m1} - p_{m2})/\dot{q}_{T1}$.

Although the procedure outlined gives a rough estimate of the resistive properties of the respiratory tract, it is rather difficult to accept that this method reflects the overall resistive effect of the system. The basic difficulty is the premise that flow in the system is zero everywhere if flow into the mouth is briefly reduced to zero. If this is true, the pressure at the mouth truly reflects the pressure of the total tract; if it is not true (and this seems likely) then the mouth pressure during interrupted flow reflects only the pressures at select portions of the tract.

These ideas are more clearly understood if reference is made to the model circuit of Fig. 7.4b. From this circuit we note if $\dot{q}_T$ is interrupted and all the branch flows are zero then p_m and p_s will be the same and p_m will indeed be a measure of p_s. The total drop across the tract prior to flow interruption might then legitimately be taken as $p_m - p_s$. If we can then claim that very little flow remains stored in the tube (that is, $\dot{q}_b$ in Fig. 7.4b is neglectable compared with $\dot{q}_s$) we might claim the pressure drop $p_m - p_s$ causes the total flow $\dot{q}_T$ and thereby consider the ratio of these variables as a definition of the overall airway resistance. However, we must also note that, even if flow is zero everywhere during interrupted flow and even if $p_m - p_s$ does represent the total drop across the airways, if $\dot{q}_b$ is not neglectable compared with $\dot{q}_s$ then the ratio $(p_m - p_s)/\dot{q}_T$ does not clearly define the resistive effects of the respiratory system; in fact, it would be difficult then to interpret the significance of this ratio.

A more critical objection to the procedure outlined is the assumption that flow is everywhere zero if $\dot{q}_T$ is zero. An examination of the circuit in Fig. 7.4b indicates clearly that this need not be the case. If $\dot{q}_T$ is zero, $\dot{q}_b$ and $\dot{q}_s$ can be nonzero and they can still satisfy Kirchhoff's law that the sum of the currents entering node a must be zero ($\dot{q}_T + \dot{q}_b + \dot{q}_c = 0$; if $\dot{q}_T = 0$, then $\dot{q}_b + \dot{q}_c = 0$. The last equation states $\dot{q}_b = \dot{q}_c$ but does not require that $\dot{q}_b$ and $\dot{q}_c$ must always be zero; it simply requires that these flows be equal). This implies that, during interrupted flow, gas from the distended airways may enter the alveoli or gas from the distended alveoli may enter the airways. Under these circumstances, pressure measurements at the mouth during interrupted flow do not represent the alveolar pressures. The mouth pressure p_{m2} in Fig. 7.26b would then most likely be the pressure represented by p_a in Fig. 7.4b. The ratio $(p_{m1} - p_{m2})/\dot{q}_T$ would then represent the resistive properties of the upper respiratory tract (R_b in Fig. 7.4b) and not the overall respiratory resistance.*

As we are interested in this particular problem only as an illustrative example, and as there is no way for us to know how to interpret the figure of 2 cm $H_2O/l/s$ for airway resistance, we shall make some arbitrary assumptions. We assume first that the value of 2 cm $H_2O/l/s$ includes only about half the resistance of the lower bronchial tree ($R_s/2$) and that the resistance of the upper bronchial tree is about twice that of the lower tree ($R_b = 2R_s$). On this basis R_b would be about 1.6 cm $H_2O/l/s$ and R_s would be about 0.8 cm $H_2O/l/s$.

* This illustration is a good example of how a system analog can be used effectively to evaluate the utility (and limitations) of measurements and their interpretation. The circuit in Fig. 7.4 cannot be considered an adequate representation of the lung and yet, crude though it may be, it can still be used to explain the significance of various measurements.

Problems

7.1   (a) Figure P7.1a is a series circuit composed of two impedances, Z_1 and Z_2. Show the total impedance e/i is $Z_1 + Z_2 = Y_1Y_2/Y_1 + Y_2$.

(b) Figure P7.1b is a parallel circuit composed of two admittances, Y_1 and Y_2. Show the total admittance $(1/e)$ is $Y_1 + Y_2 = Z_1Z_2/Z_1 + Z_2$.

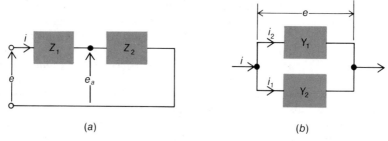

(a)   (b)

FIG. P7.1

7.2   Figure P7.1a is a form of a voltage divider (see step 18 in Appendix A).

(a) Show $e_a/e = Z_1/(Z_1 + Z_2)$.

(b) If the circuit in Fig. P7.1a is really that shown in Fig. P7.2, what is the transfer function e_a/e?

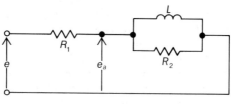

FIG. P7.2

7.3   Equation (7.24) describes the operation of a voltage divider; the pressure drop across the series impedance is proportional to the ratio of the impedance to the total impedance. A similar relationship exists for a flow divider. For Fig. P7.1b show $i_1/i = Y_1/Y_1 + Y_2$.

7.4   If the interpleural pressure which drives the model of Fig. 7.4b is given by $P_2u(t) + (P_1 - P_2)\epsilon^{-at}u(t)$ (see Fig. 7.9a) the flow response may be represented by

$$\dot{q}_T = \frac{P_1}{R_b} \frac{(s + aP_2/P_1)(s + 1/\tau_{bs})}{(s + a)[s^2 + (1/\tau_b + 1/\tau_{bs})s + 1/\tau_b\tau_s]} \qquad *$$

This expression should now describe the flow-response curve (Fig. 7.9b). To see more clearly how this response curve is related to the system and the excitation, Eq. (7.39) can be written in factored form and the result separated into partial fractions. To outline the procedure symbolically: If it is assumed the quadratic in Eq. (7.39) can be factored into the product $(s + 1/\tau_1)(s + 1/\tau_2)$, the equation becomes

$$\dot{q}_T = \frac{P_1}{R_b} \frac{(s + aP_2/P_1)(s + 1/\tau_{bs})}{(s + a)(s + 1/\tau_1)(s + 1/\tau_2)}$$

where τ_1 and τ_2 depend upon the system constants R_s, R_b, C_s, and C_b; see Eq. (7.36). This function can be expanded by a partial-fraction expansion. The equation then takes the form

$$\dot{q}_T = \frac{P_1}{R_b} \left(\frac{A}{s + a} + \frac{B}{s + 1/\tau_1} + \frac{C}{s + 1/\tau_2} \right)$$

The coefficients A, B, and C are related to the system constants and the excitation constants. If the latter result is expressed in the time domain,

$$\dot{q}_T = \frac{P_1}{R_b} (A\epsilon^{-at} + B\epsilon^{-t/\tau_1} + C\epsilon^{-t/\tau_2})$$

The constants A, B, and C are evaluated in the process of a partial-fraction expansion [see Eq. (7.41)]. Determine these constants to show $\dot{q}_{T0}$ (flow at $t \approx 0$) $= P_1/R_b$ (i.e., show $A + B + C$ reduces to unity).

7.5 If concentration gradient is related to current and solute flow to voltage, show that the circuit in Fig. 7.18a may be used to simulate the transfer function given by Eq. (7.47). If solute *weight* (not flow) is related to i and concentration gradient is related to e, show the circuit of Fig. 7.18b may be used as a simulation. Under what circumstances can the circuit in Fig. 7.18c be used?

* If we transform the expression for the interpleural pressure we have

$$p_t - p_m = \frac{P_2}{s} + \frac{P_1 - P_2}{s + a} = \frac{aP_2 + sP_1}{(s + a)s} = \frac{P_1(s + aP_2/P_1)}{s(s + a)}$$

From Eq. (7.20) we have

$$\frac{\dot{q}_T}{p_t - p_m} = \frac{s(s + 1/\tau_{bs})}{s^2 + (1/\tau_b + 1/\tau_{bs})s + 1/\tau_b\tau_s} \frac{1}{R_b}$$

Therefore

$$\dot{q}_T = \frac{P_1}{R_b} \frac{(s + aP_2/P_1)(s + 1/\tau_{bs})}{(s + a)[s^2 + (1/\tau_b + 1/\tau_{bs})s + 1/\tau_b\tau_s]}$$

We might note, incidentally, that if we use this expression to determine $\dot{q}_{T0}$ by the initial-value theorem (i.e., flow at $t \approx 0$) we obtain $\dot{q}_{T0} = P_1/R_b$, as it should be. [See footnote associated with the discussion of Eq. (7.37).]

7.6 Show that the third-order exponential of Fig. 7.21 gives rise to a three-section ladder network.

REFERENCES

1. Tuttle, David F., Jr.: "Electrical Networks: Analysis and Synthesis," Mc-Graw-Hill Book Company, New York, 1965.
2. Kuo, Franklin F.: "Network Analysis and Synthesis," John Wiley & Sons, Inc., New York, 1962.
3. Fry, D. L.: Theoretical Considerations of the Bronchial Pressure-Flow-Volume Relationships with Particular Reference to the Maximum Expiratory Flow-Volume Curve, *Phys. in Med. Biol.*, **3**(2):174 (October, 1958).
4. Tammeling, G. T., W. Chr. Berg, and H. J. Sluiter: A Comparative Study of the Value of Forced Expirograms and Flow Curves in Health and Obstructive Lung Disease. *Amer. Rev. Respiratory Disease*, **93**:238 (1966).
5. Florey, E.: "An Introduction to General and Comparative Animal Physiology," W. B. Saunders Company, Philadelphia, 1966.
6. Snell, F. M., S. Shulman, R. Spencer, and Carl Moos: "Biophysical Principles of Structure and Functions," Addison-Wesley Publishing Company, Inc., Reading, Mass., 1965.
7. Ling, G. N.: Cell Membrane and Cell Permeability, *Ann. N.Y. Acad. Sci.*, **137**:2 (1966).
8. Scheer, B. T.: "General Physiology," John Wiley & Sons, Inc., New York, 1953.
9. Dawson, H., and J. F. Danielli: "The Permeability of Natural Membranes," 2d ed., Cambridge University Press, New York, 1952.
10. Dawson, H., and M. Grace Eggleton (eds.): "Starling's Human Physiology," Lea & Febiger, Philadelphia, 1962.

CHAPTER EIGHT

PERIODIC SIGNALS

8.1 INTRODUCTORY COMMENTS

In most of the situations thus far examined, either the excitation signals have been nonrepetitious or it has been possible to express them in terms of nonrepetitious signals. In Fig. 6.19, for example, the temperature-excitation curve was synthesized by adding two ramp-functions and a decaying exponential; in Fig. 6.23 the triangular excitation waveform was synthesized by using a step-function and a series of ramps. The ramp, the decaying exponential, the step are all examples of nonrepetitive signals in the sense that the waveshape is not recurrent *during the course of an investigation*. Such signals are generally described as aperiodic signals; contradistinctively, signals that are repetitious throughout a specified observation time are described as periodic signals. EKG signals and signals obtained from the respiratory system are examples of physiological-system signals that have periodic characteristics.

Periodicity and aperiodicity depend to some degree upon the observation period. Without this restriction it would be necessary to consider all signals aperiodic. The waveform of Fig. 6.25b, for example, would be considered aperiodic if the observation period were $-1 < t < \infty$. During this interval the signal undergoes a change which is not repeated at any later time. On the other hand, if the observation period is restricted to $0^+ < t < \infty$ (and it is assumed the wave continues) there is a recurrent pattern; the signal would then be considered a periodic function. Although this distinction is often merely an academic technicality, it is sometimes important to remember that the beginning of a periodic signal could have been (one might say "must have been") aperiodic; the initial response of a system to a periodic excitation can thus be (and often is) aperiodic.

Although it is always possible to synthesize any curve with an appropriate combination of ramps, steps, and other aperiodic functions, it is sometimes a clumsy procedure. This is particularly true when the signal waveform is a smooth, continuous function (no sharp corners) with distinctly periodic properties. Intuitively, one would expect that it would be more appropriate to represent such signals by mathematical functions having natural periodic properties. The most likely choice for such representation would be the sinusoid. In this chapter we examine how periodic signals may be described by a sinusoidal representation and how sinusoidal analysis may be used in systems investigations.

8.2 THE SINUSOIDAL FUNCTION

Although the sine function is probably familiar, some of the concepts associated with sinusoidal representation may have been forgotten. In this section we shall review some of the pertinent features of the sinusoid and its representation* before we discuss periodic signals in general.

The fundamental definition of the sine function is given by the equation $y/A = \sin\theta$. This familiar trigonometric identity defines an angle between two lines in terms of some length relationship of the lines (see Fig. 8.1a). In this form the sine function is not very exciting. It becomes slightly more interesting when we write the expression as

$$y = A \sin\theta$$

In this form a function y is expressed in terms of a magnitude A (or amplitude, as it is often called) and an inclination angle θ. Evidently, as the inclination angle changes, so does the value of y. The changes of y with θ can be described by indexing the line A around in a circle, as shown in Fig. 8.1a, or by plotting the value of y for each corresponding

---

* This will include the concept of the sine as related to a rotating phasor (line segment) and the representation of the sine in complex notation. Readers familiar with this subject may proceed directly to Sec. 8.3.

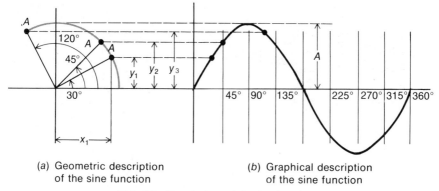

(a) Geometric description
of the sine function

(b) Graphical description
of the sine function

FIG. 8.1 *Descriptions of the sine function.*

value of θ. The latter procedure gives rise to the familiar sine graph shown in Fig. 8.1b. The maximum value of the plot is evidently A; this is often described as the amplitude of the plot.

The transition from a sinusoidal "plot" to a sinusoidal "waveform" is now just a technicality. Waveforms, as we have discussed them, are time-dependent functions. The curve of Fig. 8.1b thus becomes a waveform when the angle θ becomes related to time. To show this transformation more forcefully, we shall examine a realistic situation such as that shown in Fig. 8.2. A reciprocating piston causes a vibrating air column to be issued from the end of a funnel-shaped tube. This air motion causes an oscillating pressure distribution in the air (often described as areas of rarefaction and compression); the oscillatory pressure wave is detected by the ear as a sound wave.* The oscillating pressure wave may be measured by a pressure-detection device (not shown) and recorded as shown by the vertical plot (p_D versus t) in Fig. 8.2. The ear responds to such pressure excitations by generating cochlear potentials. These potentials have been detected in animals by implanting fine electrodes in the cochlea of a guinea pig.† A typical voltage response to a pure tone is also shown in Fig. 8.2. Both curves are obviously periodic, and both curves seem to have sinusoidal properties. As a first approximation we assume that we can represent the pressure waveform mathematically as a sinusoidal function. (No claim is made that the pressure waveform is sinusoidal. We use the sine function in the same way we used the step-function:

---

* A good description of a mechanical tone generator is given in Ref. 1, p. 61. Another good description of tone generation is given in Ref. 2, chap. 2. This chapter also includes some background on the connection between pressure and sound and has a simple pictorial illustration of the relationship between a rotating wheel and the generation of a sinusoidal wave.

† Some interesting curves are shown in Ref. 3.

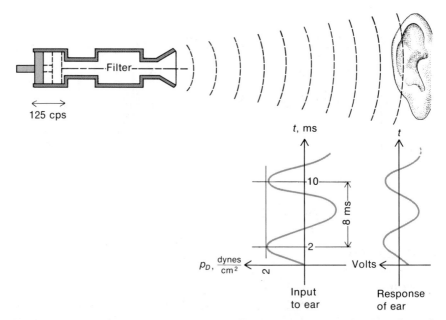

FIG. 8.2 *Generation and detection of a low tone. (Pressure measurement shown is referenced to atmospheric pressure; thus p_D is actually $p - P_{atm}$, where p is absolute pressure.)*

as a mathematical representation of a physical signal.) This implies that the pressure waveform may be described as

$$(8.1) \qquad p_D = P_0 \sin \theta = 2 \sin \theta \, \frac{\text{dyn}}{\text{cm}^2}$$

where P_0 is obviously the amplitude, or maximum value, of the excitation pressure (above or below the atmospheric pressure; see comment in Fig. 8.2). The mathematical procedure by which θ is related to time should be evident: p_D is maximum (i.e., equal to P_0) when θ is 90°, 450°, etc. [Eq. (8.1) and Fig. 8.1*b*]; it is also maximum when *t* is 2 ms, 10 ms, etc. (Fig. 8.2). From these figures it appears θ changes at a rate of 90° in 2 ms or 450° in 10 ms or 45°/ms. We might therefore write $\theta = (45°/\text{ms})t$, where *t* is in milliseconds. Equation (8.1) can then be written

$$(8.2) \qquad p_D = 2 \sin (45°/\text{ms})t$$

A better picture of the relationship between θ and time is secured by comparing the sinusoidal plot and the pressure waveform from a "cyclic" point of view. We note from Fig. 8.1*b* the sinusoidal plot repeats itself every 360°. Each repeated sinusoid is considered a cycle, and a cycle is often described as equivalent to 360°. The pressure waveform repeats periodically; it goes through a complete cycle every 8 ms. If we are to describe the pressure waveform by a sinusoidal function of the form given in Eq. (8.1) we must assume θ goes through a complete cycle of 360° every 8 ms. This implies θ changes at a rate of 360°/8 ms or 45°/ms (or 45,000°/s). At any instant, θ must therefore be given by (45°/ms)t, and p_D is then written as shown in Eq. (8.2).

We can now realize that the sine function may be used to represent (generate might be a better word) a sinusoidal waveform if the angular term (θ) is expressed as the product of an angular velocity (e.g., 45°/s) and time. Angular velocity is often designated by the Greek letter ω; θ may therefore be expressed as ωt and Eq. (8.2) would then be written

$$(8.3) \qquad\qquad p_D = P_0 \sin \omega t$$

One interpretation of the significance of θ expressed in terms of an angular rate may be obtained by reference to Fig. 8.1a. There the sine function is generated by the abscissa term y (or the projection of A on the y axis) as the line segment A is indexed around a center of rotation. The angular motion of A is described by the angle θ. If the angle θ is given in terms of an angular velocity (ω) and time, then we would picture A as a continuously rotating line segment; the angular velocity of the line would evidently be ω. Now as A goes through one revolution, y goes through one sinusoidal cycle. If the angular velocity of the line is ω, it will go through ω revolutions in 1 s; y will then evidently go through ω cycles per second (c/s) or hertz (Hz). The periodicity or frequency of y is thus equivalent to the angular velocity of the rotating line segment. Rotating line segments used to generate sinusoidal waveforms are often called phasors. (Some texts still refer to these rotating line segments as vectors. Vectors and phasors have similar properties but the former have more general significance and the use of phasors for rotating line segments is to be preferred.)

A phasor can also be used to generate a cosinusoidal waveform. If we had plotted x versus θ (the x-axis projection of A) in Fig. 8.1b instead of y versus θ, the result would be a cosine function. A cosine waveform is generated by permitting θ to be represented by ωt.

There are a number of interesting manipulations possible with phasors; as these are well described in many introductory electrical or systems engineering texts,* we shall not go into detailed explanations of the utility of phasor representation. We shall, however, use the phasor concept as an intuitive tool to establish some other representations of

* See Ref. 4. See also Probs. 8.2 and 8.3.

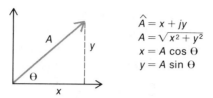

FIG. 8.3 *Complex notation.*

sinusoidal functions. The point to keep in mind is that the sinusoidal waveform may be derived from a rotating line segment. Any representation of a line segment may therefore be ultimately related to the sinusoid.

As a phasor is just a line, it can be represented by complex-number notation as shown in Fig. 8.3. The directed line A (the carot above the letter indicates a directed-line segment) is given by the complex number $x + jy$. x is the magnitude of $\hat{A}$ along the x axis (equivalent to $A \cos \theta$) and y is the magnitude of $\hat{A}$ along the x axis. The magnitude of $\hat{A}$ is given by $A(= \sqrt{x^2 + y^2})$. If we use the trigonometric functions for x and y we can write A as

$$(8.4) \qquad \hat{A} = A \cos \theta + jA \sin \theta$$

The j indicates that the y magnitude of $\hat{A}$ is $A \sin \theta$. The term without the j is the x magnitude.

There is another interpretation of Eq. (8.4) which is interesting and useful. If we factor out A, we can write

$$(8.5) \qquad \hat{A} = A(\cos \theta + j \sin \theta)$$

This equation emphasizes that $\hat{A}$ is a directed line of magnitude A inclined an angle θ with the x axis. The inclination of the line is governed by the parenthetic term in Eq. (8.5). We might, in fact, consider this term as an operator that serves to rotate a horizontal line (line along the x axis) to an inclination of $\theta°$. A rotation of $20°$ would thus be given by $A(\cos 20° + j \sin 20°)$; one of $130°$ would be $A(\cos 130° + j \sin 130°)$ (see Fig. 8.4). If we want the line segment to rotate continuously, we express θ in terms of time. We would then have $\hat{A} = A(\cos \omega t + j \sin \omega t)$, which would represent a rotating line segment with angular velocity ω and magnitude A.

The operator $\cos \theta + j \sin \theta$ has interesting implications when rotations of $90°$ are examined. Since $\cos 90°$ is zero and $\sin 90°$ is unity, the operator reduces to simply the j coefficient. We can thus interpret the j term as an operator that causes a line to rotate $90°$. As a simple illustration, a line represented by $\hat{A} = 5$ is interpreted as a line five units long along the x axis (see Fig. 8.4b); a line represented by $\hat{A} = j5$ is interpreted as a line five units long along the y axis. If we extend this idea once more, we see that a line in the positive x direction may be reoriented to lie in the negative x direction by a rotation of 180°. Thus $\hat{A} = 5$ is in the positive x direction; $\hat{A} = j5$ is in the y direction (a rotation of 90°); $\hat{A} = j(j5)$ is a line directed 90° from the y direction or 180° from the positive x direction, or, in other words, directed in the negative x direction. If we handle the operator j as an algebraic quantity then $j(j5)$ may be interpreted as $5j^2$; a line in the negative x direction may, however, also be expressed as -5. We have, therefore,

(8.6)
$$5j^2 = -5$$

or

$$j^2 = -1$$

or

$$j = \sqrt{-1}$$

By this interpretation it is somewhat meaningless to consider j a numeric (it is the familiar imaginary number designated by i in mathematics texts); j^2, however, has real significance.

The interpretation of j^2 as -1 (or -1 as j^2) permits conversion of the operator $\cos \theta + j \sin \theta$ into another form which we shall find useful in transform analysis. Cos θ may be expanded into a series* given by

(8.7)
$$\cos \theta = 1 - \frac{\theta^2}{2!} + \frac{\theta^4}{4!} + \cdots$$

* Expanded into a Maclaurin's series. The series may not be familiar but its applicability can be demonstrated by trying a few simple values: It is known $\cos 30° = \sqrt{3}/2 = 0.866$; to check this by the series representation of Eq. (8.7), we note $\theta = 30° = \pi/6$ rad. Therefore $\cos \ 30° = 1 - (\pi/6)^2/2! + (\pi/6)^4/4! + \cdots$ or $\cos \ \theta = 1 - 0.136 + 0.0027 + \cdots \approx 0.86$.

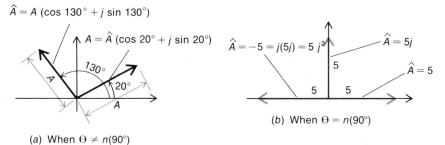

(a) When $\Theta \neq n(90°)$

(b) When $\Theta = n(90°)$

FIG. 8.4   *Operator* $\cos \theta - j \sin \theta$.

If -1 can be written j^2 and $+1$ written j^4, etc., then Eq. (8.7) may be given by

$$(8.8) \qquad \cos \theta = 1 + \frac{(j\theta)^2}{2!} + \frac{(j\theta)^4}{4!} + \cdots$$

In a similar manner $\sin \theta$ may be written

$$(8.9) \quad \sin \theta = \theta - \frac{\theta^3}{3!} + \frac{\theta^5}{5!} + \cdots = \theta + \frac{j^2\theta^3}{3!} + \frac{j^4\theta^5}{5!} + \cdots$$

Therefore

$$(8.10) \qquad j \sin \theta = j\theta + \frac{(j\theta)^3}{3!} + \frac{(j\theta)^5}{5!} + \cdots$$

Equations (8.8) and (8.10) are now added to give $\cos \theta + j \sin \theta$. Alternate terms taken from each equation form a series given by

$$(8.11) \qquad \cos \theta + j \sin \theta = 1 + j\theta + \frac{(j\theta)^2}{2!} + \frac{(j\theta)^3}{3!} + \cdots$$

Mathematicians recognize the series in Eq. (8.11) as an expansion of the term $\epsilon^{j\theta}(e^x = 1 + x + x^2/2! + x^3/3! \cdots)$. Therefore we may finally write

$$(8.12) \qquad \cos \theta + j \sin \theta = \epsilon^{j\theta}$$

This equation is technically known as Euler's identity.

At this time we can interpret Eq. (8.12) only as the definition of another form of phasor operator. Multiplication of a number by $\epsilon^{j\theta}$ converts the number to a line segment inclined $\theta°$ to the x axis. If we wish the line segment to rotate continuously, we express θ in terms of ωt. Thus $A\epsilon^{j\omega t}$ is interpreted as a line A units long that rotates continuously at an angle frequency of ω.

The use of the phasor-operator notation makes it possible to relate the phasor analytically to the sinusoidal waveform. Since a phasor is written

$$(8.13) \quad \hat{A} = A(\cos \omega t + j \sin \omega t) = A \cos \omega t + jA \sin \omega t = A\epsilon^{j\omega t}$$

the sinusoid is often defined as the imaginary part of a phasor and a cosinusoid as the real part of a phasor. In equation form this may be written

$$(8.14) \qquad A \sin \omega t = \text{Im } \hat{A} = \text{Im } A\epsilon^{j\omega t} = A \text{ Im } \epsilon^{j\omega t}$$

$$A \cos \omega t = \text{Re } \hat{A} = \text{Re } A\epsilon^{j\omega t} = A \text{ Re } \epsilon^{j\omega t}$$

These equations are simply mathematical statements of the concept established earlier, namely: The sinusoidal waveform may be represented by the y-axis projection of a rotating line segment, and the cosinusoidal waveform may be represented by the x-axis projection of a rotating line segment.

One other notation for the sinusoid should be introduced at this time. This involves a mathematical definition of the sinusoid (and cosinusoid) in terms of $A\epsilon^{j\omega t}$. We note from Eq. (8.12) if θ were a negative angle (i.e., if the phasor were to be turned clockwise instead of counterclockwise) we would write

$$(8.15) \qquad \cos -\theta + j \sin -\theta = \epsilon^{j(-\theta)}$$

or

$$\cos \theta - j \sin \theta = \epsilon^{-j\theta}$$

(since $\cos -\theta = \cos \theta$ and $\sin -\theta = -\theta$). If this second form of Eq. (8.15) is added to Eq. (8.12), the sine terms drop out and we have

$$(8.16) \qquad \cos \theta = \frac{\epsilon^{j\theta} + \epsilon^{-j\theta}}{2}$$

Similarly, if we subtract Eq. (8.15) from Eq. (8.12) the result is

$$(8.17) \qquad \sin \theta = \frac{\epsilon^{j\theta} - \epsilon^{-j\theta}}{2j}$$

Equations (8.16) and (8.17) are two additional representations for the same function which we shall find useful when we seek to transform it to the s domain.

8.3 SINUSOIDAL ANALYSIS OF AN INSTRUMENT SYSTEM

To illustrate the utility of some of the concepts established in Sec. 8.2 we examine the response of an instrument system designed to measure air flow (and, in particular, respiration). A schematic diagram of the system assembly is shown in Fig. 8.5 (see also Fig. 2.4). The system components include a pneumotachograph, a pressure transducer, and a means to record or "readout." The pneumotachograph is the device that senses the air flow. In its simplest form it consists of a tube with a wire mesh. Flow can occur only if there is a slight pressure drop across the mesh. This slight pressure drop is used to measure the flow. The pressure drop is detected by a sensitive pressure gauge, and this information is transformed to an electrical signal which can be displayed on a recording device or on an oscilloscope. The question is whether this system can be

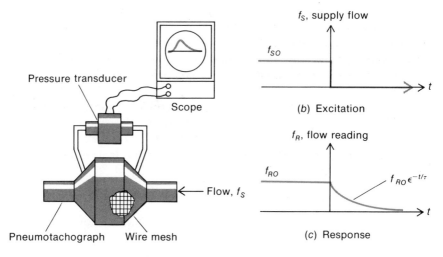

(a) Experimental arrangement

FIG. 8.5   *Instrument assembly to measure air flow.*

used faithfully to record steady breathing when the respiratory rhythm changes from low-frequency deep breathing to a high-frequency "pant." One way to answer this question would be to subject the system to a test and thereby calibrate the assembly. This is, however, easier said than done: To test such a system, one must have a known air-flow signal source that can deliver a cyclic air flow. Such equipment would be rather expensive. The easier way to handle this problem is through analysis.

To determine whether this system will faithfully record air flows at different respiratory frequencies we make the assumption that the respiratory waveform may be closely represented by a sinusoidal response when the flow excitation is sinusoidal. A block-diagram representation of the problem is shown in Fig. 8.6. A sinusoidal air flow of amplitude F_{SA} and frequency ω is to be applied to an instrument system with transfer function $T(s)$; the flow response is to be determined.

The problem evidently cannot be solved unless we can obtain an

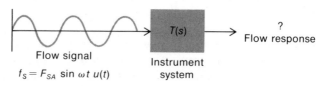

FIG. 8.6   *Block diagram of instrument system with sinusoidal stimulus.*

analytic representation of the instrument system. This can be done by applying a known excitation and recording the response. From such information we can obtain some approximate representation for the system transfer function. The procedure should be quite familiar by now. For the purposes of illustration, we assume the excitation signal shown in Fig. 8.5b is used as a test signal. This signal can be generated quite easily: A steady flow (f_{S0}) through the pneumotachograph is maintained until some predetermined cutoff time. Flow is then reduced quickly to zero by valve action. The system response to this excitation is shown in Fig. 8.5c. Before the supply flow is interrupted, the flow (or flow response) is constant at f_{R0}; after excitation cutoff the response curve decays exponentially.* In accordance with the procedure previously outlined, the transfer characteristics may be evaluated as $(1/\tau)/(s + 1/\tau)$. (See Fig. 7.17 and the associated discussion. It is assumed here that the instrument system has a scale factor or calibration factor of unity. The instrument reading, f_{R0}, and the actual flow, f_{S0}, under steady-state conditions are therefore equal and f_{R0}/f_{S0} is unity.)

With the transfer characteristics established, we can now determine the response of the system for any driver. The procedure is simply to transform the driver into a function of s and to operate on the driver with the system transfer function. In equation form, this is written

$$(8.18) \qquad f_R(s) = \frac{1/\tau}{s + 1/\tau}\, f_S(s)$$

For this particular situation the driver is a sinusoid given by $F_{SA} \sin \omega t\, u(t)$. [The $u(t)$ is included to indicate that the stimulus does not exist (or is zero) prior to $t = 0$. The excitation waveform would thus appear as shown in Fig. 8.6.] The next problem therefore is to convert the sinusoid into the s domain.

The transform of $\sin \omega t$ is available in almost all transform tables (see Appendix D). We can derive the transform easily enough, however, if we use the exponential from the sinusoid as given by Eq. (8.17). In this case θ is equivalent to ωt, and so we write

$$(8.19) \qquad \sin \omega t = \frac{\epsilon^{(j\omega)t} - \epsilon^{-(j\omega)t}}{2j}$$

The $j\omega$ term in the parentheses of the exponential term may be considered the time coefficient a as in row 4, Table 6.1. The transform of $\epsilon^{j\omega t}$ is thus

---

* The input-output characteristics shown in Fig. 8.5 represent the characteristics of the wholly assembled system. The signal decay shown in Fig. 8.5c is the result of all the component characteristics between input and output. From such information it is not possible to ascribe any characteristics to the individual components. All that can be obtained from the information given is the composite characteristics of the total system.

$\delta(t)/(s - j\omega)$, and $\delta(t)/(s + j\omega)$ is the transform of $\epsilon^{-j\omega t}$. The transform of the sinusoid may therefore be written

$$(8.20) \qquad \sin \omega t = \left(\frac{1}{s - j\omega} - \frac{1}{s + j\omega}\right) \frac{\delta(t)}{2j}$$

When the terms in parentheses are combined they become $2j\omega/(s^2 + \omega^2)$ (j^2 is interpreted as -1), and the transform of $F_{SA} \sin \omega t$ is thus

$$(8.21) \qquad F_{SA} \sin \omega t = F_{SA} \frac{\omega}{s^2 + \omega^2} \delta(t)^*$$

where the amplitude term F_{SA} is simply carried along.

The sinusoidal response of the instrument system in Fig. 8.5 can now be expressed completely in terms of the s transform. From Eqs. (8.18) and (8.21) we have

$$(8.22) \qquad f_R(s) = \frac{F_{SA}\omega}{\tau} \frac{1}{(s + 1/\tau)(s^2 + \omega^2)} \delta(t)$$

The time response of f_R must now be evaluated by reference to a transform table or by separation into recognizable forms by partial-fraction expansion. Before we obtain an exact solution for the instrument reading, we shall employ some physical reasoning to determine what kind of solution might be expected. The system is being driven sinusoidally; if the system responds with reasonable fidelity (i.e., if distortion is small) one would expect a sinusoidal response. The *shape* of the output should thus finally be a sinusoid with the same frequency as the input. On the other hand, although we would not expect the instrument system to alter the signal waveshape, we would expect magnitude and timing changes to occur between input and output. For example, we might find that the output amplitude seldom corresponds to the input amplitude and that the maximum output always occurs sometime after the maximum input is applied. We could predict that this condition must exist on the basis of the test performed to obtain the system transfer function (see Fig. 8.5). The response of a step change consisted of a variable amplitude, and a final steady state was obtained four time-constants (approximately) after the initiation of the step. Some alteration of amplitude and time base would thus also be expected with a sinusoidal driver. From a graphical point of view, therefore, if the input waveform is given by the top curve in Fig. 8.7, we might expect the output curve to appear as shown by the lower curve.

---

* The denominator of this expression has interesting implications. The summation sign between the two terms implies the terms should have the same units. This would imply that s has the same units as frequency. In fact, s is often referred to as a complex frequency. We shall see the significance of this concept in the next section.

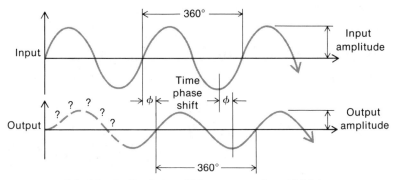

FIG. 8.7 *Predicted sinusoidal response to sinusoidal driver.*

We are now prepared to examine Eq. (8.22) for a total solution of the system response to a sinusoidal excitation. Rather than immediately handle the problem in a completely general fashion, it will be somewhat easier if we use specific numbers for a first analysis. At normal-breathing rates, flow varies from 0 to about 500 ml/s; F_{SA} will therefore be taken as 500. Normal breathing is at about 16 breaths/min; slow breathing is at about 8 breaths/min; rapid breathing is at about 60 breaths/min. As an intermediate estimate we assume the instrument should be able to handle about 30 breaths/min. Therefore ω is 30 c/min or $\frac{1}{2}$ c/s or 3 rad/s.* As an arbitrary figure we assume the instrument time-constant to be about 100 ms (this is not too far out of line for mechanical transducers). Therefore τ is 0.1 s. Equation (8.22) can now be written

$$(8.23) \qquad f_R(s) = \frac{500 \times 3}{0.1} \frac{1}{(s + 1/0.1)(s^2 + 3^2)} \delta(t)$$

$$= 15,000 \frac{1}{(s + 10)(s^2 + 9)} \delta(t)$$

To effect a solution of Eq. (8.23) we must factor the denominator into first-order terms (where factors containing j terms are now acceptable) and then separate the terms by a partial-fraction expansion:

$$(8.24) \qquad f_R = 15,000\delta(t) \frac{1}{(s + 10)(s + 3j)(s - 3j)}$$

The procedure is now mechanical. To separate into partial fractions, we

---

* In all calculations, ω, the frequency term, should have units (seconds)$^{-1}$ or 1/s. This will become clear if units are checked in the various expressions and particularly if the final answer is checked. Since radians are dimensionless, radians per second are essentially (seconds)$^{-1}$. In general, therefore, ω should be expressed in terms of radians per second.

write

$$(8.25) \quad \frac{1}{(s + 10)(s + 3j)(s - 3j)} = \frac{M}{s + 10} + \frac{N}{s + 3j} + \frac{P}{s - 3j}$$

To obtain M, we multiply through by $s + 10$ and find the limit as s approaches -10:

$$(8.26) \quad \lim_{s \to -10} \frac{s + 10}{(s + 10)(s + 3j)(s - 3j)}$$

$$= \lim_{s \to -10} \left[M + \frac{N(s + 10)}{s + 3j} + \frac{P(s + 10)}{s - 3j} \right]$$

$$(8.27) \quad \frac{1}{(-10 + 3j)(-10 - 3j)} = M = \frac{1}{100 + 9} = \frac{1}{109}$$

By a similar procedure we find

$$(8.28) \quad N = \frac{1}{(-3j + 10)(-6j)} \qquad P = \frac{1}{(+3j + 10)(+6j)}$$

We can now write

$$(8.29) \quad f_R = 15{,}000 \left[\frac{(1/109)\delta(t)}{s + 10} - \frac{\delta(t)}{6j(10 - 3j)(s + 3j)} \right.$$

$$\left. + \frac{\delta(t)}{6j(10 + 3j)(s - 3j)} \right]$$

If we interpret the s terms in a standard fashion $[A/(s + a)\,\delta(t) = A\epsilon^{-at}]$ we can write

$$(8.30) \quad f_R = 15{,}000 \left[\underbrace{\frac{\epsilon^{-10t}}{109}}_{\substack{\text{Free} \\ \text{response}}} + \underbrace{\frac{\epsilon^{+3jt}}{(10 + 3j)(6j)} - \frac{\epsilon^{-3jt}}{(10 - 3j)(6j)}}_{\substack{\text{Forced response or} \\ \text{steady state}}} \right]$$

Equation (8.30) is the solution for the response of the instrument system as a function of time when the input stimulation is sinusoidal. The first bracketed term in the solution is simply an exponential-decay term having a time-constant representative of the instrument system. It is the free response of the system which always appears, no matter what the driver is.* After four time-constants this term will be practically zero. Although the other bracketed terms in the solution are not entirely familiar, from the reasoning presented earlier (and prior information that sinusoids are related by $\epsilon^{j\theta}$) we intuitively expect that these terms will give rise to a sinusoidal function. The latter terms thus represent the forced response of the system, and we now examine how these forced-

* Unless very special pains are taken to eliminate it.

response terms may be converted to a recognizable sinusoidal function. When this is done we shall then be able to see how the forced response is related to the sinusoidal forcing function.

There are a variety of ways to convert the forced-response terms in Eq. (8.30) into a form that may be recognized as a sinusoid. Intuitively we realize that to do so we must rearrange these terms so that they form the expression given by the right side of Eq. (8.17).

$$(8.17) \qquad \sin\theta = \frac{\epsilon^{j\theta} - \epsilon^{-j\theta}}{2j}$$

To effect this transformation we first examine alternative ways of writing the complex number $10 + 3j$. One way was suggested in Sec. 8.2 where it was shown that a directed-line segment can be represented by a complex number in the form of $A(\cos\theta + j\sin\theta)$. (See the discussion relating to Fig. 8.3.) If we use this concept, then $10 + 3j$ may be written

$$(8.31) \qquad 10 + 3j = 109(\cos\theta + j\sin\theta)$$

where θ is defined as $\tan^{-1} +0.3 = 16.7° = 0.3$ rad (see Fig. 8.8). By Euler's identity [Eq. (8.12)] this can be expressed in another form:

$$(8.32) \qquad 10 + 3j = \sqrt{109}\,(\cos\theta + j\sin\theta) = \sqrt{109}\,\epsilon^{j\theta}$$

When this last form is used to replace $10 + 3j$, the first forced-response term may be written

$$(8.33) \qquad \frac{\epsilon^{+3jt}}{(10 + 3j)(6j)} = \frac{\epsilon^{+3jt}}{\sqrt{109}\,\epsilon^{+j0.3}6j} = \frac{\epsilon^{+3jt}\epsilon^{-j(0.3)}}{\sqrt{109}\,6j} = \frac{\epsilon^{+j(3t-0.3)}}{\sqrt{109}\,6j}$$

We could use exactly the same procedure to modify the last term in Eq. (8.30). However it should be evident the last term and the second term in Eq. (8.30) are essentially the same except $+j$ has been replaced by $-j$ (these terms are therefore complex conjugates). The angle will thus

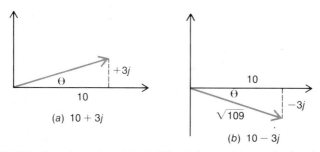

FIG. 8.8 *Transformation of $10 \pm 3j$ into trigonometric complex functions.*

change sign (see Fig. 8.8b), and we have

$$(8.34) \qquad \frac{\epsilon^{-3jt}}{(10-3j)(6j)} = \frac{\epsilon^{-j3t}}{\sqrt{109}\,\epsilon^{-j(0.13)}6j} + \frac{\epsilon^{-j3t}\epsilon^{+j(0.3)}}{\sqrt{109}\,6j} = \frac{\epsilon^{-j(3t-0.3)}}{\sqrt{109}\,6j}$$

Equations (8.33) and (8.34) can now be used to replace the last two terms in Eq. (8.30):

$$(8.35) \qquad \frac{\epsilon^{+j3t}}{(10+3j)(6j)} - \frac{\epsilon^{-3jt}}{(10-3j)(6j)} = \frac{1}{3\sqrt{109}}\,\frac{\epsilon^{j(3t-0.3)}-\epsilon^{-j(3t-0.3)}}{2j}$$

A comparison of Eqs. (8.35) and (8.17) shows the final term in Eq. (8.35) may be written [sin $(3t - 0.3)$]/3; therefore f_R may be finally written

$$(8.36) \qquad f_R = \frac{15{,}000}{109}\epsilon^{-10t} + \frac{5{,}000}{\sqrt{109}}\sin{(3t-0.3)}$$

In this form we can readily see that the total solution contains terms representing the effect of the system (the free response) and the effect of the driver (the forced response). A plot of the total response is shown in Fig. 8.9. As expected, the response ultimately becomes a sinusoid (after four time-constants) with the same frequency as the driver. There are some changes, however; the amplitude of the output sinusoid, for example, is smaller than the amplitude of the input sinusoid by a factor of $10/\sqrt{109}$. Although the difference is not great, it is significant. There has also been a time-scale shift between the input and the output; the output

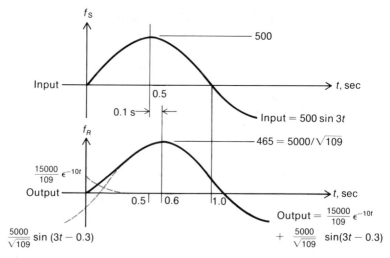

FIG. 8.9 *Sinusoidal excitation and response of the instrument system of Fig. 8.5.*

reaches a maximum point later than the input. This time shift is mathematically related to the angle shift (or phase shift, as it is sometimes called) in the sine function of the force response. The angle term of the output sinusoid is given by $3t - 0.3$ rad. This could be written $3(t - 0.1)$; the forced-response term can now be written

$$(8.37) \qquad \text{Forced response} = \frac{5,000}{\sqrt{109}} \sin 3(t - 0.1)$$

In this form it is seen that the output *sinusoid* term has the same magnitude as the input sinusoid term had 0.1 s earlier.

It will be of interest to see how the amplitude and phase changes were brought about. The amplitude change will be examined first. As mentioned above, the output amplitude appears to be related to the input amplitude by a factor $10/\sqrt{109}$. The output amplitude is $5,000/\sqrt{109}$ [Eq. (8.37)]; the amplitude of the input is $(F_{SA} =)$ 500. If we call the output (response) amplitude F_{RA} then we have

$$(8.38) \qquad \frac{F_{RA}}{F_{SA}} = \frac{5,000/\sqrt{109}}{500} = \frac{10}{\sqrt{109}}$$

or

$$F_{RA} = F_{SA} \frac{10}{\sqrt{109}}$$

The question now is: Where in the solution for the response $[f_R(t)]$ did the input amplitude (F_{SA}) appear, and from where did the multiplicative factor $(10/\sqrt{109})$ come? If we trace back in the solution for the total response, we find that the input amplitude with a multiplicative constant (ω/τ) first appears in Eq. (8.22); the coefficient of the transform solution for f_R is $F_{SA}\omega/\tau$. For the particular values chosen $(F_{SA} = 500$, $\omega = 3$, $\tau = 0.1)$ this term is evaluated to be 15,000. If we trace this term, we note in Eq. (8.30) it is still retained as a coefficient; in Eq. (8.36), however, other multiplicative factors have appeared. The amplitude of the forced-response term $(5,000/\sqrt{109})$ is now the product of 15,000 and $\frac{1}{3}\sqrt{109}$. The amplitude of the forced-response term is therefore $(F_{SA}\omega/\tau)(\frac{1}{3}\sqrt{109})$. If we trace the origin of this new factor we find $\frac{1}{3}\sqrt{109}$ appears first in Eq. (8.35); tracing back further we find the 3 is due to the excitation frequency ω and the $\sqrt{109}$ is the magnitude of the complex number $10 \pm 3j$ [see Eq. (8.32)]. Further checking will show that this number is governed by the time-constant of the system and the frequency of the excitation. Specifically,

$$(8.39) \qquad 10 \pm 3j = \frac{1}{\tau} \pm j\omega$$

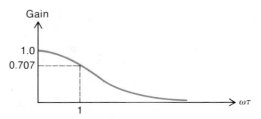

FIG. 8.10   *Gain vs. frequency plot for a first-order system.*

The number $\sqrt{109}$ is in reality $\sqrt{(1/\tau)^2 + \omega^2}$. In general terms, therefore, we can now write

$$(8.40) \qquad F_{RA} = \frac{F_{SA}\omega/\tau}{\omega\sqrt{(1/\tau)^2 + \omega^2}} = F_{SA}\frac{1}{\sqrt{1 + (\omega\tau)^2}}$$

Although this cannot be called a general derivation, it serves to show how the output amplitude of the sinusoidal response is related to the amplitude of the input excitation. For this case we see that, as the driving frequency goes up, the response amplitude goes down. This same general result will be obtained for all real systems: As driving frequencies are increased, response amplitudes must ultimately be decreased.

The ratio of output amplitude to excitation amplitude is often described as the gain or magnification of the system. The gain of the instrument system now under discussion is thus given by

$$(8.41) \qquad G = \frac{F_{RA}}{F_{SA}} = \frac{1}{\sqrt{1 + (\omega\tau)^2}}$$

A plot of the system gain function is shown in Fig. 8.10. It should be noted that the abscissa of this plot is in terms of $\omega\tau$, a dimensionless number.* Dimensionless plots can often be used to advantage when a function depends on several variables. In this case, for example, by plotting gain versus $\omega\tau$ we can readily see gain varies with ω and τ in the same manner.

The gain-frequency curve in Fig. 8.10 is characteristic of all first-order systems. At low frequencies the gain is roughly unity; as the frequency of the excitation increases, the gain decays monotonically (decays continuously, never showing an increase) until at very high frequencies

* We can easily establish that $\omega\tau$ must be dimensionless, since the denominator of Eq. (8.40) is composed of $\omega\tau$ added to a pure numeric. Such an addition may be performed only if $\omega\tau$ is also a pure numeric. This establishes also the most convenient set of units that should be used for ω. If τ, the time-constant, is given in seconds, then ω should have the units (seconds)$^{-1}$ or radians per second. See the footnote in the discussion relating to Eq. (8.22), p. 381.

the gain is zero. In physical terms this means that the output of the system is practically zero when excitation frequencies are high. For the instrument system presently under discussion we interpret this to mean the instrument will indicate true flow when respiratory frequency is about 15 breaths/min (normal breathing) but will read at least 13 percent in error when respiration is in the form of a pant (approximately 60 breaths/min).* A more revealing type of plot with this same information is generally shown as a log-gain vs. log-frequency plot. The details of the construction of these graphs and the interpretation of this information are deferred to a later section.

We can also examine the cause for the sinusoidal phase shift by tracing back through the analysis to determine the origin of the phase angle. In doing so we find immediately that the angular shift stems directly from the angle term associated with the complex number $10 \pm 3j$ (see Fig. 8.8). Specifically, we find

$$(8.42) \qquad \tan \theta = \frac{3}{10} = 0.3 \qquad (\text{See Fig. 8.8}a)$$

As we have already identified the 3 as the excitation frequency and the 10 as the reciprocal of the system time-constant $(1/\tau)$, we see that the phase angle is given by

$$(8.43) \qquad \tan \theta = \omega\tau$$

and once again we find the $\omega\tau$ product governs some feature of the output response. This is not a surprising result as it is reasonable to expect any alterations between input and output waveforms should reflect something of the driver and the system. The phase shift, like the magnitude change, represents some modification between input and output waveforms; it should therefore be governed by excitation and system, as was the amplitude ratio (gain). Phase-frequency curves can be used for system identification in the same way gain-frequency curves are used. The phase-frequency plot of Fig. 8.11 is, for example, representative of a first-order system. However, phase-frequency plots are not as sensitive as gain-frequency plots, and the latter are more frequently used for analysis. More will be said of these plots in a later section.

From the above discussion it is obvious that, for this system, the complex number $10 \pm 3j$ governs both the magnitude and phase of the

---

* At 15 c/min, $\omega = 1.5$ rad/s. Therefore gain $= 1/\sqrt{1 + [(1.5)(0.1)]^2} = 1/\sqrt{1 + (0.15)^2}$ or approximately unity. Therefore $F_{RA} \approx F_{SA}$, or the amplitude of the instrument reading is approximately equal to the amplitude of the actual flow. On the other hand, when ω is 60 c/min, $\omega = 6$ rad/s. Then gain $= 1/\sqrt{1 + (0.6)^2} \approx 0.87$. Therefore the instrument amplitude (F_{RA}) is about 0.87 of the input amplitude (F_{SA}) with a contingent error of at least 13 percent.

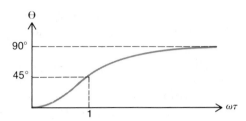

FIG. 8.11   *Phase-frequency characteristics of a first-order system.*

steady-state response.* This point has significant implications and will be analyzed more carefully after a brief "intuitive" examination of the relations governing the steady-state phase and amplitude response. The fact that the output magnitude is governed by the term $10 + 3j$ was implied by Eq. (8.40). A verbal statement of this equation is

(8.44)   Output amplitude $= \dfrac{(\text{input amplitude})/(\text{time-constant})}{\text{magnitude of complex number } (10 + 3j)}$

where the magnitude of the complex number is simply the length of the line defined by the complex number $10 + 3j$ (see Fig. 8.8). In symbolic form this may be written

(8.45) $$F_{SA} = \frac{F_{RA}/\tau}{|10 + 3j|}$$

or

$$F_{SA} = F_{RA} \frac{1/\tau}{|j\omega + 1/\tau|} = F_{RA} \left| \frac{1/\tau}{j\omega + 1/\tau} \right|^{\dagger}$$

Before we examine this any further we shall also look at the relationship between the output phase angle and the complex number $10 + 3j$. The connection between these quantities was implied in Fig. 8.8, in Eq. (8.31), and again in Eq. (8.43). A verbal statement of the relationship is

(8.46)   Output phase angle
$= $ angle associated with complex number $10 + 3j$

* Steady-state response is a term often used in sinusoidal analysis (instead of forced response). It is the response remaining after the free-response terms have disappeared.

† The manipulations required to arrive at the last relationship should be evident by inspection. The magnitude of 1/(complex number) is the same whether only the denominator or the whole fraction is considered to be complex.

$$\frac{1}{|\text{complex number}|} = \left| \frac{1}{\text{complex number}} \right|$$

In words, 1/(magnitude of complex number) is equivalent to the magnitude of 1/(complex number).

This may be written symbolically as

(8.47)
$$\theta = \left/ \frac{1}{10 + 3j} \right. = \left/ \frac{1/\tau}{j\omega + 1/\tau} \right.$$

where we see that the angle associated with a complex number is not altered by a multiplicative constant. An examination of Eqs. (8.45) and (8.47) indicates that both phase and amplitude are governed by the number $(1/\tau)/(j\omega + 1/\tau)$. This expression is immediately recognized as the system transfer function $T(s)$, where s is replaced by $j\omega$ [see Eq. (8.18) and the discussion leading to this equation].

We can now see how the phase and amplitude of the sinusoidal output are dependent upon the input signal and system properties; both phase and amplitude are governed by the system transfer function with s replaced by j operating on the frequency of the sinusoidal excitation. To use the brevity of mathematical notation, we might say, if a system is excited with a sinusoid stimulus, phase and output amplitude are functions of $T(j\omega)$, where $T(j\omega)$ is understood to mean the system transfer function with s replaced by $j\omega$. In symbols, this total statement for the instrument system may be written as follows: If

(8.48a)
$$T(s) = \frac{1/\tau}{s + 1/\tau} = \frac{f_R}{f_S} \qquad \text{and} \qquad f_S = F_{SA} \sin \omega t$$

then

(8.48b)
$$T(j\omega) = T(s)]_{s \to j\omega} = \left(\frac{1/\tau}{s + 1/\tau} \right)_{s \to j\omega} = \frac{1/\tau}{j\omega + 1/\tau}$$

and

(8.48c)
$$F_{RA} = F_{SA} \left| \frac{1/\tau}{j\omega + 1/\tau} \right| = F_{SA} |T(j\omega)|$$

(8.48d)
$$\theta = \left/ T(j\omega) \right.$$

and finally

(8.48e)
$$f_R \text{ (steady state)} = F_{SA} |T(j\omega)| \sin \left[\omega t + \left/ T(j\omega) \right. \right]$$

We see then, if the transfer function is known, it is a simple matter to determine the sinusoidal, steady-state response of the system. Although this concept was developed on the basis of a specific system, it is general, and in the next section we develop this relationship for a general case.

8.4 RELATIONSHIP BETWEEN TRANSFER FUNCTION AND SINUSOIDAL RESPONSE

In the preceding section it was shown that for a particular situation the steady-state response to a sinusoidal driver can be determined immedi-

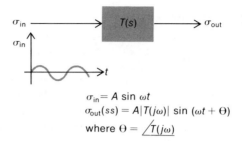

$\sigma_{in} = A \sin \omega t$
$\sigma_{out}(ss) = A|T(j\omega)| \sin (\omega t + \Theta)$
where $\Theta = \underline{/T(j\omega)}$

FIG. 8.12 *Steady-state sinusoidal response of a system.*

ately if the system transfer function is known. We shall now establish this concept for a general case. To develop the general equations we use the block diagram of Fig. 8.12 and write immediately

$$(8.49) \qquad\qquad \sigma_{out} = \sigma_{in} T(s)$$

If σ_{in} is a sinusoid of amplitude A and frequency ω, then

$$(8.50) \qquad\qquad \sigma_{in} = A \sin \omega t$$

or

$$\sigma_{in} = \frac{A\omega}{s^2 + \omega^2}$$

We therefore write

$$(8.51) \qquad \sigma_{out} = \frac{A\omega}{s^2 + \omega^2} T(s) = \frac{A\omega T(s)}{(s + j\omega)(s - j\omega)}$$

To obtain a total solution for Eq. (8.51) we would have to expand into partial fractions. This would result in two fractional terms due to the factors $s \pm j\omega$ and additional fractional terms due to the other denominator factors of $T(s)$ [see Eq. (8.25)]. This may be written

$$(8.52) \quad \sigma_{out} = \frac{A\omega T(s)}{(s + j\omega)(s - j\omega)} = \frac{M}{s + j\omega} + \frac{N}{s - j\omega}$$
$$+ \text{ other fractional terms}$$

For the steady-state sinusoidal response only the first two terms of Eq. (8.52) are of interest. All other terms are exponential terms (i.e., free-response terms) which decay to zero after long-term operation [see Eqs. (8.30) and (8.36) and the response curve of Fig. 8.9]. For the steady-state analysis, therefore, we need evaluate only the numerators of $s \pm j\omega$. To obtain M we multiply Eq. (8.52) through by $s \pm j\omega$ and allow s to approach $-j\omega$. In mathematical notation the result may be expressed as

$$(8.53) \qquad M = \left[\frac{A\omega T(s)}{s - j\omega}\right]_{s \to -j\omega} = \frac{-AT(-j\omega)}{2j}$$

By a similar procedure we have

$$(8.54) \qquad N = \left[\frac{A\omega T(s)}{s + j\omega}\right]_{s \to +j\omega} = \frac{+AT(+j\omega)}{2j}$$

With the results given by Eqs. (8.53) and (8.54) we can now write the steady-state sinusoidal solution:

$$(8.55) \quad \sigma_{\text{out}}(ss)^* = \frac{M}{s + j\omega} + \frac{N}{s - j\omega} = \frac{A}{2j}\left[\frac{-T(-j\omega)}{s + j\omega} + \frac{T(+j\omega)}{s - j\omega}\right]$$

$$= A\,\frac{T(+j\omega)\epsilon^{+j\omega t} - T(-j\omega)\epsilon^{-j\omega t}}{2j}$$

We now see $T(j\omega)$ is the same as any complex number and may be expressed in terms of a magnitude and an angle [see Figs. 8.3 and 8.8 and Eq. (8.32)]. Therefore $T(j\omega)$ may be written

$$(8.56) \qquad T(j\omega) = |T(j\omega)|\epsilon^{j\theta}$$

where $T(j\omega)$ represents the magnitude of the complex number (i.e., the length of the directed line) and θ represents its inclination. By the same reasoning, $T(-j\omega)$ may be written

$$(8.57) \qquad T(-j\omega) = |T(-j\omega)|\epsilon^{-j\theta} = |T(+j\omega)|\epsilon^{-j\theta}\dagger$$

With these last two relationships Eq. (8.55) becomes

$$(8.58) \qquad \sigma_{\text{out}}(ss) = A|T(j\omega)|\,\frac{\epsilon^{j\theta}\epsilon^{j\omega t} - \epsilon^{-j\theta}\epsilon^{-j\omega t}}{2j}$$

or

$$\sigma_{\text{out}}(ss) = A|T(j\omega)|\sin(\omega t + \theta)$$

The last equation is the general expression relating the system transfer

* Where (ss) implies the "steady-state" solution.

† The last form of Eq. (8.57) is easy to justify graphically: $T(-j\omega)$ is the same as $T(+j\omega)$ except every $+j$ has been replaced by a $-j$ (these are complex conjugates). This means if $T(j\omega)$ is a directed line with a positive inclination (Fig. 8.8a) $T(-j\omega)$ is a directed line with the same magnitude as $T(j\omega)$ but with a negative inclination (Fig. 8.8b). We can therefore write $T(-j\omega)$ as a line with magnitude $T(+j\omega)$ and a negative angle as given by Eq. (8.57).

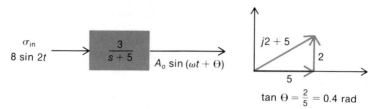

FIG. 8.13 *Steady-state response of a specific system.*

function to the steady-state sinusoidal response of the system. It is identical to Eq. (8.48e), which was developed essentially by intuitive reasoning.

In the next section we shall utilize these concepts to examine some of the flow relationships that exist during respiration. However, to illustrate the simplicity of the technique just developed, we examine the academic situation shown in Fig. 8.13. A system with a transfer function $3/(s + 5)$ is excited by a sinusoid of an amplitude of eight units and frequency of 2 rad/s. The steady-state response is required. We can write the result almost by inspection:

$$(8.59a) \quad |T(j\omega)| = \left| \frac{3}{j\omega + 5} \right| = \frac{3}{\sqrt{\omega^2 + 5^2}} = \frac{3}{\sqrt{4 + 25}} = \frac{3}{\sqrt{29}}$$

The output amplitude is therefore

$$(8.59b) \qquad A_0 = \frac{3}{\sqrt{29}} \times 8 = \frac{24}{\sqrt{29}} \approx 4.5$$

To obtain the phase angle,

$$(8.59c) \qquad T(j\omega) = \frac{3}{j\omega + 5} = \frac{3}{\sqrt{29}\ \epsilon^{j\theta}}$$

where

$$(8.59d) \qquad \tan \theta = \frac{2}{5} \approx 0.4 \text{ rad} \qquad \text{(See Fig. 8.13)}$$

Therefore

$$(8.59e) \qquad T(j\omega) = \frac{3}{\sqrt{29}}\ \epsilon^{-j(0.4)}$$

$$(8.59f) \qquad \underline{/T(j\omega)} = 0.4 \text{ rad}$$

Finally

$$(8.59g) \qquad \sigma_{\text{out}}(ss) = 4.5 \sin (2t - 0.4)$$

8.5 SINUSOIDAL ANALYSIS OF A SECOND-ORDER SYSTEM

The technique developed in Sec. 8.4 to determine the sinusoidal response of a system is not restricted by the order of the system. The transfer function $T(s)$ was not specified; the final result [Eq. (8.58)] is therefore completely general and applicable to a system of any order. The procedure used to evaluate the sinusoidal response of a first-order system [see the sequence Eqs. (8.48) to (8.48e) or Eqs. (8.58a)to (8.58g)] may therefore also be used to determine the response of systems of any order. In evaluating the sinusoidal response of second- and higher-order systems, however, it is necessary to evaluate at least second-order complex terms. The manipulation of these terms may be unfamiliar (or forgotten). To illustrate how these terms may arise and how they may be handled, we shall examine the sinusoidal response of a second-order system.

The second-order system chosen for illustration is the analog representation of a flexible tube feeding a compliant chamber, as shown in Fig. 7.4. As mentioned in the discussion of this model, the simulation may be considered a lumped representation of the lung (see Sec. 7.3). This affords a particularly appropriate example for this discussion since the lung (as part of the respiratory system) represents a system stimulated naturally by a driver that may be approximated by a sinusoid.

The model to be used is repeated in Fig. 8.14b. It is the same as that of Fig. 7.4 except in this case R_b is composed of two series resistors. One of them represents the resistive effect of the trachea; the other represents part of the resistive effect of the flexible airways. The change is not a significant one and does not affect the circuit operation as the model now stands. Later, when we wish to examine the separate operation of the right and left lungs, we shall see that the tracheal resistance must be separately represented so that additional circuitry may be included at point B.

To indicate how the sinusoidal excitation is related to the physiological system, we examine briefly some of the mechanical features of the respiratory function. A schematic diagram of the operating parts of the respiratory system is shown in Fig. 8.14a. Breathing is induced by motion of the diaphragm and rib cage. This motion creates a variable pressure in the thoracic cage (chest cavity) which causes air flow to and from the lung. When the diaphragm is "down" or flat and the rib cage distended (solid lines in Fig. 8.14a) pressure in the thorax (or the intra-

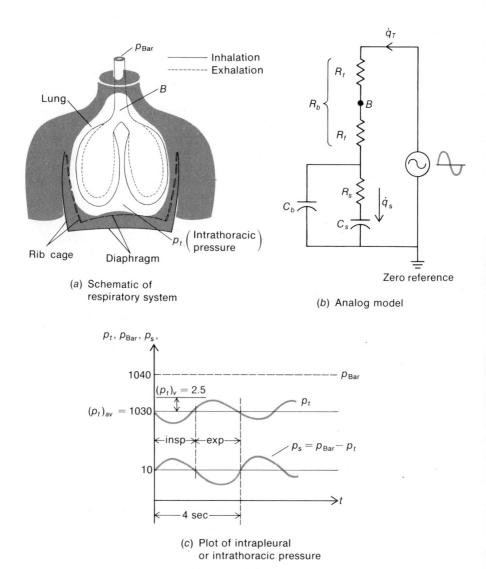

FIG. 8.14   *Representation of respiratory system.*

pleural pressure, as it is sometimes called*) is reduced below a "resting level." This reduced pressure creates a pressure differential between

---

* There is considerable interchangeable terminology in this area. Intrathoracic pressure is often designated as the pleural pressure, pressure in the pleural cavity, or the intrapleural pressure. In most cases these terms all refer to the "space" directly adjacent to the lung surface and the chest wall. See also Sec. 4.4 for a discussion of the various pressures involved.

mouth and chest cavity; the lung becomes distended and air flows into the lung (inhalation). For exhalation, the opposite effects are induced: The diaphragm moves up and "bows," and the rib cage contracts to reduce the intrathoracic volume (dashed lines in Fig. 8.14a); intrathoracic pressure is increased; the lungs are compressed and air is forced out. We see, then, that the pressure signal that drives the respiratory system is the differential pressure between the mouth (assumed to be atmospheric pressure or p_{bar}) and the intrathoracic cavity. This pressure stimulus may be expressed mathematically as

$$(8.60) \qquad\qquad p_s = p_{bar} - p_t$$

Since the mouth is normally at barometric pressure and this pressure remains ostensibly constant, normal breathing is induced by changes in p_t.* A plot of this pressure curve is shown in Fig. 8.14c. It may be noted from this plot that p_t (the intrathoracic pressure) varies around some average value [designated by $(p_t)_{av}$]. We can therefore describe p_t as an average pressure plus or minus some variable component. If we call this variable component $(p_t)_v$, then p_t may be written

$$(8.61) \qquad\qquad p_t = (p_t)_{av} - (p_t)_v$$

The pressure stimulus therefore becomes

$$(8.62) \quad p_s = p_{bar} - p_t = p_{bar} - (p_t)_{av} + (p_t)_v = (p_s)_{av} + (p_t)_v$$

The pressure terms p_{bar} and $(p_t)_{av}$ in Eq. (8.62) are constants; their difference represents the average (or dc or constant) level of the excitation signal $(p_s)_{av}$. This average pressure is important physiologically as it keeps the lung in an extended, noncollapsed condition, and it sets the operating level of the resistive and compliant properties of the lung.† For the steady-state, periodic analyses, however, only the pressure variation is of importance; the average (or dc) level can be neglected. This seems reasonable, since in the steady state the average (or constant or dc) level contributes nothing to the variable output response; only the continuous (or repeatable) variations are of interest for the steady-state analysis of systems with periodic excitations. If we ignore the steady-pressure levels, the variable component of the intrathoracic pressure

---

* Breathing is induced so long as p_s, the differential pressure, is varied. From Eq. (8.60) we see p_s may also be varied if p_{bar} is artificially cycled and p_t remains constant. This is one way to force respiration or to perform artificial respiration. See Sec. 8.11.

† Increased interthoracic pressure due to air trapped in the chest cavity results in a pathological condition known as pneumothorax. Compliance and resistance are high under these circumstances, and breathing is shallow as the rib cage and diaphragm cannot produce a sufficient pressure drop to induce a full inspiration.

$[(p_t)_v]$ is the only pressure stimulus of interest and we can write

(8.63) $$p_s = (p_t)_v$$

The waveshape of the variable intrathoracic-pressure component is shown in Fig. 8.14c. The curve is drawn closely sinusoidal. In reality, the intrapleural-pressure curve does not appear so symmetrically defined. It is, however, a relatively periodic waveform and a sinusoid serves as a first-order approximation. If we accept a sinusoidal representation, p_s may be described as

(8.64) $$p_s = 2.5 \sin 1.6t \quad \text{cm } H_2O$$

where 2.5 cm H_2O represents the amplitude of the excitation and 1.6 rad/s represents the normal-breathing frequency (1.6 rad/s is approximately $\frac{1}{4}$ c/s or $\frac{1}{4}$ Hz, which is the same as 15 breaths/min). This is the pressure stimulus used to excite the analog model of Fig. 8.14b.

We are now prepared to use the analog model to examine some of the flow relationships to be expected in the physiological system. For the purposes of illustration we shall accept the values for the circuit elements established in Sec. 7.3. Thus R_b will be taken as 1.6 cm $H_2O/l/s$, R_s as 0.8 cm H_2O l/s, C_b as 0.02 l/cm H_2O, and C_s as 0.18 l/cm H_2O. As an academic exercise we first use the model to determine the total resultant flow. The relationship between pressure and flow is given by Eq. (7.20). If we substitute the numerical values indicated above we may write

(8.65) $$\frac{\dot{q}_T}{p_s} = \frac{s(s + 70)}{s^2 + 109s + 273}\left(\frac{1}{1.6}\right) \quad \text{[See also Eq. (7.30)]}$$

If p_s is a sinusoid of amplitude 2.5 cm H_2O and with a frequency 1.6 rad/s [Eq. (8.64)], the output amplitude of $\dot{q}_T$ is given by

(8.66)   Amplitude of $\dot{q}_T$

$$= (\text{amplitude of } p_s)\left|\frac{s(s + 70)}{s^2 + 109s + 273}\left(\frac{1}{1.6}\right)\right|_{s \to j(1.6)}$$

$$= \frac{2.5}{1.6}\left|\frac{1.6j(1.6j + 70)}{(1.6j)^2 + 109(1.6j) + 273}\right|$$

[See Eqs. (8.59a) and (8.59b).] We must now evaluate the magnitude of the complex number shown in the second form of Eq. (8.66). As expected, the second-order system gives rise to complex numbers with second-order terms. To evaluate such complex numbers we have only to remember that j to a power may be replaced by a real number or by j alone. Thus j^2 is equivalent to -1, j^3 is equivalent to $-j$, etc. [see Eq. (8.6) and the discussion following]. If we use this viewpoint, Eq. (8.66)

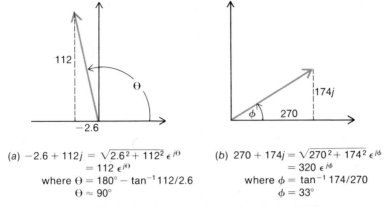

(a) $-2.6 + 112j = \sqrt{2.6^2 + 112^2}\,\epsilon^{j\Theta}$
$\qquad\qquad\qquad = 112\,\epsilon^{j\Theta}$
where $\Theta = 180° - \tan^{-1} 112/2.6$
$\qquad\qquad \Theta \approx 90°$

(b) $270 + 174j = \sqrt{270^2 + 174^2}\,\epsilon^{j\phi}$
$\qquad\qquad\qquad = 320\,\epsilon^{j\phi}$
where $\phi = \tan^{-1} 174/270$
$\qquad\qquad \phi = 33°$

FIG. 8.15   *Evaluation of complex numbers in Eq. (8.67).*

may be rewritten

(8.67) $\qquad \dot{q}_T \text{ (amplitude)} = 1.56 \left| \dfrac{-2.56 + 112j}{(273 - 2.56) + 174j} \right|$

$\qquad\qquad\qquad\quad \approx 1.56 \left| \dfrac{-2.6 + 112j}{270 + 174j} \right|$

We can now evaluate the numerator and denominator of this last complex number by the trigonometric relations indicated in Fig. 8.15. From the diagrams in this figure and the relationships derived, we see Eq. (8.67) may now be written

(8.68) $\qquad \dot{q}_T \text{ (amplitude)} = 1.56 \left| \dfrac{112\epsilon^{j90°}}{320\epsilon^{j33°}} \right| = 1.56 \left| \dfrac{112}{320}\,\epsilon^{j57°} \right|$

The magnitude of the last complex number is obviously $112/320 = 0.35$.

(8.69) $\qquad \dot{q}_T \text{ (amplitude)} = 1.56 \times 0.35 = 0.55 \text{ l/s} = 550 \text{ ml/s}$

To determine the phase angle associated with the flow, we evaluate the angle of the complex number $T(j\omega)$ [that is, $\underline{/T(j\omega)}$; see Eq. (8.59f)]. This is already available in Eq. (8.68), and we can immediately write

(8.70) $\qquad\qquad \text{Phase of } \dot{q}_T = \underline{/T(j\omega)} = 57° \approx 1 \text{ rad}$

From Eqs. (8.69) and (8.70) we can now describe the sinusoidal flow response as

(8.71) $\qquad\qquad \dot{q}_T = 550 \sin (1.6t + 1 \text{ rad}) \qquad \text{ml/s}$

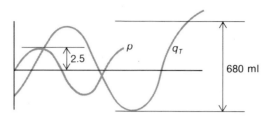

(a) p_s and q_T curves as derived

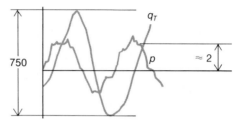

(b) p_s and q_T curves as measured

FIG. 8.16 *Plots of pressure and volume curves. [Part b is adapted from Starling and Lovatt Evans, "Principles of Human Physiology," 13th ed., H. Dawson and H. Grace Eggleton (eds.), Lea and Febiger, Philadelphia, 1962, p. 355.]*

As a basis for comparison, the pressure and volume relationships* derived are shown in Fig. 8.16a. Figure 8.16b shows pressure and volume

---

* We can obtain the volume relationships by integrating the flow equation [Eq (8.71)]. Thus,

$$q_T = \int \dot{q}_T \, dt = \int 550 \sin (1.6t + 1)$$

$$q_T = \frac{-550}{1.6} \cos (1.6t + 1) = \frac{550}{1.6} \sin (1.6t - 0.6)$$

As an alternative procedure we could write Eq. (8.65) as

$$\frac{\dot{q}_T}{p_s} = \frac{s q_T}{p_s} = T(s)$$

Therefore

$$\frac{q_T}{p_s} = \frac{T(s)}{s}$$

$$q_T \text{ (amplitude)} = \left| \frac{T(j\omega)}{j\omega} \right| = \left| \frac{T(j\omega)}{\omega \epsilon^{j90}} \right| = \left| \frac{T(j\omega)}{\omega} \right| \epsilon^{-j90}$$

$$q_T \text{ (amplitude)} = \frac{|T(j\omega)|}{\omega} = \frac{550}{1.6} = 340$$

$$q_T \text{ (phase)} = \underline{/T(j\omega)} - 90° = 1 \text{ rad} - \frac{\pi}{2} = -0.6 \text{ rad}$$

$$q_T = 340 \sin (1.6t - 0.6)$$

q_T is the volume of gas that has entered or left the lung in any period of time t; it is known as the tidal-air volume (see Fig. 4.13).

curves obtained from a resting subject. It is rather difficult to make any direct comparisons as the numerical values used in the model were chosen rather arbitrarily. In spite of these crude approximations, and in spite of the fact that the pressure curve is not really sinusoidal, the volume curves appear to agree reasonably well; the derived phase relationship between pressure and volume also agrees fairly well with the measured values. We cannot help but feel elated at the good correspondence between measured and derived results; on the basis of this correspondence, there is a great temptation immediately to accept the model as a good representation of the physiological system. Although the feeling of elation might be justified, we should keep in mind the crudity of the assumptions. The good correspondence might be coincidental, or it might indicate that a sinusoidal analysis with one frequency is too insensitive for this situation. It may be possible that a wide range of element values could be used with the proposed model and that we would still obtain "close results." In this case, for example, we could double R_s and reduce R_b in half and the amplitude and phase of the resultant flow would remain virtually unchanged. To justify the adequacy of a model representation, more testing would be required.

8.6 THE CONCEPT OF SINUSOIDAL IMPEDANCE

When only sinusoidal excitations are involved, system transfer functions can be expressed as complex numbers by replacing s with $j\omega$. These complex numbers may then be used to define the magnitude and phase of the response. As impedance functions are essentially special forms of transfer functions, these too may be expressed in terms of $j\omega$ when sinusoidal behavior is under consideration. The complex impedance can be used to define the magnitude and phase of a through- or an across-variable response. When an impedance function is written in terms of $j\omega$, the function is generally described as the sinusoidal impedance of a network (as opposed to the parametric impedance, which is the impedance function defined in terms of s; see the end of Sec. 7.2).

As a simple illustration of how this concept may be used, we note that the (parametric) impedance of a capacitor was defined by Eq. (7.11) as

$$(7.11) \qquad \text{Impedance of capacitor} = Z_c(s) = \frac{1}{sC}$$

If we wish to consider only the sinusoidal behavior of the capacitor, we replace the s by $j\omega$ and we have

(8.72)   Sinusoidal impedance of capacitor $= Z_c(j\omega) = \dfrac{1}{j\omega C}$ *

where ω is the frequency of the sinusoidal excitation signal applied to the capacitor. If the excitation signal is a voltage of amplitude E and frequency ω (see Fig. 8.17a) the amplitude of the current response is given by $E/|Z(j\omega)| = E/\omega C$, and the phase angle of the current response is $\underline{/Z(j\omega)} = -90°$. The current flow may thus be described by

(8.73) $i = \dfrac{E}{\omega C} \sin (\omega t - 90°) = \dfrac{E}{\omega C} \sin \left(\omega t - \dfrac{\pi}{2}\right)$

The relationship between the waveforms of e and i are shown in Fig. 8.17b. The voltage peak always occurs 90° [or $(\pi/2\omega)$ s; see the discussion associated with Eqs. (8.36) and (8.37)] before the current peak. This is often described by saying the voltage leads the current (or the current lags the voltage) by 90°.

As a more meaningful illustration of the utility of the impedance concept in studies of sinusoidally excited systems, we examine once more the model of the respiratory system in Fig. 8.14 and redrawn in Fig. 8.18. In this case we assume the system is stimulated by a current source, $\dot{Q} \sin \omega t$, rather than a pressure source†; we wish to examine the resultant flow into the lower bronchial tree (i.e., the alveoli at the terminations of the airways). From the diagram in Fig. 8.18 we see clearly that all the inspired air does not enter the alveolar sacs (C_s); some is stored in the flexible airways. To obtain an estimate of the flow to the alveoli we can use some of the principles of circuit analysis. From the diagram in Fig. 8.18 we realize immediately we can write the total flow as the sum of the

---

* The magnitude of this impedance is simply $1/\omega C$. If ω in radians is replaced by $2\pi f$ (where f is in cycles per second) the magnitude may be written $1/2\pi f C$. The latter form is very often used in elementary electrical studies to describe the sinusoidal impedance of a capacitor.

† This shift in viewpoint is one of the many advantages of an analytic investigation. We have the choice here of assuming the excitation is a pressure signal and the response is the flow (as on the left in the accompanying block diagram) or we may assume the flow is the

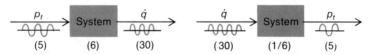

excitation and the response is the intrathoracic pressure (block diagram on the right). The question of which is "right" is not relevant, as either point of view can be used for analytic purposes. The analytic results will be correct either way; it is just as correct to say $5 \times 6 = 30$ as it is to say $30/6 = 5$.

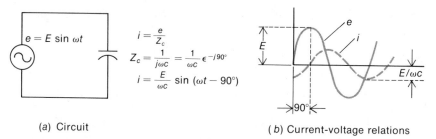

(a) Circuit   (b) Current-voltage relations

FIG. 8.17 *Sinusoidal impedance of a capacitor.*

flows to the alveolar sacs and the expanded airways. Thus $\dot{q}_T = \dot{q}_b + \dot{q}_s$. This total flow may be described by a pressure at a (p_a) acting on the admittance between point a and the ground reference (see Fig. 8.18). Since this admittance is composed of two parallel branches, it may be written as the sum of the branch admittances $Y_b + Y_s$, where Y_b is the (sinusoidal) admittance of C_b and Y_s is the (sinusoidal) admittance of the R_sC_s combination. We have thus far, therefore,

$$(8.74) \qquad \dot{q}_T = p_a(Y_b + Y_s)$$

The alveolar flow may also be directly defined by the pressure p_a acting on the admittance Y_s. Thus

$$(8.75) \qquad \dot{q}_s = p_aY_s$$

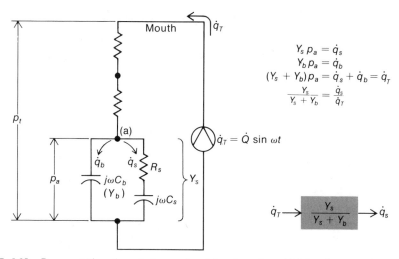

FIG. 8.18 *Representation of respiratory system driven by a sinusoidal flow (or current) source.*

The ratio of Eqs. (8.75) and (8.74) now yields the desired relation between the input flow and the alveolar flow.

$$(8.76) \qquad \frac{\dot{q}_s}{\dot{q}_T} = \frac{Y_s}{Y_b + Y_s} = \frac{1}{Y_b Z_s + 1}$$

where the last form is obtained by using $Y_s = 1/Z_s$. It is now fairly easy to define the alveolar flow when the delivered flow is given by the sinusoid $340 \sin 1.6t$ ml/s. The sinusoidal admittance $Y_b(j\omega)$ is given by $j\omega C_b$; $Z_s(j\omega)$ is the series combination of R_s and C_s. This may be written $R_s + 1/j\omega C_s$. The admittance ratio in Eq. (8.76) may therefore be expressed as

$$(8.77) \qquad \frac{1}{Y_b Z_s + 1} = \frac{1}{j\omega C_b(R_s + 1/j\omega C_s) + 1}$$

$$\frac{1}{Y_b Z_s + 1} = \frac{1}{j\omega C_b R_s + (C_b/C_s) + 1} = \frac{1}{(C_b + C_s)/C_s + j\omega C_b C_s}$$

If we use the numerical values previously established for the circuit elements ($C_b = 0.02$ l/cm H_2O, $C_s = 0.18$ l/cm H_2O, $R_s = 0.8$ cm H_2O/l/s, $\omega = 1.6$ rad/s) we have

$$(8.78) \qquad \frac{1}{Y_b Z_s + 1} = \frac{1}{(10/9) + j(1.6)(0.02)(0.18)} = \frac{1}{1.1 + j0.0064} \approx 0.9$$

$$\frac{1}{Y_b Z_s + 1} = 0.9\epsilon^{-j0.3}$$

Alveolar air flow may therefore be written

$$(8.79) \quad \dot{q}_s = (340)(0.9) \sin(1.6t - 0.3°) = 306 \sin(1.6t - 0.3°)$$

A plot of the alveolar flow is compared with the delivered flow in Fig. 8.19. As expected, the amplitude of the alveolar flow is slightly less

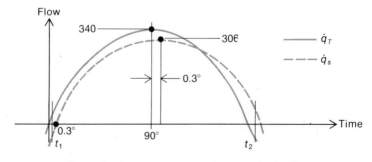

FIG. 8.19 *Comparison of total flow and alveolar flow.*

than the delivered flow; some of the delivered flow is used to fill the expanded airways. Of somewhat greater interest in this case is the phase shift between the alveolar flow and the total flow. Although this phase shift is very small numerically (0.3° or 0.005 rad) it must exist. This implies that sometime during the breathing cycle delivered flow must be positive and alveolar flow is negative (and vice versa). In Fig. 8.19, for example, at t_1 total flow is positive (inhalation at the mouth) but alveolar flow is negative (exhalation at the lower end of the bronchial tree); at t_2 the reverse situation occurs. The surprising conclusion is that during inhalation the alveoli may be actually delivering to the expanded airways and that during exhalation air may be delivered to the alveoli from the contracting airways.

If we extend the idea developed in the preceding paragraph, we see that in a breathing cycle air may be shifted between alveoli, between different lung sections, and in fact between the lungs. We can easily justify the latter claim by using an analysis similar to the one just developed except we now use a model that includes a representation for both lungs. Such a simulation is shown in Fig. 8.20; the resistance R_t represents the tracheal resistance common to both lungs. Models to represent both lungs are connected to the trachea in parallel at B; the circuit elements for the right and left lungs are assumed to be different (if they were identical, the circuit in Fig. 8.20 would reduce to that of Fig. 8.14b). From the procedure outlined earlier we see that the total flow is given by $p_B(Y_L + Y_R)$, where Y_L and Y_R represent the admittances of the left and right lung, respectively. In a similar manner the flow to the left lung ($\dot{q}_L$) is given by $p_B Y_L$. The ratio of these two flows is then

$$(8.80) \qquad \frac{\dot{q}_L}{\dot{q}_T} = \frac{Y_L}{Y_L + Y_R}$$

From Eq. (8.80) we see if $Y_L = Y_R$ (i.e., the two lungs are identical) the flow to the left lung would be one-half the total flow. This would seem to be reasonable except that it is unrealistic to expect the right and left lungs to be identical in every respect. It is more probable that the two lungs would manifest differences in dimensions as well as structural details. The chances are, therefore, Y_L and Y_R are not alike; cancellations will not be possible, and we must accept that the admittance ratio in Eq. (8.80) will give rise to a complex number when a sinusoidal analysis is performed. The complex number will have a phase angle associated with it and so we must expect that flow to the left lung (and right lung) will not be in phase with the total flow. This would mean that at some time during the breathing cycle air will be inhaled into the respiratory system while air is simultaneously expelled from the left

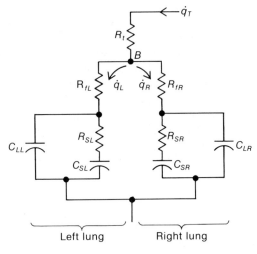

(a) Circuit model

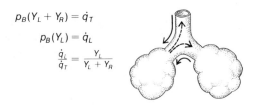

(b) Concept of pendelluft

FIG. 8.20 *Model of the respiratory system when right and left lungs have different element properties.*

lung During this period all flow will be to the right lung. At some other time during the cycle all flow will be to the left lung while the right lung exhales. Some possibilities of air flow are shown in Fig. 8.20b. It is of interest to note that the transfer of air between the lungs as predicted by the circuit analysis can be demonstrated by experiment. The phenomenon is generally described as *pendelluft* (i.e., pendulous air); one research report which includes some information about this effect was published by Hugh-Jones and West (Ref. 5).

The last circuit analysis is a good illustration of how the most general formulation can lead to meaningful results if the principles are understood. Equation (8.80) cannot be considered a detailed description of the flow relationship, yet if the implications of the admittance ratio are understood no further analysis is needed to predict some aspects of the flow relationships.

8.7 GAIN AND PHASE PLOTS AS A FUNCTION OF FREQUENCY

In Sec. 8.3, it was mentioned that the gain-frequency and phase-frequency plots of Figs. 8.10 and 8.11 are characteristic of all first-order systems. We can easily establish the validity of this claim from the definitions of gain and phase:

(8.81) $G = |T(j\omega)|$ [See Eqs. (8.41) and (8.48c)]

(8.82) $\phi = \underline{/T(j\omega)}$ [See (Eq. 8.48d)]

Since the form of the transfer function for all first-order systems is the same, the mathematical relationship between gain (or phase*) and frequency must also be the same. The plots of gain vs. frequency for all first-order systems must therefore have the same characteristic shape.

To illustrate this point with a specific example, we note if a system has a transfer function of $T_1 = 2/(s + 4)$ and another has a transfer function $T_2 = 16/(s + 8)$ then both have a gain function expressed by $A/\sqrt{\omega^2 + B^2}$. For the first system A is 2 and B is 4; for the second system A is 16 and B is 8. Since the gain functions of both systems are described by the same mathematical expression, their graphical features will be the same: At "low" frequencies the gain is approximately constant (when ω is small, $G \approx A/\sqrt{B^2} = A/B$; therefore $G_1 \approx \frac{1}{2}$, $G_2 \approx 2$); at high frequencies the gain curve is approximately hyperbolic [when ω is large (that is, $\omega \gg B$) $G \approx A/\sqrt{\omega^2}$; therefore $G\omega = A$].

These features are shown by the gain vs. frequency plots in Fig. 8.21.

* In these next two sections we deal with the gain function. The phase function will be examined in Sec. 8.9.

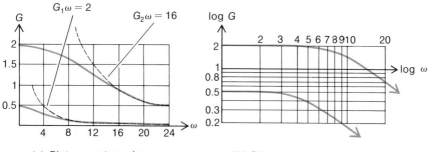

(a) Plot on rectangular axes   (b) Plot on logarithmic axes

FIG. 8.21 *Gain-frequency plots if* $T_1 = 2/(s + 4)$ *and* $T_2 = 16/(s + 8)$.

In Fig. 8.21a the plot is shown on rectangular axes; the curve starts with zero slope and ends with hyperbolic characteristics. These features are more pronounced in Fig. 8.21b where the gain-frequency curve is plotted on logarithmic coordinates. Here we can clearly see the low-frequency behavior is essentially a horizontal line, which indicates that gain remains relatively constant; the high-frequency behavior approaches a straight line with a specific negative slope, which indicates that gain and frequency are related by a product function ($G\omega = A$; therefore $\log G = \log A - \log \omega$; if $\log G$ is the ordinate and $\log \omega$ is the abscissa, this may be written $y = \log A - x$, which is a straight line with a negative slope).

From the foregoing discussion it is readily seen how gain-frequency plots may serve as a useful tool for system identification. The system is sinusoidally excited by a spectrum (multiplicity, variety) of driving frequencies. The gain function at each frequency is measured. A plot of the gain-frequency function then serves to identify something about the system transfer function. The procedure is diagrammatically illustrated in Fig. 8.22. Even with limited experience, one can readily say that the system shown [that is, $T(s)$] is definitely not a first-order system; the gain-frequency plots in no way resemble the first-order plots of Fig. 8.21. We shall see later that the gain-frequency plots of Fig. 8.22 represent second-order systems with underdamped (or resonant) characteristics and, in fact, see how these plots may be used to determine system transfer functions. The point to be established here, however, is that frequency-response curves are related to system transfer functions, and the gain-frequency plot may be used for system identification.

To develop a technique that may be used for system identification, we must relate the frequency-response curves to system gain functions. One obvious way to do this is to use some formal curve-fitting techniques. The frequency-response curve can then be described by a gain function expressed in terms of frequency (ω). To obtain the overall transfer function it is then necessary to replace the frequency terms with appropriate functions of s. This overall procedure can be rather complex; curve fitting is not an especially easy process for most situations and in this case it does not even lead to a final result. After a mathematical expression to describe the response curve has been obtained, this expression must be converted from the "frequency domain" to the "s domain." The process that must be used to effect this conversion is not as straightforward as it might appear and is, in fact, beyond the scope of this text. We shall use a less exacting technique to relate transfer functions to frequency-response curves. Although the procedure does not have as general application as that just outlined, it is adequate for most applications of interest here.

To develop this identification technique we examine first how the

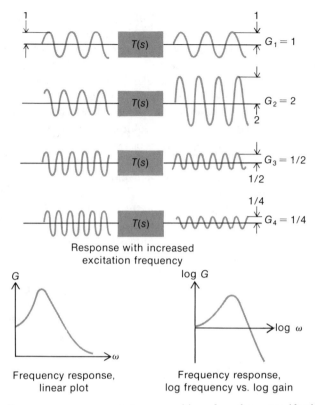

FIG. 8.22 *Frequency response as a pattern-recognition scheme for system identification.*

form of the transfer function affects the shape of the gain-frequency curve. It will then be relatively clear how the gain-frequency curve may be used to propose an appropriate transfer function. We have already seen that, for a first-order transfer function, the gain curve starts at a constant level and then rapidly decays to zero. We shall examine briefly why the first-order gain curve behaves in this fashion. The system gain of a first-order system is defined by the expression $A/\sqrt{\omega^2 + B^2}$. This function may be rewritten as

$$(8.83) \qquad G = |T(j\omega)| = \frac{A/B}{\sqrt{(\omega/B)^2 + 1}}$$

In this form we can easily see how the value of ω (the excitation frequency) affects the frequency response. If ω is small compared with B

(some system constant, e.g., the time-constant) then $(\omega/B)^2$ is small compared with unity. As an approximation, therefore, we might assume, when ω is less than B, $(\omega/B)^2$ may be neglected compared with unity. The gain function is then constant at A/B, and we see for $\omega \ll B$ that gain is essentially independent of excitation frequency. On the other hand, when ω is larger than B, $(\omega/B)^2$ is larger than unity. In this range ($\omega \gg B$) we assume unity may be neglected compared with ω/B, and the gain function becomes A/ω. In this range, therefore, the gain varies inversely with ω; as $\omega \to \infty$; $G \to 0$. The separation between the high and low range for ω is evidently governed by B; this value for ω (that is, $\omega = B$) is generally described as the break frequency of a gain-frequency plot. The reason for this terminology will become self-evident later.

It was noted earlier the characteristics of the frequency-response curve are more clearly displayed when logarithmic rather than rectangular coordinates are used. When the frequency axis is logarithmic, the low-frequency section of the axis is spread out (when $\omega = 0$, $\log \omega = -\infty$). This emphasizes the constancy of the gain-frequency curve at low frequencies. At high frequencies the gain function is proportional to a power of frequency (in this case to the -1 power; $G = A\omega^{-1}$). Logarithmic plots of such functions result in straight-line graphs which are readily identified and convenient to use. This feature is, no doubt, one of the reasons for the popularity of the log-gain vs. log-frequency plots.

Before we use numerical examples to illustrate the construction and interpretation of the gain-frequency curve on logarithmic coordinates, one other point regarding the conventional description of gain should be mentioned. Gain is defined as the ratio of an output amplitude to an input amplitude. In some cases the output amplitude may be very much smaller than the input amplitude. Gain is then a very small number. In other situations (particularly in electronic amplifier equipment) the output amplitude is very much higher than the input amplitude, and gain is very large (as much as 1 million to 1). To avoid such a wide range of numbers, gain is often defined on a logarithmic basis and is described in terms of "decibels" [or "dee-bee's" (dB's), as it is frequently called]. Absolute gains are therefore expressed in terms of exponents of 10, which eliminates the need for high numerical values. As a specific mathematical description, gain in decibels is defined as

$$(8.84) \qquad \text{Gain in decibels} = 20 \log \text{gain}^*$$

---

* The evolution of this terminology and relationship is interesting but confusing. To recount this history may allay difficulties of interpretation for those involved in auditory research. The original description of the bel (named after Alexander Graham Bell) does *not* involve input-output ratios. It relates to the logarithm of power ratios and was used to describe sound intensity. As a crude illustration: If a sound that can just barely be heard delivers

As an illustration of how Eq. (8.84) may be used: If a system has a gain of 2,000 at some frequency, its gain in decibels (G_{dB}) is given by 20 log 2,000 = 66 dB. In popular language one would say the system is "up 66 dB," indicating that the gain is higher than unity (= 0 dB). If a system has a gain of 1/2,000 at some frequency, then its gain in decibels is given by $20 \log \frac{1}{2} = -20 \log 2,000 = -66$ dB. In popular language this would be read "down 66 dB" to indicate the gain is lower than unity. For most situations the simple logarithmic ratio is adequate; the need for the coefficient of 20 is questionable. However, the decibel definition has been continuously used and it is now a well-established convention.

If we apply the definition given by Eq. (8.84) to the gain function for a first-order system [e.g., Eq. (8.83)] we have

$$(8.85) \qquad G(\mathrm{dB}) = 20 \log \frac{A}{B} - 20 \log \left[\left(\frac{\omega}{B} \right)^2 + 1 \right]^{1/2}$$

$$= 20 \log \frac{A}{B} - 10 \log \left[\left(\frac{\omega}{B} \right)^2 + 1 \right]$$

If A were 16 and B were 8, we would have

$$(8.86) \quad G(\mathrm{dB}) = 20 \log 2 - 10 \log \left[\left(\frac{\omega}{8} \right)^2 + 1 \right]$$

$$= +6 \text{ dB} - 10 \log \left[\left(\frac{\omega}{8} \right)^2 + 1 \right]$$

$$= +6 \text{ dB} - 10 \log \left[\left(\frac{\omega}{8} \right)^2 + 1 \right]$$

A plot of Eq. (8.86) is shown in Fig. 8.23. We note that at low frequencies

---

power p_1 to the ear and a more intense sound delivers twice the power ($p_2 = 2p_1$), the power ratio is 2:1. One would then describe the power of the second sound as 0.301 bel above audibility (log p_2/p_1 = bels = log 2 = 0.301) or 3 *deci*bels above audibility. This is not an input-output relationship but simply magnitude above a reference level (p_1 is considered the threshold of audibility in this case; therefore p_1 is at 0 dB).

When the electrical engineer developed *signal* amplifiers, he was interested in the output-signal power compared with the input-signal power (as a reference). He utilized the decibel as a scale to rate the effectiveness of his amplifier but in this case the decibel compared the output of a device with the input. In this application, then, the decibel involves input-output ratios. As amplifier design became more sophisticated, it became important to rate amplifiers on the basis of voltage ratios rather than power ratios (i.e., output voltage/input voltage rather than output power/input power). In a rather arbitrary fashion the electrical engineer assumed that input and output power could be defined by the square of the voltage ($P = E^2/R$) and on this basis assumed p_2/p_1 could be written $(E_2/E_1)^2$. The bel thus became log $(E_2/E_1)^2 = 2 \log (E_2/E_1)$ and the decibel was then 20 log (E_2/E_1). Finally, in systems work one is generally interested in the amplitude of the sinusoidal variables and now E_2 is considered the output amplitude of interest and E_1 is the input amplitude of interest. The ratio of E_2/E_1 then becomes "gain," and the gain in decibels is as given by Eq. (8.84).

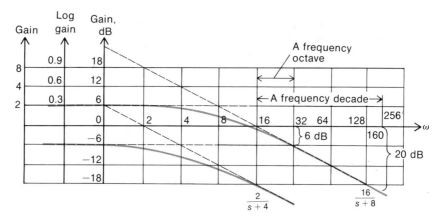

FIG. 8.23 $G(dB)$ *versus log-frequency plots for first-order systems.* $T_1 = 16/(s + 8)$, $T_2 = 2/(s + 4)$.

($\omega \ll 8$) the gain remains essentially constant at 6 dB (a gain of 2). At high frequencies ($\omega \ll 8$) the gain curve decays at a constant slope. We can uniquely describe this slope if we note that at $\omega = 16$ the gain in decibels is approximately 6 dB $-$ 10 log $(16/8)^2 = $ 6 dB $-$ 20 log 2 $= 0$; if we double the frequency, $\omega = 32$ and the gain is

$$6 \text{ dB} - 20 \log 4 = 6 \text{ dB} - 40 \log 2 = -6 \text{ dB}$$

From these figures we can see the gain curve "falls off" at a rate of 6 dB for each twofold increase in frequency ($\omega = 64$, $G_{dB} = -12$, etc.). A twofold increase in frequency is generally described as a frequency octave.* If we use this terminology, we would say the gain curve falls off at 6 dB/octave.

The slope of the first-order gain curve is sometimes also expressed in terms of tenfold increases in frequency (i.e., decades). From Eq. (8.86) (or the curve in Fig. 8.23), when $\omega = 160$ Hz (a decade above 16 Hz) the gain is -20 dB. For a change of frequency from 16 to 160 Hz, the gain changes from 0 to -20 dB. The slope of the curve may thus also be described as a gain loss of 20 dB/decade.

We now examine how the low-frequency and high-frequency approximations compare with the actual gain (decibels) vs. frequency (log) curve as plotted from Eq. (8.85). The solid curves in Fig. 8.23 are the curves for two first-order systems as plotted directly from the defining equation. The dashed lines show the curves that are obtained if we use

---

* An octave in the musical scale consists of eight notes. Thus from middle A to high A there are eight tones. Middle A has a frequency of 440 Hz; the frequency of high A is 880 Hz. A musical octave therefore includes a twofold increase in frequency.

the low- and high-frequency approximations.* The approximate curves evidently represent the asymptotic behavior of the actual curves at low and high frequencies. As can be seen from the plots, the asymptotic lines intersect at what has been described as the "break frequency." Above this frequency the high-frequency asymptote most nearly approximates the actual frequency-response curve; below this frequency the low-frequency asymptote is the better approximation.

The meaning of the descriptive term "break frequency" should now be apparent: It is the frequency at which it may be assumed that the frequency response breaks from low-frequency behavior to high-frequency behavior. At this frequency, the error between the actual curve and the low-frequency asymptote is the same as the error between the actual curve and the high-frequency asymptote. (See Prob. 8.6.) At all other points the error between the appropriate asymptotes and the actual curve is less. As a final comment about the break frequency, it should be evident that the break frequency and the system time-constant are reciprocal functions. Thus if ω_B is the break frequency and τ is the first-order time-constant, then $\omega_B = 1/\tau$. (See Prob. 8.6.)

From the foregoing discussion we see that there is a very simple way to construct the gain (decibels) vs. frequency (log) characteristics of any first-order system. We determine the magnitude of the low-frequency gain and consider the gain to be constant up to $\omega = 1/\tau$ (the break point). At this point we draw a line with slope -20 dB/decade (or -6 dB/octave). These straight-line (or asymptotic) approximations represent the true-gain curve with small error for high and low frequencies. The maximum error between the asymptotic approximations and the true curve is 3 dB at the break point. This procedure is illustrated in Fig. 8.23. To construct the frequency-response curve for a system with transfer function $2/(s + 4)$ we draw a horizontal line at the -6 dB level [G(dB) at low frequencies is $20 \log \frac{2}{4} = -6$ dB] to $\omega = 4$. At this point we draw another line which decays at a rate of 20 dB/decade (or 6 dB/octave). The true-gain curve follows the low-frequency asymptote to about $\omega = 2$; it passes through a point 3 dB below the intersection of the asymptotes and follows the high-frequency asymptote from about $\omega \geq 8$.

Now that we have learned how to construct the logarithmic frequency-response curve for a first-order system, we can construct the

---

* If $\omega \ll B$, Eq. (8.85) becomes G(dB) $= 20 \log (A/B)$; when A is 16 and B is 8, G_{dB} is $+6$ dB. When $A = 2$ and $B = 4$, $G_{dB} = \log \frac{1}{2} = -6$ dB. In both cases the approximate curves are frequency-independent and the plots are therefore horizontal lines.

If $\omega \gg B$, Eq. (8.85) becomes G(dB) $= 20 \log (A/B) - 20 \log (\omega/B) = 20 \log A - 20 \log \omega$. If we write G(dB) $= y$ and $\log \omega = x$, this last form may be written $y = 20 \log A - 20x$, which is easily recognized as a straight line with slope 20 dB per unit change in $\log \omega$ (per decade change in ω).

response curve for any system composed of first-order factors. The use of the logarithm is particularly advantageous when dealing with higher-order systems as it permits adding first-order effects to obtain the response of higher-order systems. To illustrate the procedure: If a system has a transfer function $T(s) = 3/(s + 2)(s + 5)$, the gain function in terms of frequency is written

$$(8.87) \qquad G = \frac{3}{\sqrt{\omega^2 + 4}\,\sqrt{\omega^2 + 25}} = |T(s)|_{s \to j\omega}$$

The gain in decibels is

$$(8.88a) \qquad G(\text{dB}) = 20 \log \frac{3}{\sqrt{\omega^2 + 4}\,\sqrt{\omega^2 + 25}}$$

$$(8.88b) \qquad G(\text{dB}) = 20 \log \frac{3}{\sqrt{\omega^2 + 4}} + 20 \log \frac{1}{\sqrt{\omega^2 + 25}}$$

$$(8.88c) \qquad G(\text{dB}) = 20 \log \frac{3/2}{\sqrt{(\omega/2)^2 + 1}} + 20 \log \frac{1/5}{\sqrt{(\omega/5)^2 + 1}}$$

In the last two forms of Eq. (8.88) the logarithmic frequency response of the total system may be considered the sum of the logarithmic frequency response of two first-order systems. The low-frequency gain of the first term is $20 \log 3/2 = +3.6$ dB, and the break point is $\omega = 2$; the low-frequency gain of the second term is $20 \log 1/5 = 14$ dB, and the break point is at $\omega = 5$. The gain curves of the two first-order components are shown by the broken lines in Fig. 8.24. The solid curve is the overall decibel-gain curve; it is the sum of the two first curves. The low-frequency behavior for the overall system is taken to be the response below $\omega = 2$ Hz. At such values the gain is $+3.6$ dB $- 14$ dB $= -10.4$ dB. At some intermediate range of frequencies ($2 < \omega < 5$) the overall-gain curve decays as a first-order system at the rate of 6 dB/octave. At the high-frequency range ($\omega > 5$), the gain curve decays at the rate of 12 dB/octave, or twice the rate of a first-order system. This decay rate (12 dB/octave or 40 dB/decade) is characteristic of all second-order systems and is due basically to the s^2 term in the denominator of the transfer function. At high frequencies this term dominates in the denominator and causes the gain to behave as $1/\omega^2$ instead of $1/\omega$, as in a first-order system. It is easy now to extrapolate and state that a third-order system must ultimately decay at a rate of 18 dB/octave (or 60 dB/decade), a fourth-order at 24 dB/octave, etc.

With a little practice it is easy to become proficient at constructing the decibel-gain curve directly from the transfer function. One can quickly recognize, for example, that a system with a transfer func-

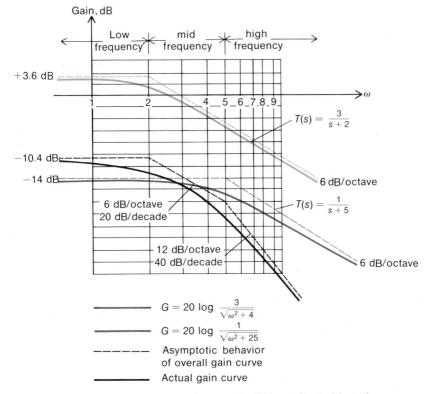

FIG. 8.24 $G(dB)$ versus log frequency for $T(s) = 3/(s+2)(s+5)$.

tion $6(s+3)/(s+6)$ is really a zero-order system and at high frequencies (s and $\omega \to \infty$) this system behaves as if it had a constant gain of six units or 15.6 dB; at low frequencies (s and $\omega \to 0$) the gain is constant at three units (or $+9.6$ dB). For intermediate frequencies ($3 < \omega < 6$) the gain rises at 6 dB/octave. The sketch of the gain curve is shown in Fig. 8.25. In the same fashion it is easy to recognize that, for a transfer function $T(s) = 6(s+3)/(s+6)(s+4)$ when $\omega < 3$, $G(dB)$ is -2.5 dB; when $3 < \omega < 4$, gain rises at 6 dB/octave; when $4 < \omega < 6$, gain is again constant at 0 dB; finally, when $\omega > 6$, gain drops off at 6 dB/octave.

From the foregoing examples it should now be evident that it is often possible to obtain some estimate of a system transfer function if one has a frequency-response characteristic with which to work. By drawing asymptotes with slopes that are multiples of ± 6 dB/octave, one can obtain some indication of the order of a system and the various

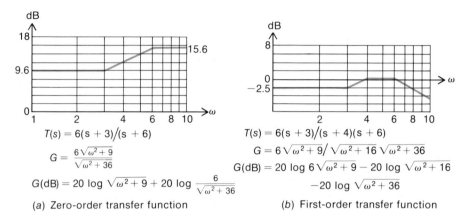

(a) Zero-order transfer function

$T(s) = 6(s + 3)/(s + 6)$

$G = \dfrac{6\sqrt{\omega^2 + 9}}{\sqrt{\omega^2 + 36}}$

$G(dB) = 20 \log \sqrt{\omega^2 + 9} + 20 \log \dfrac{6}{\sqrt{\omega^2 + 36}}$

(b) First-order transfer function

$T(s) = 6(s + 3)/(s + 4)(s + 6)$

$G = 6\sqrt{\omega^2 + 9}/\sqrt{\omega^2 + 16}\sqrt{\omega^2 + 36}$

$G(dB) = 20 \log 6\sqrt{\omega^2 + 9} - 20 \log \sqrt{\omega^2 + 16}$

$-20 \log \sqrt{\omega^2 + 36}$

FIG. 8.25 *G(dB) versus log frequency for other simple systems.*

break points. From this information a preliminary guess can be made as to the form of the system transfer function. In the next section we shall examine some specific problems to illustrate the procedure.

To close this section, it should be mentioned that it is possible to use phase-frequency plots to estimate transfer functions in somewhat the same manner as gain-frequency plots are used. This procedure is not as sensitive or well defined as the gain-frequency technique, and it is therefore not generally used to define the system transfer function. However, the phase-frequency information is important for determining whether a system has a time delay associated with its response. Such information cannot be obtained from the gain-frequency or phase-frequency plot alone but must be obtained by using both forms of frequency-response data. The use of gain and phase vs. frequency plots (often called a Bode diagram or simply Bode plots) will be examined in Sec. 8.9.

8.8 EVALUATION OF A TRANSFER FUNCTION
FROM FREQUENCY–RESPONSE INFORMATION

A rather interesting application of the use of frequency-response analysis to determine a system transfer function for design purposes was recently reported in *Medical and Biological Engineering* (Ref. 6). The overall objective of the investigation was to obtain a quantitative estimate of muscle response to an electrical excitation to assess the feasibility of electrically controlled muscular motion. It is hoped this information can be used ultimately as a guide toward the design of assistive devices for victims of certain types of paralysis.

To describe how the frequency-response technique was used to characterize the muscle transfer function it will be helpful to examine first some of the qualitative relationships between electrical stimulation and muscle response. Since Galvani's classic discovery that frog muscle may be induced to twitch when electrically excited, much experimentation has been performed to obtain data about this physiological peculiarity. It has now been fairly well established that, to obtain a muscle response, the electrical stimulation must exceed a minimum threshold value (sometimes described as a subminimal stimulus). Any excitation above this threshold value will cause the muscle to exhibit some form of mechanical response. If the muscle is rigidly anchored (see Fig. 8.26) the muscle will generate a characteristic *force* vs. time curve when a short suprathreshold electrical pulse is applied; this is generally classified as an isometric, constant-length twitch response. If the muscle sustains a constant load but can move freely, it will generate a characteristic *displacement* vs. time curve when a short suprathreshold electrical pulse is applied; this is generally described as an isotonic (constant-tone) twitch response. Some typical response curves are shown in Fig. 8.26. The numerical values associated with the response curves depend upon the muscle tested. Latency periods (delay period between application of stimulus and inception of response) may vary from 1 to 50 ms. Contraction times for isometric responses (see graphical characteristics in Fig. 8.26a) may be as short as 8 ms (for fast muscles such as the medial rectus of the eye) and as long as 100 ms (for slow muscles such as the soleus in the leg).

For gross demonstrations of muscle reaction or demonstrations involving a small number of muscle fibers, the shape, strength, and duration of the electrical-stimulation curve are not very important. The only criterion to be satisfied is that a "minimal excitation stimulus" be applied. This minimal signal is just sufficient to deliver adequate charge to the muscle membrane to ensure membrane conduction. Once conduction is established, a muscle fiber will deliver the maximal response that it can generate. This is often described as an all-or-nothing response characteristic of nerves and muscles.*

Since a complete muscle is composed of many fibers, an applied electrical signal may deliver a minimal excitation charge to one section of the muscle while the rest of the muscle remains unaffected. A pictorial representation of this situation is shown in Fig. 8.27. A short current pulse delivers sufficient charge to depolarize (trigger) only a small

---

* An interesting general description and illustration of an all-or-nothing response are given in Ref. 7. The authors use the familiar toilet-flushing mechanism as an analog of an all-or-nothing response (p. 39). This same excellent text also includes a good discussion of the minimal excitation stimulus and related topics.

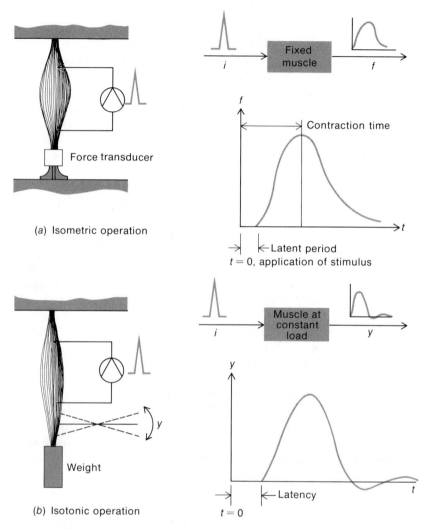

FIG. 8.26 *Muscle-response modes due to electrical-pulse stimulation.*

fraction of the available muscle fibers (represented by three blocks as shown in Fig. 8.27b). Because a different amount of time is required for the depolarizing charge to reach the various fibers, each fiber may have a different latency period. The total effect is the summation of the separate effects, as shown in Fig. 8.27c. It is easy now to appreciate the statistical nature of muscle energization: as more charge is delivered more fibers are energized; the resultant response is increased with in-

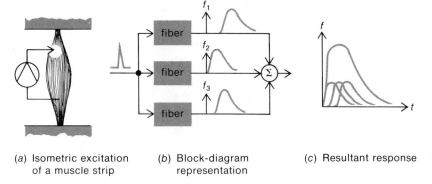

(a) Isometric excitation   (b) Block-diagram   (c) Resultant response
of a muscle strip   representation

FIG. 8.27 *Pulse excitation of muscle strip.*

creased charge delivery until sufficient charge is delivered to excite all fibers. After this point, further increases in the delivery of charge will not induce increased strength of response. Evidently, therefore, it is possible to grade muscle response (up to a point) by varying the magnitude of a current pulse, the duration of the pulse, or the number of pulses delivered; any of these will alter the quantity of delivered charge and hence the muscle response. These ideas are diagrammatically presented in Fig. 8.28.

The discussion in the foregoing paragraph deals only with the short-time response of muscle. From the manner in which the material was presented one might assume the short-time response is due to the short-term duration of the pulse excitation; to obtain a prolonged response, a prolonged excitation should be used. Although this appears to be a reasonable supposition, it is not supported experimentally. An increased pulse width increases response duration (as well as amplitude; see Fig. 8.28*b*), but a limit is soon reached beyond which the response cannot be extended by additional pulse width. It appears that increased pulse width gives rise to electrode polarization and muscle fatigue (continuous energization of all fibers does not allow time for elimination of waste products). These factors operate against prolonged response during the application of steady excitation.

It is possible to generate a sustained contraction, however, by applying the same kind of excitation signal the physiological system uses to energize muscular tissue. In vivo muscles are excited by a train of pulses (see Fig. 8.29) delivered from the controlling neural connection. If a muscle is stimulated by a similar train of pulses generated by an electronic stimulator, it can be made to exhibit a sustained contraction for an appreciable amount of time (i.e., with relatively little fatigue).

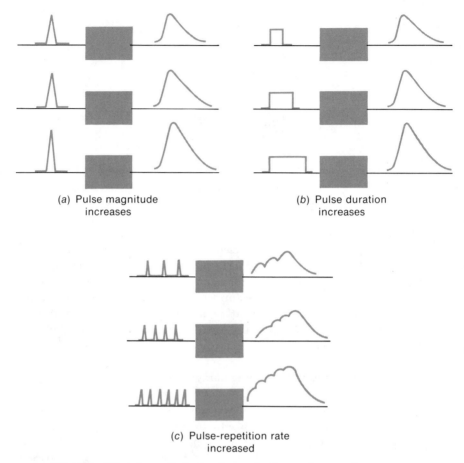

(a) Pulse magnitude
increases

(b) Pulse duration
increases

(c) Pulse-repetition rate
increased

FIG. 8.28   *Effect of magnitude, duration, and pulse count on muscle response*

An explanation of how a pulse train may produce a sustained response is outlined in Fig. 8.29. If a pulse train (of suprathreshold amplitude) is applied to a muscle, the muscle responds with a series of twitches. If the repetition rate of the pulses is low, the muscle exhibits a series of superimposed but distinctly separate twitch responses, as shown in Fig. 8.29a. As the repetition rate is increased the twitch responses become superimposed sooner and a more nearly continuous contraction is obtained (Fig. 8.29b). When the repetition rate is increased further, the twitch responses merge and the muscle response appears to be a sustained contraction. The continuous response due to fused twitch responses (i.e., due to high-frequency pulse excitations) is generally described as a state of tetany, or a tetanized response. Further increases in pulse repetition rate cause the tetanic force to increase; a saturation limit

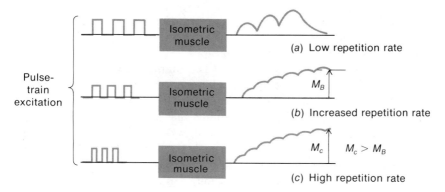

FIG. 8.29   *Muscle responses to pulse train.*

is soon reached, however. Higher repetition rates do not cause improved performance and may, in fact, cause performance to deteriorate. At very high repetition rates a pulse train begins to act as a steadily applied excitation and the muscle begins to show signs of fatigue. Fatigue does not appear to be a problem at lower repetition rates; at lower rates some of the muscle fibers apparently have a chance to recover from fatigue effects while others carry the load. The effects of fatigue are thus delayed for a substantial amount of time.

Although the foregoing discussion deals only with the in vivo response of muscle it provides some insight into the in vivo operation and stimulates some thoughts about the feasibility of muscle control through external excitation. We now know muscle force can be varied by adjusting the frequency of a current pulse train, the amplitude of the train, or the width of the individual pulses. In the physiological system, muscles are controlled by nerves, and nerve-generated pulses are known to be generally of fixed amplitude and fixed pulse width. Muscle control in the physiological system is thus governed by pulse-rate variations (i.e., frequency-modulated pulse trains). It is known also that muscle tissue can be energized by electrical stimulation delivered through the skin (as everyone who has experienced an electrical shock will testify). This now raises the interesting question: Can an appropriate pulse train be delivered through the skin to muscle so that the muscle may be controlled by external stimulation? Should this procedure prove feasible, it should then be possible to excite and control "functional but inoperative" muscle (such as a muscle paralyzed because of a nonfunctioning nerve).

Crochetiere et al. (Ref. 6) have investigated one aspect of the problem of artificial, in vivo control of muscle. These researchers elected to study the feasibility of the amplitude-modulated pulse train as a mode of muscle control, and as a first step in the study they obtained a repre-

sentation for the muscle transfer function (force-excitation characteristics) by the frequency-response technique. A description of this investigation is given in the following paragraphs.

The muscle chosen for test was the human biceps. An electrode
placed over the muscle motor point delivered the stimulating pulse train;
the excitation caused motion of the forearm with respect to the upper arm
around the elbow joint. By constraining the arm within a mechanical
fixture equipped with instrumentation it was possible to measure the
isometric elbow torque generated for each signal waveform. The experimental arrangement is pictured in Fig. 8.30.

To obtain the desired information, a stimulating pulse train of
50 pulses/s and a pulse width of 0.4 ms was applied to the biceps motor
point. These signal parameters were chosen as they provided a tetanic-
torque level with no discomfort to the subject. A current train with a
pulse amplitude of 15 mA was sufficient to generate a tetanic torque of
4.5 ft-lb (see Fig. 8.31); these are considered the quiescent operating
levels. To obtain the frequency-response characteristics of the system
the pulse amplitude was varied sinusoidally, as shown in Fig. 8.31a. The
amplitude of the sinusoidal variation (I_a) was 5 mA, and the frequency
of the amplitude variation was between 1 rad/s (about $\frac{1}{6}$ Hz) to 20 rad/s
(3 Hz). For each excitation frequency, the amplitude of the isometric

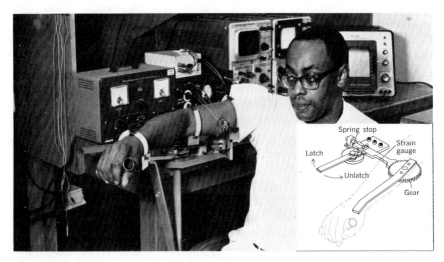

FIG. 8.30 *Experimental apparatus to study pulse-excitation characteristics of biceps. (Reproduced with permission of Medical and Biological Engineering.)*

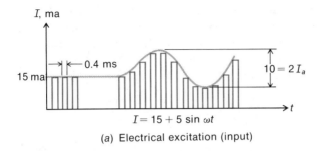

(a) Electrical excitation (input)

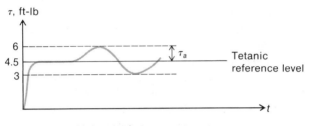

(b) Isometric torque response
at low frequencies.

FIG. 8.31 *Electrical stimulus and isometric torque response for biceps muscle studies. [Adapted from W. J. Crochetiere et al., Electrical Stimulation of Skeletal Muscle: A Study of Muscle as an Actuator, Med. Biol. Eng., **5**: (1967). Numerical values shown were taken from data in published article, figs. 17, 19.]*

torque response (τ_a) was obtained. The decibel gain of the system [20 log (τ_a/I_a)] was then plotted against log frequency (Fig. 8.32); from these results it was possible to define a transfer-function representation for the biceps muscle when driven by an amplitude-modulated-current pulse train.

A frequency-response plot for one of the subjects tested is shown in Fig. 8.32.* The smooth curve through the points represents the frequency characteristics of the pulse-train–muscle system. To obtain an *s*-function representation for the system we note that at high frequencies (above 15 rad/s) the curve falls off at 12 dB/octave. It appears that a second-order representation will be required. As a first guess we assume the

---

* In the published data, the frequency-response curve was normalized to read 0 dB at low frequencies. The curve in Fig. 8.32 has been modified to include the low-frequency gain. To obtain this information it was noted in the published data (Fig. 17) that during static testing a change of 10 mA in stimulus amplitude resulted in a torque change of 3 ft-lb. As static testing is equivalent to zero-frequency testing, the low-frequency gain is considered to be ³⁄₁₀ ft-lb/mA. In decibel notation this is 20 log 0.3 or −9.5 dB.

system has only one time-constant but is second-order. Such a system would have a perfect square in the denominator of the transfer function and would be written $A/(s + a)^2$. For the particular case at hand the break frequency is at 9 rad/s; the transfer-function denominator would therefore be $(s + 9)^2$. The total transfer function could then be written

$$(8.89) \qquad T(s) = \frac{0.3 \times 81}{(s + 9)^2} = \frac{24.3}{(s + 9)^2}$$

where the constants are chosen to satisfy the gain at the low-frequency end of the curve. A gain (decibel) vs. log-frequency plot of Eq. (8.89) is included in Fig. 8.32 for comparison.

The correspondence between the proposed gain function and the actual data is fairly good. As the data are rather limited and represent only a single experimental test, it would be pointless to seek further refinement. As an academic exercise, however, we shall examine another possible representation. We may feel (on the basis of other information, possibly) that the system might more appropriately be represented by a transfer function with two different time-constants. This would imply that the s-function representation should be of the form $A/(s + a)(s + b)$; this indicates that in some intermediate frequency range $(a < \omega < b)$ there will be an asymptotic gain curve that falls at the rate of 6 dB/ octave. A pair of possible asymptotic lines is shown in Fig. 8.33. From the figure we can see that, if we assume one break frequency to be at

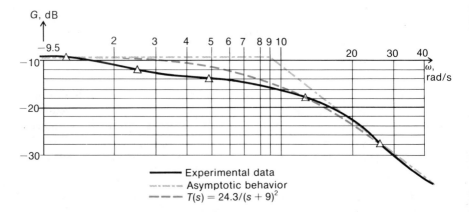

FIG. 8.32 $G(dB)$ *versus log-frequency plot for test of biceps frequency response.* [*Modified from W. J. Crochetiere et al., Electrical Stimulation of Skeletal Muscle: A Study of Muscle as an Actuator, Med. Biol. Eng.,* **5**: (*1967*).]

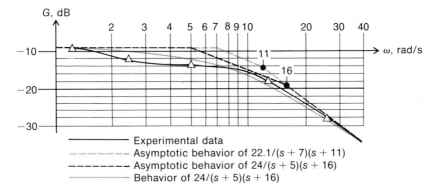

FIG. 8.33 *Possible asymptotic representations of biceps-frequency-response characteristics.*

5 rad/s, the other should be at 16 rad/s (see diagrammatic inserts in Fig. 8.33). On the other hand, if one break frequency is at $\omega = 7$ rad/s, the other should be at 11 rad/s. In short, by using these constructions we see the system transfer function might be taken as

$$(8.90) \qquad T(s) = \frac{0.3 \times 77}{(s + 7)(s + 11)} = \frac{22.1}{(s + 7)(s + 11)}$$

or

$$(8.91) \qquad T(s) = \frac{0.3 \times 80}{(s + 5)(s + 16)} = \frac{24}{(s + 5)(s + 16)}$$

The gain curves for Eqs. (8.90) and (8.91) have been plotted in Fig. 8.33 where the original data curve is included for comparison. It is noted that Eq. (8.91) is a slightly better fit than Eq. (8.90), and if it were established the system operated with two different time-constants, we might accept Eq. (8.91) as a close representation of the physical system.

8.9 RELATIONSHIP BETWEEN PHASE LAG AND TIME DELAY

At the end of Sec. 8.7 it was noted that phase vs. frequency plots can also be used to evaluate system transfer functions. It was pointed out, however, that this procedure is not as sensitive as the gain-frequency technique and for this reason phase-frequency plots are not generally used to define system functions. Many investigators, in fact, do not bother to obtain such data; it is assumed phase-frequency characteristics can be derived (if needed) from the transfer function; the transfer function may be obtained from the gain-frequency characteristics. Although this view-point is often correct, it is incorrect for a very important situation often encountered in physiological systems: If a system has a time delay* associated with response, the phase-frequency characteristics cannot be obtained from the gain-frequency information. Conversely, and more importantly, if the phase-frequency characteristics are not available, it is impossible to determine from the gain characteristics alone whether the system has a time lag. The connection between time delay and phase characteristics may be sensed intuitively, but it is not evident how they are related quantitatively or how one may be obtained from the other.

Probably the best way to gain an understanding of how these quantities are related is through an illustrative example. In this section we shall examine the frequency-response characteristics of a specific system with a view toward examining the relationship between time delay and phase angle. To avoid the danger of a purely academic discussion and to minimize the possibility of monotony through repetition, we shall examine a realistic physiological system which has not yet been discussed. It will, however, be necessary to establish first some background about the system. The first part of this section is devoted to a brief introductory description of the physiological system and the apparatus to make experimental measurements. At the end of the section we shall analyze the results of the experimental data obtained.

The system to be examined is the pupillary control system of the eye. We are not yet equipped to investigate the overall feedback system associated with pupillary control; feedback systems are the subject of Chap. 10. At this stage, however, we can examine one aspect of the pupillary system without considering the overall control function. In the words of the servosystem analyst, we can investigate the open-loop characteristics of the system without considering the closed-loop behavior. To appreciate more fully what this investigation involves (and what this terminology means) it will be helpful first to discuss briefly the overall pupillary control system.

A schematic diagram of the essential optical mechanical features of the eye is shown in Fig. 8.34. Light passes through the clear cornea of the eye to the plane of the iris. The iris is the colored, circular portion of the eyeball which has a dark center. Aside from the aesthetic

---

* That is, if the response does not start when the stimulus is applied or if there is an appreciable period before a detectable response can be obtained. This is sometimes also called a time lag or transportation lag. The latter terminology is particularly appropriate in a situation such as dye-dilution, where dye must be transported from the injection site to the sampling site before a measurable signal may be obtained.

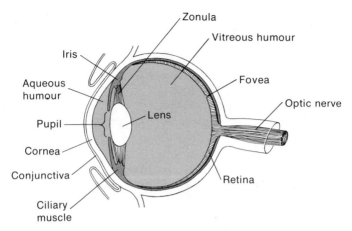

FIG. 8.34 *Schematic diagram of essential morphological features of the eye.*

value of the pigmentation in the iris (beautiful brown eyes, etc.), the pigment serves to make the iris periphery opaque to light. Light can therefore enter the eye only through the black center portion of the iris. This black center is in actuality an adjustable aperture (the pupil) leading directly to the lens of the eye. The muscle fibers in the iris adjust the diameter of the pupil to changes in the focal length of the lens system and to limit the light delivered to the interior of the eye. The iris evidently corresponds to the diaphragm of a camera. After the light passes through the pupil it enters the lens system which focuses the image on the light-sensitive retina. The light-sensitive elements in the retina convert light information into nerve impulses; these are collected, processed, and sent to the brain by the optic nerve.

Of particular interest for this discussion is the variation of pupil diameter with the variation of incident light. Under normal operating conditions, the size of the pupil governs the amount of light entering the eye. As the intensity of light delivered to the eye is increased, the pupil diameter is decreased. The amount of light that reaches the retina is thus diminished to reduce the possibility of damage. The inherent feedback characteristics of this arrangement are evident from the verbal description of the operation: The light reaching the retina controls the diameter of the pupil, which controls the amount of light that reaches the retina. A pictorial representation of the feedback property of the system is shown in Fig. 8.35a; Fig. 8.35b shows a more schematic representation of the situation. Both diagrams are primarily qualitative in nature. To obtain a somewhat more quantitative representation we could use the

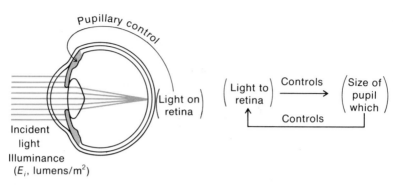

(a) Pictorial representation of pupillary control system

(b) Diagrammatic representation of closed loop

FIG. 8.35   *Closed-loop aspects of pupillary control system.*

very simple equations

(8.92)   Light flux to retina = (incident illuminance)* × (pupil area)

or

$$\mathfrak{F}_R = E_i A_p$$

and

(8.93)   Area of pupil = function of light flux to retina

or

$$A_p = f(\mathfrak{F}_R)$$

Equation (8.92) may be diagrammatically represented as shown in Fig. 8.36a; E_i (the incident illuminance) and A_p (the pupil area) enter a

* A point light source with an intensity of 1 candle emits a radiant flux of 12.57 lumens (lm) over a solid angle of 4π steradians (sr). (An intensity of 1 candle is thus equivalent to an intensity of 1 lm/s.) The flux incident per unit area is defined as the illuminance. The units of illuminance are lumens per square meter and foot-candles. The latter is the illuminance that would be obtained on the surface everywhere 1 ft from 1 candle at its center. (Therefore 1 foot-candle is an illuminance of 1 lm/ft².)

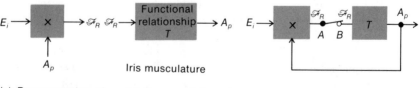

(a) Representation of Eq. (8.92)

(b) Representation of Eq. (8.93)

(c) Simultaneous solution by closing the loop

FIG. 8.36   *A more quantitative description of the closed-loop aspect of pupillary control system.*

multiplying device. The output of the device is the product of retinal flux $\mathfrak{F}_R$. Equation (8.93) may also be diagrammatically represented as shown in Fig. 8.36b. Thus far the variables $\mathfrak{F}_R$ and A_p are only functionally related; therefore, we must use an undefined block to associate A_p and $\mathfrak{F}_R$. If we combine the diagrammatic representations of Eqs. (8.92) and (8.93) we obtain the closed-loop configuration in Fig. 8.36c. In a sense, the latter configuration represents a simultaneous solution of Eqs. (8.92) and (8.93); each equation is used to eliminate one of the unknowns of the other equation.

Although the closed-loop diagram of Fig. 8.36 cannot yet be considered a quantitative representation of the pupillary control system, it clearly indicates what variables must be examined to obtain a more quantitative representation. The block relating $\mathfrak{F}_R$ and A_p should be analytically defined if any further work is to be done with this control system. From the diagram (or from the equations, if one prefers) it can be seen there are two ways to define the functional blocks relating $\mathfrak{F}_R$ and A_p. One way is to work directly with the closed-loop configuration; measure E_i and A_p; and from this information define $\mathfrak{F}_R$ in terms of A_p. As an alternative procedure, we can seek to make $\mathfrak{F}_R$ an independent variable (not affected by A_p). We can then adjust $\mathfrak{F}_R$, measure the resultant A_p response, and thereby define A_p as a function of $\mathfrak{F}_R$ directly. This procedure requires that the closed-loop system be converted into an open-loop one: The influence of the block output (A_p) on its input $(\mathfrak{F}_R)$ must be eliminated, and the control aspects of the system must be destroyed or made inoperative. Diagrammatically, what is implied is that switch AB in Fig. 8.36c be opened and $\mathfrak{F}_R$ be supplied to block T (the retinal system) without interference of A_p.

Although it is always easy to open a closed-loop system on paper, it is not always easy (or possible) in practice. In this particular situation, however, there is a relatively easy way to open the loop. The procedure involves the use of a special optical arrangement that permits light to be introduced to the retina without interference from the pupil.* A schematic representation of the apparatus is shown in Fig. 8.37a. Light from a source is directed to the eye through a collimating-lens system. Before the light enters the eye, it is focused at the plane of the iris, as shown in the diagram. The light thus enters the eye through an effective aperture which is significantly smaller than the smallest pupil diameter. This aperture is constant and independent of the diametral changes of the

---

* This procedure is described by L. Stark in Ref. 8. Stark has been responsible for much of the recent work dealing with the human pupillary control system and has published many articles on the subject. Many of the data used in this section have been taken from his published works.

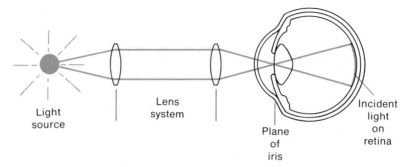

(a) Optical arrangement to provide
for open-loop operation

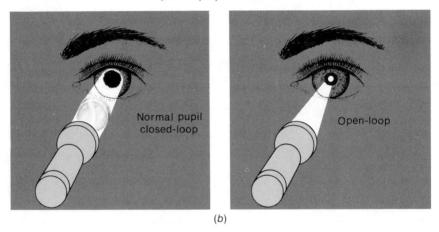

(b)

FIG. 8.37 *Opening the pupillary control loop by optically decoupling the entering light flux from the pupil aperture.*

pupil. The light reaching the retina is thus independent of pupillary adjustments, and the system can operate effectively as an open loop.

It is now easy to see how a functional representation for the flux-aperture system may be obtained by direct measurements: Known flux signals are applied to the eye in an open-loop mode; the response to these signals is simultaneously recorded. From the input-output data an analytic description of the system may be obtained. A diagram of the experimental arrangement to obtain the required information is shown in Fig. 8.38a. A light source is focused on the plane of the pupil as described previously. A recording for the incident light is obtained by incorporating a beam-splitting mirror into the incident light beam. Part of the incident light is thus deflected to a photocell pickup which converts the light to an equivalent electrical signal. A recording for the response is obtained by using an infrared pupillometer arrangement (refer to Fig. 8.38a): An

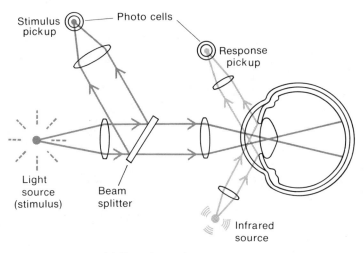

(a) Experimental arrangement

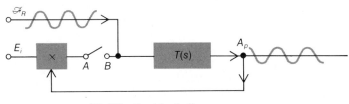

(b) Effective block diagram

FIG. 8.38 *Arrangement of apparatus to obtain open-loop stimulus-response data of the pupillary-retinal system.*

infrared source is collimated to illuminate the iris of the eye. (As the retina is not sensitive to infrared, this light does not interfere with the stimulus light.) The infrared is reflected from the iris and focused on an infrared-sensitive photocell. The amount of reflected infrared depends upon the pupil area. If the pupil is large, much of the infrared is absorbed and little is reflected; the electrical response signal thus is small. As the pupil size decreases, reflection increases and the electrical signal increases.*

Having established that meaningful stimulus and response data can be obtained, we can now excite the system with any stimulus, and from the response we should then be able to propose some model representation. Stark (Ref. 8) chose to use a sinusoidal excitation as he felt by this technique he could obtain a given level of accuracy with small-

---

* Acquisition of meaningful data sometimes requires more work and ingenuity than the interpretation and analysis of the data!

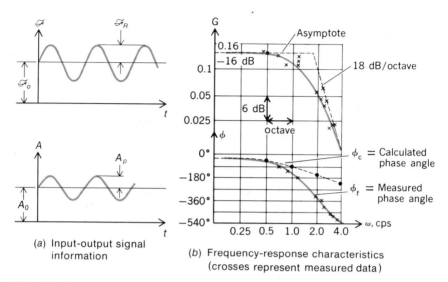

(a) Input-output signal
information

(b) Frequency-response characteristics
(crosses represent measured data)

FIG. 8.39   *Open-loop data and characteristics of pupillary-retinal system.* (*Part b is from Ref. 8.*)

signal excitations. (Small-signal levels permit the use of linearized techniques as a first-order approximation.) Rotating Polaroid film (and later electronic techniques) was used to provide a sinusoidal light excitation. The excitation and response curves for various frequencies were obtained from signals generated by the photocell pickups. The information acquired was then presented in the form of frequency-response plots as shown in Fig. 8.39.* Such frequency-response plots (i.e., gain and phase vs. frequency) are often called Bode plots after H. W. Bode, who first

---

* Some comments regarding the gain function shown in Fig. 8.39 are required. We have defined gain as the ratio of input amplitude to output amplitude. From Fig. 8.38b we would therefore assume gain to be $\mathcal{F}_R/A_p$, where $\mathcal{F}_R$ and A_p are the *amplitudes* of the input flux and output area (see Fig. 8.39a). However, the significance of the gain function is not lost if a known multiplicative constant is used in the gain function. Stark (Ref. 8) chose to define gain as the ratio of percent change in area to the percent change in flux, or $G = (A_p/A_0)/(\mathcal{F}_R/\mathcal{F}_0)$, where A_0 and $\mathcal{F}_0$ are the quiescent (constant) levels of area and flux (see Fig. 8.39a). This may be rewritten $G = (A_p/\mathcal{F}_R)(\mathcal{F}_0/A_0)$. Since $\mathcal{F}_0$ and A_0 are constants, this gain function is the same as that defined earlier to within a multiplicative constant.

One further comment is necessary: Light flux $\mathcal{F}$ and light illuminance (E) are related by the area of the light beam [Eq. (8.92)]; similarly, light intensity (I in lumens per steradian) and light flux are proportionately related. The percent change in flux must therefore be equal to the percent change in illuminance, which must also be equal to the percent change in intensity. Gain can therefore be defined in terms of any of these variables, depending upon the measurements made. As Stark defines his light measurements in terms of light intensity, he defines gain as the ratio percent change in area to percent change in intensity, or $G = (A_p/I)(I_0/A_0)$ [or $(dA/A)/(dI/I)$].

showed the close relationship between gain and phase for amplifiers and electrical networks in 1938.

We use the data in Fig. 8.39b in the manner outlined in Sec. 8.8. To obtain an s-function representation of the system we note first that the approximate smooth curve through the experimental points seems to decay 18 dB/octave. The pupil-retinal system thus appears to be a third-order system. As a first approximation we might guess the form to be

$$(8.94) \quad T(j\omega) = \frac{k}{(j\omega + a)(j\omega + b)(j\omega + c)}$$

$$= \frac{k/abc}{[(j\omega/a) + 1][(j\omega/b) + 1][(j\omega/c) + 1]}$$

or

$$T(s) = \frac{k}{(s + a)(s + b)(s + c)} = \frac{k/abc}{[(s/a) + 1][(s/b) + 1][(s/c) + 1]}$$

where a, b, and c are the various break frequencies and k/abc is the low-frequency gain constant. The latter term can be evaluated immediately from the gain curve. At very low frequencies the gain curve may be approximated by the horizontal asymptote at 0.16 (or -16 dB); k/abc may therefore be taken as 0.16.

By using a, b, and c to represent the break frequencies we accept the possibility that the system has three separate time-constants. It is possible that the system have only two time-constants (and hence only two break frequencies) and still be third-order. Under these circumstances the transfer function would contain two factors; one of them would be second-order, and the other first-order. The transform would then become

$$(8.95) \quad T(s) = \frac{0.16}{[(s/a) + 1]^2[(s/b) + 1]}$$

where now a and b represent the two break frequencies. We can extend this viewpoint one step further and note it is possible also that the system have only one time-constant; the system function would then contain a single cubic term in the denominator and would be written

$$(8.96) \quad T(s) = \frac{0.16}{[(s/a) + 1]^3}$$

As this last form is the easiest one to "check out," we examine it first as an initial guess for the system function. To obtain the break frequency we extrapolate the high-frequency asymptote back until it intersects the low-frequency asymptote (dashed lines in Fig. 8.39). From the intersection point we find the break frequency to be 1.6 Hz or 10 rad/ s; s/a is therefore $0.1s$, and Eq. (8.96) can be written

$$(8.97) \qquad T(s) = \frac{0.16}{(0.1s + 1)^3}$$

To check quickly the accuracy of this representation we note for a first-order system the true-gain curve falls 3 dB below the intersection of the asymptotes (see Prob. 8.6); for the squared second-order system the true curve is 6 dB below the intersection; for the cubic system the curve should be 9 dB below the break frequency point. An examination of the gain plot in Fig. 8.39b shows that a curve that follows the high- and low-frequency asymptotes and is 9 dB below the break point appears to fit the plotted points with fair accuracy. For the purposes of a preliminary investigation we might therefore accept Eq. (8.97) as a first estimate for the functional representation of the pupillary-retinal system.

The transfer-function representation given by Eq. (8.97) was based upon gain considerations only. If the system behaves in a linear fashion with no time lag and the expressions in the numerator and denominator of the transfer function have roots with negative real parts only,* then the gain-frequency characteristics can be used alone to derive the transfer function. If the phase-frequency curve is available, it can be used to check the results obtained from the gain-frequency characteristics; if the phase-frequency curve is not available it can be derived from the derived system function. If, however, the system has some inherent time delay (or is nonminimum-phase; see the first footnote) then the transfer function derived from the gain-frequency characteristics will not adequately represent the system function; the phase-frequency characteristics must then be available to indicate how the transfer function must be modified for a more satisfactory representation.†

We can use the transfer function just derived to illustrate this point. If the transfer function given by Eq. (8.97) is an adequate representation of the system characteristics, the phase function would be given by the angle associated with $T(j\omega)$ [see the discussion in Sec. 8.3 and Eq. (8.48d) at the end of the section; see also Sec. 8.5 and Eqs. (8.59c) to (8.59f)]. In particular, the phase angle would be given by $\underline{/T(j\omega)} = -3\theta$, where

---

* That is, the roots are of the form $s = -a \pm jb$, where a is a positive number only. Such systems are sometimes called minimum-phase systems.

† Evidently, therefore, unless there is definite information to the contrary, phase information should always be obtained with gain information to ensure a total representation will be possible.

$\tan \theta = 0.1\omega$.* A plot of the phase angle vs. log frequency is shown by the dashed line labeled ϕ_c on the phase-frequency characteristics of Fig. 8.39b. As can be seen, the calculated phase characteristics always lie above the measured phase curve shown by the solid line through the experimental points. Evidently, the transfer-function representation of Eq. (8.97) in its present form does not properly characterize the system operation.

We can use the difference between the calculated (minimum) phase characteristics and the experimental phase curve to learn how the transfer function should be modified to give proper results. We might note, for example, that the difference between the two curves increases with frequency.

To see how the difference between the true phase curve and the calculated phase curve varies with frequency, we can plot this difference vs. frequency. Such a plot is shown in Fig. 8.40. Although the relationship between $\phi_c - \phi_t$ and frequency is not exactly linear, we might accept a straight line as an adequate approximation for this situation. In doing so, we in essence accept that the difference between the calculated and actual results is proportional to frequency. The result may be expressed

* $$T(j\omega) = \frac{0.16}{(j0.1\omega + 1)^3} = \frac{0.16}{[\sqrt{(0.1\omega)^2 + 1^2}\ \epsilon^{j\theta}]^3}$$

where $\tan \theta = 0.1\omega/1 = 0.1\omega$. The denominator can be expanded:

$$T(j\omega) = \frac{0.16}{[\sqrt{(0.1\omega)^2 + 1^2}]^3 \epsilon^{j3\theta}} = \frac{0.16}{[(0.1\omega)^2 + 1]^{3/2}}\ \epsilon^{-j(3\theta)}$$

Therefore $\underline{/T(j\omega)} = -3\theta = $ calculated phase angle ϕ_c.

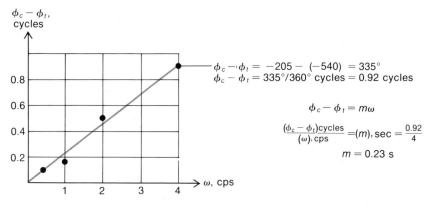

FIG. 8.40   *Plot of difference between calculated and true phase angles vs. frequency.*

analytically as

$$(8.98) \qquad\qquad \phi_c - \phi_t = m\omega$$

where m is the slope of the curve in Fig. 8.40. As the phase angles may be measured in radians and ω may be expressed in radians per second, it is evident the constant m has the dimensions of time. To understand the significance of this time term, we review the true meaning of the phase angle. It will be recalled the overall phase angle [or phase shift, as it is often called; see Sec. 8.3 and, in particular, the discussion associated with Eq. (8.37)] causes the time axis of a sinusoid to be shifted. As a specific illustration, if a system has a sinusoidal response with a phase shift ϕ_t, the response term is written $A \sin (\omega t + \phi_t)$, where A is the amplitude of the response. This term may be rewritten as $A \sin \omega (t + \phi_t/\omega)$. In this form we see the phase angle causes a shift in the time axis; the direction of the shift will depend upon the sign of ϕ_t. In most practical cases (though not all) the phase angle is negative; the phase angle, therefore, represents a time lag; events at the output occur after similar events have occurred at the input.

For the particular situation presently under investigation, ϕ_t is given by $\phi_c - m\omega$ or $-3\theta - m\omega$. The sinusoidal response term thus becomes $A \sin \omega (t - m - 3\theta/\omega)$. In this form we can clearly see how the response function is shifted in time by the various phase-angle components. The last term represents a frequency-dependent time shift related to the system properties. As the frequency or the system properties change, so does the time shift. This is the familiar time shift due to the system s function, and it has significance for periodic (frequency-dependent) functions only. The m term represents a time shift which is independent of the signal frequency. In fact, this term exists even if the signal excitation has no periodic content (i.e., it is aperiodic). This time shift is an inherent delay within the system; it can be due to the system properties or physical construction but is generally independent of the signal excitation.

A good example of such an inherent delay within a system is the transportation lag associated with dye-dilution testing. A dye signal injected into the circulatory system at one point is detected at a remote sampling point only after a finite time. The delay is, of course, due to the time required for the blood to carry the signal from the injection site to the sampling site. This is an obvious case of time delay; the cause is self-evident. In many situations the delay exists but the cause is not obvious. The pupillary-retinal system presently under discussion is an illustration of this situation. A signal is delivered to the retina; a finite time elapses before the pupil responds. From the available data this delay may be evaluated to be approximately 0.2 s (from the slope of the curve in Fig. 8.40). The specific cause of the delay, however, is not clear and would probably have to be determined as a separate investigation.*

From a block-diagram point of view, it is relatively easy to indicate a system has a time lag. A simple but adequate descriptive representation is shown in Fig. 8.41a. Delay blocks can be realized physically through the use of various devices with inherent transport-lag characteristics. For example, electrical delay lines (long lengths of coiled, specially constructed wire) can provide transport delays of the order of microseconds. Acoustical delay lines (acoustical conductors) can provide delays in the milliseconds range. Magnetic-tape transports can be used to obtain delays of the order of seconds; by special techniques delays of hours or even days can be obtained.

To show a system has a time delay in an analytic fashion, the transfer-function representation must be modified so that this information may be included. To determine what modifications may be required, we assume the true transform is represented by $T_T(s)$. In complex form this would be written

$$(8.99) \qquad T_T(j\omega) = |T_T(j\omega)|\epsilon^{j\phi_t}$$

where $T_T(j\omega)$ is the magnitude of the true (complex) transfer function and ϕ_t is the angle associated with the function. The magnitude of the true transfer function is obtained from the gain curve and can be determined from Eq. (8.97):

$$(8.100) \quad |T_T(j\omega)| = |T(j\omega)| = \left| \frac{0.16}{(0.1s + 1)^3} \right|_{s \to j\omega} = \left| \frac{0.16}{(0.1j\omega + 1)^3} \right|$$

* We could, however, guess that part of this delay is due to the transduction (conversion) of light to a nerve pulse of sufficient magnitude to be transmitted from the optic sensors, and part of the delay is due to the transmission lag of signal information from eye to brain and from brain to pupillary muscle system. The existence of this lag is easily verified by other types of investigations. Pulse excitation of the eye, for example, has shown that there is approximately a 0.2-s lag before the diameter of the pupil begins to change. Work along these lines is described in Ref. 9.

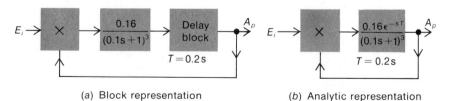

(a) Block representation   (b) Analytic representation

FIG. 8.41 *Pupil-retinal system with delay function included.*

From Eqs. (8.99) and (8.100) we can write

$$(8.101) \qquad T_T(j\omega) = \left| \frac{0.16}{(0.1j\omega + 1)^3} \right| \epsilon^{j\phi_t}$$

Since ϕ_t is given by Eq. (8.98) as $\phi_c - m\omega$ and since ϕ_c is -3θ (see earlier footnote) we have $\phi_t = -3\theta - m\omega$; therefore Eq. (8.101) becomes

$$(8.102) \qquad T_T(j\omega) = \left| \frac{0.16}{(0.1j\omega + 1)^3} \right| \epsilon^{j(-3\theta - m\omega)} = \left[\left| \frac{0.16}{(0.1j\omega + 1)^3} \right| \epsilon^{-j3\theta} \right] \epsilon^{-j\omega m}$$

The parenthetic term in Eq. (8.102) is the transfer function derived from the gain characteristics when the time delay was not considered [see the discussion associated with Eqs. (8.94) to (8.96)]. Equation (8.102) may therefore be written

$$(8.103) \qquad T_T(j\omega) = T(j\omega)\epsilon^{-j\omega m}$$

If we replace $j\omega$ by s to obtain the more general form, we have

$$(8.104) \qquad T_T(s) = T(s)\epsilon^{-sm}$$

Equation (8.104) indicates finally how time-delay information may be included in the transfer function: The transfer function is obtained as if there were no delay and then the result is multiplied by the factor ϵ^{-sm} to signify a signal transmission delay of m units of time. The block diagram of Fig. 8.41b is thus identical to that of Fig. 8.41a except that the delay is represented as part of the system transfer function instead of a separate block. The new notation does nothing more than does the added block; it is just another way to describe the same thing. The delay term (ϵ^{-sm}) does have additional significance and mathematical utility for some applications. For our purposes here, however, it will be sufficient to consider this notation as simply an indication of transmission delay.

8.10 FOURIER EXPANSION OF A PERIODIC SIGNAL

The discussions to this point have been concerned primarily with the use of the sinusoidal function as a representation for special types of naturally and artificially generated signal waveforms. From this point of view the study of sinusoidal excitations and responses merits at least as much consideration as the study of other excitations and responses. If, for example, a system is naturally excited by a signal with sinusoidal characteristics, then sinusoidal stimulus and response data must be used to analyze the system. The respiratory system is an illustration of a system that operates naturally with an approximate sinusoidal rhythm. One should certainly have some facility to handle the sinusoidal analysis associated with such a system. When the signal excitation of a system is under the control of the experimenter, he may choose a simple aperiodic function as a stimulus; he may, however, have equipment and instrumentation that will give greater accuracy if a sinusoidal excitation is used (ac amplifiers are still somewhat better than dc amplifiers, and they minimize drift problems). In this instance too, therefore, some facility with sinusoidal analysis and data interpretation would be required.

Aside from the reasons stated above, there is a much more basic reason for understanding the sinusoidal signal. Although pure sinusoidal signals may be the exception rather than the rule, periodic signals (repetitious but not necessarily sinusoidal) are a familiar part of everyday life and of every physiological system, and such signals may be expressed in terms of sinusoids. The vibration of the commuter's railroad, the bouncy bus ride, the physiological electrocardiograph (EKG) signal, the musical note from an instrument are all illustrations of periodic but not necessarily sinusoidal signals; all, however, can be represented by some combination of sinusoidal components. In this section we determine how periodic waveforms may be represented by sinusoidal functions.

We can easily appreciate that periodic waveforms may be described by sinusoids if we take the reverse approach and construct waveforms from sinusoids. As a trivial first illustration, we note that if two sinusoids of the same frequency are added the result is a periodic function which is also a sinusoid of the same frequency. This result can be obtained mathematically or by phasor considerations, as shown in Fig. 8.42. We can extend this reasoning and state that the summation of any number of sinusoids of one frequency must yield a sinusoid with the same frequency. Then the obvious question is: What waveforms can be expected if sinusoids of different frequencies are added? The variety of possibilities

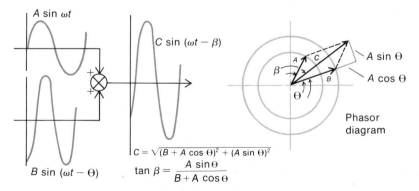

FIG. 8.42 *Summation of sinusoids of the same frequency.*

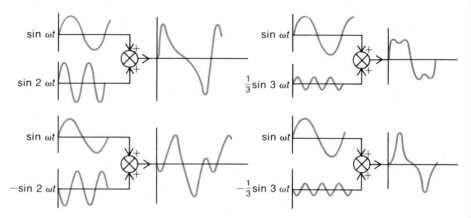

FIG. 8.43 *Representative examples of the summation of two sinusoidal signals that have different frequencies.*

in this case is infinite; a few representative examples are shown in Fig. 8.43. The only general statement that can be made about the summation is that, although the result will probably not be a sinusoid, it must be periodic.* The shape of the resultant waveform is altered by the amplitude of the sinusoidal component, by the frequency of the components, by the phase relation between the components, and by the number of different frequency components included. In short, as mentioned a short while ago, the variety of possible shapes is infinite.

It now becomes clear how sinusoidal functions may be used to study systems subjected to nonsinusoidal but periodic stimuli. We first find a suitable set of sinusoids that can be used to represent the periodic waveform (i.e., we find its frequency content, its spectral content, or its Fourier components; these terms are often used to describe sinusoidal representations of a periodic function). We then assume each of the components gives rise to its own response, independent of the effect caused by the other components. The total response to the periodic excitation is then considered the sum of the responses of the separate component responses. A diagrammatic description of the process is given in Fig. 8.44. The periodic signal is decomposed into sinusoidal components; in the diagram, this is shown by assuming the signal is subjected to a selective "filtration" system which allows only signals of specific frequencies to be transmitted. (Such filter devices are available commercially or can be built.) The sinusoidal signals are then assumed to act

---

* This statement should be intuitively obvious. If two waveforms start at the same instant and recur periodically, sooner or later both must *simultaneously* reach the value at which they started. The waveform of the sum must therefore be repeated at this point, and it must therefore be periodic also.

individually as excitations of the systems under study. The sinusoidal responses of all the excitations are summated to obtain the output response. This process of decomposition and recombination presupposes signal magnitudes are sufficiently small so that linear operation may be presumed and superposition may be employed.

We are already familiar with one of the processes depicted in Fig. 8.44; we can describe the system responses to the various sinusoidal excitations. However, we have not yet examined how periodic functions may be represented by sinusoidal combinations and we shall now do so.

Fourier* first developed the mathematical procedure to express

* Jean Baptiste Joseph Fourier (1768–1830) was a versatile scientist and mathematician whose diverse capabilities were recognized by Napoleon Bonaparte. In 1798 Fourier accompanied Napoleon to Egypt and was appointed governor of lower Egypt. Much of Fourier's work was devoted to studies of heat conduction, and in conjunction with these studies he developed the theory of the trigonometric series which bears his name. The development of the series was based upon work of earlier mathematicians and physicists (Bernoulli and Euler, around 1750) who similarly examined the trigonometric mathematical series representation in conjunction with the study of a physical problem.

We have frequently mentioned sinusoidal representation; we could just as well have used the term sinusoidal and/or cosinusoidal representations. To include these possibilities, the term "trigonometric representation" is often used to indicate that both sinusoid and cosinusoid terms may be present.

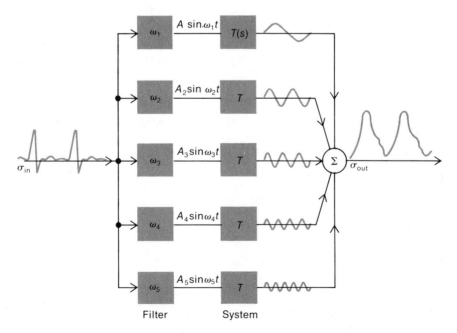

FIG. 8.44 *Principle of analysis of periodic excitation by decomposition into sinusoidal components.*

periodic waveforms in terms of trigonometric components. A formal statement of the theory can be found in almost any mathematical text that discusses mathematical series. We shall not develop the theory in any formal manner but instead shall use an intuitive approach to evolve the appropriate relationships. As a start we use the knowledge that the finite sum of sinusoidal (and/or cosinusoidal) waveforms gives rise to periodic waveforms. On the basis of this observation, we arbitrarily assume for most practical purposes the converse is also true; i.e., periodic waveforms may be approximately represented by a finite sum of sinusoids and cosinusoids.* This statement can be expressed mathematically by

$$(8.105) \quad p(t) = a_1 \sin \omega_1 t + a_2 \sin \omega_2 t + \cdots + a_n \sin \omega_n t$$
$$+ b_1 \cos \omega_1 t + b_2 \cos \omega_2 t + \cdots + b_n \cos \omega_n t$$

The sinusoidal components are called harmonics of the given waveform. As we do not know how many harmonics may be needed, we designate some arbitrary number of n components. Obviously, the frequencies of the harmonics must be related to each other: During one cycle of the periodic waveform $p(t)$, each harmonic must go through an integral number of complete cycles (e.g., two or five cycles, not $2\frac{1}{2}$ or $3\frac{1}{2}$ cycles). If this were not the case, the given periodic waveform $p(t)$ would begin a second oscillation when all the components have not yet simultaneously reached their starting value. Under these circumstances the sum of the harmonics would not give rise to a curve with the same periodicity as that of the given function $p(t)$.

To ensure that the harmonics are correctly related to the period of the original waveform we can set ω_1 equal to the period of $p(t)$; ω_2 can then be 2ω, ω_3 can be 3ω, etc. The first frequency is often called the first harmonic or the fundamental frequency; the second frequency is called the second harmonic, and so on. If we use this new notation, Eq. (8.105) may be written

$$(8.106) \quad p(t) = a_1 \sin \omega t + a_2 \sin 2\omega t + a_3 \sin 3\omega t + \cdots + a_n \sin n\omega t$$
$$+ b_1 \cos \omega t + b_2 \cos 2\omega t + b_3 \cos 3\omega t + \cdots + b_n \cos n\omega t$$

As Eq. (8.106) is now written, it implies that $p(t)$ has no average value (or the average value is zero, or the dc level is zero); the average value of all the sine and cosine terms is zero. If $p(t)$ has an average value, a

---

* There are situations where mathematical periodic functions cannot be represented by a finite series but require a nonending or infinite series for representation; even then, it is possible the representation will not be valid as the series will not converge to a final answer. However, for most practical purposes where realistic signal waveforms are encountered, these problems will not occur and the statement above may be accepted.

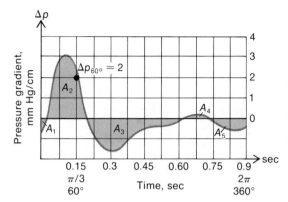

Harmonic	ω	a (sin)	b(cos)
1	$7\frac{rad}{sec}$	0.24	1.024
2	14	1.346	−0.126
3	21	0.477	−0.819
4	28	0.002	−0.305
5	35	−0.144	−0.26

Avg or dc level = 0.335

FIG. 8.45 *Harmonic analysis of femoral artery pulse.* [*From Donald A. McDonald, "Blood Flow in Arteries," p. 292, Edward Arnold (Publishers) Ltd., London, 1960.*]

constant term must be included on the right side of Eq. (8.106), and we have

$$(8.107) \quad p(t) = b_0 + b_1 \cos \omega t + b_2 \cos 2\omega t + \cdots + b_n \cos n\omega t$$
$$+ a_1 \sin \omega t + a_2 \sin 2\omega t + \cdots + a_n \sin n\omega t$$

This equation is one form of the Fourier trigonometric series used to represent periodic functions. In the formal definition n need not be finite but could be infinite.

To show how numerical values for the various terms may be obtained, we shall use a specific example for illustration. Figure 8.45 shows one cycle of the pressure gradient across a centimeter of the femoral artery. For a heartbeat of about 67 strokes/min the duration of one cycle is about 0.9 s. The period of the pressure pulse is thus 0.9 s, and its frequency would be taken as 7 rad/s [$(1/0.9) \times 2\pi \approx 7$]. The term "frequency" of the pressure pulse is somewhat of a misnomer as the pulse is not sinusoidal. A more appropriate description would be to state the fundamental or first harmonic of the pulse is 7 rad/s; the second harmonic is then 14 rad/s, etc. This establishes the value of the ω term in Eq. (8.107). The b_0 is a simple term to evaluate. It represents the average value of the pulse waveform. To determine b_0, therefore, we simply find the average gradient over the pulse period. This can be obtained by averaging a number of pulse magnitudes taken at equal intervals over the cycle. A typical value at 60° ($\pi/3$ rad or 0.15 s) is shown in Fig. 8.45. The smaller the intervals, the more accurate is the average; as a limit we would use the area under the curve over the period. This can be expressed mathematically as

$$(8.108) \qquad b_0 = \frac{-A_1 + A_2 - A_3 + A_4 - A_5}{0.9}$$

where A_1, A_2, etc., are areas under the pressure-gradient curve (in mmHg-s) as shown in Fig. 8.45. If we carry out the operations indicated by Eq. (8.108) we obtain the average gradient of 0.335 mmHg.

An interpretation of Eq. (8.108) in a more general fashion leads to an understanding of the systematic process by which the other constants in Eq. (8.107) may be evaluated. The area under a curve within an interval is ordinarily expressed as an integral. Thus the area under the pulse curve over a single cycle can be expressed as $\int_0^T p(t)\, dt$. If we use this notation, Eq. (8.108) becomes

$$(8.109) \qquad b_0 = \frac{\int_0^{T=0.9} p(t)\, dt}{T(=0.9)}$$

or

$$b_0 T = \int_0^T p(t)\, dt$$

In this form we see b_0 may be defined by performing a mathematical operation on the pulse waveform over a cycle; in this case the operation is integration. To illustrate this point very simply, we integrate both sides of Eq. (8.107) over the period T; the result is

$$(8.110) \quad \int_0^T p(t)\, dt = \int_0^T b_0\, dt + \int_0^T b_1 \cos \omega t\, dt + \int_0^T b_2 \cos 2\omega t + \cdots$$
$$+ \int_0^T a_1 \sin \omega t\, dt + \int_0^T a_2 \sin 2\omega t + \cdots$$

The integral of the trigonometric terms is interpreted as the area under the trigonometric curves for one period. This area is, of course, zero, and the right side of Eq. (8.110) becomes simply $\int_0^T b_0\, dt$ or $b_0 \int_0^T dt$ or $b_0 T$. Equation (8.110) thus reduces to Eq. (8.109).

The technique to evaluate the Fourier constants (a_1, a_2, b_1, b_2, etc.) suggested by the foregoing discussion can be described verbally in a fairly simple manner: Operate on Eq. (8.107) in a fashion that causes all but one term on the right side of the equation to reduce to zero and causes the left side to reduce to a numerical value. From the results of such an operation, one Fourier coefficient may be evaluated. Although the concept is easily stated, it is not at all obvious what operations can be used to obtain the desired results. Integration has attractive possibilities, as the integral of sinusoidal (or cosinusoidal) functions yields zero.

This concept was used to evaluate b_0, however; if we now seek to evaluate any of the other constants by some integration, we must modify the b_0 term so that upon integration this term will reduce to zero; e.g.

$$\int_0^T b_0 \sin \omega t \, dt = 0$$

To see how this idea may work out, we multiply Eq. (8.107) by $\sin \omega t$ and integrate the result. Upon performing the operations, we get

$$(8.111) \quad \int_0^T p(t) \sin \omega t \, dt = \int_0^T b_0 \sin \omega t \, dt + \int_0^T b_1 \cos \omega t \sin \omega t \, dt$$

$$+ \int_0^T b_2 \cos 2\omega t \sin \omega t \, dt + \cdots$$

$$+ \int_0^T a_1 \sin^2 \omega t \, dt + \int_0^T a_2 \sin 2\omega t \sin \omega t \, dt + \cdots$$

The first term on the right side of the equation is zero. We can evaluate the second term by using formal integration* or by examining a graphical interpretation of the results of the operations indicated.

The solid curve in Fig. 8.46a is a cosine curve, and the dashed curve is a sine curve; the second term in the series expression of Eq. (8.111) is the area under the product of the sine and curve functions. The product curve is shown in Fig. 8.46b. From the symmetry of the curve we can immediately see that the area (and hence the integral) under the resultant product is zero. In a similar manner we can recognize $\int_0^T b_2 \cos 2\omega t \sin \omega t \, dt$ is zero (Fig. 8.46c; during the first half period the cosine curve is multiplied by the positive half of a sine wave; during the second half the same cosine curve is multiplied by the negative half of a sine wave; the result-

---

$* \, b_1 \int_0^T \cos \omega t \sin \omega t \, dt = b_1 \int_0^T \frac{-\cos \omega t}{\omega} \, d(\cos \omega t) = \frac{-b_1}{\omega} [\cos^2 \omega t]_0^T = 0$

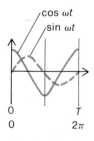

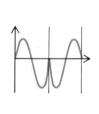

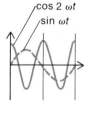

(a) sin ωt and cos ωt (b) Product of sin ωt and cos ωt (c) Cos 2 ωt and sin ωt (d) Sin² ωt

FIG. 8.46 *Graphical interpretation of sine and cosine products.*

ant curve must therefore be symmetrical about the x axis and the area must therefore be zero). We can, in fact, show that every term on the right side of Eq. (8.111) reduces to zero except the term that includes the square of the sinusoid. We can see by inspection (Fig. 8.46d) that the area under the $\sin^2 \omega t$ function is not zero; the integral is therefore not zero and may be evaluated. If we carry out the required integration (or use integral tables), we find $a_1 \int_0^T \sin^2 \omega t \, dt$ may be evaluated to be $a_1 T/2$. Equation (8.111) may thus be written

(8.112)
$$\int_0^T p(t) \sin \omega t \, dt = a_1 \frac{T}{2}$$

or

$$a_1 = \frac{2}{T} \int_0^T p(t) \sin \omega t \, dt$$

or

$$a_1 = \frac{1}{\pi} \int_0^{2\pi} p(t) \sin \omega t \, d\omega t$$

The last form is obtained by using $2\pi/\omega$ to replace the period T and changing the variable of integration to ωt instead of t.

Equation (8.112) conveniently relates the Fourier coefficient a_1 to the original periodic waveform $p(t)$ and, in fact, indicates how the coefficient may be evaluated: The product $p(t) \sin \omega t$ is formed. The area under the product curve is equivalent to $a_1 \pi$; a_1 can therefore be determined (see Fig. 8.47). If we carry out the indicated operations, the a_1 term is evaluated to be 0.240.

The procedure to evaluate the other Fourier coefficients for $p(t)$ now seems to be established. To obtain b_1 we use $\cos \omega t$ as a multiplier of Eq. (8.107) and integrate over one cycle; all terms on the right side of the equation should reduce to zero except the one containing b_1; b_1 can then

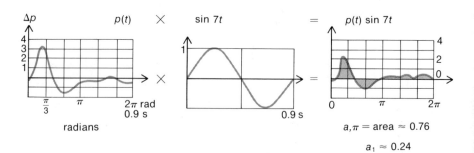

FIG. 8.47   *Evaluation of the coefficient for first-harmonic sine term.*

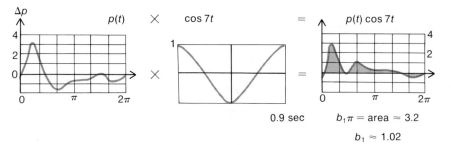

$$b_1\pi = \text{area} \approx 3.2$$
$$b_1 \approx 1.02$$

FIG. 8.48 *Evaluation for the coefficient of first-harmonic cosine term.*

be evaluated. As an outline of the procedure,

(8.113a) $\displaystyle\int_0^T p(t) \cos \omega t \, dt = \int_0^T b_0 \cos \omega t \, dt + \int_0^T b_1 \cos^2 \omega t \, dt$

$$+ \int_0^T b_2 \cos \omega t \cos 2\omega t \, dt + \cdots, \text{ etc.}$$

$$+ \int_0^T a_1 \cos \omega t \sin \omega t + \cdots, \text{ etc.}$$

(8.113b) $\displaystyle\int_0^T p(t) \cos \omega t \, dt = \int_0^T b_1 \cos^2 \omega t \, dt = b_1 \frac{T}{2}$

(8.113c) $\displaystyle\int_0^{2\pi} p(t) \cos \omega t \, d\omega t = b_1\pi$

or

$$b_1 = \frac{1}{\pi} \int_0^{2\pi} p(t) \cos \omega t \, d\omega t$$

If we perform the operations indicated, the b_1 term is evaluated to be 1.024 (Fig. 8.48).

From the foregoing discussion, we should now be willing to accept that the general Fourier coefficients (a_n or b_n, where $n \neq 0$) may be obtained from

(8.114a) $\displaystyle a_n = \frac{1}{\pi} \int_0^{2\pi} p(t) \sin n\omega t \, d\omega t$

(8.114b) $\displaystyle b_n = \frac{1}{\pi} \int_0^{2\pi} p(t) \cos n\omega t \, d\omega t \qquad n \neq 0$

Although it has not been shown that the outlined procedure can be generalized for all coefficients, we should intuitively expect that the method has general applicability. [It is, in fact, possible to justify mathematically the validity of Eq. (8.114) but we do not do so here.] If we accept Eq. (8.114) as a general procedure we can evaluate as many har-

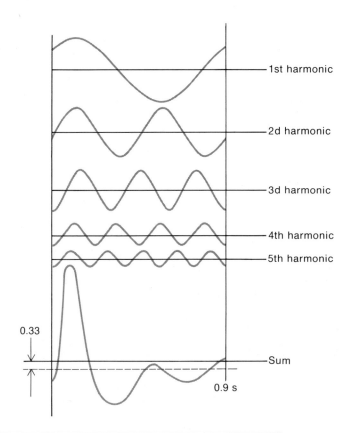

Harmonic	Amplitude	Phase
1	1.052	13.2°
2	1.352	93.23°
3	0.948	149.79°
4	0.305	179.62°
5	0.380	209.33°

FIG. 8.49 *Fourier components of p(t) in Fig. 8.45. [Values taken from Donald A. McDonald, "Blood Flow in Arteries," Edward Arnold (Publishers) Ltd., London, 1960. Phase angles were derived for cosinusoidal reference rather than sinusoidal reference; see Eq. (8.116).]*

monic coefficients as are necessary to provide an adequate representation of the given periodic waveform. The method can be mechanized, and computer programs are available which can automatically calculate harmonic coefficients.

To complete this discussion we examine the representation of the given waveform after five harmonics have been defined. A table for the sine and cosine coefficients is given in Fig. 8.45. From these values we can construct an approximate representation of the given waveform in terms of the harmonic components. If $p_a(t)$ represents the approximate representation, we may write

(8.115)
$$\begin{aligned}
p_a(t) = 0.335 &+ 1.024 \cos\ \ 7t + 0.24\ \ \sin\ \ 7t \\
&- 0.126 \cos 14t + 1.346 \sin 14t \\
&- 0.819 \cos 21t + 0.477 \sin 21t \\
&- 0.305 \cos 28t + 0.002 \sin 28t \\
&- 0.26\ \ \cos 35t - 0.144 \sin 35t
\end{aligned}$$

To examine the final result we would have to plot each component and add all the results. This would require 10 plots and considerable tedious work. We can simplify the procedure somewhat if we combine terms of the same frequency in the manner outlined by Fig. 8.42. In this case only sine and cosine terms are involved and so θ is 90° and the relationship is reduced to $C = \sqrt{A^2 + B^2}$ and $\tan \phi = A/B$. If we use this viewpoint, each harmonic will have an amplitude and phase as shown in the table of Fig. 8.49. Equation (8.115) may then be written

(8.116)
$$\begin{aligned}
p_a(t) = 0.335 &+ 1.05\ \ \cos\ (7t\ \ - 13.2°) \\
&+ 1.352 \cos\ (14t - 93.23°) \\
&+ 0.948 \cos\ (21t - 149.79°) \\
&+ 0.305 \cos\ (28t - 179.6°) \\
&+ 0.380 \cos\ (35t - 209.33°)
\end{aligned}$$

It is now necessary to construct only five sinusoids to represent $p(t)$. These curves and their sum are shown in Fig. 8.49. One can readily see that the sum of the Fourier components approaches the actual waveform almost exactly in some places. For better representation more terms would be needed.

8.11   USE OF FOURIER EXPANSION

In a previous section we established that the sinusoidal output of a system depends upon the frequency and amplitude of the sinusoidal driver. In Sec. 8.10 it was established that periodic waveforms can be represented by a series of sinusoidal terms of different amplitudes and frequencies. We now examine a situation in which both concepts can be used to study a system stimulated by a periodic excitation.

A diagrammatic sketch of the system to be studied is shown in Fig. 8.50. A respirator is used to provide artificial ventilation for resuscitation

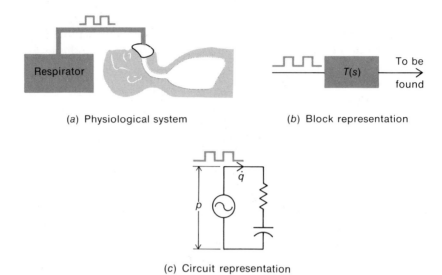

(a) Physiological system

(b) Block representation

(c) Circuit representation

FIG. 8.50 *Periodic excitation of a physiological system.*

or possibly for sustenance when the respiratory function has been temporarily impaired. The amount of air delivered depends on the characteristics of the patient and on the pressure delivered by the respirator. For illustrative purposes it will be assumed that the respirator can supply a square pressure wave as shown in Fig. 8.51.* An estimate of the air flow to and from the patient is to be made.

In this study the system excitation is known and the system response is to be determined. The block-diagram configuration is shown in Fig. 8.50b. To obtain a solution, the system characteristics and proper-

---

* "Constant"-pressure respirators are available commercially. Such units may be considered pressure sources that deliver constant pressure for some portion of the cycle and then vent the patient to the atmospheric pressure for another portion of the cycle. Other kinds of respirators are also available: Constant-flow units maintain a steady inspiratory flow; servo or demand-type units work in accordance with command signals generated by the patient.

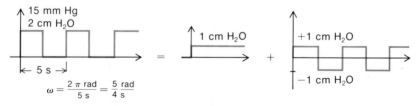

FIG. 8.51 *Square-wave pressure excitation signals.*

ties [that is, $T(s)$] must be defined. As a crude approximation, we assume the pulmonary system may be considered a simple resistive-compliant system. The network shown in Fig. 8.50c may thus be taken as an analog representation of the physiological system. The lumped pulmonary resistance is 1.5 cm $H_2O/l/s$; the lumped compliance is 0.2 l/cm H_2O. With this diagram and the use of the impedance concept we can write immediately

$$(8.117) \qquad \dot{q} = \frac{p(t)}{Z} = \frac{p(t)}{R + 1/sC} = \frac{s}{s + 1/RC} \frac{p(t)}{R}$$

If the excitation $p(t)$ can be expressed in operator form (i.e., as a function of s) we could find a solution for Eq. (8.117) by using standard techniques.* As an approximate procedure we assume the periodic square wave $p(t)$ can be represented by a Fourier expansion of the form given by Eq. (8.107). In block-diagram form this can be represented as shown in Fig. 8.52a. The given periodic waveform is represented by the sum of a dc or constant-level component (b_0) and a series of oscillating components $\left[\sum_{n=1}^{\infty} (b_n \cos n\omega t + a_n \sin n\omega t) \right]$. The system operates on the sum of all the components.

An alternative viewpoint is shown in Fig. 8.52b. In this diagram it is assumed the system operates individually on the constant-level component and on the sinusoidal components. $\dot{q}_1$ represents the flow due to the steady average pressure of 1 cm H_2O (positive pressure breathing) and $\dot{q}_2$ represents the flow due to the periodic portion of the excitation pressure. The total flow is evidently the sum of the separate flows generated by the component excitations. (This presumes the system may be considered linear and that superposition is valid.) From this point of view it is easy to see the dc pressure component does not contribute anything to the steady-state flow. To justify this contention we simply determine $\dot{q}_1$, the flow due to a step-pressure excitation of 1 cm H_2O. Since

---

* This is sometimes more easily said than done. In this case, for example, we can determine the transform of a square wave to be $\delta(t)/s(1 + \epsilon^{-sT/2})$ (see Prob. 8.8). A formal statement of the solution for $\dot{q}$ would be

$$\dot{q} = \frac{1}{s + 1/RC} \frac{\delta(t)}{1 + \epsilon^{-sT/2}} \frac{1}{R}$$

It is not possible, however, to translate this expression into a closed-form solution for $\dot{q}$ in the time domain; instead, the solution must be expressed as the sum of time-displaced signals $\left[\text{i.e., a series of the form} \sum_{n=0}^{\infty} f(t - t_n)u(t - t_n) \right]$. More information about such methods of solution can be found in texts dealing with transform techniques. See, for example, Ref. 10.

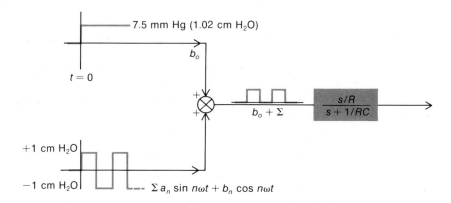

(a) System excited by total signal

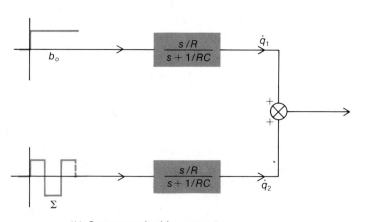

(b) System excited by separate
signal components

FIG. 8.52   *Alternative viewpoints of system excitation.*

the transform of such a signal would be $1/s$, we have

$$(8.118) \qquad \dot{q}_1 = \frac{1}{s} \frac{s/R}{s + 1/RC} = \frac{1/R}{s + 1/RC} = \frac{1}{R} \epsilon^{-t/RC}$$

In the steady state, $t \to \infty$ and $\dot{q}_1 \to 0$. This seems intuitively sensible. A steady pressure would inflate the patient but it could not sustain flow. The choice of the average pressure to be applied to the patient is, however, important as it affects the physiological parameters of R and C (i.e., it affects the quiescent operating level; see also the discussion in Sec. 8.5). It may therefore affect the total flow.

From the foregoing discussion we can now see that to determine air flow during artificial ventilation, we need only consider the alternating components of the pressure waveform. These are the trigonometric components of the square wave, and when the components have been defined we can assume each acts individually on the patient to elicit a flow response. The total response will be the summation of the responses due to each component. A diagrammatic outline of the proposed procedure is shown in Fig. 8.53. The A_1, A_2, etc., coefficients are the Fourier coefficients of the component excitation waveform. The $\dot{q}_n$ terms are the resultant flow components due to each of the excitation components; $\dot{Q}_n$ represents the amplitude of the flow components, and ϕ_n represents the phase angles associated with the flow components.

To carry out the proposed analysis we must first define the Fourier coefficients of the excitation waveforms. We can determine these coefficients in a formal manner by using Eqs. (8.114a) and (8.114b), or we can use the graphical technique. In this case the waveform is so simple and easily defined we can almost immediately obtain the general coefficients by using the equations. Since the alternating portion $p(t)$ is $+1.02$ cm H_2O for 2.5 s and -1.02 cm H_2O for 2.5 s (see Fig. 8.54) Eq. (8.114a) may be written

$$(8.119) \quad a_n = \frac{1}{\pi} \left(\int_0^\pi 1.02 \sin n\omega t \, d\omega t + \int_\pi^{2\pi} - 1.02 \sin n\omega t \, d\omega t \right)$$

where the integral has been split into an integral between 0 and π rad (0 and 2.5 s) and another between π and 2π rad (2.5 and 5 s). By integration (or tables or counting squares to get areas), we find

$$(8.120) \quad a_n = \frac{1}{\pi} \left\{ \left[\frac{-1.02 \cos n\omega t}{n} \right]_0^\pi + \left[\frac{+1.02 \cos n\omega t}{n} \right]_\pi^{2\pi} \right\}$$

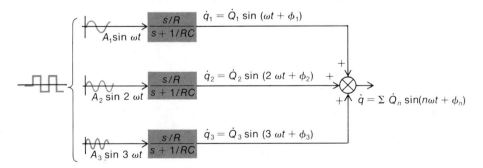

FIG. 8.53 *Technique to determine flow response by Fourier expansion.*

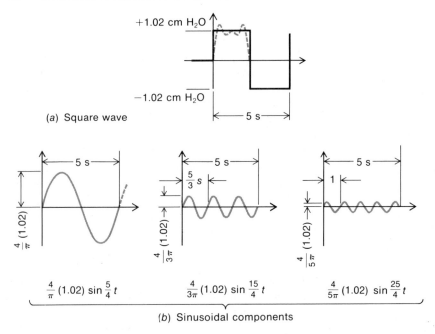

(a) Square wave

(b) Sinusoidal components

FIG. 8.54 *First three Fourier components of a square wave.*

If we work out the algebra we find when n is an even number a_n reduces to zero. If n is an odd number, the Fourier sine coefficients are given by $a_n = 4.08/n\pi$, where n can be 1, 3, 5, ... This implies that the given square wave has only odd harmonics. If the fundamental frequency is $\frac{1}{5}$ Hz or $\frac{5}{4}$ rad/s (see Fig. 8.51), the frequency of the next component will be $\frac{3}{5}$ Hz or $\frac{15}{4}$ rad/s, etc. [This conclusion can easily be justified by graphical construction of the integral in Eq. (8.114a). The resultant plots will have nonzero areas only when odd harmonics are used.] The Fourier expansion of the variable portion of the applied pressure [$p_v(t)$] may thus far be written

$$(8.121) \quad p_v(t) = \frac{4.08}{\pi} \sin \frac{5t}{4} + \frac{4.08}{3\pi} \sin \frac{15t}{4} + \frac{4.08}{5\pi} \sin \frac{25t}{4} + \cdots$$

$$+ \sum_{n=1}^{\infty} b_n \cos n\omega t$$

The cosine coefficients must now be evaluated. However, by graphical construction or by using Eq. (8.114b) we would find all these coefficients must be zero,* and Eq. (8.121) then reduces to a sine series of odd har-

---

* The integral of Eq. (8.114b) in this case yields the same expression as Eq. (8.120) except the cosine terms are replaced by the sine. But $\sin 0 = \sin n\pi = \sin n\, 2\pi = 0$; therefore $b_n = 0$.

monics which may be written

(8.122) $p_v(t) = 1.3 \sin 1.25t + 0.435 \sin 3.75t + 0.26 \sin 6.25t + \cdots$

The results are shown diagrammatically in Fig. 8.54. The sum of the first three components of the square-wave excitation is shown by the gray curve in Fig. 8.54a.

We are now prepared to use the format suggested by Fig. 8.53 to obtain a solution for the patient flow response. From the experience gained earlier regarding the response to sinusoidal excitations we can see the coefficient of the flow due to the first harmonic of the pressure excitation is given by

(8.123) $\dot{Q}_1 = A_1 \left| \dfrac{s/R}{s + 1/RC} \right|_{s \to jw} = \dfrac{A_1}{R} \left| \dfrac{jw}{jw + 1/RC} \right|$

The amplitude A_1 of the first harmonic is given as 1.3 [see Eq. (8.122); the fundamental frequency is 1.25 rad/s and the system parameters are such that RC is 0.3 s ($R = 1.5$, $C = 0.2$)]. With these values, the response amplitude of the first harmonic becomes

(8.124) $\dot{Q}_1 = \dfrac{1.3}{1.5} \left| \dfrac{j1.25}{j1.25 + 10/3} \right| = 0.87 \dfrac{1.25}{\sqrt{(1.25)^2 + (3.3)^2}}$

or

$$\dot{Q}_1 = 0.305$$

The phase angle of this term is obtained from the transfer function by evaluating the angle associated with the complex form of the system function.

(8.125) $\phi_1 = \underline{/T(j\omega)}$ = angle associated with $\left| \dfrac{j\omega/R}{j\omega + 1/RC} \right|$

The numerator of the complex form of the transfer function may be written

(8.126) $$\dfrac{j\omega}{R} = \dfrac{\omega}{R} \epsilon^{j90°}$$

The denominator can be written

(8.127) $$j\omega + \dfrac{1}{RC} = \sqrt{\omega^2 + \left(\dfrac{1}{RC}\right)^2} \, \epsilon^{j\theta_1}$$

where

(8.128) $$\tan \theta_1 = \omega RC = 1.25 \times 0.3 = 0.375$$

or

$$\theta_1 = 20.5°$$

From Eqs. (8.126) and (8.128) we can see the phase angle of the first harmonic of the response may be taken as

(8.129) $$\phi_1 = 90 - 20.5° = 69.5° = 1.2 \text{ rad}$$

The fundamental of the flow response may thus be written

(8.130) $$\dot{q}_1 = 0.305 \sin (1.25t + 1.2)$$

The procedure to evaluate the constants for the first harmonic can serve as a pattern for the evaluation of the constants of successive harmonics. In fact, this same procedure can be used to evaluate the constants for the general (or nth) harmonic. The amplitude of the nth harmonic can be written

(8.131) $$\dot{Q}_n = \frac{A_n}{R} \left| \frac{j\omega n}{j\omega n + 1/RC} \right|$$

where ωn is the frequency of the nth (odd) harmonic. Since A_n is given by $4.08/n\pi$ [see Eq. (8.120) and the associated discussion] and RC is 0.3 s we can write

(8.132) $$\dot{Q}_n = \frac{4.08}{\cancel{n}\pi} \frac{1}{1.5} \frac{\omega \cancel{n}}{\sqrt{(\omega n)^2 + (1\frac{9}{3})^2}} = \frac{4.08}{1.5\pi} \frac{5\frac{1}{4}}{\sqrt{(5\frac{1}{4})^2 n^2 + 11.1}}$$

or

$$\dot{Q}_n = \frac{1.08}{\sqrt{1.56n^2 + 11.1}}$$

In a comparable fashion, ϕ_n may be determined to be

(8.133) $$\phi_n = 90° - \theta_n \qquad \text{where } \tan \theta_n = \omega nRC = 0.375n$$

From Eqs. (8.132) and (8.133) we can evaluate the constants for the third harmonic to be

$$\dot{Q}_3 = \frac{1.08}{\sqrt{1.56 \times 9 + 11.1}} = 0.216$$

and $\phi_3 = 90° - 48.5° = 0.7$ rad. The third harmonic component can therefore be written

(8.134) $$\dot{q}_3 = 0.216 \sin (3.75t + 0.7)$$

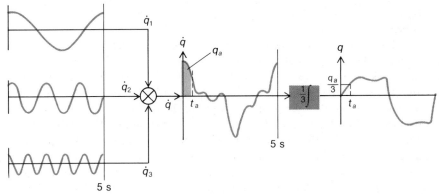

FIG. 8.55 *First three components of flow response and their results.*

In a similar fashion we find the fifth harmonic to be given by

$$\dot{q}_5 = 0.145 \sin (6.25t + 0.49)$$

The flow response, if only the first three harmonics of the excitation are considered, can now be written

(8.135) $\dot{q} = 0.305 \sin (1.25t + 1.2) + 0.216 \sin (3.75t + 0.7)$
$$+ 0.145 \sin (6.25t + 0.49)$$

A plot of the harmonic components and their sum is shown in Fig. 8.55. As can be seen from the figure [or Eq. (8.135)], the amplitude contribution of the fifth harmonic (third term) is less than one-half that of the first harmonic. A check on the seventh harmonic will show that the amplitude of this term is 0.113 or about one-third of the first harmonic. Thus, as the frequency of the excitation harmonic increases, its contribution becomes less and less; in other words, the higher harmonics of the squared-wave excitation do not produce significant movements of air.

To determine the volume of air moved into and out of the patient, we need only to integrate the flow information. We can obtain a general result by integrating the generalized flow component

$$\dot{q}_n = \dot{Q}_n \sin (n\omega t + \phi_n)$$

where $\dot{Q}_n$ is given by Eq. (8.131) and ϕ_n is given by Eq. (8.133).

$$q_n = \int_0^T \dot{q}_n \, dt$$

One can obtain the result for a specific case by integrating the flow curve as shown in Fig. 8.55. Either technique could be used to study the variations of delivered volume with changes of excitation waveform or system properties.

The foregoing analysis is somewhat academic as it serves more as an illustration of how Fourier expansions and sinusoidal analyses are jointly used rather than the usefulness of these techniques. However, something can be gleaned even from this academic discourse. From the analysis of the particular situation we note the first harmonic of flow is related to the first harmonic of the pressure excitation and the system properties. Conversely, this implies that, if we have the first-harmonic pressure and flow, we should know something about the system. It is, in fact, easy to see that, if we accept a simple RC representation for the physiological system (Fig. 8.50c), the phase and amplitude information of the stimulus and the response first harmonics can be used to define these constants. From the measured fundamental frequency and phase angle, the system time-constant [Eq. (8.128)] can be obtained; this in conjunction with the amplitude datum [Eq. (8.131)] can be used to define R (C is then automatically available). If we are not willing to accept a simple RC representation but the circuit in Fig. 7.4 seems to be adequate, we would need the first and second harmonics to determine the four unknown circuit elements.

One might extrapolate further and note that, since the nonsinusoidal periodic input contains a spectrum of frequencies, it may be possible to plot gain-frequency curves from the input-output harmonic data. From this representation one might be able to determine an appropriate model rather than using cut-and-try guesswork. It is to be noted no mention has been made of square-wave excitation in this discussion. The same comments could be made regardless of the kind of periodic excitation used. This implies that if the periodic pressure excitation attendant with respiration could be recorded (as with an esophageal balloon) and the periodic flow could be recorded simultaneously, the harmonics of these waveforms could be used in the manner outlined. In a practical situation the stimulus and response waveforms are not as periodic as one might desire (and the system is certainly nonlinear). To perform such an analysis, therefore, it might be necessary to average a number of cycles to obtain a representative waveform or, alternatively, one might make an analysis with a number of different waveform sections and average the results. In either case, it appears that such studies could be used to assess some of the dynamic properties of the respiratory system.

Problems

8.1   The two phasors shown in Fig. P8.1 represent the pressure waveform of a sound and the voltage response at the cochlea.

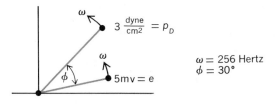

FIG. P8.1

(a) Sketch the two waveforms on the same set of axes. Indicate when the pressure wave reaches its peak.

(b) If ϕ (called a phase angle) is given as 30°, what is the time lapse between the pressure maxima and voltage maxima?

8.2 Two pressure waves are delivered to the ear from two different points (see Fig. P8.2). The pressure waveforms at the ear may be given by $p_{D1} = 2 \sin (256\ c/s)t$ and $p_{D2} = 3 \sin [(256\ c/s)t + 30°]$. (See Prob. 8.1.)

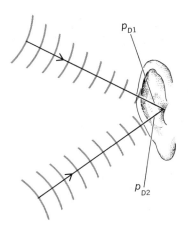

FIG. P8.2

(a) What is the amplitude of the resultant pressure waveform (can phasors be added like vectors)? What is the phase angle (time lapse) between the resultant pressure waveform and the reference pressure waveform (i.e., the one with zero phase angle)? In this case this will be the pressure wave with smaller amplitude.

(b) If the voltage amplitude of the cochlea is assumed to be approximately proportional to pressure in this range, write an expression for the voltage response of the cochlea and draw a phasor to represent it.

8.3   When an instrument converts a physiological signal to an electrical signal, some modification of signal waveform may be expected, because of the instrument properties (transfer function). In some cases the modification may be so extreme that the output waveform may bear relatively little resemblance to the input waveform (the instrument is then said to have distorted the input signal). The degree of alteration depends upon the signal-frequency content and the system properties. To illustrate this point, assume the input to an instrument is of the form

$$\sigma_{\text{in}} = \sin t + \frac{1}{2} \sin 4t$$

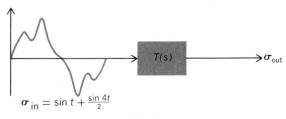

$$\sigma_{\text{in}} = \sin t + \frac{\sin 4t}{2}$$

FIG. P8.3

   (a) What will be the shape of the output waveform (σ_{out}) if $T(s)$ is given by $2/(s + 2)$?
   (b) Suppose the instrument in (a) is replaced by one having a transfer function $8/(s + 8)$. What will be the output waveform now?
   (c) What will be the output waveform when $T(s)$ is $16/(s^2 + 2s + 16)$?

8.4   If a light of fixed illuminance E_{i0} impinges on a fixed aperture of area A_{p0}, the light flux that passes through the aperture is $E_{i0}A_{p0}$ [$= \mathfrak{F}_{R0}$; see Eq. (8.92) and Fig. P8.4a]. If the incident light varies around some fixed level (as does an ordinary incandescent lamp) the illuminance may be represented by $E_{i0} + E_{iv}$, where E_{iv} represents the variable component of the illuminance around the fixed illuminance level of E_{i0} (see Fig. P8.4b). If this light falls on a fixed aperture of area A_{p0}, the delivered light is given by $(E_{i0} + E_{iv})A_{p0} = E_{i0}A_{p0} + E_{iv}A_{p0} = \mathfrak{F}_{R0} + \mathfrak{F}_{Rv}$, where $\mathfrak{F}_{Rv}$ is the variable component of the delivered flux.

   (a) If the illuminance varies sinusoidally around some fixed level with an amplitude equivalent to 2 percent of the fixed amplitude (E_{i0}) and at a frequency of 60 Hz, write an expression for the delivered flux. What is the amplitude variable flux component?
   (b) If the illuminance varies and the aperture also varies, the delivered flux may then be written

$$(E_{i0} + E_{iv})(A_{p0} + A_{pv}) = \mathfrak{F}_{R0} + \mathfrak{F}_{Rv}$$

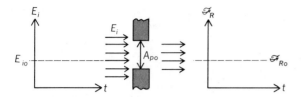

(a) Fixed aperture and illuminance

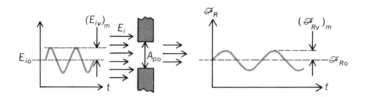

(b) Fixed aperture, variable illuminance

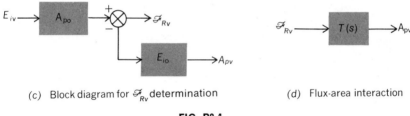

(c) Block diagram for $\mathscr{F}_{Rv}$ determination   (d) Flux-area interaction

FIG. P8.4

where A_{pv} represents the variable component of the aperture variation. Write an expression for the constant level of delivered flux and an expression for the variable component of the delivered flux. If increased illuminance results in decreased aperture area, what would the expression then be?

(c) If the variable component of the illumination has a maximum amplitude of 2 percent of the fixed illuminance and the variable component of the aperture area (A_{pv}) has a maximum amplitude of 1 percent of the fixed area (A_{po}), what is the maximum amplitude of the variable flux component $\mathscr{F}_{Rv}$?

(d) On the basis of the foregoing development, it should seem reasonable that the variable flux component can be represented approximately by

$$\mathscr{F}_{Rv} = A_{po}E_{iv} - E_{io}A_{pv}$$

(The negative sign implies an area reduction for increased illuminance.) This last equation may be represented by the block diagram in Fig. P8.4c. If the delivered flux ($\mathfrak{F}_{Rv}$) has an effect on the aperture area as suggested by Fig. P8.4d [i.e., if the model represents the eye and the interaction is as suggested by Eq. (8.93)], draw a completed diagram to show how the light illuminance affects the pupillary area of the eye and how this area affects the delivered flux.

(e) In Chap. 10 we shall learn that the diagram constructed for part d of this problem is a closed-loop system. The overall transfer function for E_{iv} as input and $\mathfrak{F}_{Rv}$ as output may be given by

$$\frac{\mathfrak{F}_{Rv}}{E_{iv}} = \frac{A_{p0}}{1 + E_{i0}T(s)} = \frac{\mathfrak{F}_{R0}/E_{i0}}{1 + E_{i0}T(s)}$$

If $T(s) = 0.16/(0.1s + 1)^3$ and $E_{iv} = 0.2E_{i0} \sin \omega t$ (where ω is 25 c/s), derive an expression for the amplitude of the delivered flux in terms of E_{i0}. When E_{i0} is very small, what is $\mathfrak{F}_{Rv}/\mathfrak{F}_{R0}$? Compare with (a). What happens when E_{i0} is very high? How are the results affected when $\omega = 60$ cs?

8.5 A mechanical representation of a muscle would consist of a spring to represent the elastic property of the muscle, a dashpot to represent the internal viscous friction (see Sec. 4.12) within the muscle tissue, and a mass to represent the inertial property of the muscle. One possible schematic arrangement is shown in Fig. P8.5a. The muscle mass is lumped into a single mass M_m; the friction and elastic properties are represented by two springs and dashpots.

If the velocity-voltage and force-current analogs are used, the spring and dashpot are analogous to an inductor and resistor, respectively, and a mass would be analogous to a capacitor.* The analogous electrical representation is shown in Fig. P8.5b. (The velocity of the mass is always measured with respect to earth; points A_1 and A_2 in Fig. P8.5a must be the same velocity and are represented by node m in Fig. P8.5b.)

(a) If the top of the muscle is driven by a velocity input V_p, use the impedance concept and the concept of a voltage divider to derive the transfer function of the muscle where V_w is the system output (i.e., find V_w/V_p as a function of s).

(b) If V_p is a sinusoid of amplitude V_{pa} and frequency ω, what will be the amplitude of the weight oscillation V_{wa}? (This method has been used by King and Lawton (see Ref. 1, Chap. 4) to assess some of the muscle elastic and viscous

---

* If ground or earth is used as a velocity reference, the velocity of the mass with respect to ground is $V_m - V_g$, or simply V_m. The force to drive the mass is given by Newton's law $f = ma$ or

$$f = m\frac{dv}{dt} = m\frac{d(V_m)}{dt}$$

If $f \sim i$ and $v \sim e$, we have

$$i = m\frac{de_m}{dt}$$

From this last relationship it is easy to see mass and capacitance may be considered analogous components.

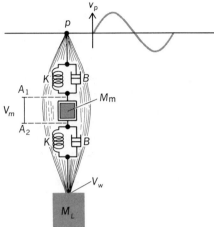

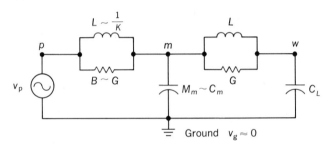

(a) Mechanical model

(b) Circuit analog

FIG. P8.5

properties. If V_{pa} and V_{wa} are measured and ω, M_m, and M_w are known, some information about the muscle viscoelastic properties may be obtained by this procedure.)

8.6 If a system has a transfer function $T = \sigma_{\text{out}}/\sigma_{\text{in}} = T(s) = 10/(s + 2)$, the gain for "low"-frequency input signals will be 5.

(a) Express the gain in decibels.

(b) Draw the low-frequency gain curve G_{dB} versus log frequency (i.e., draw the low-frequency asymptote).

(c) What is the true gain (in decibels) when the input signal has a frequency 1 rad/s? What is the difference (in decibels) between the low-frequency asymptote and the true-gain curve when the signal frequency is 1 octave below the break frequency? (Note 1 rad/s is one-half the break frequency of 2 rad/s; it is therefore 1 octave below the break frequency.)

(d) When the input signal has a frequency of 2 rad/s, the true gain is given by

$$G = \frac{10}{|j2 + 2|} = \frac{10}{\sqrt{8}} = \frac{5}{\sqrt{2}}$$

Express this gain in decibels and show the true gain at the break frequency is 3 dB below the low-frequency asymptote.

(e) The high-frequency asymptote passes through the break-frequency point on the low-frequency asymptote and has a slope 6 dB/octave. Show the true-gain curve at 1 octave above the break frequency (in this case at $\omega = 4$ rad/s) is 1 dB from the high-frequency asymptote.

(f) On the basis of the above discussion, show that if the transfer function is $10/(s + 2)^2$ the true-gain curve and asymptotic approximations differ by 6 dB at the break frequency and by 2 dB 1 octave above and below the break frequency.

8.7 In a physiological test the gain-amplitude ratio is determined as a function of the excitation frequency. The results are shown below in tabular form. The gain is given in terms of absolute ratio (*not* decibels) and frequency in radians per second.

ω(rad/s)	0.1	0.2	0.5	1	2	3	4	5	8	10
Gain (absolute)	1	1	1.05	1.1	1.1	0.95	0.7	0.5	0.24	0.14

(a) Plot the frequency-response characteristics of gain (decibels) vs. log ω. What is the high-frequency rate of decay?

(b) Could either of the transfer functions given below be used to represent the gain-frequency characteristics of the system?

$$T(s) = \frac{23.1(s + 1)}{(s + 2)(s + 3.4)^2}$$

$$T(s) = \frac{9}{s^2 + 3s + 9}$$

(c) Are the phase characteristics of the two transfer functions given in (b) closely alike or distinctly different? Would the phase characteristics be of any use to help decide which transfer function might more closely represent the given system?

8.8 The square wave shown in Fig. P8.8 may be described by the mathematical expression

$$p(t) = u(t) - 2u(t - t_1) + 2u(t - 2t_1) - \cdots$$

where the term $u(t - t_1)$ is a step delayed t_1 sec, etc.

(a) Show that the summation indicated yields the desired square wave.

(b) Since the transform of a step is $1/s$, that of a delayed step is ϵ^{-st_1}/s. Show that the transform of the square wave is $(1 - \epsilon^{-st_1})/(s)(1 + \epsilon^{-st_1})$.

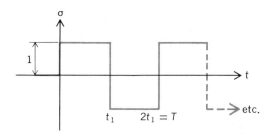

FIG. P8.8

(c) If t_1 is 1 min and this signal is applied to a system with a transfer function $5/(s + 5)$ (i.e., the system time-constant is 12 s), draw the block diagram and determine the system response as the summation of successive step-function excitations (see the discussion associated with Fig. 6.23). Compare this result with that obtained by using a Fourier representation of the square wave.

(d) If this same signal is applied to a system with transfer function $3/(s + 3)$, what will be the *steady-state* response of the system?

(e) Compare the average response obtained in (c) and (d) and the corresponding values that would have been obtained by using a Fourier spectrum for the excitation signal.

8.9 At the end of Chap. 6 we examined the response of a thermometer which was excited by a triangular waveform (see Sec. 6.10). To analyze this situation the excitation was represented by a sum of ramp-functions. This problem can also be handled by using a Fourier representation of the triangular function. A block diagram of this procedure is shown in Fig. P8.9. The triangular waveform is

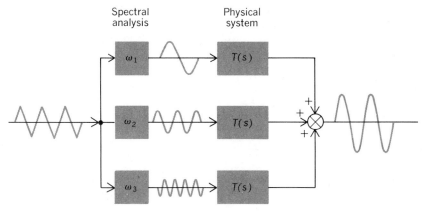

FIG. P8.9

separated into its sinusoidal components; each component is operated upon by the system transfer function; the sinusoidal responses of all the components are added to produce the final response.

(a) Write expressions for the first three sinusoidal components of the temperature waveform shown in Fig. 6.25.

(b) If the transfer function of the thermometer is $5/(s + 5)$, what would be the response for each of the first three components?

(c) If it is assumed the first three components of the triangular waveform are adequate to approximate the true input, what will be the thermometer output? Compare this result with the steady-state response curve shown in Fig. 6.26.

REFERENCES

1. Von Bekesy, Georg: "Experiments in Hearing," p. 61, McGraw-Hill Book Company, New York, 1960. (Translated and edited by E. G. Wever.)

2. Hirsch, I. J.: "The Measurement of Hearing," chap. 2, McGraw-Hill Book Company, New York, 1952.

3. Tasaki, I., H. Davis, and J. P. Legouix: The Space-Time Pattern of the Cochlear Microphonics (Guinea Pig) as Recorded by Differential Electrodes, *J. Acoust. Soc. Am.*, **24**:502.

4. Lynch, W. A., and J. G. Truxal: "Introductory System Analysis," McGraw-Hill Book Company, New York, 1961.

5. Hugh-Jones, P., and J. B. West: *Thorax*, **15**:154 (1960).

6. Crochetiere, W. J., L. Vodovnik, and J. B. Reswick: Electrical Stimulation of Skeletal Muscle: A Study of Muscle as an Actuator, *Med. Biol. Eng.*, **5**: (1967).

7. Ruch, T. C., and H. D. Patton: "Physiology and Biophysics," W. B. Saunders Company, Philadelphia, 1965.

8. Stark, L.: Stability Oscillations and Noise in the Human Pupil Servomechanism, *Proc. IRE*, **47**(11): 1925–1939 (November, 1959).

9. Baker, Frank H.: Pupil Response to Double Pulse Stimulation; a Study of Nonlinearity in Human Pupil System, *J. Opt. Soc. Am.*, **53**(12):1430–1436 (December, 1963).

10. Aseltine, J. A.: "Transform Methods in Linear System Analysis," McGraw-Hill Book Company, New York, 1958.

CHAPTER NINE

TRANSIENT OSCILLATIONS AND RESONANCE

9.1 INTRODUCTORY COMMENTS

Two common modes of system behavior have not yet been encountered in our examination of system responses. They might be described as the oscillatory transient evoked by an aperiodic excitation and the resonant response evoked by a periodic excitation. The first is exemplified by the familiar vibration when a system is given a nonrepetitive stimulus. The bounce sometimes experienced when gratefully falling into bed is one example of a decaying oscillatory response induced by an aperiodic excitation. The response of an instrument to a sharp excitation (such as a tap or other type of abrupt stimulation) is another example of this kind of reaction; the reflex reaction of the leg when a blow is applied to the patellar tendon (the common knee-jerk response) is still another illustration of an oscillatory transient.

The resonant response is the familiar behavior pattern exhibited by systems that are excited by periodic signals. These responses are characterized by increased vibratory amplitudes as the excitation frequency

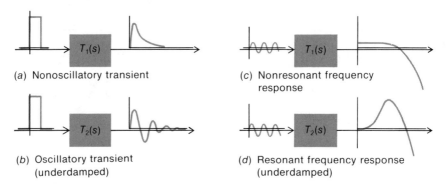

(a) Nonoscillatory transient

(b) Oscillatory transient
(underdamped)

(c) Nonresonant frequency
response

(d) Resonant frequency response
(underdamped)

FIG. 9.1 *Overdamped and underdamped responses of systems.*

passes through some critical range. The vibration noted in a car as it passes through the velocity range of about 14 to 16 mi/h (the resonant range of the author's car) is an illustration of resonant behavior. The bounce experienced by commuters in a railroad car running at some fixed speed is another illustration of the resonance phenomenon. Above and below the critical speed the vibration effect is reduced considerably; the critical speed or the critical frequency at which resonance is noted is described appropriately as the resonant point.

To compare graphically the kinds of systems to be discussed in this chapter with those discussed previously, we refer to Fig. 9.1. Figure 9.1a shows the kind of response we have thus far obtained from systems stimulated with an aperiodic excitation. The output is unidirectional in the sense that it approaches the final operating state from one direction with no overshoot; the slope of the response curve changes sign only if the excitation curve changes sign. In this chapter we shall examine systems that exhibit the type of behavior shown by Fig. 9.1b. The transient-response curve is not unidirectional but exhibits oscillatory characteristics. It overshoots its final operating state and then may possibly undershoot; in general, it exhibits a decaying oscillatory response. In a comparable manner, Fig. 9.1c typifies the kind of frequency-response curve thus far obtained from systems stimulated with a periodic signal.* The curves decay steadily with stimulation frequency; the slope of the curve never varies in sign (such curves are often described as monotonic). This implies the gain decreases (output amplitude decreases for a fixed input amplitude) as excitation frequency is increased. The systems to be exam-

---

* There are exceptions; see Fig. 8.25b, for example. This situation was due to terms in the numerator of the transfer function. The difference between this system and those to be examined here will be evident as the discussion progresses.

ined in this chapter will have frequency-response curves typified by Fig. 9.1*d*. For some range of excitation frequencies the gain curve rises to a peak and then falls. The rise and fall may occur over a very narrow range (or band) of frequencies or they may be spread over a wide band. In any case, the gain curve is generally not monotonic. Physically this implies the gain of the system will increase for a range of frequencies rather than decrease, as might ordinarily be expected; in short, the system has a resonant point.

Although it may not be evident, the resonance phenomena just described and the transient oscillatory response previously described are manifestations of the same system characteristics. Thus, if we find that a system can exhibit resonant behavior when excited with periodic stimuli, we would also expect this same system to exhibit a transient oscillation when aperiodically excited. Systems which show such behavior patterns are generally described as underdamped systems. In this chapter we shall see that these systems are at least second-order, they have resistive properties below a critical maximum level, and they contain more than one type of storage element (capacitive and inductive or compliance and mass, etc.). In the next chapter we shall see that this same behavior may indicate the existence of some feedback mechanism.

9.2 TRANSIENT RESPONSE OF AN UNDERDAMPED SECOND–ORDER SYSTEM

We could develop the concepts associated with underdamped systems by examining illustrative physical examples first and, from these, develop the mathematical and other model representations; alternatively, we can use the mathematical approach directly and then show application to physical systems. For persons not familiar with systems viewpoints and mathematical manipulation, the former procedure is probably more effective pedagogically. However, the latter procedure is more efficient and, with the level of competence which has (hopefully) now been established, we shall use the academic approach to introduce this new material.

The first comment to be made about systems exhibiting transient oscillations or resonance is they must be more than first-order systems. The first-order system has a single time-constant and must respond exponentially to an aperiodic excitation; such systems cannot oscillate. The systems to be discussed here must therefore be second- or higher-order systems.

We have examined a number of second-order systems in previous discussions but have not yet encountered the type of response indicated in Fig. 9.1*b* or *d*. On this basis one might discount the possibility of underdamped (i.e., oscillatory or resonant) responses for second-order systems.

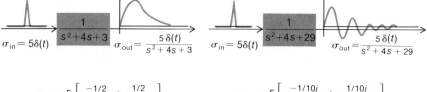

$$\sigma_{\text{in}} = 5\delta(t) \qquad \frac{1}{s^2+4s+3} \qquad \sigma_{\text{out}} = \frac{5\,\delta(t)}{s^2+4s+3}$$

$$\sigma_{\text{in}} = 5\delta(t) \qquad \frac{1}{s^2+4s+29} \qquad \sigma_{\text{out}} = \frac{5\,\delta(t)}{s^2+4s+29}$$

$$\sigma_{\text{out}} = 5\left[\frac{-1/2}{s+3} + \frac{1/2}{s+1}\right]$$

$$\sigma_{\text{out}} = 5\left[\frac{-1/10j}{s+2+5j} + \frac{1/10j}{s+2-5j}\right]$$

$$\sigma_{\text{out}} = \frac{5}{2}\left[\epsilon^{-t} - \epsilon^{-3t}\right]$$

$$\sigma_{\text{out}} = \epsilon^{-2t}\sin 5t \text{ (see text)}$$

(a) Overdamped second-order system   (b) Underdamped second-order system

FIG. 9.2 *Damped and underdamped second-order systems.*

However, if we reexamine the second-order systems discussed, we find them to be rather specialized; in all cases the second-order terms were factorable into real first-order terms. In particular, the second-order systems thus far discussed have been of the form shown in Fig. 9.2a. The denominators of such systems were factorable so that the response could be expressed as the equivalent of two first-order systems operating together. We have not yet examined a second-order system in which the denominator factors into two complex terms, as shown in Fig. 9.2b. From our experience with complex transforms, we should expect these factors to give rise to some form of sinusoidal expression. This would lead us intuitively to expect that underdamped systems may be represented by functions that are (or include) quadratic expressions with complex roots. This intuitive viewpoint proves to be correct, and in the following discussion we shall examine this aspect of second-order systems in a more quantitative fashion.

To establish analytically how underdamped systems and second-order representations are related we examine first the impulse response of the specific second-order system shown in Fig. 9.2b. The procedure to determine the response is handled in the standard manner; we first write the response as the product of the transfer function and the system excitation. For the situation shown in Fig. 9.2b this is

$$(9.1) \qquad \sigma_{\text{out}} = \frac{5}{s^2 + 4s + 29}\,\delta(t)$$

If we had no further interest in examining the significance of the various terms and were interested only in the final response, we could use transform tables to evaluate the output function. However, we should like ultimately to relate the various constants in the response and the transfer function and shall therefore follow the formal procedure of partial-fraction expansion and conversion to the time domain. From the quadratic formula, we find the factors of the denominator of the transfer function to be $s + 2 + 5j$ and $s + 2 - 5j$. Equation (9.1) is thus written

$$(9.2) \qquad \sigma_{\text{out}} = \frac{5}{(s + 2 + 5j)(s + 2 - 5j)} \, \delta(t)$$

If we expand into partial fractions, we then have

$$(9.3) \qquad \sigma_{\text{out}} = 5 \left(\frac{-1/10j}{s + 2 + 5j} + \frac{+1/10j}{s + 2 - 5j} \right) \delta(t)$$

Each of the fractional terms is of the form $A/(s + a)$, which transforms to ϵ^{-at} in the time domain. Equation (9.3) in the time domain therefore becomes

$$(9.4) \qquad \sigma_{\text{out}} = 5 \left(\frac{1}{10j} \, \epsilon^{-(2-5j)t} - \frac{1}{10j} \, \epsilon^{-(2+5j)t} \right)$$

Each of the exponential terms can be expressed as product functions:

$$(9.5) \qquad \sigma_{\text{out}} = 5 \left[\frac{\epsilon^{-2t}\epsilon^{5jt}}{5(2j)} - \frac{\epsilon^{-2t}\epsilon^{-5jt}}{5(2j)} \right] = 5 \, \frac{\epsilon^{-2t}}{5} \left(\frac{\epsilon^{5jt} - \epsilon^{-5jt}}{2j} \right)$$

The final term is the complex representation for the sine function with frequency 5 rad/s. Equation (9.5) can therefore be written

$$(9.6) \qquad \sigma_{\text{out}} = \epsilon^{-2t} \sin 5t$$

In this form we can readily picture the nature of the response. The sine function indicates the response will be oscillatory. The coefficient of the sine term is, however, an exponential function that decays with time. The amplitude of the sinusoidal oscillation thus continually decreases with time; the transient response to an aperiodic excitation must therefore ultimately be zero. A plot of the response as given by Eq. (9.6) is shown in Fig. 9.3. This simple analysis in a sense proves that second-order systems with complex roots always give rise to transient oscillations. If there are complex roots in the transfer function they give rise to exponential terms with complex time-constants. Such functions must give rise to sinusoidal terms in the world of real time. On the basis of these introductory comments we may confidently state: If a system shows a transient oscillation, the transfer function of the system can be expected to contain a quadratic factor with complex roots.

We shall introduce here some of the terminology associated with underdamped second-order systems. From the solution for the impulse

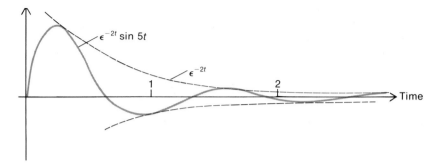

FIG. 9.3 *Transient oscillation of waveform given by $\epsilon^{-2t} \sin 5t$.*

response of this particular system [Eq. (9.6) and Fig. 9.3] the oscillatory frequency of the transient is 5 rad/s. This frequency is generally called the damped natural frequency of the system. The reason for this terminology will become somewhat clearer when we examine the second-order system in more general terms. We can, however, even now guess the implication of this terminology. Because of its construction or properties, the system has some natural tendency to oscillate when excited. Frictional (or resistive) effects diminish this tendency by reducing the oscillatory frequency and by causing the oscillation to decay. The decay of oscillation is governed by the time-constant of the exponential amplitude term ϵ^{-2t}. If the coefficient of t were greater than 2, the oscillation would decay more rapidly and the system would be considered more heavily damped. The time-constant of the exponential term thus contains information about the system damping (or damping factor, as it is called); the oscillatory frequency and amplitude decay thus contain the information about the system properties. In an identification problem the response curve may be used to evaluate the frequency and decay terms, and the system properties may be defined from them. It will become clearer later how these terms may be used to evaluate system properties.

9.3 GENERAL DESCRIPTION OF NATURAL FREQUENCY AND DAMPING

To obtain a better picture of how the constants of the system transfer function are related to the response waveshape, we examine the impulse response of a general second-order system as shown in Fig. 9.4. The procedure to be followed is identical to that employed in Sec. 9.2 except we now use literal values instead of numbers. The output response to an impulse of a general second-order system is given by

$$(9.7) \qquad \sigma_{\text{out}} = \frac{A}{s^2 + as + b} \, \delta(t)$$

The factors of the denominator are obtained from the quadratic formula $[-a/2 \pm \sqrt{(a/2)^2 - b}]$; Eq. (9.7) can thus be written

$$(9.7a) \quad \sigma_{\text{out}} = \frac{A \, \delta(t)}{[s + (a/2) + \sqrt{(a/2)^2 - b}][s + (a/2) - \sqrt{(a/2)^2 - b}]}$$

Before proceeding further, we note first the radical that appears in both factors. If the constants a and b are such that $(a/2)^2$ is greater than b, the expression under the radical is positive; the quadratic factors in this case are real. This would be the situation for the system shown in Fig. 9.2a. Here a is 4 and b is 3; therefore $(a/2)^2 - b$ is 1. The factors in Eq. (9.7a) are thus $[s + (\frac{4}{2}) + 1][s + (\frac{4}{2}) - 1]$ or $(s + 3)(s + 1)$. Such systems are considered overdamped; there is no oscillatory transient. (See the top curve in Fig. 9.4.)

If a and b are related so that $a = 2\sqrt{b}$ [that is, $(a/2)^2 = b$] the second-order denominator becomes a perfect square and there are two equal factors $(s + a/2)$. Equation (9.7a) then becomes

$$(9.8) \qquad \sigma_{\text{out}} = \frac{A \, \delta(t)}{(s + a/2)^2}$$

This is a new form of transfer function. It is already in its simplest form, and so it cannot be expanded into partial fractions; yet it is not a recognizable function. To convert this expression to the time domain we could take advantage of published tables or we can perform some mathe-

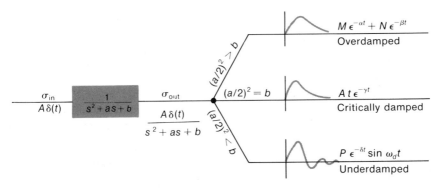

FIG. 9.4 *Impulse responses of a general second-order system.*

matical gymnastics.* Whatever technique is used the result will be

$$(9.9) \qquad \sigma_{out} = At\epsilon^{-at/2}$$

Evidently the response is still not oscillatory; it is still "too heavily damped." (See the middle curve in Fig. 9.4.)

The condition $(a/2)^2 = b$ is known technically as the critically damped condition for a second-order system. If the constant a is reduced any further (or b is increased) the system becomes frequency-sensitive and has a natural tendency to oscillate. Specifically, if $(a/2)^2 < b$, the system becomes underdamped and will exhibit oscillatory behavior during transient periods. The latter comments follow directly from the discussion in Sec. 9.2. When $(a/2)^2 < b$, the factors of the second-order system are complex; such factors give rise to sinusoidal representations in the time domain. Thus when $(a/2)^2 < b$ the system will have a transient oscillation. To show this more precisely we rewrite the radical in Eq. (9.7a) so as to include a j coefficient:

$$(9.10) \qquad \sqrt{\left(\frac{a}{2}\right)^2 - b} = j\sqrt{b - \left(\frac{a}{2}\right)^2}$$

where $b > (a/2)^2$. Equation (9.7a) may now be written

$$(9.11) \quad \sigma_{out} = \frac{A\,\delta(t)}{[s + a/2 + j\sqrt{b - (a/2)^2}][s + a/2 - j\sqrt{b - (a/2)^2}]}$$

Equation (9.11) is the same as Eq. (9.2) except the former is written as a general expression with literal terms instead of numbers. We can handle Eq. (9.11) in the same manner as Eq. (9.2) to determine the final expression in the time domain. If we simply compare the terms of Eqs. (9.2) and (9.11), however, we can use the solution for Eq. (9.2) [see Eq. (9.6)] to state the general solution for Eq. (9.11):

$$(9.12) \qquad \sigma_{out} = \frac{A}{\sqrt{b - (a/2)^2}}\, \epsilon^{-at/2} \sin\sqrt{b - \left(\frac{a}{2}\right)^2}\, t$$

[If A is 5, a is 4 and b is 29. Equation (9.12) reduces to Eq. (9.6).] The damped natural frequency is $\sqrt{b - (a/2)^2}$, and the time-constant of the amplitude-decay term is $a/2$.

The general form of Eq. (9.12) permits us to understand the terminology associated with the various terms a little more clearly and to

---

* One possibility is to expand $(s + a/2)^{-2}$ into a series and let each term operate on $\delta(t)$, as was done when the operator concept was first introduced. An alternative procedure is to assume $1/(s + a/2)^2$ can be represented by $1/[(s + a/2 + \Delta)(s + a/2 - \Delta)]$, where Δ is very small. A partial-fraction expansion is now possible, and when the result is determined the limit can be found as Δ becomes neglectable.

relate the terms of the transfer function with the time-domain solution. We note, for example, as a decreases, the oscillatory frequency increases and the amplitude-decay term has a longer time-constant; in the limit (if a can be reduced to zero) Eq. (9.12) would then become

$$(9.13) \qquad \sigma_{\text{out}} = \frac{A}{b} \sin \sqrt{b}\, t$$

That is, if a is zero (or, practically speaking, very small) the system will respond with a sustained oscillation of frequency given by $\sqrt{b}$ and with very little amplitude attenuation. This frequency is the highest natural (or transient) oscillation the system can have and is often described as the natural frequency of the system. To emphasize this point, the $\sqrt{b}$ is often written ω_n; b is therefore ω_n^2, and one now can easily see that the system natural frequency may be obtained from the constant term of the quadratic factor in a system transfer function (e.g., the natural frequency of the system shown in Fig. 9.2b is $\sqrt{29}$).

From the foregoing discussion it is apparent the term a is related to the damping within the system. As a increases, the tendency to oscillate decreases; the frequency of oscillation decreases and the amplitude decays rapidly. The critical value for a is given by the relation

$$a_{\text{crit}} = 2\sqrt{b} = 2\omega_n$$

At this value for a the system is just critically damped. Thus, if the coefficient of the s term in the system of Fig. 9.2b were $2\sqrt{29} = 10.7$ instead of 4, the system would be critically damped. The value of 4 rather than 10.7 indicates the system is $4/10.7 = 37$ percent of critical damping. This is often expressed by the statement that the system has a damping factor of $\zeta = 0.37$ (where the Greek letter ζ is generally employed to indicate the damping factor). To put this into more general form: If a is the actual s coefficient in a denominator quadratic factor of a transfer function and a_{crit} is the value for a at which the system becomes critically damped, then the system damping factor is defined by

$$(9.14) \qquad \text{Damping factor} = \zeta = \frac{a}{a_{\text{crit}}}$$

Since a_{crit} is defined by $2\omega_n$, the damping factor may be written

$$(9.15) \qquad \zeta = \frac{a}{2\omega_n}$$

In this form it is easy to see how the s coefficient a is related to the system damping. Because of the connection between the s coefficient and damping, this coefficient is often written $2\zeta\omega_n$ rather than a.

If we use the more descriptive terms of damping factor and natural frequency for the formulation of a second-order transfer function, the a and b terms in Fig. 9.4 and Eq. (9.7) are replaced by $2\zeta\omega_n$ and ω_n^2, respectively. The response function [Eq. (9.7)] is then written

$$(9.16) \qquad \sigma_{\text{out}} = \frac{A}{s^2 + (2\zeta\omega_n)s + \omega_n^2}$$

The time-domain solution [Eq. (9.12)] then becomes

$$(9.17) \qquad \sigma_{\text{out}} = \frac{A}{\sqrt{\omega_n^2 - (\zeta\omega_n)^2}} \epsilon^{-\zeta\omega_n t} \sin \sqrt{\omega_n^2 - (\zeta\omega_n)^2}\, t$$

or

$$\sigma_{\text{out}} = \frac{A}{\omega_n \sqrt{1 - \zeta^2}} \epsilon^{-\zeta\omega_n t} \sin \omega_n \sqrt{1 - \zeta^2}\, t$$

These results are valid only if the system is underdamped or $\zeta < 1$. This last expression is sometimes further simplified to read

$$(9.18) \qquad \sigma_{\text{out}} = \frac{A}{\omega_d} \epsilon^{-\zeta\omega_n t} \sin \omega_d t$$

where ω_d is defined as the damped natural frequency. Obviously, the damped natural frequency is defined by $\omega_n \sqrt{1 - \zeta^2}$. From the relationship we can easily see the damped natural frequency or the oscillatory frequency of the system will be closely equal to the natural frequency when the damping factor is low. (For a damping factor of about 0.3 the damped natural frequency ω_d is approximately 95 percent of the natural frequency.)

As some very simple illustrations of the utility of the notation developed in the preceding paragraph, if we are given a system with a transfer function $1/(s^2 + 12s + 16)$, we can state immediately:

$$(9.19a) \qquad \omega_n = \sqrt{16} = 4$$

$$(9.19b) \qquad 2\zeta\omega_n = 12 \qquad \text{or} \qquad = \frac{12}{2\omega_n} = \frac{12}{8} = 1.5$$

This system is overdamped ($\zeta > 1$), it will not oscillate, and its impulse-response will be the sum of two decaying exponentials. If the transfer function is given by $1/(s^2 + 8s + 16)$

$$(9.20a) \qquad \omega_n = \sqrt{16} = 4$$

$$(9.20b) \qquad 2\zeta\omega_n = 8 \qquad \text{or} \qquad = \frac{8}{2\omega_n} = \frac{8}{8} = 1$$

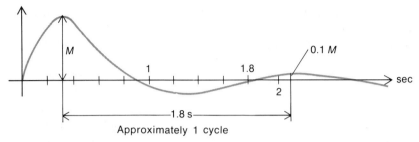

FIG. 9.5 *Estimated response of system with transfer function $1/(s^2 + 4s + 6)$.*

This system is critically damped ($\zeta = 1$); it will not oscillate but will respond to an impulse according to the function $t\epsilon^{-4t}$. If the transfer function is given by $1/(s^2 + 4s + 16)$,

(9.21a) $\omega_n = 4$

(9.21b) $2\zeta\omega_n = 4$ or $= \dfrac{4}{2\omega_n} = \dfrac{4}{8} = 0.5$

This system is underdamped with a damping factor 0.5; it will oscillate at its damped natural frequency of

(9.21c) $\omega_d = \omega_n \sqrt{1 - \zeta^2} = 4\sqrt{1 - (0.5)^2} = 3.5$ rad/s

The amplitude-decay coefficient ($\epsilon^{-\zeta\omega_n t}$) has a time-constant of $\frac{1}{2}$ s. Since the oscillatory frequency is 3.5 rad/s or about 0.56 c/s it will take approximately 1.8 s for the system to execute one cycle of oscillation. This is about four time-constants of amplitude decay; in one cycle, therefore, the amplitude will undergo approximately a 90 percent decay. From this information we can roughly sketch the impulse-response of the system as shown in Fig. 9.5.

9.4 PHYSICAL SIGNIFICANCE OF UNDERDAMPED RESPONSES

Although the mathematical analysis presented in the previous sections indicates how the constants of the transfer function are related to the shape of the transient-response curve, it does not provide insight into the components of underdamped systems. To realize that the s coefficient of the quadratic factor relates to the damping factor of the system, which relates to the amplitude decay of the curve, is helpful analytically; this information alone, however, does not help to understand how these terms (and the transient response) are affected by the system properties.

To gain insight into implications of underdamped responses we examine some systems that are known to give rise to such responses. A fertile source of such systems is in the field of instrumentation.* In some cases, instruments are designed to be highly responsive and as a result become particularly susceptible to transient oscillations; in other cases, an inherent transient oscillation may be undesirable, and pains are taken to eliminate this effect. In any case, the second-order underdamped characteristics of most instruments are relatively easy to explain and to relate to the physical properties of the system. We use these to illustrate the relationship between system properties and underdamped responses. In the next section we shall apply these ideas to a physiological system.

Figure 9.6 shows schematic diagrams of some very elementary instruments. These are common devices and should be familiar. The U-tube manometer is simply a glass tube filled with a fluid; pressure applied to one leg causes the fluid to be displaced, and the difference between the liquid levels in the two legs measures the pressure. The compass is a balanced magnet needle that seeks the earth's magnetic north. The pan scale is simply a pan suspended from a spring. The deflection of the spring is measured to determine the weight of material in the pan.

Although these instruments are not very sensitive or accurate (a fair degree of accuracy may, however, be obtained with a manometer),

---

* Most instrumentation texts include a wealth of information about physical systems analysis, static and dynamic responses, and related material. Some good texts in this area, which deal with both theory and application, are Refs. 1 to 3.

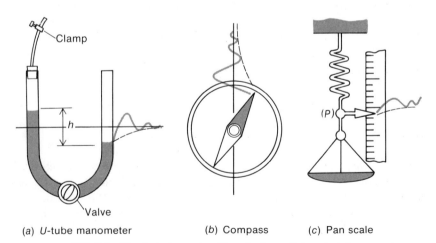

(a) U-tube manometer   (b) Compass   (c) Pan scale

FIG. 9.6 *Simple instruments with underdamped characteristics.*

they can be used to illustrate qualitatively some of the features common to underdamped systems. We note first that a transient oscillation could be expected from all these devices. The sudden application of a signal, a tap, or a displacement from equilibrium with a sudden subsequent release could result in an oscillation as shown by the solid curves in each of the diagrams in Fig. 9.6. Of the three, the oscillation of the pan scale is the most easily understood. If the pan scale is pulled down and released, the stretched spring tends to contract and delivers an unbalanced force to the pan. The pan accelerates upward and may attain a velocity high enough to drive the spring into compression. The compressed spring drives the mass down and the cycle repeats. Ultimately, the motion dies down and the pan comes to rest at some equilibrium point. As the pan scale has all the features of an underdamped system and as the properties are practically obvious by inspection, we shall use this system to relate the transient characteristics to the system properties.

We begin by noting the spring is one of the important elements that govern the frequency of oscillation. We can see intuitively a more rigid or a stiffer spring (i.e., higher spring-constant) would cause the oscillatory frequency to increase; we would expect no oscillation if there were no spring. Therefore, it seems reasonable to expect that the spring determines, to some extent, the natural tendency to oscillate or the natural frequency of the system. From the reverse point of view, we see the natural frequency of the system [ω_n or b; see the discussion relating to Eq. (9.16)] reflects some aspect of the springiness of the system. As springiness is recognized as a system-storage property (it stores potential energy; see also the discussion in Sec. 4.10), we see the natural frequency reflects a storage property of the system.

Equally important (though possibly not quite as obvious) is the effect of the mass on the transient oscillatory frequency. If the pan is a heavy one (or laden with heavy material) a low persistent oscillation would be expected; with a light mass the oscillation would probably be rapid although it might be short-lived. It appears, then, that the natural frequency also reflects the "massiness" of the system. As massiness is also a system-storage property,* we see the natural frequency reflects a combination of two different kinds of storage properties within the same system.

The viewpoints presented in the last two paragraphs may be combined into a single observation: A system can be expected to have two

---

* Mass stores kinetic energy. If the acceleration produced by forces applied to a body is measured and the values obtained are plotted, the slope of the resultant curve has units of mass only. The ratio of force to acceleration thus has time units to an even power (T°), and it defines a storage property.

(or more) dissimilar storage elements if it exhibits an underdamped response. This is not as positive a conclusion as one would like; it does not state there must be two dissimilar storage elements—only that this may be the situation. The reason for this nondefinitive conclusion is that underdamped responses can also be the result of a feedback configuration within the system under consideration. This aspect of second-order underdamped responses will be more clearly understood when feedback systems are examined in the next chapter. On the basis of the latter comments we note, however, if a system is composed of only one type of storage element, it cannot have an underdamped response unless feedback is present. This comment has not been justified analytically, and it is stated here without proof. It does not seem unreasonable, however, since a thermal system, for example, has only heat capacity as a storage element, it will not oscillate unless some feedback condition exists (such as a thermostat, by which the temperature obtained is fed back to control the heat delivered). A similar observation may be made about diffusion systems which have only volume-storage elements and about electrical networks which are composed only of R's and C's (i.e., inductance is too small to be considered). On the other hand, electrical systems that include both storage elements (inductance and capacitance) can have underdamped characteristics.

With these introductory ideas established, we might well ask: Where are the two storage elements in the devices shown in Fig. 9.6a and b? One of the elements in each of the devices is fairly obvious. In the manometer the fluid mass constitutes one storage element; in the compass, the inertia of the pointer represents a storage element. Mass (or some aspect of mass) almost always appears as a storage element in mechanical systems.

Although the other storage element in each of the systems in not self-evident, it is apparent these elements are related to position. None of the systems will oscillate unless displaced from a positional equilibrium point. This hint is sufficient to prompt us to examine the displacement aspects of the system more carefully. In the manometer system the displacement is caused by pressure applied to one leg of the U tube; the more the pressure, the greater is the displacement. A check on the ratio of pressure to displacement (i.e., the slope of the pressure-displacement characteristics) indicates the units include time to an even power. The pressure and displacement measurements therefore define a storage property of the system. Further study would reveal that this property may be attributed to the density (not weight or mass) of the fluid in the tube. By employing a similar procedure to the compass system we find the displacement of the compass needle is dependent upon the torque applied to the needle. The torque-displacement characteristics define a storage property of the system. Further checking will show this property is related to the magnetic strength in the pivoted needle.

Although the spring and the mass of the pan-scale system are the fundamental elements that govern the frequency of the system, they do not entirely determine the oscillatory behavior of the system. This becomes patently clear when we realize we can change the response frequency (and can, in fact, eliminate oscillation) even though the spring and mass remain the same. One simple way to produce such a change would be to immerse the whole assembly in a pail of water. The tendency to oscillate would be considerably reduced (and probably eliminated; this would be true even if the pan were constructed as a sieve so that water could flow freely through the pan). We can produce a similar effect in a more realistic fashion by simply bending the pointer so that it rubs against the scale. In either case we have increased the friction of the system and have thereby decreased (or eliminated) its oscillatory frequency and have caused the oscillation to decay more rapidly. In short, by increasing the friction (or mechanical resistive property of the system) we have increased the damping in the system. As the amplitude decay and the oscillatory frequency depend on both the natural frequency (ω_n) and the damping factor ζ [see Eq. (9.17)] and as the natural frequency was not altered in this case, the damping factor ζ is related directly to the resistive property of the system.

We can also use the other instruments in Fig. 9.6 to show the qualitative relationship between resistive properties and damping. By increasing the existing resistive property (or introducing other resistive properties) the system damping factor can be increased in the same way the pan-scale factor was increased. To reduce the transient oscillation of compasses, the needle mechanism is often floated in a bath of kerosene or light oil; most ship and aircraft magnetic compasses are constructed in this fashion. To damp the oscillatory behavior of a manometer, a valve (or clamp, if flexible hose is used) is included between the legs of the manometer, or a restrictive orifice may be included at the outlet of the legs (see Fig. 9.6a). In the first case, liquid viscous friction is used to increase damping, and in the second case air friction causes the damping factor to increase.

It should be pointed out that although viscous friction changes the dynamic character of a system it does not affect the static operation. In the manometer system, for example, the added air or liquid damping may cause the system to exhibit an overdamped response rather than underdamped; the final magnitude of the response, however, remains unchanged. Thus the application of a step pressure will cause the same

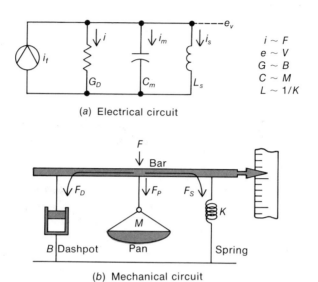

(a) Electrical circuit

(b) Mechanical circuit

FIG. 9.7   *Analogous systems with two dissimilar storage elements.*

liquid-level displacement in the manometer legs with low or high damping. The static accuracy of the instrument is thus not affected by the addition of viscous damping. The dynamic accuracy may, however, be critically affected as increased friction makes the instrument more sluggish and slower responding. The effect would be the same as an increased time-constant in the thermometer problem of Sec. 6.10.

To show more quantitatively how system properties are related to underdamped responses, we examine a simple electrical system which includes a resistive property and two dissimilar storage properties. Figure 9.7a shows a parallel RLC circuit driven by a current source. This circuit is easily seen to be an analog of the pan-scale system as redrawn in Fig. 9.7b. A comparison between Figs. 9.7b and 9.6c shows these diagrams are mechanically equivalent: The bar in Fig. 9.7b couples the spring and pan together rigidly, as does the pin P in Fig. 9.6c; motion of the spring is prevented at one end by fastening it to a fixed structure (ground or ceiling). The dashpot in Fig. 9.7b represents schematically the viscous friction in the system.* (A brief review of Sec. 4.12 is in order here.) The

---

* The dashpot in Fig. 9.7b is a linear lumped representation of the friction in the system. This includes the pointer rubbing against the scale, air friction opposing the motion of the pan, molecular friction in the spring, etc. Inclusion of the dashpot does not imply the system is truly linear, as none of the aforementioned effects is always linear and these effects may not always act parallel with the spring as shown. The dashpot is included in the manner shown only to provide some estimate or "feel" for the effect of friction on the system.

analogy between the electrical and mechanical systems is straightforward: If we use the force-current and velocity $(\dot{x})$–voltage analog the resistor represents the dashpot, the inductor represents the spring $(L = 1/K$; see Sec. 4.11), and the capacitor represents the mass.* The current delivered to the parallel configuration of electrical elements is distributed in the same fashion as the force is applied to the mechanical elements.

To obtain an expression that relates the resultant voltage (velocity) to an applied current (force) we must determine the system transfer function when e_v is the output and i_f is the input. Since the required transfer function (e_v/i_f) is simply the impedance of the parallel circuit, we can write immediately

$$(9.22) \qquad \frac{e_v}{i_f} = T(s) = Z = \frac{1}{Y}$$

or

$$\frac{e_v}{i_f} = \frac{1}{G_D + sC_m + 1/sL_s} = \frac{sL_s}{s^2 C_m L_s + sL_s G_D + 1}$$

or

$$\frac{e_v}{i_f} = \frac{s/C_m}{s^2 + (G_D/C_m)s + 1/L_s C_m}$$

From the denominator of the last form of Eq. (9.22), we can see the system may have underdamped characteristics; the system is second-order, and the denominator of the transfer function may contain complex factors. A comparison of this denominator with that of Eqs. (9.7) and (9.16) shows that the natural frequency (ω_n) of the system can be related to $1/\sqrt{L_s C_m}$ (or, in mechanical elements, $\sqrt{K/M}$; see Fig. 9.7); this is in agreement with the intuitive conclusion that the transient oscillation of a system is governed by the storage elements in the system. The damping in the system is related to G_D/C_m (or $2\zeta\omega_n = G_D/C_m$); again, as anticipated, the damping in the system is related to the friction [the greater the friction, the greater the mechanical coefficient B (see Sec. 4.12) and the greater the electrical conductance G] in the system.

9.5 THE UNDERDAMPED RESPONSE OF A PHYSIOLOGICAL SYSTEM

As an illustration of how some of the ideas presented in the previous section may be used to gain insight into and understanding of a physiological system we shall examine an elementary model of an aortic length

---

* A force applied to a mass causes the mass to accelerate according to the relationship $f = m\, dv/dt$. If $f \sim i$ and $v \sim e$, the analogous representation would be $i = m(de/dt)$. m is apparently a capacitor of magnitude m.

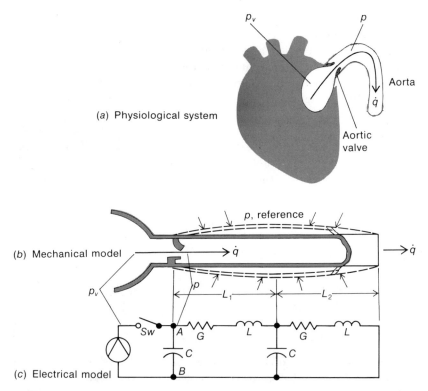

(a) Physiological system

(b) Mechanical model

(c) Electrical model

FIG. 9.8 *System and model representations of post-systolic operation in aortic arch.*

proximal to the aortic valve of the heart. We shall seek to understand the pressure response that may be expected in the aorta when the flow through the valve is suddenly reduced to zero. Physiologically this is the transient effect that is induced when the aortic valve closes just at the end of systole and is the effect that causes the familiar dicrotic notch to appear in the recordings of aortic pressure pulses.

A diagrammatic representation of the system to be examined is shown in Fig. 9.8a. A mechanical model of the system is proposed in Fig. 9.8b. The aortic valve is represented by a simple flapper check valve, and the aorta is represented by a length of flexible hose; a corresponding electrical model is shown in Fig. 9.8c. In the latter model it is assumed the hose length may be lumped into two representative lengths L_1 and L_2 with identical properties. As this analysis is only to gain some insight into the parameters that govern the system operation, we make some additional oversimplifications and propose that the valve is not pressure-actuated but programmed to close as we command. In reality the aortic valve closes when ventricular pressure falls below aortic pressure. Since we are concerned here not with why the valve closes (see the supplementary discussion at the end of Chap. 2) but only with what happens to the aortic pressure after the valve closes, we propose that valve closure is an independently controlled function.

The model operation as we see it, therefore, is: Flow is delivered through a valve to a flexible hose. At some time ($t = 0$) the valve is closed (in the electrical system the switch sw is opened) so that the delivered flow is reduced to zero. The question to be resolved is: How does the pressure directly behind the valve respond? A diagrammatic representation of the problem as posed is shown in Fig. 9.9. The graphical description in Fig. 9.9a indicates the system is excited by a step flow which reduces flow from a quiescent level to zero; the pressure response is to be evaluated. The same information is shown in block form in Fig. 9.9b. To obtain a quantitative picture of the pressure response we can use the electrical analog shown in Fig. 9.8c.

Before doing so, however, we shall justify the model and describe the significance of the elements. This model is essentially the same as

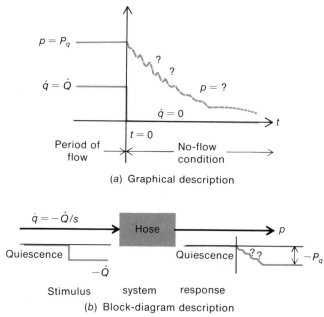

(a) Graphical description

(b) Block-diagram description

FIG. 9.9 *Graphical and block-diagram representation of post-systolic operation as described in text.*

that derived in Sec. 4.5. The capacitors represent the aortic compliance per length of section; the resistors represent the fluid friction per length of section encountered by the fluid flow through the tube length. (It is assumed L_1 and L_2 have identical properties.) The need for the inductor was implied in Sec. 4.5 but not justified. To justify the need for this element we propose a rigid frictionless pipe with flow as shown in Fig. 9.10. The pressure needed to accelerate the fluid can be defined by Newton's law $f = m\dot{v}$. The accelerating force in this case is the product of the pressure differential across the ends and the tube across the sectional area. Since the fluid velocity multiplied by the tube area is equivalent to the volumetric flow ($\dot{q}$), we can describe $f = m\dot{v}$ by $p_1 - p_2 = (m/A^2)(d\dot{q}/dt)$ (see Fig. 9.10). If we consider p and $\dot{q}$ analogous to e and i, respectively, we see m/A^2 may be represented by an inductor. The model for a rigid frictionless tube would be simply an inductor. If friction is included, the model becomes an RL series arrangement*; if the tube is flexible the capacitor is introduced to represent the compliance. The model representation we have evolved is only a first approximation to the true physiological system. Yet in spite of this we can use the model to learn how the various physiological elements affect the system operation.

The pressure response to a step-flow excitation can now be written almost by inspection. The transfer function desired is $p/\dot{q}$; this function is seen to be the circuit impedance behind the switch (valve). We can write therefore

$$(9.23) \qquad \frac{p}{\dot{q}} = Z_{AB}$$

* When steady-state flow is reached, fluid inertia is unimportant but fluid resistance still offers opposition to flow; when steady electrical flow is achieved through a series RL circuit, the opposition of the inductor is zero but the resistor still offers resistance to flow.

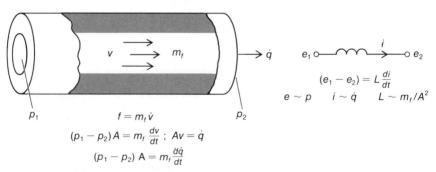

FIG. 9.10 *Relationship between fluid inertia and inductance.*

where Z_{AB} is the total impedance "looking into" points A-B in Fig. 9.8c. To evaluate Z_{AB} we work backward and first determine the impedance of the tube length L_2. This impedance is given by

$$(9.24) \qquad Z_{L_2} = \frac{1}{sC + 1/(R + sL)} = \frac{R + sL}{(R + sL)sC + 1}$$

This impedance is in series with the G-L section of L_1. The combined impedance of G, L, and Z_{L_2} is therefore

$$(9.25) \qquad Z_{G,L,L_2} = (R + sL) + \frac{R + sL}{(R + sL)sC + 1}$$
$$= \frac{(R + sL)^2 sC + 2(R + sL)}{(R + sL)sC + 1}$$

The compliance of the L_1 section must now be included. If we use the admittance form we have

$$(9.26) \qquad Y_{AB} = sC + \frac{(R + sL)sC + 1}{(R + sL)^2 sC + 2(R + sL)}$$

Therefore

$$(9.27) \qquad Z_{AB} = \frac{(R + sL)^2(sC)^2 + 2(R + sL)}{[(R + sL)sC]^2 + 3(R + sL)sC + 1}$$

To obtain the pressure response to a step-flow excitation $(-\dot{Q}/s)$ we use Eqs. (9.23) and (9.27) and write

$$(9.28) \qquad p = -\frac{\dot{Q}}{s} \frac{(R + sL)^2 sC + 2(R + sL)}{M^2 + 3M + 1} \delta(t)$$

where $M = sC(R + sL)$.

We can make one more simple modification: If we factor the denominator of Eq. (9.28) we can then write

$$(9.29) \qquad p = -\frac{\dot{Q}}{s} \frac{(R + sL)^2 sC + 2(R + sL)}{[M + (3 + \sqrt{5})/2][M + (3 - \sqrt{5})/2]} \delta(t)$$

or

$$p = -\frac{\dot{Q}}{s} \frac{(s + L/R)^2(s/G) + 2(s + 1/R)/(LC^2)}{(s^2 + sR/L + 2.62/LC)(s^2 + sR/L + 0.38/LC)} \delta(t)$$

(This last expression was obtained by replacing M with its equivalent expression in R, L, and C and then collecting terms.) It is possible further to modify this last expression by separating into partial fractions and attempting a general solution; for the purpose of understanding the results, however, this is not necessary as we can see what the form of the solution will be without actually obtaining it. If we had used a partial-fraction expansion we could have obtained an expression of the form

$$(9.30) \quad p = \frac{A\delta(t)}{s} + \frac{(B_1 s + D_1)\delta(t)}{s^2 + sR/L + 2.62/LC} + \frac{(B_2 s + D_2)\delta(t)}{s^2 + sR/L + 0.38/LC}$$

where the A, B's, and D's are constant terms. In this form we see the solution for p will consist of some constant or steady-state term* and two second-order terms. The latter terms give rise to transient effects, and it is these which concern us now, as we know the steady-state pressure must finally be zero. To determine what these transient responses might be, we examine the damping factor of each second-order term. From earlier discussions we see immediately the damping factor of the first second-order term may be written

$$(9.31) \quad \zeta_1 = \frac{R/L}{2\omega_{n1}}$$

The natural frequency (ω_{n1}) of this first term is

$$(9.32) \quad \omega_{n1} = \sqrt{\frac{2.62}{LC}} = \frac{1.62}{\sqrt{LC}}$$

The damping factor is therefore defined by

$$(9.33) \quad \zeta_1 = \frac{R/L}{2(1.62/\sqrt{LC})} = \frac{R\sqrt{C/L}}{3.24}$$

From this relationship we see the first second-order term will give rise to a transient oscillation only if $R\sqrt{C/L}$ is less than 3.24. If this condition is not satisfied, the denominator of the second term in Eq. (9.30) will be factorable into two first-order terms (which may be equal). The second transform of Eq. (9.30) would then give rise to a second-order decay with no oscillatory characteristics (as, for example, the sum of two decaying exponentials).

By a similar analysis we can show the third term in Eq. (9.30) will give rise to a transient oscillation only if $R\sqrt{C/L} < 1.23$.† For values of $R\sqrt{C/L}$ equal to or greater than 1.23 the third transform in Eq. (9.30) will also respond with critically or overdamped characteristics. We see, therefore, three different kinds of responses can be expected, depending

---

* This term must obviously be equal to the quiescent pressure P_q. A/s transforms into $Au(t)$ in the time domain. A must be equal to $-P_q$ to reduce the pressure from the quiescent level $+P_q$ to a final level of zero. See Fig. 9.9a.

† $\zeta_2 = (R/L)/2\omega_n$ but $\omega_{n2} = \sqrt{0.38/LC} = 0.615/\sqrt{LC}$. Therefore $\zeta_2 = R/2L\omega_{n2} = R\sqrt{LC}/2L$ (0.615) = $(R\sqrt{C/L})/1.23$.

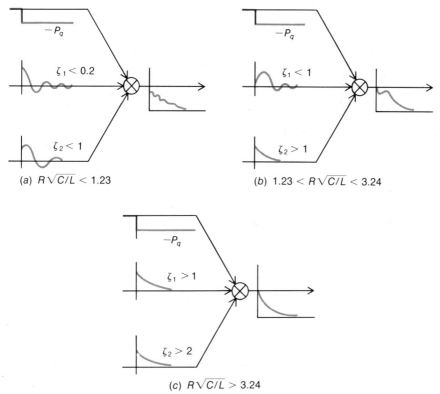

FIG. 9.11   *Transient responses expected for three ranges of* $R\sqrt{C/L}$.

upon the magnitude of the function $R\sqrt{C/L}$. If $R\sqrt{C/L} < 1.23$, both second-order terms in Eq. (9.30) will be underdamped ($\zeta_1 < 0.4$ and $\zeta_2 < 1$) and both terms will give rise to a transient oscillation. As the first second-order term is always less damped than the second, ζ_1 is always less than ζ_2. This term will give rise to a more persistent and higher-frequency oscillation. A relative representation of the oscillations that might be expected when $R\sqrt{C/L} < 1.23$ is shown by the lower two curves in Fig. 9.11a. If $R\sqrt{C/L}$ is greater than 1.23 but less than 3.24, one of the second-order terms in Eq. (9.30) is underdamped ($\zeta_1 < 1$) and the other is overdamped ($\zeta_2 > 1$). Under these circumstances one of the terms will give rise to a transient oscillation; the other will not. This situation is shown by the lower two curves in Fig. 9.11b. Finally, if $R\sqrt{C/L} > 3.24$, both second-order terms will be overdamped and neither will give rise to a transient oscillation. Both will decay monotonically as shown in Fig. 9.11c.

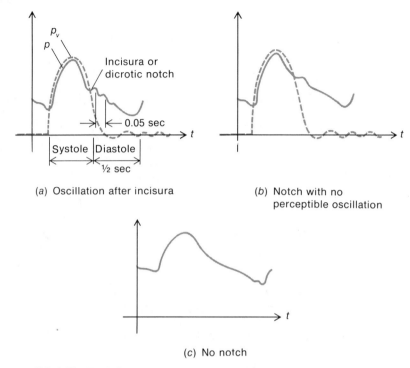

(a) Oscillation after incisura

(b) Notch with no perceptible oscillation

(c) No notch

FIG. 9.12 *Typical curves for ventricular and aortic pressures adjacent to valve.*

The total pressure response of the system to a step decrease in flow is the sum of the three components of Eq. (9.30). A qualitative description of the resultant response for each range of $R\sqrt{C/L}$ is shown by the outputs of the signal-flow diagrams in Fig. 9.11. In the totally underdamped case (Fig. 9.11a) the pressure response in the aorta proximal to the aortic valve would be distinctly oscillatory (this is sometimes described by the statement that the response "rings"). If the response is partly underdamped, the damping factor of the oscillating component will be in the range $0.4 < \zeta_1 < 1$. This response could therefore still be fairly oscillatory (though not as much so as the first response) but it could also be rather sluggish with only little tendency to oscillate (Fig. 9.11b). For the overdamped condition there are no oscillatory components, and total response decays monotonically (Fig. 9.11c).

We shall now see how this study and these conclusions may be used to gain insight into the original physiological system and typical recorded data. Some representative recordings of the aortic pressure at the aortic valve (point p in Fig. 9.8) are shown by the solid curve in Fig. 9.12. The dashed curve in Fig. 9.12 is the ventricular pressure directly adjacent to the aortic valve (point p_v in Fig. 9.8).* In actual operation, the ventricular pressure during systole exceeds the aortic pressure to keep the valve open and maintain flow. When the ventricular muscles relax, the ventricular pressure falls rapidly; the aortic pressure then exceeds the ventricular pressure and the valve closes. From the point of view of the previous study, we are concerned not with why the valve closes but only with what happens when it does. In Fig. 9.12 this would be directly after systole (when $p_v < p$).

The three recordings shown in Fig. 9.12 were chosen to illustrate the strong temptation one has to extrapolate *qualitative* data when a familiar pattern is recognized and to show how this can lead to incorrect results. Because of the good correspondence between the *qualitative* results predicted by theoretical considerations and the actual data, there is a tendency to transfer the conclusions derived from the model to the actual physical system with little or no further analysis. From the response curves in Fig. 9.11, for example, one might well assume that the curve in Fig. 9.12a indicates that the resistive and storage properties of the physiological system are related by $R \sqrt{C/L} < 1.23$, that the curve in Fig. 9.12b indicates the properties are related by $1.23 < R \sqrt{C/L} < 3.24$, and that in Fig. 9.12c the relationship between the properties is $R \sqrt{C/L} > 3.24$. Although these conclusions seem reasonable from a qualitative point of view, they cannot be supported by a quantitative analysis.

To justify the contention just made, we use $R \sqrt{C/L}$ as an indication of the damping factor and use rough approximations of the system properties to obtain some estimate of the magnitude of this term. In particular, we use a consistent set of units and define†

$$(9.34a) \qquad \text{Flow resistance} = R = \frac{8\mu\lambda}{\pi r^4}$$

where μ is fluid viscosity, λ is tube length, and r is average tube radius.

$$(9.34b) \qquad \text{Fluid inertance} = L = \frac{\rho_B\lambda}{\pi r^2}$$

---

* Such in vivo recordings may be obtained by inserting catheters, tipped with pressure transducers, into the peripheral arterial and venous systems (e.g., the brachial vein in the arm, the femoral artery in the leg, etc.). The catheters are then pushed through these vessels to the site of the heart where the measurements are to be made. The procedure is not unlike that described to obtain dye-dilution measurements (see Sec. 5.8).

† A brief description of how these relationships may be derived is given in chap. 25 of Ref. 4. However, there is a typographical error in this reference. In the relationship given for the compliance of an arterial wall, Eq. 6 on p. 842 of the reference indicates C is given by $2\pi r^3/Ew\lambda$, where the literal values are as described above. A check of the units and of the basic definition of C shows C should be written as Eq. (9.34c).

where ρ_B is the density of the blood.

$$(9.34c) \qquad \text{Tube compliance} = C = \frac{2\pi r^3}{Ew}$$

where E is the modulus of elasticity of the tube wall (i.e., the elastic property of the wall *material*) and w is the wall thickness.

If these relationships are used in the expression $R\sqrt{C/L}$, we have

$$(9.35) \qquad R\sqrt{\frac{C}{L}} = \frac{8\mu\lambda}{\pi r^4} \sqrt{\frac{2\pi r^3 \lambda}{Ew} \frac{\pi r^2}{\rho_B \lambda}}$$

or

$$R\sqrt{\frac{C}{L}} = \frac{11.3\mu\lambda}{\sqrt{r^3 Ew\rho_B}}$$

To obtain an approximate numerical result, we can use rough figures for the properties of the constituents of the physiological system.* The blood viscosity μ (at normal hematocrit) may be taken as 0.035 dyn-s/cm^2 and its density as 1.05 g/cm^3. Normal aortas have an average diameter of 2.5 cm and a wall thickness 0.2 cm. The modulus of elasticity (E) of the aortic wall varies considerably, depending on which section is under consideration. In the table provided by McDonald (see Ref. 5) the values for Young's modulus (E) range from 1×10^6 to 50×10^6 dyn/cm^2; as a tangential value for E we shall use 20×10^6. (It does not matter which coefficient is used in this case; the 10^6 order of magnitude is the significant number here.) The value to be used for tube length also presents somewhat of a puzzle. There is a small ascending aorta and a rather long descending one; of the latter portion, part is in the thoracic cage and part in the abdomen. To obtain some range of values, however, we shall assume the aortic length is no less than 1 ft and no greater than 2 ft. We can therefore say $30 \text{ cm} < \lambda < 60 \text{ cm}$. When these values are used in Eq. (9.35), we have

$$(9.36) \qquad R\sqrt{\frac{C}{L}} = \frac{(11.3)(0.035)(30)}{\sqrt{(2.5)^3(20 \times 10^6)(0.2)(1.05)}} = 1.46 \times 10^{-3}$$

The larger damping factor of the system would then be $1.46 \times 10^{-3}/1.23$ (see the second footnote on page 486) or 0.0012. Even if we use values in Eq. (9.36) to maximize this expression, $R\sqrt{C/L}$ would not exceed 0.015, and the larger damping factor would then increase to about 0.012. In any case, we can see from these figures the system is grossly underdamped but this is not the interpretation we would get if we used only the qualitative results of Fig. 9.11 to explain the results of Fig. 9.12.

* See, for example, Ref. 5. Another good source of pertinent material is Ref. 6.

The knowledge that the aortic system is probably a very lightly damped system raises some questions that are not easily answered by analysis alone. We must now ask: Where are the expected oscillations in Fig. 9.12b and c, and why does the oscillation in Fig. 9.12a not persist for a longer time instead of decaying after one or two cycles? It may be these anomalies are due to pathological conditions (the last curve, in fact, is due to a low-blood-pressure condition). To answer these questions, additional laboratory work and a reevaluation of the data may be required. We can, however, use the analysis to make one observation about the data which may be useful in further experimental work. We note, for example, the oscillation in Fig. 9.12a is probably not due to the physiological system. This contention can be supported from two points of view: The physiological system is theoretically highly underdamped; if the oscillation at the beginning of the response were due to the physiological system, it would persist for the full diastolic period. From another point of view we note the frequency of the oscillation shown in Fig. 9.12a is about one cycle in 0.05 s or about 20 Hz. The physiological system should show oscillatory frequency components of no less than 2 Hz and no more than 15 Hz.*

These observations suggest that the oscillations shown in Fig. 9.12a may be derived from some source other than the physiological system. The most likely possibilities in this case are spurious signal stemming from stimuli external to the physiological system (artifacts) or the inherent response characteristics of the instrument system. The instrumentation used to acquire the aortic-pressure recordings is generally

---

* This can be established from the relationship between the damping factor and the damped natural frequency as mentioned at the end of Sec. 9.3:

$$\omega_d = \omega_n \sqrt{1 - \zeta^2}$$

Since ζ is only 0.015 for this situation, $\omega_d \approx \omega_n$. From the discussion relating to Eq. (9.32) and the associated footnote,

$$\frac{0.615}{\sqrt{LC}} < \omega_n < \frac{1.62}{\sqrt{LC}}$$

If we accept the earlier definitions for L and C as given by Eqs. (9.34), we can then write

$$\frac{0.615}{\lambda} \sqrt{\frac{Ew}{\rho_B r}} < \omega_d < \frac{1.62}{\lambda} \sqrt{\frac{Ew}{\rho_B r}}$$

If we use the numerical values previously suggested ($r = 2.5$, $w = 0.2$, $1 \times 10^6 < E < 50 \times 10^6$ and $30 < \lambda < 60$) we have

$$12 \text{ rad/s} < \omega_d < 110 \text{ rad/s}$$

or

$$2 \text{ Hz} < \omega_d < 16 \text{ Hz}$$

fairly sensitive and often has a low damping factor; an abrupt signal from the system (as when the valve closes) may cause the instruments to deliver a transient oscillation that was not present in the applied signal.* In any case, we see the brief analysis can serve to alert the experimenter to the need for careful attention to acquisition and interpretation of data if he plans further experimentation to obtain information about the dicrotic wave.

We conclude this section with one final observation about the dicrotic wave. If the oscillatory transient has a frequency 2 Hz $< \omega_d <$ 15 Hz we note that at the low frequency only one cycle of oscillation would appear during the diastolic period ($\frac{1}{2}$ s). This may be what Fig. 9.12 shows: a low-amplitude, low-frequency transient. The comment about "low amplitudes" may also be true for transients of higher frequency; it may be the instrumentation has not been chosen to detect these oscillations. The questions cannot be resolved by analysis alone but must be supplemented by additional laboratory and/or library research work.

9.6 SIGNAL CONSTANTS OF AN UNDERDAMPED SECOND–ORDER RESPONSE

Second-order systems may have two general modes of behavior. They may exhibit an oscillatory response to an aperiodic excitation (in which case they are classified as underdamped systems) or they may respond to an aperiodic excitation in a monotonic manner (and so be classified as critically or overdamped systems). We have seen that in most cases when a system exhibits the latter response two constants that relate to the system properties can be determined from the signal data. These are the signal time-constants; the way these constants relate to the system properties depends upon the system configuration (see, for example, the discussions of Secs. 5.6, 7.3, and 7.5). It seems entirely reasonable to expect therefore that, as a second-order system changes from one with overdamped characteristics to one with underdamped characteristics, there will still be associated with the signal data two constants that relate to the system properties. This is indeed the case, and for underdamped systems the damping factor (ζ) and the frequency of the transient oscillation (ω_d) are the signal constants that relate to the system properties.

To obtain the signal constants (ζ, ω_d) associated with data from underdamped systems we must learn how these constants are related to the recorded-response curves. This is relatively straightforward as we

---

* A brief discussion of some of the aspects associated with blood-pressure measurements and instrumentation is presented in Ref. 7, chap. 5. Additional information can be found in Refs. 5 and 6.

already see intuitively that the rate of decay of the oscillation must be related to the damping factor of the system and the cyclic period must be related to the oscillatory frequency. To show this in a more quantitative manner we examine the analytic expression for an underdamped oscillatory response and compare this with the resultant waveform that would be derived from this expression. By this procedure, it will then be relatively easy to see how the waveform can be used to define the signal constants.

The analytic description of an underdamped oscillatory response is given by Eq. (9.18):

$$(9.18) \qquad \sigma_{\text{out}} = \frac{A}{\omega_d} \epsilon^{-\zeta\omega_n t} \sin \omega_d t$$

As ω_n and ω_d are related by $\omega_d = \omega_n \sqrt{1 - \zeta^2}$ [see the discussion leading to Eq. (9.18)] we see that Eq. (9.18) may be written as

$$(9.37) \qquad \sigma_{\text{out}} = \frac{A}{\omega_d} \epsilon^{-(\zeta/\sqrt{1-\zeta^2})\omega_d t} \sin \omega_d t$$

In this form we can see the response is governed by two constants: the damping factor and the damped natural frequency ω_d. A characteristic plot of this expression when damping is "light" is shown in Fig. 9.13. The amplitude of the response ($y_1, y_2, y_1, \ldots$) decays gradually, and the oscillations persist for at least two or more cycles. For more heavily damped systems the amplitude decay would be more rapid and the oscillations would disappear sooner.

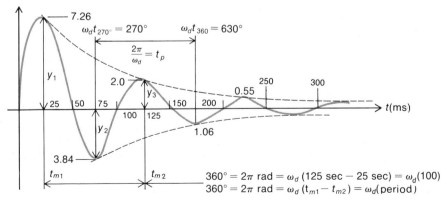

FIG. 9.13 *Characteristic decay curve for transient response of underdamped second-order system.*

To relate the amplitude decay of the response curve to the system damping and to show more forcefully how the response waveform is related to the analytic expression given in Eq. (9.37), we make the assumption that the curve of Fig. 9.13 is the recorded response from a "perfect," linear second-order instrument impulsively excited. It will be assumed the signal magnitudes are in volts (any measurable variable will suffice) and time is in milliseconds. The problem to be resolved is how we may determine the damping factor ζ and the oscillatory frequency ω_d from the recorded data. One obvious way to handle this problem is through simple substitution. If Eq. (9.37) describes the response curve, then points on the response curve must automatically satisfy Eq. (9.37). By substituting points from the curve into Eq. (9.37) we can set up a number of simultaneous equations from which the various curve constants can be evaluated. As a specific illustration, we note from Fig. 9.13 at 25 ms ($t_{m1} = \frac{1}{40}$ s) the signal amplitude (y_1) is 7.26; Eq. (9.37) may therefore be written

$$(9.38a) \qquad 7.26 = y_1 = \frac{A}{\omega_d} \epsilon^{-(\zeta/\sqrt{1-\zeta^2})(\omega_d/40)} \sin \frac{\omega_d}{40}$$

At 125 ms ($t_{m2} = \frac{1}{8}$ s) the signal amplitude (y_3) is 2.0 V. Equation (9.37) now becomes

$$(9.38b) \qquad 2.0 = y_3 = \frac{A}{\omega_d} \epsilon^{-(\zeta/\sqrt{1-\zeta^2})(\omega_d/8)} \sin \frac{\omega_d}{8}$$

If we did this for a third point we would then have three equations from which it should be possible to evaluate the three curve constants A, ω_d, and ζ. The coefficient A is the amplitude of the impulsive excitation; ω_d and ζ are curve constants related to the system resistive and storage properties.

The procedure outlined in the previous paragraph provides one possible technique to evaluate the signal constants. It is, however, somewhat clumsy as the expressions can become unwieldy, particularly if any arbitrary points are used for substitution. The procedure can be simplified considerably by taking advantage of one's knowledge of periodicity and using this concept to evaluate the damped natural frequency (ω_d) first. Then, by choosing discrete points for substitution in Eq. (9.37), very simple expressions can result and solutions are easily obtained. To carry out this suggested approach we recall first that sinusoidal functions repeat every 360° (2π rad). Since Eq. (9.37) has a sinusoidal term, it must have the same general wave characteristics every 360°. Thus, if the curve described by Eq. (9.37) reaches a maximum point at some time t_{m1} (see Fig. 9.13) and another maximum point at t_{m2}, in the time interval $t_{m2} - t_{m1}$ the sine term must have gone through a 360° cycle. Evidently, $t_{m2} - t_{m1}$ must represent the period of oscillation. For the waveform of Fig. 9.13 this is 100 ms or 0.1 s. The damped natural frequency (ω_d) must therefore be 10 c/s (10 Hz) or 20π rad/s.

Since we know ω_d, we may now use Eqs. (9.38a) and (9.38b) to solve for the two unknown constants A and ζ. Some simplifications are possible, however, as these equations were developed by the substitution of a special set of points taken from the response waveform. We may note, for example, that a relative maximum point of the response curve was used to develop the first equation and a relative maximum point one cycle removed from the first maximum was used to develop the second equation. These special points permit us to simplify Eqs. (9.38a) and (9.38b) as we know at the maximum points the sine function must be unity.* Equations (9.38a) and (9.38b) may therefore be written

$$(9.39) \qquad 7.26 = y_1 = \frac{A}{\omega_d}\,\epsilon^{-(\zeta/\sqrt{1-\zeta^2})(\omega_d/40)}$$

$$(9.40) \qquad 2.0 = y_3 = \frac{A}{\omega_d}\,\epsilon^{-(\zeta/\sqrt{1-\zeta^2})(\omega_d/8)}$$

As the amplitude A is not of interest in this case, we can eliminate it by dividing Eq. (9.39) by Eq. (9.40). The result is a single equation in which the damping factor is the only unknown:

$$(9.41) \qquad \frac{7.26}{2.0} = \frac{y_1}{y_3} = \frac{\epsilon^{-(\zeta/\sqrt{1-\zeta^2})(\omega_d/40)}}{\epsilon^{-(\zeta/\sqrt{1-\zeta^2})(\omega_d/8)}}$$

As ω_d is already known, this expression can be used to solve for ζ, the damping factor. Before we do so we simplify the expression further by noting $\omega_d/40$ is equivalent to $\pi/2$ (see footnote below) and $\omega_d/8$ is $5\pi/2$. If we use these equivalent values and combine the exponentials by performing the indicated division, the simplified expression becomes

$$(9.42) \qquad 3.63 = \frac{y_1}{y_3} = \epsilon^{+2\pi\zeta/\sqrt{1-\zeta^2}}$$

---

* If we use maximum points in Eq. (9.37), $\sin \omega_d t$ is taken as unity and the equation may be written

$$\sigma_{\text{out}}(\text{max}) = y_{\text{max}} = \frac{A}{\omega_d}\,\epsilon^{-(\zeta/\sqrt{1-\zeta^2})(\omega_d/t)}$$

As an alternative viewpoint, we now know ω_d is 10 Hz; $\omega_d/40$ is therefore 10/40 cycles; a quarter of a cycle is 90°; $\sin \omega_d/40 = \sin 90° = 1$. Equation (9.38a) therefore reduces to Eq. (9.39). Similarly, $\omega_d/8$ is $1\frac{0}{8}$ or $1\frac{1}{4}$ cycles; therefore $\sin \omega_d/8 = \sin 450° = 1$ and Eq. (9.38b) reduces to Eq. (9.40).

In this form we can see the solution for ζ is now simply a routine matter:

$$(9.43a) \qquad \ln 3.63 = \ln \frac{y_1}{y_2} = 2\pi \frac{\zeta}{\sqrt{1 - \zeta^2}}$$

$$(9.43b) \qquad 1.29 = \ln \frac{y_1}{y_2} = 2\pi \frac{\zeta}{\sqrt{1 - \zeta^2}}$$

$$(9.43c) \qquad (1.29)^2 = \left(\ln \frac{y_1}{y_2} \right)^2 = (2\pi)^2 \frac{\zeta^2}{1 - \zeta^2}$$

$$(9.43d) \qquad \left(\frac{1.29}{2\pi} \right)^2 = \left[\frac{\ln (y_1/y_2)}{2\pi} \right]^2 = \frac{\zeta^2}{1 - \zeta^2}$$

$$(9.43e) \qquad \zeta = \sqrt{\frac{(1.29/2\pi)^2}{1 + (1.29/2\pi)^2}} = \sqrt{\frac{[\ln (y_1/y_2)/2\pi]^2}{1 + [\ln (y_1/y_2)/2\pi]^2}}$$

$$(9.43f) \qquad \zeta = \sqrt{\frac{0.042}{1.042}} = \sqrt{0.04} = 0.2$$

The procedure to determine ζ as outlined above may be generalized by Eq. (9.43d). The general amplitudes y_1 and y_3 are two successive relative maxima of the underdamped-oscillatory-response curve. The ratio of these maxima (y_1/y_3) is sometimes described as the decrement of the curve, and the logarithm of this ratio is known as the "log decrement." If the log decrement is represented by δ, Eq. (9.43d) may be written

$$(9.44) \qquad \delta^2 = (2\pi)^2 \frac{\zeta^2}{1 - \zeta^2}$$

Therefore ζ would be written*

$$(9.45) \qquad \zeta = \frac{\delta/2\pi}{\sqrt{1 + (\delta/2\pi)^2}}$$

This expression makes it possible to evaluate the damping factor of any underdamped second-order response curve. One needs only the amplitudes of two successive maxima (scaled off with a ruler if necessary; the true significance of the measured amplitudes is not important as only the scaled magnitude is of importance, because the amplitude ratio is to be used). The logarithm of this ratio is the log decrement of the curve. This value substituted into Eq. (9.45) yields the damping factor.

---

* In some cases it may be desirable to use a maximum point and a successive minimum point to evaluate ζ (for example, y_1 and y_2 in Fig. 9.13). This would be appropriate when curves are heavily damped. The log decrement is then simply the logarithm of the absolute ratio $|y_1/y_2|$. Equation (9.45) is then rewritten (see Prob. 9.3) so that ζ is given by

$$\zeta = \frac{\delta/\pi}{\sqrt{1 + (\delta/\pi)^2}}$$

When ω_d and ζ have been determined from an underdamped-response curve (or, phrasing it more realistically, when one is satisfied or willing to assume that a response curve may be considered an underdamped second-order response and an appropriate ω_d and ζ have been evaluated), these values can then be used to describe the second-order transfer function of the system. The transfer function may then be used for any of the various applications described previously. For example, it may be used to evaluate the system properties of a system if the system configuration is known (or if one has been proposed as in Sec. 7.3) or it may be used to propose the existence or identification of system properties if the configuration is not known (by network synthesis as in Sec. 7.4); finally, it may be used to predict performance or serve as a guide for the design of experiments of equipment. The latter application of the transfer function has been implied but not demonstrated by illustration.

We shall digress here to outline one situation in which knowledge of a system transfer function can be used to improve or correct system performance. If the decay curve of Fig. 9.13 were the impulse response of an instrument we would estimate the instrument transfer function to be $1/(s^2 + 2.56s + 4{,}096)$ (Fig. 9.14).* Since the transfer function of

* The evaluation of the transfer function is outlined in Fig. 9.14. As a brief explanation of the steps: The response curve of Fig. 9.13 appears to be that of an underdamped second-order system. The response curve would thus be described mathematically as $A\epsilon^{-\zeta\omega_n t}\sin \omega_d t$, where $\omega_d = \omega_n \sqrt{1 - \zeta^2}$. The transform of the response curve is then $1(s^2 + 2\zeta\omega_n s + \omega_n^2)$. From the response curve we find ω_d is 20π rad/s and $\zeta = 0.2$ [see Eq. (9.43f) and the discussion related to that equation]. The transform of the response curve is then $1/(s^2 + 2.56s + 4{,}096)$. Since this is the impulse response of the system, the transform of the response curve is also the transfer function of the instrument. $[T(s)$ is therefore $1/(s^2 + 2.56s + 4{,}096).]$

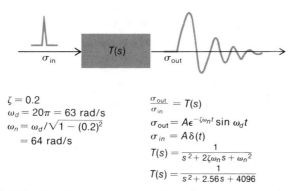

$\zeta = 0.2$

$\omega_d = 20\pi = 63$ rad/s

$\omega_n = \omega_d / \sqrt{1 - (0.2)^2}$

$\quad = 64$ rad/s

$\dfrac{\sigma_{out}}{\sigma_{in}} = T(s)$

$\sigma_{out} = A\epsilon^{-\zeta\omega_n t}\sin \omega_d t$

$\sigma_{in} = A\delta(t)$

$T(s) = \dfrac{1}{s^2 + 2\zeta\omega_n s + \omega_n^2}$

$T(s) = \dfrac{1}{s^2 + 2.56s + 4096}$

FIG. 9.14 *Use of impulse response to determine system transfer function.*

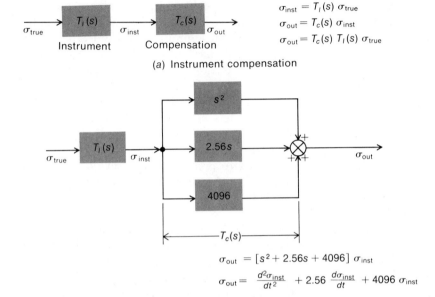

$$\sigma_{inst} = T_I(s)\,\sigma_{true}$$
$$\sigma_{out} = T_c(s)\,\sigma_{inst}$$
$$\sigma_{out} = T_c(s)\,T_I(s)\,\sigma_{true}$$

(a) Instrument compensation

$$\sigma_{out} = [s^2 + 2.56s + 4096]\,\sigma_{inst}$$
$$\sigma_{out} = \frac{d^2\sigma_{inst}}{dt^2} + 2.56\,\frac{d\sigma_{inst}}{dt} + 4096\,\sigma_{inst}$$

(b) A specific compensation scheme

FIG. 9.15 *Compensation for instrument system.*

the instrument is not a pure-gain constant, the instrument does not act as a pure amplifier. The output of the instrument is thus not an accurate magnified version of the true signal input; instead, the output will be a modified (distorted) representation of the input signal. In a later section we shall see for periodic signals of "low" frequency content this distortion will be low. If the frequency spectrum of the applied signal is wide, however, distortions may be significant.

To develop an instrumentation system that delivers a more accurate representation of the input signal, one can mechanically or analytically compensate for the effect of the instrument. The first technique would involve the construction of a device that could cancel the instrument transfer function (at least for a limited range); the second would be analytically to reverse the operations performed by the instrument. These techniques are diagrammatically represented by the block diagrams shown in Fig. 9.15.* The instrument output is delivered to a compensation block (or block assembly) which cancels the effect of the instru-

* Both techniques are the same; the compensation device required [$T_c(s)$] is a mechanical or hardware realization of the analytic computations that must be performed to effect a correction.

ment. From Fig. 9.15a it can be seen the connection block $T_c(s)$ must have a transfer function $1/T_I(s)$ [where $T_I(s)$ is the instrument transfer function] if it is to act as a compensator. In some simple cases it is possible to build the compensation block*; often, however, construction is impractical. In any case a description of the compensation block indicates what mathematical computations must be performed if a correction is to be obtained.

If, for example, the instrument has a transfer function $1/(s^2 + 2.56s + 4,096)$ then the compensation block should operate on the instrument signal according to the function $s^2 + 2.56s + 4,096$ (see Fig. 9.15b). This implies the true signal is the instrument reading increased by a factor of 4,096 (4,096 σ_{inst}) added to the product of 2.56 and the derivative of the instrument reading, which is then added to the second derivative of the instrument reading. A block diagram to represent the operations is shown in Fig. 9.15b. If the mathematical operations are performed (manually, or by machine, if possible), the resultant waveform will be a closer representation of the true signal delivered to the instrument (see Prob. 9.4).

9.7 RESONANCE

From the previous discussions and the subject organization of the chapter, the connection between resonance and underdamped systems should be intuitively evident. Resonance is the response that may be expected of underdamped systems when a periodic signal of a characteristic frequency is applied to the system. The response is characterized by a "surprisingly" large output amplitude for relatively small input amplitude (i.e., the gain is large). A translation of these comments by illustration is afforded by the annoying vibrations developed in a house when certain periodic stimuli are applied. A truck passing a house generally causes a vibration of an almost imperceptible amplitude; some dishes, however, may be "tuned" to the induced vibration and may begin to oscillate rather wildly and may (to the housewife's dismay) even shatter or fall. A faucet that delivers water at high or low rates may operate quite satisfactorily; at one delivery rate, however, it may excite the valve mechanism or pipes to resonate and thereby generate an unwelcome whine.

To show how resonance may be described quantitatively and how

* Its purpose is to provide compensation for at least a limited range. This was done for a simple instrumentation system used for dye-dilution studies. The catheter-oximeter portion of the system (see Sec. 5.8) was found to have first-order characteristics. An electrical circuit was devised to compensate for the effect of the catheter-oximeter combination. As a result of the compensation, the output curves from the system more nearly represented the signal delivered to the instrument (as checked by more expensive equipment). See Ref. 8.

this phenomenon can be used to gain insight into some system properties, we examine the response that can be expected of a second-order under-damped system when a periodic excitation is applied. Before we begin this discussion, however, it will be helpful to review the knowledge already available about the sinusoidal response of second-order systems. In Sec. 8.7 we examined the sinusoidal response of second-order systems by considering them the product of two first-order systems. This procedure provided a convenient method to construct the log-gain vs. log-frequency plots (frequency-response characteristics) of second- and higher-order systems. However, second- and higher-order systems need not be considered the product of first-order systems; the frequency-response curves can be constructed directly from a transfer function without factoring the polynomials of the function into first-order terms. The procedure is simply to replace s with $j\omega$ (a review of Sec. 8.7 is appropriate here) in the unmodified transfer function. Specifically, therefore, if we wish to plot the frequency-response characteristics of a system with a transfer function $T(s) = 3/(s^2 + 7s + 10)$ we write

$$(9.46) \qquad G(\text{dB}) = 20 \log \left| \frac{3}{s^2 + 7s + 10} \right|_{s \to j\omega}$$

which can then be written

$$(9.47a) \qquad G(\text{dB}) = 20 \log \left| \frac{3}{(j\omega)^2 + 7j\omega + 10} \right|$$

$$(9.47b) \quad G(\text{dB}) = 20 \log \left| \frac{3}{-\omega^2 + 7j\omega + 10} \right| = 20 \log \left| \frac{3}{(10 - \omega^2) + j7\omega} \right|$$

$$(9.47c) \qquad G_{\text{dB}} = 20 \log \frac{3}{\sqrt{(10 - \omega^2)^2 + (7\omega)^2}}$$

We can now plot the gain in decibels as a function of frequency and obtain directly the frequency-response characteristics of the second-order system given by Eq. (9.46). [This result can then be compared with the result obtained in Sec. 8.7 when the factored form of Eq. (9.46) was used to evaluate the frequency-response characteristics. It should be obvious the two procedures must give identical results, as Eqs. (9.47c) and (8.88a) are equivalent. This may be established by factoring Eq. (9.47c) or combining the first-order terms of Eq. (8.88a).]

The reason for taking the trouble to establish that frequency-response curves can be derived directly from second-order transfer functions (without factoring) becomes evident when one seeks to plot the decibel-gain characteristics of an underdamped second-order system: Underdamped second-order transfer functions cannot be factored into

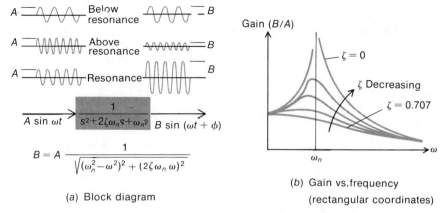

(a) Block diagram

(b) Gain vs.frequency
(rectangular coordinates)

FIG. 9.16 *Sinusoidal excitation of a general second-order system.*

real first-order terms. Therefore, the summation of first-order gain curves cannot be used (as in Fig. 8.24). The most practical way to handle such situations is to use the procedure outlined by the sequence of Eqs. (9.47a) to (9.47c).

To show how damping and resonance phenomena are related, we examine the frequency response of the general second-order system shown in Fig. 9.16. Since the gain is given by the absolute value of the complex transfer function, we have

$$(9.48a) \qquad G = \frac{B}{A} = \left| \frac{1}{s^2 + 2\zeta\omega_n s + \omega_n^2} \right|_{s \to j\omega}$$

$$(9.48b) \qquad G = \left| \frac{1}{(\omega_n^2 - \omega^2) + j(2\zeta\omega_n\omega)} \right|$$

$$(9.48c) \qquad G = \frac{1}{\sqrt{(\omega_n^2 - \omega^2)^2 + (2\zeta\omega_n\omega)^2}}$$

Equation (9.48c) is the system-gain function expressed in terms of the excitation frequency. Sometimes this expression is further modified to show the gain function in terms of the frequency ratio ω/ω_n. Equation (9.48c) then becomes

$$(9.48d) \qquad G = \frac{1/\omega_n^2}{\sqrt{[1 - (\omega/\omega_n)^2] + (2\zeta\omega/\omega_n)^2}}$$

In terms of decibels, this gain function is written

$$(9.49) \quad G(\text{dB}) = -40 \log \omega_n - 10 \log \left\{ \left[1 - \left(\frac{\omega}{\omega_n} \right)^2 \right]^2 + \left(\frac{2\zeta\omega}{\omega_n} \right)^2 \right\}$$

Although the second-order gain function is not as compendious as those obtained from first-order expressions, it is still possible to examine the gain expression for asymptotic behavior. For example, if ω is sufficiently smaller than ω_n, $(\omega/\omega_n)^2$ and $(2\zeta\omega/\omega_n)^2$ can be neglectably small compared with unity. Equation (9.49) then reduces to

$$(9.50) \qquad\qquad G(\text{dB}) = -40 \log \omega_n$$

Equation (9.50) may be considered a description of the low-frequency asymptote for this situation (see Fig. 9.17). In a similar fashion, if $\omega \gg \omega_n$, Eq. (9.49) may be written

$$(9.51) \qquad G(\text{dB}) \approx -40 \log \omega_n - 40 \log \left(\frac{\omega}{\omega_n}\right)^4 = -40 \log \omega$$

Equation (9.51) may be considered the description of the high-frequency asymptote; as might have been expected, this line falls at the rate of 40 dB/decade or 12 dB/octave.

From Eq. (9.48d) [or Eq. (9.48c) or (9.49)] we can easily begin to appreciate how the damping factor ζ affects the frequency-response curve. If ζ were very small (i.e., if the system were highly underdamped) $2\zeta\omega/\omega_n$ could probably be neglected compared with the other terms of the gain equations.* The gain equations therefore reduce to

$$(9.52a) \qquad\qquad G = \frac{1/\omega_n^2}{1 - (\omega/\omega_n)^2}$$

$$(9.52b) \qquad G(\text{dB}) = -40 \log \omega_n - 20 \log \left[1 - \left(\frac{\omega}{\omega_n}\right)^2 \right]$$

These equations indicate clearly the gain function will rise as the excitation frequency (ω) approaches the natural frequency of the system (ω_n). In fact, mathematically Eqs. (9.52a) and (9.52b) indicate the gain will become infinite if ω/ω_n becomes unity. If we use this information and our knowledge of asymptotic behavior of the gain curve we can quickly construct an approximate representation of the underdamped-frequency

---

* If $\zeta \approx 0.1$,

$$G = \frac{1/\omega_n^2}{\sqrt{[1 - (\omega/\omega_n)^2]^2 + (0.2)^2(\omega/\omega_n)^2}}$$

or

$$G = \frac{1/\omega_n^2}{\sqrt{1 - 2(\omega/\omega_n)^2 + (\omega/\omega_n)^4 + 0.04(\omega/\omega_n)^2}}$$

The term $0.04(\omega/\omega_n)^2$ may be neglected compared with $2(\omega/\omega_n)^2$.

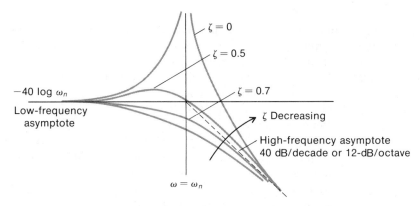

FIG. 9.17 $G(dB)$ *versus log frequency for second-order systems.*

characteristics when ζ is very low: At low frequencies ($\omega \ll \omega_n$) the curve follows the low-frequency asymptote; as ω approaches the system natural frequency, the gain curve rises sharply; when ω exceeds the natural frequency, the gain curve drops, and as ω becomes large ($\omega \gg \omega_n$) the curve approaches the high-frequency asymptote. A sketch of the $G(dB)$ versus log-frequency curve for small ζ ($\zeta \rightarrow 0$) is shown in Fig. 9.17 (a sketch on rectangular coordinates is shown in Fig. 9.16c).

Resonance phenomena now become easy to understand: If an under-damped system is driven by a sinusoid that is of a frequency close to the system natural frequency, the amplitude of the system output will be significantly larger than the input amplitude (i.e., the system will tend to resonate). From another point of view, if an underdamped system is excited by a periodic signal with a harmonic content equivalent to the system natural frequency, the system will be induced to resonate. A schematic representation of the resonance effect is shown in Fig. 9.16a. The first two curves above the block describe operation far removed from the resonant frequency; the third set of curves describes resonant operation. As can be seen, for "high-" or "low-" frequency excitation signals the output amplitudes are either equal to or less than the input amplitudes. Signals in the range of the resonant frequency (near ω_n) cause magnified (or amplified) output harmonics.

For somewhat more quantitative information about the resonant condition, we examine some of the implications of the general second-order gain relationship given by Eq. (9.48d). We have already established the gain will be "very large" at $\omega \approx \omega_n$ when the damping is very small. The exact mathematical statement is: Gain will be infinite when $\omega = \omega_n$ if ζ is zero. This is a mathematical figment of the imagination; practically speaking, ζ cannot be zero,* and, physically speaking, a system cannot reach infinite amplitudes; it would shake itself to destruction as high amplitudes are reached. However, this does indicate the intimate relationship between maximum gain, damping, and resonant frequency. The maximum-gain condition can be expressed in a more quantitative manner if the gain function of Eq. (9.48d) is maximized; gain is maximized when the denominator of the equation is minimized; the term under the radical is minimized when $(\omega/\omega_n)^2$ is $1 - 2\zeta^2$.† This conclusion has interesting implications; it indicates first the resonant frequency is not always the natural frequency of the system. The resonant frequency is given by $\omega_n \sqrt{1 - 2\zeta^2}$; this indicates a system will resonate at its natural frequency only if the damping factor (ζ) is neglectably small. As ζ increases, the resonant frequency decreases.

A second implication of the maximum-gain relationship is the limited range of underdamped systems that can resonate. Not all underdamped systems have resonant characteristics; if the damping factor is $\sqrt{2}/2$ (or 0.707), the system "resonates" at zero frequency. In effect, therefore, if ζ is equal to or greater than 0.7, the system will not have a resonant point. This indicates, therefore, that a system may be underdamped $(\zeta < 1)$ yet not have a resonant point.

9.8 SOME OBSERVATIONS RELATED TO RESONANT BEHAVIOR

From a design point of view, the condition of resonance can sometimes be used to advantage, and sometimes it can be a decided disadvantage. Filter systems (electrical and others) and tuning systems are devices in which resonance phenomena have been used to advantage. Resonant configurations in filter systems eliminate or impede the transmission of undesirable frequencies through the system; in selective-tuning systems (radio, TV) resonance is used to accentuate one specific frequency above

---

* Even if it were possible, gain would not be infinite (practically or mathematically) as soon as $\omega = \omega_n$. It would take an infinite amount of time for a zero-damped system to reach infinite gain at $\omega = \omega_n$. Passing through the resonant frequency would therefore not create any difficulties even for a zero-damped system.

† This minimum condition is determined by setting the derivative of the expression under the radical equal to zero:

$$\frac{d \text{ (radicand)}}{d\omega} = 2 \left[1 - \left(\frac{\omega}{\omega_n} \right)^2 \right] \left(-2 \frac{\omega}{\omega_n} \right) \frac{1}{\omega_n} + 2 \left(2\zeta \frac{\omega}{\omega_n} \right) \frac{2\zeta}{\omega_n} = 0$$

Therefore

$$\left(\frac{\omega}{\omega_n} \right)^2 = 1 - 2\zeta^2$$

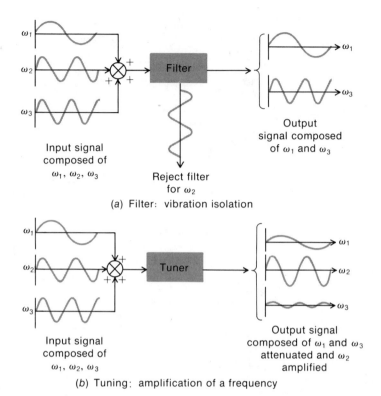

FIG. 9.18 *Block diagrams of filter and tuning applications of resonant phenomena.*

all other entering frequencies. These applications of resonance are schematically shown by the block diagrams in Fig. 9.18.

In some applications resonance is not a useful feature of the system; in fact, it may affect the operation of the system in an adverse manner. This is particularly true in instrument systems where the output signal is expected to be a true replica of the input signal. If an instrument can resonate to one of the spectral components of the signal it will selectively amplify this component. The total resultant output of the instrument will thus consist of frequency components that have undergone different degrees of amplification. The instrument output will thus be a distorted version of the signal input.

We can readily appreciate the significant effect resonance can have on signal waveforms if we examine the very simple illustrative example in Fig. 9.19 (see also Prob. 8.3). It is assumed a periodic signal with two frequency components is delivered to an instrument with a natural

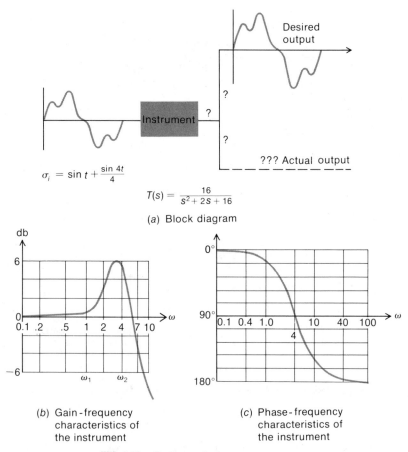

$$\sigma_i = \sin t + \frac{\sin 4t}{4}$$

$$T(s) = \frac{16}{s^2 + 2s + 16}$$

(a) Block diagram

(b) Gain-frequency
characteristics of
the instrument

(c) Phase-frequency
characteristics of
the instrument

FIG. 9.19   *Resonance in instrument systems.*

frequency $\omega_n = 4$ rads/s, a damping factor 0.25, and a low-frequency gain of unity. [The instrument transfer function is thus $16/(s^2 + 2s + 16)$.] The gain-frequency characteristics of the instrument are shown in Fig. 9.19*b*; the phase-frequency characteristics are shown in Fig. 9.19*c*. The frequency components of the applied signal are $\omega_1 = 1$ rad/s and $\omega_2 = 4$ rad/s; the amplitudes of these components are 1 and $\frac{1}{4}$ unit, respectively (see Fig. 9.19*a*). If the measuring instrument were to respond properly, the instrument output would be of the waveform as the input signal within a multiplicative constant (i.e., scaling factor); all components of the input signal should thus be amplified by the same factor and undergo a similar time shift. In this case, however, the frequency of one of the entering signal components matches the natural frequency of the instrument; this signal is amplified more than the other component and is shifted in phase with respect to this component. The total output, therefore, cannot be a scaled version of the input.

To show this in a quantitative fashion we note that the low-frequency component passes through the instrument with practically no gain at all (at $\omega_1 = 1$, the gain provided by the instrument is about $\frac{1}{4}$ dB, which is practically unity gain; see Fig. 9.19b). Similarly, this component has very little phase shift (see Fig. 9.19c). The low frequency component of the instrument output may thus be given by

$$(9.53) \qquad \sigma_{01} = \sin t \qquad \text{(See Fig. 9.20)}$$

The high-frequency component ($\omega_2 = 4$), on the other hand, undergoes appreciable amplification as it passes through the instrument. From the gain-frequency characteristics we see this component is magnified by a factor of 4 (that is, 6 dB; see Fig. 9.19b). The output amplitude of this component is thus four times the input amplitude. Furthermore, this component undergoes a 90° shift in phase during transmission through the instrument. It is therefore delayed with respect to the low-frequency component, as shown in Fig. 9.20. The expression for the high-frequency component of the output can therefore be given by

$$(9.54) \qquad \sigma_{02} = (4 \times \tfrac{1}{4}) \sin (4t + 90°) = \cos 4t$$

The total instrument output is now the sum of Eqs. (9.53) and (9.54). This summation is shown in Fig. 9.20. As can be seen, the output waveform of the instrument bears little resemblance to the input waveform; the instrument has distorted the applied signal beyond recognition.

On the basis of the foregoing discussion it might be concluded that instruments having resonant points cannot be used for measurements. However, the resonance of an instrument is not the only criterion by which its utility should be judged. If the signal components of the last

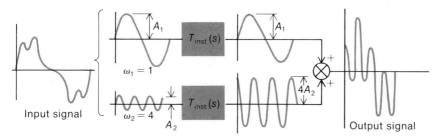

FIG. 9.20 *Effect of resonance on signal transmission.*

example were $\omega_1 = \frac{1}{4}$ rad/s and $\omega_2 = 1$ rad/s, the instrument would reproduce the input waveform with good fidelity, as both components would then be amplified equally and undergo approximately the same time shift. The criterion for faithful reproduction, therefore, is that the signal components of the input lie within a flat portion of the gain-frequency characteristics and a linear portion of the phase-frequency characteristics of the instrument. For the foregoing example, this implies the frequency components of the input wave must not be greater than about 2 rad/s, as the instrument has flat characteristics up to this frequency. (This is sometimes expressed by saying the response is flat for a bandwidth of 0 to 2 rad/s.)

Evidently, the wider the flat portion of the frequency characteristics (i.e., the greater the bandwidth of the instrument) the greater are the utility and fidelity of reproduction. However, as it is more difficult to build an instrument with a wide bandwidth, one also generally finds the greater the bandwidth, the greater is the price (note, for example, the price difference between high-fidelity music systems and ordinary sound-reproduction equipment). It is often worthwhile, therefore, to investigate one's needs before purchase of an expensive instrument. If the signals to be measured are of relatively low-frequency content, high-frequency capability is not necessary, and it is foolish to invest in an expensive instrument when a cheaper one can do an adequate job.

Problems

9.1 The transfer function of an overdamped system has a denominator of $s^2 + as + b$.

(a) Determine a relationship that describes the time-constants of the system.

(b) The constant a is continually reduced until the system becomes underdamped. What is the relationship between the damping factor, the oscillatory frequency, and the constants a and b?

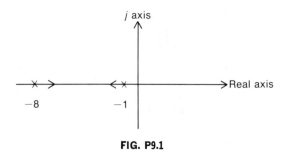

FIG. P9.1

(c) The transition from overdamped systems to underdamped systems is sometimes displayed on the complex plane with real and imaginary axes (see Fig. P9.1). The roots of the second-order denominator are plotted on this plane. Thus if the denominator is $s^2 + 9s + 8$ the roots are at -8 and -1, and these are plotted on the axis as shown. The time-constants of this system are $\frac{1}{8}$ and 1. When the number 9 is reduced (say to 7), the roots of the denominator are changed, and on the complex plane these roots appear to approach each other. A plot of the migration of roots as the system constants change is sometimes called a root locus. Plot the root locus for $s^2 + as + 8$ as a varies $0 < a < 9$.

9.2 Assume an aortic length is representable by the circuit model in Fig. P9.2.

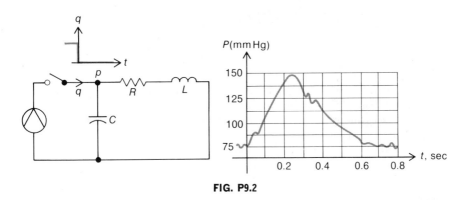

FIG. P9.2

(a) Obtain an expression in s for the pressure p when $\dot{q}$ is a negative step as shown.

(b) Write an expression for p in terms of time. What are the natural frequency of the system and the damping factor?

(c) If R, L, and C are as described by Eqs. (9.34) and an aortic-pressure curve is as shown in Fig. P9.2, obtain an estimate of the thickness and modulus of elasticity of this patient's aortic wall if it is assumed average values for μ, ρ_B, r, and λ may be used.

9.3 In some cases the transient response of an underdamped system is of such short duration that a complete response cycle cannot be clearly seen on a recording (i.e., the second positive peak is too small to be measured). This is particularly true when damping factors exceed about 0.5. In such cases one can use the first positive peak and first negative peak to obtain an estimate of the system damping factor. In Fig. 9.5 this would imply that we would use the curve magnitudes at 0.3 s and 1.2 s instead of 0.3 s and 2.1 s. The time between the peak amplitudes in this case would thus be related to π instead of 2π. Show that if these two peaks are used, the damping factor is then given by

$$\zeta = \frac{\delta/\pi}{\sqrt{1 - (\delta/\pi)^2}}$$

where δ is the log of the ratio of the two peak amplitudes.

9.4   An instrument with a transfer function $4,096/(s^2 + 2.56s + 4,096)$ delivers a signal that can be represented by $3\epsilon^{-50t}$.

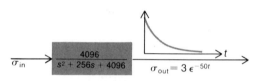

FIG. P9.4

(a) What is the maximum error between the true signal input and the reading obtained?

(b) Does the error between the true signal and the recorded signal ever become zero?

(c) Are the conclusions of (a) and (b) valid if the output signal is of the form $2(1 - \epsilon^{-13t} \cos 63t)$?

REFERENCES

1. Doebelin, E. O.: "Measurement Systems: Application and Design," McGraw-Hill Book Company, New York, 1966.
2. Cook, N. H., and E. Rabinowicz: "Physical Measurement and Analysis," Addison-Wesley Publishing Company, Inc., Reading, Mass., 1963.
3. Kay, R. H.: "Experimental Biology—Measurement and Analysis," Reinhold Publishing Corporation, New York, 1964.
4. Dow, P., and W. F. Hamilton (eds.): "Handbook of Physiology," Sec. 2: Circulation, vol. II, American Physiological Society, Washington, D.C., 1963.
5. McDonald, Donald A.: "Blood Flow in Arteries," chap. III and Table III, p. 174, Edward Arnold (Publishers) Ltd., London, 1960.
6. Burton, Alan C.: "Physiology and Biophysics of the Circulation," The Year Book Medical Publishers, Inc., Chicago, 1966.
7. Rushmer, R. F.: "Cardiovascular Dynamics," W. B. Saunders Company, Philadelphia, 1961.
8. Mitzner, W. A., W. B. Blesser, and E. Glassman: "On Line Correction of Dye Dilution Curves by a Compensation System," *Digest 7th Intern. Conf. Med. Biol. Eng., Stockholm, 1967.*

CHAPTER TEN

FEEDBACK*

10.1 INTRODUCTORY COMMENTS

In this chapter we examine some of the features and properties of the regulatory systems that enable a living organism to maintain a living state. Various terms have been applied to this mode of control behavior: The engineer describes this form of system organization as feedback; persons in the life sciences generally describe the regulatory function as homeostasis.

The terms feedback and homeostasis are so closely related that often they can be used interchangeably. Thus the physiological temperature-regulating system may be referred to as a feedback control system. However, this same system might also be described as a homeostatic system. If the terms feedback and homeostasis were truly synonymous and if these were the only two terms used to describe the regulatory behavior of living systems, this interchangeability would not cause confusion. Unfor-

* The author would like to thank Dr. John G. Truxal of the Polytechnic Institute of Brooklyn for his contributions to this chapter.

tunately they are not truly synonymous, and to confuse the issue even further one often finds biological control systems described by various other terms such as servosystem, servomechanism, or cybernetic system (to mention a few). To place these terms into proper perspective we shall first discuss what these terms imply.

The word "homeostasis" has strictly biological connotations; it refers to the ability of a living organism to maintain a fixed internal environment compatible with the living state in spite of adverse external and internal effects. The temperature-regulating mechanism in mammals and birds is one illustration of a homeostatic system. These animals are homeotherms which must maintain a relatively constant internal temperature in spite of external-temperature variations. The glucose-control system in man is another illustration of a homeostatic system. To maintain the living state, blood-sugar levels in human beings must be held to approximately 100 mg of glucose per 100 mg of blood (generally described as 100 mg percent) even if sugar is introduced internally through oral or intravenous feeding. Malfunction of this system leads to the pathological state known as diabetes. If artificial or external control is not introduced, this condition leads to death.

The concept of homeostasis and the introduction of the term to describe this condition have an interesting history.* Although we shall not discuss the details of this evolution, it should be mentioned that the concept of fixed internal environment (*milieu intérieur*) was introduced by the colorful and ingenious French physiologist Claude Bernard. The idea first appeared in Bernard's writings in about 1859 and then appears later in the introduction of his book (Ref. 2) first published in 1865. The term homeostasis to describe this condition of internal constancy was introduced by the brilliant American physiologist Walter B. Cannon. He used the term first in his article Organization for Physiological Homeostasis published in 1929; he later elaborated upon the significance of the term and the concept in his book "The Wisdom of the Body" (Ref. 3). We should also mention that homeostasis stems from the terms homeo and stasis. The first implies sameness; the second, a condition of standing or stopping. The combination is intended to describe the mechanisms or processes that function to oppose change and retain a life-sustaining environment.

The concept of feedback has more general implications than homeostasis from at least two points of view. First, feedback is not restricted to a description of biological systems; second, feedback is a generic term describing a process to achieve a condition rather than the description

---

* An interesting account of the development of the word is given in Ref. 1. This book also includes a good general discussion of feedback and descriptions of a number of homeostatic processes.

of a specific condition. To put the terms feedback and homeostasis in proper relationship to one another, one may state: Homeostasis may be (and generally is) achieved by some application of feedback. Thus a homeostatic system is generally an illustration of the concept of feedback; a feedback system, however, is not necessarily a homeostatic system.

The process implied by the term feedback is almost self-explanatory: Signal information is delivered to a system; the input signal is processed by the system, and the processed information is returned (feedback) to the signal source to modify the effectiveness of the signal stimulus. In essence, feedback implies a closed-loop configuration such as that developed in Sec. 8.9 (see Figs. 8.35b and 8.36c). In this system the input light signal is sampled by retinal sensors which transfer the signal to the pupillary muscles; these translate the electrical signals into mechanical effects which adjust the lights delivered to the retina.

To examine feedback from a schematic point of view we can translate the general verbal description of the preceding paragraphs into a block diagram such as is shown in Fig. 10.1. This is not the only kind of block-diagram configuration that can be used to represent a feedback system; however, it is the simplest one that contains all the essential features of such a system, and it includes the "set-point" concept which can be used to explain how a homeostatic process may operate. The set point is some reference level (σ_s in Fig. 10.1) which the system strives to maintain in spite of external disturbances; it is generally fixed by some internal agent of the system, the location of which may or may not be known. For example, the location of the set-point mechanism that regulates the light to the retina is not known. However, the set-point

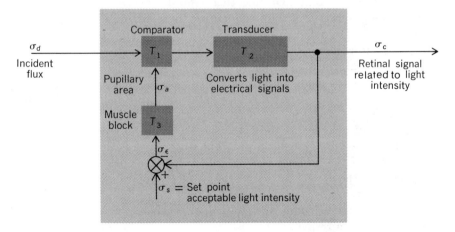

FIG. 10.1 *Representation of a homeostatic system.*

mechanism that fixes the body-temperature reference level for human beings is known to be located at the base of the brain and, in particular, in the hypothalamus. (The hypothalamus could thus be compared to the electrical thermostat that regulates or controls the house temperature in a home heating system.)

To explore the implications of the general feedback diagram in Fig. 10.1, we trace the signal paths through the diagram. The input signal σ_d (i.e., the driving signal or signal disturbance, which for the case of the eye would be the incident light flux; the diagram is labeled parenthetically to illustrate its applicability to the pupillary control system) enters the system and, through the action of the operator blocks, causes a response signal σ_c (the retinal signal indicating light intensity) to be generated. The response signal is compared with a reference signal level, σ_s, fixed by the set-point mechanism. In the diagrammatic representation, a differencing device (sometimes called a comparator) is used to effect the comparison operation. The generated signal σ_c and the set-point signal are delivered to the comparator; if these signals are the same ($\sigma_c = \sigma_s$) the output of the comparator ($\sigma_s - \sigma_c$) is zero. If the response signal is not the same as the set-point value, the comparator delivers an "error" signal σ_ϵ which is equal to the difference between the actual generated response (σ_c) and the desired signal level (σ_s) fixed by the set-point mechanism. In short, the comparator delivers an error signal given by $\sigma_\epsilon = \sigma_c - \sigma_s$.

Up to this point the description initiated in the preceding paragraph could be applied to almost any system. The basic feature that distinguishes a feedback system from other systems is how the error information is utilized. If the error information is used simply to indicate that an error exists, the system is open-looped (no feedback) with no self-correcting capability. On the other hand, if the error signal is used to modify the generation of the system error (by affecting the system components or the signal excitation) then the system may be considered a feedback (closed-loop) system with possible self-correcting capability.

In general, the diagrammatic representation of a feedback system is drawn to show the characteristic closed-loop feature of the system. Thus, in Fig. 10.1 the block-diagram configuration includes a signal path which closes on itself. The signal representing the generated error (or some signal related directly to the error signal) is fed into the signal path that propagates the system output. The diagram thus shows the closed-loop output of the feedback system where the error has a specific effect on itself. For the particular case of the pupillary control system, the error signal generates muscle motions which adjust the orifice area of the pupil. These area changes modify the effect of the incoming light signal and so affect the system response.

It was noted in the previous paragraphs that error signals in feed-back systems generally affect or modify the production of the system error. Often this effect acts to reduce continuously the generated error to some minimum or zero; such feedback systems have self-correcting capability and are stable. It seems intuitively obvious, however, that not all feedback systems need be self-correcting or stable. If the error infor-mation somehow causes the error to increase steadily (or periodically) it is evident the system is unstable and it must ultimately destroy itself if the condition persists. In physiological systems this could lead to de-struction of function and/or death. Although physiological systems are generally of the stable self-correcting type, it is conceivable that under some conditions system elements may change (e.g., through disease or accident) and the system may be converted from a stable one to an unstable one.* In a later section we shall see how modification of a block function may lead to an unstable-loop configuration.

The system shown in Fig. 10.1 has been described as a represen-tation of both a regulatory system and a homeostatic system. The engi-neer views a regulating system as any system in which the output remains at a fixed level in spite of disturbing influences; the desired level of output is set by a gauge mechanism. The thermostatically controlled home heat-ing system that maintains a relatively constant room temperature in spite of external-temperature variations is one example of a regulated system. The automatic pilot that maintains the heading of an aircraft in spite of wind disturbances is another illustration of a regulatory system. As the term "regulating system" is intended to include all systems that main-tain a fixed level of output, homeostatic systems may be considered a special class of regulators that pertain specifically to biological processes.

There is another slight difference between the implications of the terms regulatory and homeostatic. The set-point mechanism in homeo-static systems is generally located internally; it is not easily accessible nor is it ordinarily adjustable at will. In regulatory systems the set point is not necessarily internal, inaccessible, or unalterable. As an illustration of the latter situation, the set point for beam guidance systems for air-craft is not included within the system, but it is accessible and alterable.

---

* A homely qualitative example of such a situation is the relation between tiredness and sleep. Under normal circumstances a level of tiredness induces sleep to reduce tiredness (at least in the young!). When one becomes ill, as with a sore throat or nasal congestion, the sleep mechanism is impaired; lack of sleep causes more tiredness, which causes the patient's illness to become worse, which in turn makes sleep even more difficult, etc. This unstable condition persists until the body invokes other homeostatic systems to cope with the emergency situation, or drugs are used to alter the physiological components (e.g., nose drops which clear the nose) and thereby reestablish stability. If neither measure is effective the illness becomes progressively worse and leads to other pathological states which may ultimately result in death.

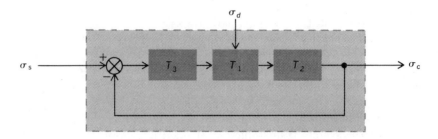

FIG. 10.2 *Representation of a regulator.*

As the reference level for a regulator system may be delivered from external sources, such systems are often drawn as shown in Fig. 10.2. This diagram is essentially the same as Fig. 10.1, except the location of the blocks has been rearranged and the set-point signal is now delivered to the system from a point external to the system. This change is evidently not of fundamental significance as it does not alter the basic signal relationships and block operations. Figures 10.1 and 10.2 thus represent the same kind of system: if σ_s is some constant-level signal, the system would be considered a regulator; if, in particular, the diagram pertains to a biological system, where σ_s is an internally fixed signal, the diagram would be considered a representation of a homeostatic system.

By one small change in viewpoint, we can now use the diagram of Fig. 10.2 to describe the concept of the servomechanism. If we remove the restriction that σ_s is a fixed signal and permit it to be continuously variable, we have what is generally considered a servomechanism representation in which the output is "slaved" to respond to the input. Ordinarily, the input and output signals of a servomechanism (or servosystem, as it is sometimes called) relate to the same kind of measurement (e.g., both represent displacement measurements). The system then operates as a master-slave unit where the output is slaved to follow the input.

As an illustration of a human being functioning as a servosystem, we shall examine the representation of an individual involved in a simple tracking task. When a person uses his finger to track a moving object, he "slaves" the position of his finger to the position of the object. As the object moves, so must the tracker's finger. If we consider p_o, the position of the object, to be the input signal and the finger position p_f to be the output signal (see Fig. 10.3), the system performs an efficient tracking task when p_o and p_f are equal. Under these conditions the positional error detected by the eye and brain is zero (that is, $p_o - p_f = 0$; it is of interest to note that in this case the eye and brain combination act as the comparator). If the comparator detects an error between the finger position (p_f)

and the desired position (p_o) (that is, $p_o - p_f \neq 0$) then signals are delivered to the appropriate muscles to reposition the finger.* The block
diagram of this servo operation, shown in Fig. 10.3b, may be compared
with the general diagram of Fig. 10.2. Although the diagrams differ in
particulars, they are of the same general configuration.

The finger tracking task is but one of the many situations in which
a physiological system functions as a servomechanism. The uncanny
ability of a bloodhound to follow an assigned scent and the ability of a
human being to direct a vehicle along a winding road are other illustrations of physiological servosystems.

Physiological systems represent a class of systems having inherent

* It might appear that the system must run in error in order for the arm continuously to
track a moving target. Intuition and experience indicate that this is not necessarily the
case; one can often track a moving target and maintain zero error. This intuitive conclusion
will be supported later when we see that the mathematical form of the blocks determine
how well a system can perform as a servo control system.

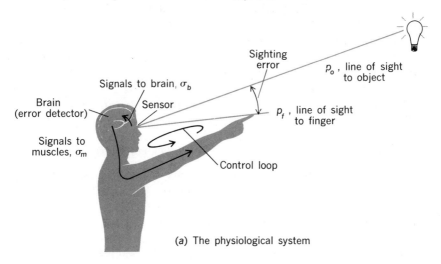

(a) The physiological system

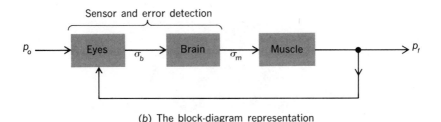

(b) The block-diagram representation

FIG. 10.3 *Human being functioning as a servosystem during a tracking operation.*

servomechanism characteristics. Nonliving systems do not generally show such characteristics although they may be designed to do so. Man-made servosystems are in evidence everywhere: Telewriter devices in business organizations and machine tools slaved to follow template guides or computer programs (to mention only two illustrations) are examples of such servo equipment. Servo principles have also been used to develop such diverse corporeal aids as myoelectrically operated prosthetic devices, respirators slaved to the patient's breathing cycle, cardiac pacemakers which can provide a variable heartbeat subject to the patient's demands, and various other man-assist devices. Although man-made servosystems are not always as versatile or adaptable as physiological systems, they can very often be designed to outperform physiological systems for specific tasks.

We conclude this preliminary discussion with a brief review of the implications of the various ideas and terms that have been introduced. The most general idea presented was that of feedback, which connotes an output effect that has an effect on itself. The most general type of feedback system is a servosystem or servomechanism; such systems deliver an output signal that follows an input signal in spite of disturbing influences. If the input signal of a servosystem remains at a fixed level (or set point), the servosystem may be classified as a regulator. Physiological regulatory systems having inherent, nonalterable set points are called homeostatic systems.

10.2   THE CHARACTERIZATION OF A PHYSIOLOGICAL FEEDBACK SYSTEM

One may speak glibly of feedback systems if the discussion is limited to qualitative descriptions of known control configurations. For very simple configurations one may even go somewhat farther: If the system can be represented by a simple block diagram, it may be possible to obtain mathematical descriptions of the blocks and then it may even be possible to obtain a rather complete quantitative analysis of the system. For the hypothetical system shown in Fig. 10.4, for example, it is easy

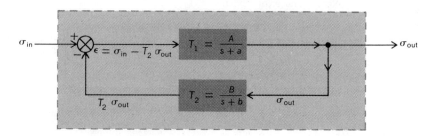

FIG. 10.4   *A single-loop feedback system.*

to determine that the overall transfer function of the system may be given by

$$(10.1)^* \qquad \frac{\sigma_{\text{out}}}{\sigma_{\text{in}}} = \frac{A(s+b)}{s^2 + (a+b)s + (ab + AB)}$$

If the form of the input signal is specified, it is possible to predict the form of the output signal by using the methods described in earlier chapters.

However, the quantitative analysis of realistic feedback systems is not so simple or straightforward as one might be led to believe by the previous illustrative example. In the first place, it is not always easy to establish beyond doubt that a system has a feedback loop. This may appear to be surprising until one considers that simple stimulus-response relationships can be obtained from all systems. One cannot conclude from such data alone that a system has feedback properties. If we excite a system with a step-function $4u(t)$, for example, and find the response may be given by $(10\frac{2}{3})u(t) + 2\epsilon^{-4t} - (16\frac{2}{3})\epsilon^{-3t}$ we could then describe the system transfer function by $2(s+5)/(s+4)(s+3)$.† We cannot know,

---

* To obtain the transfer function one simply performs the operations indicated by the diagram: Thus $\sigma_{\text{out}} = \epsilon T_1$, where ϵ is the error signal delivered by the comparator. From the diagram, $\epsilon = \sigma_{\text{in}} - T_2\sigma_{\text{out}}$. When these two relationships are combined, the result is the overall transfer function for the simple, single-loop feedback system:

$$\sigma_{\text{out}} = \epsilon T_1 = (\sigma_{\text{in}} - T_2\sigma_{\text{out}})T_1 = T_1\sigma_{\text{in}} - T_2 T_1\sigma_{\text{out}}$$

or

$$\sigma_{\text{out}} + T_2 T_1\sigma_{\text{out}} = T_1\sigma_{\text{in}}$$

or

$$\sigma_{\text{out}}(1 + T_1 T_2) = T_1\sigma_{\text{in}}$$

Therefore

$$\frac{\sigma_{\text{out}}}{\sigma_{\text{in}}} = \frac{T_1}{1 + T_1 T_2}$$

If we use $T_1 = A/(s+a)$ and $T_2 = B/(s+b)$ the result is the transfer function given by Eq. (10.1).

In most engineering texts on control systems, T_1 is described as the forward-loop gain, and it is generally designated by the product term KG (that is, $T_1 = KG$). The block T_2 in the feedback path, the feedback factor, is often designated by H. When these symbols are used, the transfer function of a single-loop system becomes

$$\frac{\sigma_{\text{out}}}{\sigma_{\text{in}}} = \frac{KG}{1 + KGH}$$

† As an outline of the algebra:

$$\text{Input signal} = \sigma_{\text{in}} = \frac{4}{s}\,\delta(t)$$

$$\text{Output signal} = \sigma_{\text{out}} = \frac{10\frac{2}{3}}{s}\,\delta(t) + \frac{2}{s+4}\,\delta(t) - \frac{16\frac{2}{3}}{s+3}\,\delta(t)$$

$$\sigma_{\text{out}} = \frac{8(s+5)}{s(s^2 + 7s + 12)}\,\delta(t)$$

$$\text{Transfer function} = \frac{\sigma_{\text{out}}}{\sigma_{\text{in}}} = \frac{2(s+5)}{(s+4)(s+3)}$$

however, whether the system is composed of a single block with the calculated properties or whether it is a feedback system of the form given by Fig. 10.1. (See Prob. 10.1.) To make this determination we must somehow enter the system and perform other tests.

The next point regarding real systems is that even though it can be shown feedback exists it may not always be easy (or even possible) to determine the description of the blocks within the closed loop. Such a determination may require that the control loop be opened at some point and experiments be performed to assess the open-loop response of the system (see, for example, the procedure used in the investigation of the pupillary control system in Sec. 8.9). Although such a procedure may be possible in some cases, often it is not possible to open the loop without destroying the characteristics of the system components. This is particularly true in physiological systems where surgery is required to open the loop.

If it can be established that feedback exists, if the system configuration can be described, and if the blocks of the system can be quantitatively described, the next step in the procedure is to analyze or study the system. This again may present problems. If the mathematical description of the blocks is relatively simple and linear and the loop configuration does not include delay functions (Sec. 8.9) or nonlinear operations (such as multiplication of variables instead of addition of variables), a mathematical analysis may then be possible. In general, however, these conditions cannot be fulfilled and a mathematical analysis

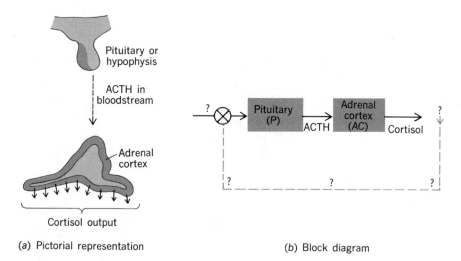

(a) Pictorial representation   (b) Block diagram

FIG. 10.5 *Hypophysis-adrenal system.*

is not feasible. As an illustration of this point, the pupillary control system has actually a relatively simple system configuration (Fig. 8.41*b*). Yet the equation relating the input to the output is a formidable expression not readily amenable to a general solution. A computer simulation is, of course, possible, and such a procedure can be used to examine a particular solution and some of the system trends.

To illustrate some of the points outlined in the previous paragraphs we shall examine the evolution of a control-system configuration which was developed to describe the operation of a particular endocrine system.* The system to be discussed deals with the cortisol- (also known as alphahydroxycorticosterone, hydrocortisone, or compound F) regulating function which involves the adrenal glands (located above the kidneys) and the pituitary gland (or hypophysis, as it is now called) located at the base of the brain. We begin by noting the following:

1. It is known the adrenal cortex secretes cortisol in response to signal secretions (adrenocorticotropin hormone or ACTH) from the hypophysis (see Fig. 10.5*a*) but it is not known exactly what function the cortisol serves.

2. It is known that ACTH and, therefore, cortisol secretions are increased when a physiological system is subjected to some stress stimulus (such as intense heat or cold or laparotomy, etc.). It is also known that if cortisol is not present the animal does not respond well to stress and may quickly expire, but it is not known how the cortisol helps the animal to resist stress.

Although this is not a great deal of information, it can be used to start the development of a diagrammatic description of the system. A representation of some of the information given by statements 1 and 2 is shown in Fig. 10.5*b*. Cortisol is the output of the adrenal cortex;

---

* The endocrine systems are chemical control systems that govern some of the specialized physiological processes associated with living systems (such as metabolism, growth, etc.). These endocrine systems may be compared to the neuromuscular systems in the sense that both provide homeostatic control functions. The two groups of systems differ, however, in their mode of operation: Neuromuscular systems exercise regulatory control of physiological processes through muscular responses to neural signals, whereas the endocrine systems exercise regulatory control through the secretion of chemical substances (hormones) in response to chemical and/or neural signals. Signals to an endocrine gland (or ductless gland, as it is sometimes called) cause the specialized cells in the gland to release hormones into the bloodstream. These secretions are distributed throughout the physiological system, and so they may act to influence the operation of the total physiological system or they may tend to act on some specific target organ in the system.

It is not possible here to provide an adequate discussion of endocrine systems. For more information the reader is referred to Ref. 5, chap. 72. This provides a good introductory background to endocrinology in general. For a more detailed discussion of such material see Refs. 6 and 7.

stress (used here as a generic term to represent an insult to the physiological system) somehow stimulates the pituitary system to activate the secretory cells in the adrenal cortex.

As the cortisol levels in the body are regulated within fairly close limits, it is evident some homeostatic mechanism is at work. This is schematically indicated by the phantom closed-loop configuration shown in Fig. 10.5b. It is not possible in this instance to indicate specifically how the closed-loop system must appear because, although it is clear some homeostatic mechanism is at work, it is not clear whether the homeostatic system is designed to control the level of cortisol or whether the function of the mechanism is to control an effect produced by the cortisol. This point might be more clearly understood if we attempt to postulate a homeostatic configuration for each situation under normal operating conditions (i.e., no external stress).

One such diagram is shown in Fig. 10.6a. In this diagrammatic representation the effect of the cortisol is considered the controlled variable. The set-point level for the system is designated by σ_{SPE}; this set point represents the desired level of an effect that is to be maintained (a temperature effect or a metabolic rate, possibly, or blood-pressure

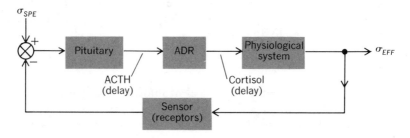

(a) Homeostatic control of physiological effect

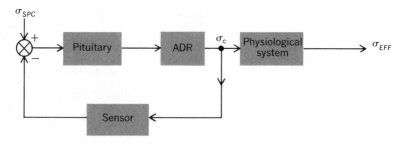

(b) Homeostatic control of adrenocortical secretion

FIG. 10.6 *A simplified homeostatic representation of pituitary-adrenal complex.*

level or volume level, or any combination). The difference between the desired effect and the actual effect produced (i.e., the error signal) actuates the pituitary gland which delivers the ACTH into the blood-stream. After a transport delay the ACTH reaches the adrenal cortex, which is activated to release cortisol. Again, after a lag the cortisol acts on the physiological system to produce the effect desired; this is represented as the output of the system. To determine whether the effect produced is at the required level, the physiological system "senses" the effect through a sensor which generates a signal to be compared with the set-point level. Thus, if the desired effect is pressure regulation, the sensor block in Fig. 10.6a will represent the pressoreceptors of the physiological system; if the desired effect is the regulation of the osmotic concentration of the extracellular fluid, the block will represent an osmoreceptor,* etc.

In contrast to a system that controls the effect of cortisol, Fig. 10.6b represents a homeostatic system in which the secretion of cortisol is the controlled variable. In this instance the set point (σ_{SPC}) represents the level at which cortisol is to be maintained. If the actual level of cortisol and the desired level are not the same, an error signal is generated to excite the adenohypophysis (i.e., the section of the pituitary gland that delivers the ACTH; this section is also sometimes described as the anterior pituitary). ACTH is delivered to excite the adrenal cortex which then operates to adjust the cortisol to the desired level.

The difference between the two control viewpoints shows up rather clearly in the diagrammatic representations. In the first case (Fig. 10.6a) the cortisol level is a secondary effect of the desired result; the physiological effect of the cortisol is the primary controlled variable. If the physiological effect of the cortisol is proportional to the level of cortisol present (i.e., if the block labeled "Physiological system" performs a simple constant multiplicative operation) the level of cortisol is regulated automatically (and simultaneously) with the regulation of the effect. If such proportionality does not exist, then only the effect of cortisol will be controlled; the cortisol level will vary accordingly. In the second configuration, however, the level of cortisol is the primary controlled variable; the effect of the cortisol is also directly regulated if the physiological system acts as a simple multiplier. If such proportionality does

* Biological sensors (or transducers) are sometimes divided into two general categories: the somatic senses and the special senses. The first group includes the sensors that detect mechanical properties (mechanoreceptors), body positions (kinesthetic senses), thermal sensations (thermoreceptors), etc. The second group includes the familiar sensors that detect light (sight), sound (hearing), odors (olfactory senses), and flavor (taste). The two groups of sensors are not completely and uniquely separable; there is often overlap, and one should not consider one sensor to be exclusively in one group or another.

not exist, only the cortisol level will be controlled. The effect of the cortisol could be anything, as the configuration of Fig. 10.6*b* implies that cortisol acts on the physiological system in an open-looped fashion; the effect produced by cortisol is not sensed or controlled physiologically.

It is not possible to resolve completely the question of which model is more valid. Yates and Urquhart* cite data that seem to support the existence of a regulatory system to control the level of cortisol. Later data presented by Smelick,† however, seem to raise questions about the conclusions drawn by Yates and Urquhart. To the author's knowledge, the question is still an open one and experimentation is still in progress to establish whether the regulatory system involving the pituitary-adrenal complex controls only the level of cortisol or only the effect produced.

It is possible both types of control functions are included in the physiological system. Such a configuration might then appear as in Fig. 10.7. By this arrangement the body would not have to depend upon a programmed relationship between cortisol and its effect to control both these variables. Each variable would be separately controlled, although the variables are interrelated.

* F. Eugene Yates and John Urquhart, Control of Corticosteroid Concentrations, *Physiol. Rev.*, **42**:359–443 (July, 1962).

† P. G. Smelick, Failure to Inhibit Corticotropin Secretion by Experimentally Induced Increases in Corticoid Levels, *Acta Endocrinol.*, **44**:36–46 (1963). See also the discussion Control of Pituitary-Adrenocortical System in Ref. 8.

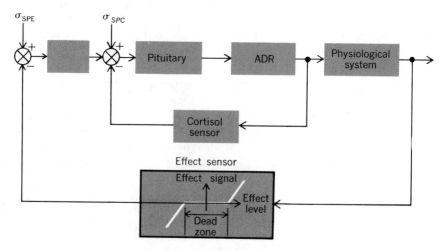

FIG. 10.7 *Feedback configuration in which effect and level of cortisol are controlled.*

The representation in Fig. 10.7 is more general than those of Fig. 10.6a and b as it includes the properties of both these representations and allows also for the possibility of separate or combined operation, depending upon the characteristics of the sensors. For example, if the cortisol sensors are nonexistent or inoperative, Fig. 10.7 represents a system in which only the cortisol effect is homeostatically controlled and Fig. 10.7 then reduces essentially to Fig. 10.6a. Similarly, if the sensors that detect the cortisol effect operate only when the effect exceeds a predetermined level, the sensor then acts as a form of switch or diode: When the effect is below a threshold value, the sensor remains inoperative; when the effect is above threshold, the sensor operates normally and transmits a control signal. Systems components having such properties are said to have dead-zone characteristics; these are represented diagrammatically by a block with a graphical description of dead-zone operation, as shown in Fig. 10.7.* If the cortisol sensor has simple linear-gain characteristics and the effect sensors have dead-zone characteristics, then below some threshold level the system in Fig. 10.7 will act as that shown in Fig. 10.6b; above the threshold level both homeostatic mechanisms will be operative.

There is no way of knowing which of the proposed control configurations most nearly represents the true system. One must accept what appears the most reasonable representation and then attempt experimentally to describe the components indicated to lend support to or refute the proposed representation. In the pituitary-adrenal system all proposed simulations include a block to represent the adrenal cortex; a description of this block is important no matter what the configuration of the system may ultimately be. One might therefore begin a quantitative description of the pituitary-adrenal complex by attempting to develop a mathematical description of the adrenal-cortex characteristics. This immediately presents a difficulty: In general, one cannot easily determine the characteristics of a system component while it is within a closed-loop configuration. Unless a significant time delay or transport lag is present in a loop, an applied signal and the resultant response do not measure the separate properties of a component in a feedback configuration. Such an experimental procedure measures the overall properties of the closed-loop system, not an individual component of the loop.

To appreciate the significance of these comments we use the configuration in Fig. 10.6b as an illustrative example and assume this diagram may serve as a representation of the physiological system (obviously a gross, naïve oversimplification). We seek to define the properties

---

* Such control components may be analogously represented through the use of function generators. See the discussion in Sec. 3.7 and Probs. 3.4 and 10.4.

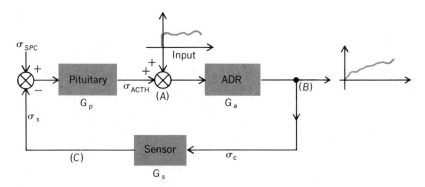

FIG. 10.8 *A step signal applied to the hypothetical representation of pituitary-adrenal system.*

of the ADR block and plan to do so by injecting a known concentration of ACTH as an input signal to the block and measuring the cortisol response as the output of the block. The block diagram of Fig. 10.8 indicates the proposed procedure.* The input signal is delivered at point A and the response is measured at point B. From a qualitative examination of the diagram we can immediately see that for closed-loop operation the signal delivered to the ADR block cannot be simply the applied input signal. The signal entering this block must be the sum of the applied signal and the signal delivered from the pituitary block; however, the latter signal is not known, nor has it been measured. The input to the ADR block is thus not actually available; therefore the ADR block cannot be defined in the manner proposed. (See Prob. 10.6.)

Further examination of Fig. 10.8 indicates clearly how the description of the ADR block should be determined:

1. If an excitation signal at A is to be used as the input signal to the block we must somehow open the loop to ensure no signal is delivered to A through other paths, such as BCA.†

---

* We assume this procedure is possible surgically. Very often what appears to be fairly simple diagrammatically proves to be grossly impractical or totally impossible from a physical point of view.

† Or, what amounts to the same thing, we must delay transmission around the loop until the ADR block has had the opportunity to respond to the applied excitation. The latter condition may already be conveniently built into the system as a result of transportation lag or other transmission delays. It should be established, however, that if such a delay exists it is of sufficient duration to permit the assumption that the system is open-looped during the period of investigation. A good illustration of this point was given in Sec. 5.8 where the dye-dilution system was discussed. It was shown that, if the transport lag (recirculation time) is sufficiently long, the system could be considered open-looped to allow an input-output study of a block within the loop.

2. If it is not possible to open the loop or depend upon transportation lag, we must measure the total input and output signals of the block under investigation as continuous functions of time. From such data it may be possible to use piecewise-linearized approximation to obtain a first estimate of the block properties. Once again, it should be mentioned that this may be easier said than done. It may not be simple to measure all delivered signals to and from a particular physiological component, nor may the data acquired by such a procedure be simple to analyze. This does not imply we should forget the whole thing. It does imply we must understand the limitations of our techniques and procedures and realize we may not have measured what we think we have measured. As a result of these possible pitfalls, great significance should not be attached to the results obtained from a single experiment if they are not supported by other information, experiments, or data.

It is not the intent here to examine fully the detailed aspects of the pituitary-adrenal control configuration. This system was used merely as a representative example of the problems and difficulties that may be expected in an investigation of physiological feedback systems. To complete this discussion, however, we shall examine a recent model that has been proposed* for this particular system. The overall block-diagram representation is shown in Fig. 10.9.

It can be seen the model is far more elaborate than the simple configuration first postulated in Figs. 10.5 and 10.6. Not only does this model include additional control loops not previously mentioned but it includes also the effects of a physiological stress (hemorrhage in this case). In this model it is assumed some of the cortisol effect is included in the cortisol-control loop (as implied by the block labeled "Distribution system").

The comparator section includes the set-point concept with a variable set-point feature. This viewpoint is to some extent the same as a constant set point with a superimposed variable excitation signal that depends upon stress stimuli. The constant set point maintains the normal level of cortisol; the actual level is governed by normal requirements plus additional requirements which may be due to internally or externally induced emergency excitations.

The comparator section in Fig. 10.9 also shows a more elaborate breakdown of the signal routing in the hypothalamic-median eminence complex. Nerve signals are shown to affect the reticular-activity system of the brain. This area is believed to be the site of neurohormonal pro-

* A detailed description of the evolution of this model is given by Donald S. Gann, Systems Analysis in the Study of Homeostasis, with Special Reference to Cortisol Secretion, *Am. J. Surg.*, **114**(1):95–102 (July, 1967).

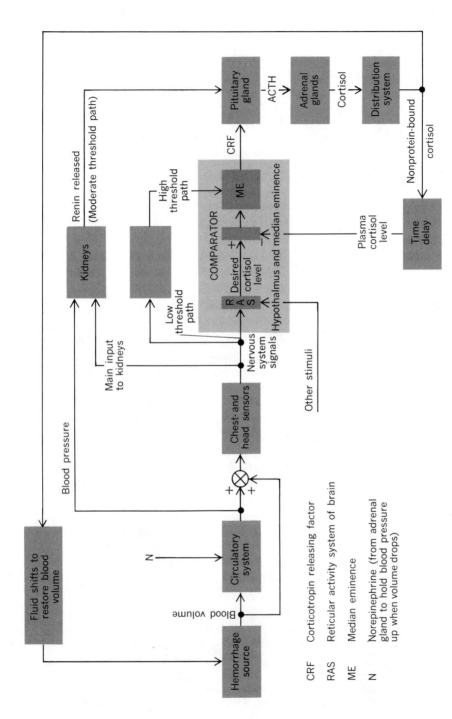

FIG. 10.9 Complete model for control of cortisol secretion after hemorrhage.

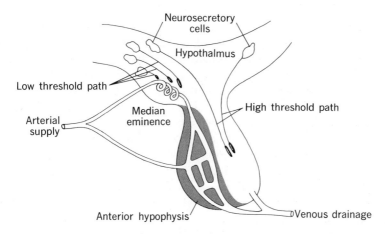

FIG. 10.10 *Anatomic description of control configuration in hypothalamic-hypophyseal complex.*

duction.* The neurohormones are carried (along or within the nerves) to the vicinity of the median eminence (see Fig. 10.10) and are then ejected to influence the arterial supply to this area. The flow from this supply (with the neurohormonal substances) enters the interior pituitary to mix with the arterial supply for this section which also contains chemical information about the cortisol level. This combined information triggers the pituitary to deliver ACTH.

The latter description may be considered the low-threshold mode of operation. For emergency situations the first section of this operational sequence may be bypassed, and the nerve stimuli sent to the hypothalamic area may trigger nerve cells which deliver neurohormones directly to cells of the anterior hypophysis to induce immediately ACTH production. This form of operation has been described as a high-threshold path in Fig. 10.9 (see also Fig. 10.10).

In addition to the two modes of behavior just described, it has been shown that, when the hypothalamus is deactivated or blocked (by drugs or saturation), an alternative control path through the kidneys becomes significant. The kidneys release the chemical renin which apparently also stimulates the pituitary gland to deliver ACTH. This effect is described as the moderate-threshold path in Fig. 10.9. The excitation to the kidneys which energizes this moderate control loop appears to be

---

* The neurohormones are thought to be chemical agents which are ejected at the nerve axons and operate to release other hormones. They are sometimes described as "releasing factors," for example, thyrotrophin-releasing factor (TRF), corticotrophin- (or ACTH-) releasing factor (CRF), etc.

due to the actual mechanical effects of blood pressure and to various pressoreceptor signals (i.e., neural signals) which respond to blood-pressure variations. Experimentation seems to indicate the latter signal is the dominant excitation that triggers the renin response of the kidneys.

It is of interest to note that chronologically the identification of the various control loops did not follow the sequence as outlined; the moderate-threshold path was discovered before the high-threshold path. When it was determined the kidneys were involved in the cortisol-control system (by hypothalamic block) experiments on animals were performed to assess the significance of this loop. In one set of experiments the loop was opened (by surgical means; i.e., the kidneys were removed) and it was then shown that some direct effect on ACTH production was still operative. Further investigation indicated that a path directly from pressure sensors to the hypothalamus probably exists, as shown by Fig. 10.9.*

The model representation in Fig. 10.9 is by no means complete. Experimentation is still necessary to verify some of the links proposed and to describe the blocks in a more quantitative manner. Even in this form, however, the diagram shows vividly how homeostasis may be realized by a variety of feedback subsystems. The various paths from the hemorrhage input signal to the pituitary gland permit the system to respond rapidly and easily to small changes in blood volume (the low-threshold path) and to utilize additional channels (the moderate- and high-threshold paths) in the case of large losses of blood.

The procedure by which the model has been derived also deserves comment as it is in itself significant. A crude model based upon early scientific knowledge is first proposed. This model suggests experiments; the results of these experiments demand modification of the original model; these modifications suggest other experiments, etc. This evolutionary process, a feedback process in its own right, leads to closer and closer representations of the true situation. Feedback might thus also be considered a way of thought as well as a means for control.

The most important use of a model such as that shown in Fig. 10.9 is the guidance it provides for medical research. If we simplify the model by combining blocks that are closely associated, we find that the portion from the hemorrhage source to the output can be represented as shown in Fig. 10.11. There is only a relatively small number of critical elements. In feedback terms, this forward portion of the overall system

---

* Gann (*op. cit.*) also reported that, when all links shown between the hemorrhage and the pituitary gland were blocked chemically and surgically, there was no response for 15 min after hemorrhage. After 30 min, there was a slow increase in cortisol secretion—an observation that would indicate an additional path in the block diagram with long-time delay.

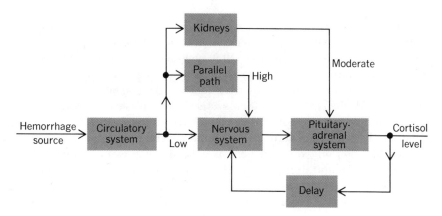

FIG. 10.11 *Essential elements of model.*

is characterized by the following:

1. An input sensor (the circulatory system)
2. The three paths in parallel to the pituitary-adrenal system
3. The pituitary-adrenal system
4. The feedback path

The two block diagrams indicate the data we should like to secure from research in order to achieve a quantitative understanding of the system. The long-time-constant element is the pituitary-adrenal system (with response times of the order of 7 to 8 min, compared with the time-constants of fractions of a second within the central nervous system). Hence the dynamics of the overall system is determined primarily by the transfer function of the pituitary-adrenal components, and accurate measurements should be made on the dynamic response here (by excitation of the pituitary, to measure ACTH and cortisol). For dynamic analysis of the low-threshold path, these measurements and the characteristics of the circulatory system should be sufficient.

In the study of the moderate-threshold path (through the kidneys), the dynamic analysis requires determination of ACTH response to blood-pressure changes with the low-threshold path blocked. Evaluation of the effect of this feedforward path on system performance characteristics requires a picture of the relative dynamics of the two paths (through the recticular-activity system and the kidneys). In engineering systems, such feedforward (from the input to a later point in the forward path, as through the kidneys in Fig. 10.9) is often used to realize system performance that is unusually insensitive to changes in the characteristics of the output elements (the pituitary-adrenal system). Is this same characteristic achieved by the feedforward in the present system?

The block diagram of Fig. 10.9 suggests that both blood pressure and volume actuate the sensors that generate signals to the recticular-activity system (as indicated by the adder). If the response characteristics of the block labeled "Circulatory system" are measured quantitatively, understanding of the complete system also requires determination of whether the volume is actually measured by the nervous-system sensors or whether the only measurement is actually of pressure, a study that would require maintaining constant pressure during a change in volume.

Finally, the block diagram indicates that, if the performance here is analogous to that of common engineering systems, the inner feedforward path (with a high threshold) is perhaps used for switching purposes in the hypothalamus, to augment the excitation of the pituitary gland in case of saturation-level operation. In order to determine whether the same function is performed here, we should have to consider the inter-relationship of this system with the numerous control systems that share the hypothalamus, pituitary, and adrenals. Hence the understanding of this system ultimately requires a more complete block diagram showing the interconnections with other physiological functions.

10.3 LINEARIZATION AND ANALYSIS OF A SIMPLE SYSTEM

In the preceding section it was noted that physiological feedback systems can seldom be analyzed in a mathematical sense, as most are far too complicated and involve nonlinear operations not readily amenable to analytic treatment. In such instances analog simulations and computer studies provide the only practical means by which a quantitative examination of system performance may be obtained.

In some instances, however, when a system is relatively simple or when a loop of a system can be examined separately (because of high-speed operation compared with the rest of the system) it may be possible to linearize the system operations (if nonlinear operations are present) and perform a simplified analysis of the system. By such procedures it is often possible to gain some insight into the properties of the system and to assess its performance characteristics. The pupillary control system discussed earlier (see Sec. 8.9) affords a good illustration of how this suggested procedure can be used to gain understanding of the system operation.

It will be recalled the pupillary control system operates to adjust the pupil diameter, thereby limiting the amount of light delivered to the retina of the eye. A diagrammatic representation of the system configu-

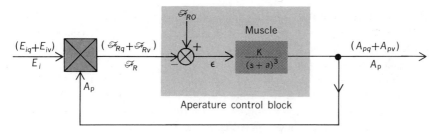

(a) Nonlinear representation

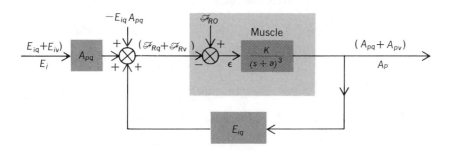

(b) Linearized approximation

FIG. 10.12   *Closed-loop configuration for the pupillary control system.*

ration is shown in Fig. 10.12.* This diagram is essentially the same as the earlier representations of this system (see Figs. 8.36 and 8.38) except that Fig. 10.12 includes a set-point mechanism and the control muscles as separate elements. The set-point level is designated as $\mathcal{F}_{RO}$; this essentially sets the quiescent operating level of the system variables (see Probs. 10.5 and 10.7). The transfer function for the muscle component of the system was derived by studies of the system sinusoidal response during open-looped operation (which was achieved through the use of special optical techniques); the procedure and derivation used to evaluate this operational representation were described in Sec. 8.9. It may be noted for this discussion no time lag has been associated with the operative control block. We might justify this omission by noting the transmission delay was determined to be about 0.2 s; this delay is probably due to a lag in the neural transmission. The operative control response is appreciably slower than the signal lag; it has a time-constant of about 10 s.

---

* A brief review of Sec. 8.9 would be appropriate at this point.

[The transfer function for this block was determined in Sec. 8.9 to be $0.16/(0.1s + 1)^3$; this is equivalent to $160/(s + 10)^3$.] In this case, therefore, we might guess the signal lag can be neglected compared with the sluggish reaction of the operative muscle control.

The system is somewhat simplified by the omission of the transmission lag but it is still not amenable to a mathematical analysis. The product operation (which involves the product of the *variables* E_i and A_p) within the loop configuration gives rise to a nonlinear mathematical representation for the overall system. A general solution for the system operation still cannot be obtained, therefore, by standard techniques (see Prob. 10.8). If we can make the further approximation that the system is subjected to small signal excitations, it will be possible to obtain an approximate, linearized representation for the system. This linearized model can then be used to study the general operating characteristics of the system in the range of the applied excitation and provide some insight into the physical properties of the system.

To show how small signal considerations can lead to a linearized representation of a system, we shall assume the input excitation in Fig. 10.12a is composed of a quiescent component (E_{iq}) and a small variable component (E_{iv}).* The input signal may thus be written

$$(10.2) \qquad E_i = E_{iq} + E_{iv} \qquad \text{(See Fig. 10.12)}$$

The small variable excitation component induces corresponding variational components in all the system variables. The aperture area A_p thus becomes $A_{pq} + A_{pv}$, and the incident flux becomes $\mathfrak{F}_{Rq} + \mathfrak{F}_{Rv}$. If we now use these representations for the system variables, the product operation required by the configuration of Fig. 10.12 may be written

$$(10.3) \qquad E_i A_p = (E_{iq} + E_{iv})(A_{pq} + A_{pv})$$
$$= E_{iq}A_{pq} + E_{iq}A_{pv} + A_{pq}E_{iv} + A_{pv}E_{iv}$$

The multiplicative operation is thus equivalent to the summation of four terms. Three of them are the product of two constants ($E_{iq}A_{pq}$) or the product of a variable and a constant ($E_{iq}A_{pv}$ and $A_{pq}E_{iv}$); only the last term is the product of two variables. On a comparative basis, however, we note if E_{iv} is about 1 percent of E_{iq} and A_{pv} is about 2 percent of A_{pq} (i.e., small variations are assumed) then the maximum value of $A_{pv}E_{iv}$ will be given by $2 \times 10^{-4} A_{pq} E_{iq}$. When we compare this number with the maximum value of $A_{pq} E_{iv}$ ($= 10^{-2}A_{pq}E_{iq}$) and $E_{iq}A_{pv}$ ($= 2 \times 10^{-2} A_{pq}E_{iq}$) we see that the product of the two variables can be neglected, compared with the product of the constant and the variable. This approx-

* A review of Prob. 8.4 would be helpful at this point.

imation permits us to write Eq. (10.3) as

(10.4)
$$E_i A_p \approx E_{iv} A_{pq} + A_{pv} E_{iq} + A_{pq} E_{iq}$$

Equation (10.4) is a simple linear relationship involving only the sum of variables. The equation can be represented diagrammatically as shown in Fig. 10.12b (see also Fig. P8.4c).*

Figure 10.12b is now a linearized representation of the pupillary control system, and standard techniques can be used to derive a relationship between the input and output. Since the system is linear, we could calculate the response (A_p) for each separate input (while other signals are ignored) and summate the results to find the overall result. Thus we could determine A_p when $\mathcal{F}_{RO}$ acts alone and then find A_p when E_{iq} acts alone, etc. The overall response would, of course, be the sum of the separate responses. A somewhat easier procedure results from the fact that all quiescent levels in the system are either known or available. We can therefore disregard these values and examine only how the system behaves around the quiescent levels (the absolute results will be the sum of the quiescent values and their variable components). If we use this concept, Fig. 10.12b may be reduced to the simpler diagram in Fig. 10.13.

---

* The validity of the representation may be checked by carrying out the indicated operations. In Fig. 10.12b we note

$$E_i A_{pq} = (E_{iq} + E_{iv}) A_{pq} = E_{iq} A_{pq} + E_{iv} A_{pq}$$

Similarly,

$$A_p E_{iq} = E_{iq} A_{pq} + A_{pv} E_{iq}$$

The sum of these two equations is

$$E_i A_{pq} + A_p E_{iq} = E_{iv} A_{pq} + A_{pv} E_{iq} + 2 E_{iq} A_{pq}$$

If we now deliver a term $-E_{iq} A_{pq}$ to the adder, as is shown in Fig. 10.12b, the result will be Eq. (10.4) as required.

The diagram of Fig. 10.12b is not the only way to represent the operations required by Eq. (10.4); other configurations are possible. The important point here is that the nonlinear multiplicative operation can be replaced by a linear additive operation if small signal behavior is postulated.

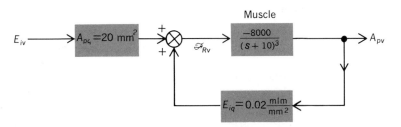

FIG. 10.13 *Pupillary control system when only small signal variations are considered.*

In this diagram all quiescent levels have been excluded; only the variable components are shown. (This procedure may be compared to the technique used in Chap. 5, Fig. 5.5.)

Before the system shown in Fig. 10.13 is examined, some comments regarding the description of the blocks should be made. First, it may be noted the muscle block has a negative sign associated with it. This sign is required to compensate for the positive signal that enters the block when $\mathfrak{F}_{RO}$ is neglected (it can be seen in Fig. 10.12b, when $\mathfrak{F}_{RO}$ is neglected, that the signal entering the muscle block should be negative). We can readily appreciate the physical significance (and need) for this negative relationship when we consider that increases (above quiescence) in the flux delivered to the eye should cause decreases (i.e., a reduction below quiescence) of the pupil diameter.

The transfer function for the muscle block was derived earlier when the open-looped sinusoidal response of the pupillary system was considered (Sec. 8.9). However, although the form of the transfer function is the same here as that derived earlier, the numerical values are not in agreement. To explain this discrepancy we must recall the basic definition of the transfer function derived in Sec. 8.9; the input and output variables were not simply signals but percentage variation. Specifically, the transfer function was defined as the ratio of percentage change in flux to percentage change in area (see the discussion in the footnote of Sec. 8.9). If we translate this definition into a mathematical expression that involves the variable used in this discussion we have

$$(10.5) \qquad T_m = \frac{A_{pv}/A_{pq}}{\mathfrak{F}_{Rv}/\mathfrak{F}_{Rq}} = \frac{0.16}{(0.1s + 1)^3}$$

where A_{pv} is the variational component of the pupil area, $\mathfrak{F}_{Rv}$ is the variational component of the applied flux, and A_{pq} and $\mathfrak{F}_{Rq}$ are the quiescent levels of the pupil area and flux, respectively. For the purposes of this discussion we are not interested in the ratio of percentage variations but rather in the ratio of actual variations. The transfer function we require, therefore, is defined by $A_{pv}/\mathfrak{F}_{Rv}$. From Eq. (10.5) we see this transfer function may be given by

$$(10.6) \qquad \frac{A_{pv}}{\mathfrak{F}_{Rv}} = \frac{A_{pq}}{\mathfrak{F}_{Rq}} \frac{0.16}{(0.1s + 1)^3}$$

To describe completely the function given by Eq. (10.6) we must now specify numerical values for the quiescent levels. From Stark's articles (*op. cit.*) it appears he worked with a quiescent pupil area of about 20 mm². From other graphical data (see, for example, Prob. 10.7) this corresponds to an illuminance (E_i) of about 0.02 mlm/mm²; the quiescent incident flux $(\mathfrak{F}_{Rq})$ would therefore be about 0.4 mlm. With these numerical values and after algebraic rearrangement, the muscle-control block can be described by

$$(10.6a) \qquad \frac{A_{pv}}{\mathfrak{F}_{Rv}} = \frac{8,000}{(s+10)^3} \quad \text{mm}^2/\text{mlm}$$

It is now a very simple task to describe the pupil response to illuminance excitation. We need only perform the indicated operations and form the ratio A_{pv}/E_{iv} to define the total transfer function. If we use the general formula for the overall description of a single-loop feedback configuration (see the first footnote on page 519) we have

$$(10.7) \qquad \frac{A_{pv}}{20E_{iv}} = \frac{-8,000/(s+10)^3}{1 - 0.02[-8,000/(s+10)^3]}$$

or

$$(10.7a) \qquad \frac{A_{pv}}{E_{iv}} = \frac{16 \times 10^4}{(s+10)^3 + 160} = \frac{-16 \times 10^4}{s^3 + 30s^2 + 300s + 1,160}$$

Equation (10.7a) provides the desired transfer function; if E_{iv} is specified, A_{pv} can be determined.

The true significance of Eq. (10.7a) is not that we can calculate the output for a given input; this is an almost incidental consideration. The more important usefulness of such a derived relationship is that, even without an exact analysis, it gives some idea what responses may be expected and it also indicates how the responses may be changed by component variations. Examination of Eq. (10.7), for example, indicates the transient pupil response will have either three time-constants or one time-constant and a damped natural frequency. (The transfer function is of third order; there must therefore be at least one real root and at most three real roots.) The pupil may therefore respond to a flicker of light (an impulse excitation) with a gradual change in diameter and a gradual monotonic decay to equilibrium, or its initial response may be relatively quick and the return to equilibrium may be gradual and oscillatory. The actual response depends upon the numerical values of the transfer function, and these depend upon the system constants.

The specific function given in Eq. (10.7) has a single real root and two complex roots. (The denominator may be factored into $s + 15.4$ and $s^2 + 14.6s + 75$.) The flicker response will thus be the sum of an exponential function which decays very rapidly (time-constant $= 1/15.4$ s) and an oscillatory term which has a natural frequency about 1.4 c/s $(\omega_n^2 = \sqrt{75}/2\pi)$ and a damping factor about 0.8 (see Sec. 9.3 and Prob.

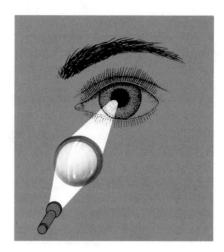

FIG. 10.14   *Arrangement to induce pupillary hippus.*

10.9). This system is rather heavily damped, and although some oscillation may be present in the transient portions of a response it is not likely to be observed.

The possibility of an observable pupil oscillation as implied in the previous paragraph is, however, real. In fact, ophthamologists can induce a persistent oscillation of the pupil by locating a small circle of light on the periphery of the pupil (as shown in Fig. 10.14, the oscillatory condition thus induced is technically known as pupillary hippus). This procedure changes the properties of the system to cause an unstable-feedback configuration; the result is a sustained oscillation even though the incident light is of a steady, invariant nature.

It is easy to explain qualitatively the nature of this oscillation. The illuminated area of the pupil is small; the illumination causes the pupil area to decrease; the decrease in pupil area (diameter) causes the flux delivered to the eye to be reduced effectively to zero; the pupil dilates to permit more light to enter, and so the cycle continues. To explain the operating characteristics of this system in quantitative terms, however, and to understand which physical mechanisms of the system may be involved (and how they may be involved) we must develop a deeper understanding of some of the properties of closed-loop systems, particularly with regard to the stability consideration of such systems. These concepts will be developed in the following sections.

10.4   STABILITY CONSIDERATIONS DUE TO CHANGE IN GAIN

To develop some of the concepts that relate to the stability of closed-loop systems we shall examine the specific illustration of Sec. 10.3 in closer detail. In this case, however, we shall use literal (instead of numerical) values to describe the component blocks. This will provide more generality and yet retain the familiarity of a system already examined. For this discussion, therefore, we use the closed-loop system defined by Fig. 10.15.

The overall transfer function for the system shown in Fig. 10.15 is derived in the standard fashion (see Sec. 10.2). The output-input ratio can be written

$$(10.8) \qquad \frac{\sigma_{\text{out}}}{\sigma_{\text{in}}} = \frac{K/(s+a)^3}{1 + KK_f/(s+a)^3}$$

or

$$(10.8a) \qquad \frac{\sigma_{\text{out}}}{\sigma_{\text{in}}} = \frac{K}{(s+a)^3 + KK_f} = \frac{K}{s^3 + 3as^2 + 3a^2s + a^3 + KK_f}$$

If the input signal is specified, Eq. (10.8a) may be used to define the output.

In the previous section, when a was 10 and KK_f was 160, it was shown this system could exhibit oscillatory characteristics in the transient response. When the literal terms had these numerical values, the denominator of the transfer function had complex roots; these complex roots gave rise to damped sinusoidal oscillations in the transient response (see Sec. 9.3 and Prob. 10.9). From a more general point of view, we may state this system will exhibit oscillatory transient-response characteristics when $s^3 + 3as^2 + 3a^2s + a^3 + KK_f$ contains complex factors. In an even more general sense, any closed-loop system (in fact, any system) will exhibit damped oscillatory transient behavior when the transfer function has a denominator with complex roots. These complex roots govern both the frequency of the oscillatory behavior and the damping in the system. The lower the system damping, the higher is the oscillatory frequency and the longer oscillation persists.

The comments in the preceding paragraph are simply a recapitulation of ideas previously presented and are included here only to emphasize that the roots of the denominator of a transfer function are the prime

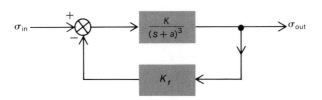

FIG. 10.15 *General representation of the system in Fig. 10.13.*

TABLE 10.1

a	KK_f	Factors of denominator	$1/\tau$	ζ	ω_n	ω_D
10	15,600	$(s+35)(s-5s+475)$ $(s+35)(s-2.5+j21.7)(s-2.5+j21.7)$	$+35$	-0.115	21.8	21.7
10	3,375	$(s+25)(s+5s+175)$ $(s+25)(s+2.5+j12.9)(s+2.5-j12.9)$	$+25$	$+0.19$	13.2	12.9
10	160	$(s+15.4)(s^2+14.6s+75)$ $(s+15.4)(s+7.3+j4.75)(s+7.3-j4.75)$	$+15.4$	$+0.84$	8.65	4.75
10	-125	$(s+5)(s^2+25s+175)$ $(s+5)(s+12.5+j0.42)(s+12.5-j0.42)$	$+5$	0.95	13.2	0.42
10	$-3,375$	$(s-5)(s^2+35s+475)$ $(s-5)(s+17.5+j13.1)(s+17.5-j13.1)$	-5	0.8	21.8	13.1
5	160	$(s+10.4)(s^2+4.6s+27)$ $(s+10.4)(s+2.3+j4.6)(s+2.3-j4.6)$	$+10.4$	0.44	5.2	4:61
2	160	$(s+7.4)(s^2-1.4s+22.4)$ $(s+7.4)(s-0.7+j4.7)(s-0.7-j4.7)$	$+7.4$	-0.15	4.74	4.7

factors governing the form of the system response. If we next note that the roots of the denominator depend entirely upon the constant coefficients in the denominator, it is easy to realize the system response may be changed radically when the constant terms in the denominator are altered.

As a specific illustration of how the variation of system constants can affect the response characteristics, we note once again that, when a in Fig. 10.15 was taken as 10 and KK_f as 160, the impulse-response of the system consisted of an exponential term with a time-constant of $1/15.4$ s and an oscillatory term with a (damped) frequency of 4.75 rad/s* and a damping factor 0.84. If a remains 10 but KK_f is increased to 3,375, the impulse-response consists of a transient term with a longer time-constant (25 s; see Table 10.1) but the oscillatory term now has a higher frequency (12.9 rad/s) and the damping factor is reduced to about 0.2. If KK_f is further increased to 15,600, the transient term in the impulse-response has an even longer time-constant (35 s) but now the oscillatory term not only has a higher frequency; it also has a negative damping

* Actual frequency $= \omega_n \sqrt{1-\zeta^2} = 8.65 \sqrt{1-0.84^2} = 8.65 \sqrt{0.3} = 4.75$.

factor. The latter condition is rather significant as it implies the transient response will contain a term of the form $\epsilon^{+0.1t} \sin 21.7t$. The positive exponential term indicates that the amplitude of the sinusoidal oscillation will continually increase (rather than decrease) with time. The transient response will therefore grow with time, and unless the response is limited by other factors (e.g., nonlinear variation of the system constants) the system will destroy itself. This situation is one illustration of an unstable system.*

The unstable condition described in the foregoing discussion was induced by increasing the gain-constants, K and/or K_f, of the system in Fig. 10.15. It is also possible to create an unstable condition in this system by reducing the gain-product. It is noted in Table 10.1, for example, if KK_f becomes sufficiently negative† one of the roots of the system denominator will be positive. This implies the transient response will contain a term of the form ϵ^{+5t}; it implies further the system transient response will grow without bounds until the system destroys itself or it will grow until other factors tend to limit the system output. Once again, therefore, we see adjustment of the gain-constant can change a stable system into an unstable one.

Although we have used only gain changes to illustrate how system responses may be altered, it should be obvious that adjustments of any other system constants may also change the system response from stable to unstable operation. The mathematical explanation for this effect involves the denominator of the transfer function; if the change of system constants alters the system denominator so that it includes factors with positive real roots, the system will exhibit unstable characteristics. Although the relationship between positive real roots and stability has not been explicitly stated prior to this point, the meaning of the phraseology should be clear. A brief examination of Table 10.1 will show that an unstable condition is created when the denominator contains factors of

---

* The exact definition of "unstable" varies from text to text. In general, systems having a tendency to respond with oscillations of increasing amplitude are consistently classified as unstable systems in all texts. Some texts, however, also consider systems that respond with sinusoidal oscillations of fixed amplitude (i.e., with zero damping; see Prob. 10.10) as unstable; other texts consider such systems marginally stable. This distinction is largely a question of semantics and application. This borderline situation (of zero damping) may be acceptable and even desirable for some situations and completely intolerable for others. An oscillator, for example, depends upon the zero-damped condition to provide a useful sinusoidal output signal; such systems would generally not be classified as unstable. On the other hand, a power-steering servosystem with zero damping would most certainly be considered an unstable system as an incessant oscillation of any magnitude could not be tolerated for such an application.

† A negative gain simply implies the output of a block will increase as the input decreases, or vice versa. This concept was encountered in an earlier discussion.

the form $s - a$ or $s - a \pm j\omega$. These factors give rise to positive roots and to exponential terms with positive time coefficients. Such terms indicate the system response will include signal components that grow with time.

10.5   TECHNIQUES USED TO ESTIMATE SYSTEM STABILITY

As noted repeatedly in the preceding section, the fundamental criterion for judging the stability of a system is the sign of the denominator roots in the transfer function. This basic concept is easy to state and understand and often is relatively easy to apply, particularly if one can enlist the aid of a computer to facilitate the evaluation of the roots of higher-order polynomials. In any case, the problem of system stability reduces essentially to the location and evaluation of the roots of polynomial expressions by any of the standard algebraic techniques (Descartes' rule of signs, Newton's method, etc.).

In many instances, the denominator of a transfer function is a relatively complex expression not readily amenable to straightforward factoring techniques. In such cases it is helpful to know what alternative techniques may be used to test stability and to understand how these techniques may be employed. The discussion to follow will introduce some of these stability tests; a brief explanation of the application of these tests is also included. It must be emphasized, however, this discussion should be considered only an introductory exposition; for a more detailed and thorough explanation of stability, the reader should consult any of the many excellent texts that deal with servo theory.*

The simplest and most direct test to determine whether an algebraic polynomial has roots in the right-hand half plane† is the Routh test. This test involves arithmetic operations with the polynomial coefficients when these are specially arranged in what is called a Routh array. The results of these operations yield information about the existence of roots with positive real parts. In more specific terms, if a polynomial has the form

$$(10.9) \quad \text{Denominator} = a_n s^n + a_{n-1} s^{n-1} + a_{n-2} s^{n-2} + \cdots + a_0$$

* Milhorn's book (Ref. 9) might be appropriate for those with life-science backgrounds.

† This is simply another way to describe roots with positive real parts. The terminology is derived from the location of such roots on the complex plane. It will be recalled the complex plane consists of a real (horizontal or x) axis and an imaginary (vertical or y) axis. Complex numbers are located in this plane according to their real and imaginary parts (see the discussion of complex notation and phasors in Sec. 8.2). If a complex number has a negative real part and an imaginary part (for example, $-a + bj$), it would be located above the real axis and to the left of the imaginary axis; all complex numbers with positive real parts are located to the right of the imaginary axis, or in the right-hand half of the complex plane, or in the right-hand half plane.

the beginning of the Routh array is formed from the polynomial coefficients a_n, a_{n-1}, etc., arranged in two rows as shown in the following:

$$\left.\begin{array}{cccc} a_n \kern-0.5em \diagdown & \kern-0.5em \nearrow a_{n-2} & a_{n-4} & \cdots \\ & \kern-0.5em \diagdown \kern-1em \diagup & & \\ & {}_1 \kern0.5em {}_2 \kern0.5em & & \\ a_{n-1} & a_{n-3} & a_{n-5} & \cdots \end{array}\right\}$$ coefficients from given polynomial

$$\left.\begin{array}{ccc} b_1 & b_2 & b_3 \quad \cdots \\ c_1 & c_2 & c_3 \quad \cdots \end{array}\right\}$$ constructed terms

The first coefficient and every other one thereafter form the first row; the second coefficient and every one thereafter form the second row. A third row (b_1, b_2, b_3, etc.) of the array is now constructed from the coefficients in the first two rows. The first term in the third row has the value

$$(10.10) \qquad b_1 = \frac{a_{n-1}a_{n-2} - a_n a_{n-1}}{a_{n-1}}$$

The second term of the third row has the value

$$(10.10a) \qquad b_2 = \frac{a_{n-1}a_{n-4} - a_n a_{n-5}}{a_{n-1}}$$

This pattern is continued until all the polynomial coefficients have been used to develop the third row. When the third row has been completed, a fourth row (c_1, c_2, c_3, etc.) is constructed in a comparable fashion. Thus

$$(10.11) \qquad c_1 = \frac{b_1 a_{n-3} - b_2 a_{n-1}}{b_1}$$

$$(10.11a) \qquad c_2 = \frac{b_1 a_{n-5} - b_3 a_{n-1}}{b_1} \qquad \text{etc.}$$

The array is complete when no more rows can be formed. The Routh criterion now states that if all the terms in the first *column* (that is, a_n, a_{n-1}, b_1, c_1, etc.) are of the same sign the original polynomial has no roots in the right-hand half plane. If, however, there is one change in sign the polynomial has one root in the right-hand plane; two changes indicate two such roots, etc. There is no indication where the roots lie in the right-half plane; all that is known is that such roots exist. This is sufficient for a check of stability, however, for, as we already know, a system will be unstable if the polynomial in the denominator of the transfer function has roots with positive real parts.

To show how this technique may be employed, we can use it to examine the stability of the system shown in Fig. 10.15 when the values of a and KK_f are specified. Thus, if a is taken as 10 and KK_f is taken as 3,375, the transfer function for the system is

$$(10.12) \qquad \frac{\sigma_{out}}{\sigma_{in}} = \frac{K}{s^3 + 30s^2 + 300s + 4,375}$$

To determine if this system is stable we must examine the denominator for positive real parts. The Routh array for the denominator is

$$1 \qquad\qquad\qquad\qquad 300$$
$$30 \qquad\qquad\qquad\qquad 4,375$$
$$\frac{(30)(300) - 4,375}{30} = +154 \qquad 0$$
$$\frac{(154)(4,375) - 0}{154} = +4,375 \qquad 0$$

As there are no changes in the signs of the terms in the first column, this denominator has no roots in the right-half plane and hence this system is stable (as expected; see Table 10.1).

On the other hand, if a is taken as 10 and KK_f is 9,000, the system denominator becomes $s^3 + 30s^2 + 300s + 10,000$. The Routh array is

$$+1 \qquad\qquad\qquad\qquad 300$$
$$+30 \qquad\qquad\qquad\qquad 10,000$$
$$\frac{9,000 - 10,000}{30} = -\frac{100}{3} \qquad 0$$
$$+\frac{100}{3}$$

Here there are two changes in sign (from $+30$ to $-100/3$ and from $-100/3$ to $+100/3$); two roots of the denominator are thus in the right-half plane. This system is unstable (comparable to the first entry in Table 10.1).

The Routh test as outlined is academically satisfying, but its usefulness to the life scientist is limited. In its present form, the test might tell an equipment designer he can assemble a particular group of known system components and can (or cannot) expect stable operation. The test is thus useful as a tool for problems of synthesis. The life scientist, however, is concerned with an assembled system, and the components of this system are seldom quantitatively defined (if they are defined at all). His problem is primarily analysis, and as presently outlined the Routh test does not serve his ends.

It is possible, however, to use the Routh test for some investigative work when it is of interest to know what conditions might drive a system into instability. To understand how the Routh test can be employed for this application we note that in the previous illustrative examples the system changed from stable to unstable as one of the terms in the first column changed from a positive number to a negative number ($+154$ changed to -10%; the cause of this transition was, of course, some change in the value of one or more system constants). As the transition from a positive value (stability) to a negative value (instability) should involve passage through zero, it seems reasonable to suspect that a zero in the first column is related to borderline stability. This is indeed partly the case; the total criterion for borderline stability is the existence of a row of zeros. The total statement is: If a set of system coefficients causes a row of zeros to exist in the Routh array, this set will cause the system to operate on the verge of stability (or instability). This concept is often useful as it permits an investigator to determine what conditions must be present for a system to become unstable.

To illustrate how the ideas developed in the preceding paragraph may be employed, we utilize once more the system shown in Fig. 10.15. In this case, however, we shall retain the literal values for the system constants as given in the figure. The system transfer function is now written

$$(10.13) \qquad \frac{\sigma_{out}}{\sigma_{in}} = \frac{K}{s^3 + 3as^2 + 3a^2s + a^3 + KK_f}$$

The Routh array for the system denominator is constructed in the standard manner:

$$
\begin{array}{cc}
1 & 3a^2 \\
3a & a^3 + KK_f \\
\dfrac{9a^3 - (a^3 + KK_f)}{3a} & 0 \\
a^3 + KK_f & 0
\end{array}
$$

The system will be at borderline stability if the constants a and KK_f can be adjusted to create a row of zeros. Examination of the array shows that a row of zeros can be created if $9a^3 - (a^3 + KK_f)$ becomes zero; this condition is satisfied if $KK_f = 8a^3$. This equality now indicates what conditions must exist for stable operation; in particular, if the system is stable (or is to be constructed so that it will be stable) KK_f must be less than $8a^3$.

Information obtained about borderline stability of a system has application in both the analysis and synthesis of systems. From an analytic point of view, such information can be used to set bounds on the values of the system constants. As noted above, for example, the system in Fig. 10.15 can be stable only if KK_f is less than $8a^3$. Thus, if KK_f is somehow evaluated to be 160, we should know that a must be greater than 2.7. (See the last two entries in Table 10.1.) As applied to a situation involving synthesis, a control system that is to be constructed with the configuration shown in Fig. 10.15 should not have a closed-loop gain (KK_f) greater than 8,000 when a is fixed at 10 ($KK_f < 8a^3 = 8,000$; see the first two entries in Table 10.1).

The Routh test is easy to use; this is the prime justification for including it here. It should be repeated, however, its utility for the life scientist is limited, as the applicability of the test is restricted to linear systems with little transport lag (or where transport lag can be ignored). The former restriction might be acceptable if a piecewise-linear analysis is to be used to examine a real system (and if the limitations of such an analysis are recognized). The latter restriction may also be acceptable in some cases but, in general, transport lag in physiological systems is often significant and should not be ignored.

A more elegant and more informative test for the stability of closed-loop systems involves plots of the denominator roots of the system transfer function as the system constants (generally gain) are varied. These plots are made on the complex plane and show the location of all possible denominator roots (for a given system configuration) as the system gain is varied. From such diagrams (called root-locus plots) it is easy to see whether any roots can "migrate" (because of gain change) into the right-hand half plane and the conditions under which this migration can take place. At a glance one can then make some estimate about the probability of instability and the case of (or how to avoid) such problems.

Although this root-locus technique (as it is often called) is of great use in engineering design work, it has the same limitations for the life scientist as has the Routh test; viz., it is difficult to include the effects of transport lag. Because of this limited applicability and because of the relatively detailed discussion required to develop this technique we shall not examine this procedure in any depth. Instead, only a brief introduction to the principles behind root-locus plotting will be offered here. For a more comprehensive discussion of this method of analysis, texts dealing with control should be consulted (e.g., Ref. 9).

We can learn what a root-locus plot implies simply by plotting the location of a few roots for a particular system under different conditions. In Table 10.1 we note, when KK_f is fixed at 160, the system has three roots at $s = -15.4$, $s = -7.3 + 4.75j$, and $s = -7.3 - 4.75j$. These three roots can be plotted in the complex plane as shown by the circles

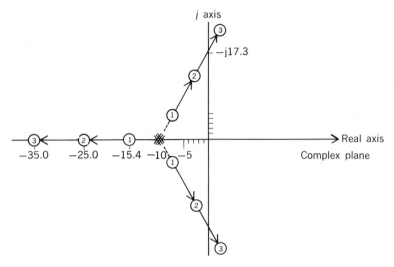

FIG. 10.16 *Root migration in the complex plane.*

(with the number 1 inscribed) in Fig. 10.16. It is of interest now to note how the positions of the roots change as the gain-constant KK_f is varied. From Table 10.1 we can see the roots are at $s - 25$, $s = -2.5 \pm 1.29j$ when KK_f is 3,375 (plotted in Fig. 10.16 as circles with 2s inscribed) and $s = -35$, $s + 2.5 \pm 21.7j$ when KK_f is 15,600 (plotted in Fig. 10.16 as circles with 3s inscribed). From these three sets of points the migration pattern of the roots (for this case) is easily perceived. As the system gain is increased, the roots for this particular system move along clearly defined straight-line paths: One path is along the negative real axis; the lines (or curves) defined by the root migrations are the aforementioned root loci. Obviously when any of the loci intersect the imaginary axis this indicates there are roots on this axis, and this can be interpreted as the condition of borderline or marginal stability; instability can be expected when the gain increase causes the loci to enter the right-half plane.

The foregoing discussion establishes what root loci are but does not indicate how these plots may be constructed. One cannot construct a table of roots (as Table 10.1) every time the root loci are to be sketched. Such a process would be self-defeating (and not always possible; higher-order polynomials are not generally easily factorable). The purpose of the loci is to estimate the roots; if the roots are evaluated by another technique there would be no need for root loci. It would be desirable, therefore, to be able to construct the loci without evaluating specific roots for chosen values of gain-constants. Such a procedure is possible, and it depends upon a change in viewpoint that recognizes gain-constants can sometimes be adjusted after certain roots are chosen. This comment will be more clearly understood as the discussion progresses.

To illustrate how to get some idea of the shape and location of the curves that define the root loci, we return to the system denominator given in Eq. (10.13) and examine how the roots of this denominator are determined. The first step is to equate the denominator to zero; this is the basic definition of the root of a polynomial. For this situation the equation is

$$(10.14) \qquad s^3 + 3as^2 + 3a^2s + a^3 + KK_f = 0$$

The next step is to find values of s that satisfy the equality. At this point a number of different approaches are possible. The one leading to the description of the root loci is to find values of s that satisfy a restatement of Eq. (10.14). The revised form of Eq. (10.14) is

$$(10.15)^* \qquad \frac{KK_f}{(s+a)^3} = 1/180$$

From this relationship we can see two mathematical properties must be satisfied if a value of s is to be a root of the system denominator. The first is that the value of s must be chosen such that the magnitude of $KK_f/(s+a)^3$ is to be unity. If the gain-constant KK_f can be varied as

---

* The steps leading from Eq. (10.14) to Eq. (10.15) should be self-evident:

$$s^3 + 3as^2 + 3a^2s + a^3 + KK_f = 0$$
$$s^3 + 3as^2 + 3a^2s + a^3 = -KK_f$$
$$(s+a)^3 = -KK_f$$
$$\frac{KK_f}{(s+a)^3} = -1$$

where -1 is equivalent to a magnitude of unity at a displacement angle of $180°$ ($1/180°$; see Chap. 8).

A number of comments can be made here: (1) The first is the obvious statement that values of s that satisfy any of the latter equations must satisfy all of them. Values of s that satisfy the last equation must therefore be roots of the first equation. (2) Another point is that Eq. (10.15) could have been derived directly from the basic relation that governs closed-loop operation. If we examine Eq. (10.8), for example, we note when we equate the denominator to zero (to find the denominator roots) we have in effect written Eq. (10.15). (3) As a final point, from the third equation it is easy to see how the denominator roots for *this* system may be determined immediately:

$$s = (-a + \sqrt[3]{KK_f})/60°, 180°, 300°$$

This is due to the perfect cubic which appears as part of the operator characteristic of one of the blocks. As this is a specific property of a particular operator block we shall proceed as if no such simplification were possible.

desired, this requirement is easy to satisfy. Thus, if in a particular system a is 10, one can always find a value of KK_f for a chosen value of s that will make the magnitude of the fraction $KK_f/(s + 10)^3$ be unity. The second constraint on s is a more stringent one: Eq. (10.15) points out that not only must the ratio $KK_f/(s + a)^3$ be unity but also s must be chosen to make this ratio a negative number (i.e., a number with 180° angular displacement in the complex plane). The latter constraint is the prime restriction on the values of s that form the root loci of the given system; in other words, the values of s that satisfy the negative-sign requirement of Eq. (10.15) form the root loci of the given system.

The last comment might be considered a definition of the root loci and also might be taken as an indication of how to find points that lie on the root loci. In the latter sense it suggests that a trial-and-error procedure can be used to locate the root loci; it proposes that points on the complex plane be chosen and checked with Eq. (10.15) (or a more general version of this equation such as $1 + KGH = 0$; see Sec. 10.2 and the second comment in the footnote on page 548) to see whether these points yield a negative number (or a 180° displacement angle). If this condition is satisfied, the chosen point lies on the root loci. This may seem a fairly haphazard procedure but it is, in fact, a relatively straightforward technique, as rules can be developed to limit the choice in the complex plane and some instruments are available (the Spirule* for one) to facilitate locations of roots.

As a brief illustration of how the foregoing comments may be used to estimate the location of the root loci, we shall assume a in Eq. (10.15) is 10 and KK_f is a positive number. (The latter constraint is more of a convention than an actual requirement. In most cases the signs of the systems equations are adjusted so that the gain-constants may be taken as positive numbers.) We now ask what values of s will cause Eq. (10.15) to become negative. We might suggest any value for s greater than zero; for such a choice we can easily see Eq. (10.15) can never be satisfied. In fact, it should now be obvious that only real values less than -10 can be used for s if we are to satisfy the sign requirements of Eq. (10.15). We have now located one locus of points (of the three possible loci for the three roots of the third-order polynomial) that may be roots of the system denominator. This root locus can be shown as a heavy line on the complex plane, as in Fig. 10.17.

An examination of Eq. (10.15) reveals there are no other real roots for this system; the other roots must be complex conjugates. As a first

---

* Developed by W. R. Evans and mentioned in various papers and in Evans' book "Control-system Dynamics," McGraw-Hill Book Company, New York, 1954. Available from the Spirule Co., Whittier, Calif.

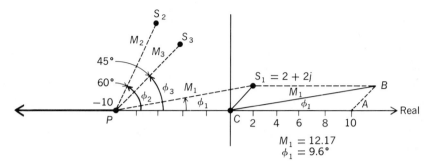

FIG. 10.17   *Location of root loci for $KK_f/(s - 10)^3 = -1$.*

guess we might try $s = 2 + 2j$; $s + 10$ then becomes

$$12 + 2j = \sqrt{148}/\tan^{-1} 0.17 = 12.17/9.6°$$

and $(s + 10)^3$ is thus $(12.17)^3/3(9.6)°$ or $(12.17)^3/28.8°$. The displacement angle of $(s + 10)^3$ is not $180°$; therefore $2 + 2j$ is not on the root locus (it is not a denominator root).

We could continue this algebraic process but it is somewhat easier to perform the calculations graphically as shown in Fig. 10.17. The line CS_1 represents the choice of $s = 2 + 2j$; the line CA represents the line segment $a = 10$. The sum of these two quantities is obviously the line CB in Fig. 10.17. If point S_1 ($2 + 2j$) were on the root locus, it should now be obvious that angle ϕ_1 in Fig. 10.17 would be $60°$.

The graphical procedure just outlined simplifies the location of the root loci to some extent as the process now degenerates to measuring the angle of the resultant of a complex summation. It is possible to simplify the procedure even further since the graphical summation of s and $+10$ may also be represented by the line connecting the -10 point (on the real axis) to the assumed value for s. This construction can be easily justified by examining Fig. 10.17, where it can be seen the line PS_1 is the same as $CB = M_1/\phi_1 = 12.17/9.6°$. We now see we can check whether a point lies on the root locus by checking the angle of the line connecting this point to the -10 point. If the angle is $60°$, the point is on the root locus; if not, the s value chosen cannot be a denominator root of the given system. Thus from Fig. 10.17 we see S_3 cannot be on the root locus of the system but S_2 can. In fact, it should now be evident the root loci of the complex roots for this system are straight lines that originate at -10 and are inclined $\pm 60°$ to the real axis (see Fig. 10.16).

The above discussion is somewhat deceptive as the results for this particular example are very simple and easily obtained. In general, root loci are not straight lines or simple geometric curves and ordinarily are not easily defined. This is particularly true when the system blocks do not contain repeated factors [as $(s + a)^3$, for example; Prob. 10.11 is an illustration of this point]. As pointed out earlier, however, there are some rules and computational aids that simplify the construction of the loci considerably. We shall not discuss these labor-saving schemes as they require additional detailed explanations and, for the reasons put forth earlier, such complete coverage is not necessary. If further information is required there is adequate reference material available.

The primary disadvantages of the stability tests discussed thus far are that these techniques can be applied only to linear systems and the effects of transport lag cannot be easily included in the analysis.* Although the former objection might be tolerable (particularly if small excitations are used) the latter cannot generally be accepted. Transmission delays can be significant even for small signal excitations, and if these are known to exist, they should be included in the system analysis. Unfortunately, however, there is no easy way to incorporate such information in either the Routh test or the root-locus investigation. (At least no easy way is known to the author; one possible procedure is given in Prob. 10.12 but this does not lead to a simple determination, nor is it, in general, very accurate.) For these reasons these tests have only limited application in physiological investigation.

The stability test to be discussed next has the same disadvantage as the Routh and root-locus tests as it cannot (or at least it should not) be used unless a linear representation is applicable. This method, however, can be applied to systems that include transport lag, and to that extent it has more utility than either of the tests previously discussed. The basic concept behind this check of system stability depends upon frequency-response plots (Bode plots). The mechanics of constructing such plots were discussed in Chap. 8; it remains here to show how these ideas may be applied to the investigation of closed-loop stability.

We can understand how frequency plots may be used to study system stability if we review briefly the implications of marginal stability. Such systems have denominator roots on the imaginary axis (see the footnote in Sec. 10.4); this implies the denominator of a marginally stable

* We have treated transportation lag as something separate from nonlinear behavior. This is not entirely valid; transport lag is a nonlinear effect in the sense that systems including such delay must be described by a nonlinear differential equation. However, if we think of nonlinear systems as those in which the system constants are altered by signal excitation levels, it may then be reasonable to separate the concepts of system nonlinearity and transport lag. This is largely a question of semantics, but it should not create any difficulties here as the use of the terms will be clear; any ambiguities will be separately discussed.

system has second-order factors of the form $s^2 + \omega^2$.* Such factors indicate some portion of the system output will be a second-order response with no damping. From earlier considerations (see Sec. 9.7) we know a second-order system with no damping will theoretically reach infinite amplitude if driven continuously at the resonant frequency. (In numerical terms this implies the output of the system in Fig. 10.15 will ultimately become infinite if the systems constants are $KK_f = 8,000$ and $a = 10$ and the input is a sinusoid of finite amplitude with frequency $\omega = 17.3$ rad/s.) From this simple discussion we can now see how two separately developed concepts must be related. The concepts stated separately are (1) marginally stable systems have imaginary roots in the denominator of the system transfer function and (2) the output of a system with imaginary roots in the denominator of its transfer function tends toward infinite amplitudes when driven at its resonant frequency. The single statement combining these concepts is: The output amplitudes of marginally stable systems tend toward infinity if such systems are driven at their resonant frequencies.

The foregoing discussion establishes to some extent the connection between system stability and frequency analysis, and it provides in addition a hint of how such an analysis may be used to study system stability. If we invert the last statement of the previous paragraph we have: Systems delivering outputs that tend toward infinity when driven at their resonant frequencies are marginally stable. Although this statement is minimally true,† it is sufficient for our purposes, as this idea can be used to determine whether a system is on the verge of unstable behavior. The procedure is to excite a system with a spectrum of frequencies (from a mathematical point of view) and to determine whether the amplitude becomes infinite at some frequency. If the system shows such response it may be considered to be at least marginally stable.

The ideas just presented may be expressed in a somewhat more analytic fashion if we use the concepts and notations developed in Chaps.

---

* As an example: From the plot in Fig. 10.16 and from the values obtained by the Routh test following Eq. (10.13), we note, when KK_f is 8,000, the system has roots at

$$s = \pm 17.3j = \pm 10 \sqrt{3} \, j$$

This implies the system has two complex factors $s + j10 \sqrt{3}$ and $s - j10 \sqrt{3}$; these combine to form a second-order factor $s^2 + 300$. Evidently ω^2 is 300 for this case.

† A more complete statement would be such systems are *at least* marginally stable systems. This implies these systems may in fact be unstable. Although we shall not prove the validity of this extended viewpoint, it seems intuitively reasonable, since unstable systems tend to destroy themselves no matter how they are excited. We have now shown that marginally stable systems tend to destroy themselves when excited at a specific frequency. It seems safe, then, to say that systems that tend to destroy themselves when driven at a specific frequency are either marginally stable or unstable.

8 and 9. In these chapters it was shown the response amplitude of a sinusoidally driven system may be given by

$$(10.16) \qquad \sigma_{out} = A|T(j\omega)|$$

where ω is the excitation frequency, $T(j\omega)$ is the magnitude of the transfer function when $j\omega$ replaces s, and A is the (finite) input amplitude of the system [see Eq. (8.48c)]. From the previous discussion it should now be evident how the criterion for stability may be phrased in terms of Eq. (10.16). In particular, we can now state that, if σ_{out} in Eq. (10.16) becomes infinite at some ω, the system under investigation must be at least marginally stable. Since A is considered finite, we can now also see from Eq. (10.16) that, if $|T(j\omega)|$ becomes infinite at some ω, the system having the transfer function $T(s)$ must be at least on the verge of instability.

The criterion established in the preceding paragraph is applicable to all systems and can be used to check the stability of any system. To use this criterion, the $|T(j\omega)|$ for the system is evaluated as ω changes from zero to infinity. If $|T(j\omega)|$ becomes infinite, the system is marginally stable.

This stability test as outlined is adequate but it may be clumsy to carry out. For closed-loop systems, however, it is possible to simplify the procedure to some extent and thereby reduce much of the computational labor. To show how this simplification may be achieved, we return first to the general expression for the transfer function of a simple single-loop feedback system such as that shown in Fig. 10.18. This transfer function is given by

$$(10.17) \qquad T(s) = \frac{KG}{1 + KGH}$$

From the criterion established, this system is marginally stable if $|T(j\omega)| \rightarrow \infty$ at some value for ω. An examination of Eq. (10.17) indicates this would be equivalent to the condition

$$(10.18) \qquad |1 + KGH|_{s=j\omega} = 0$$

which is in turn equivalent to

$$(10.19) \qquad (KGH)_{s=j\omega} = -1 = 1\underline{/180°}$$

This equation is a relatively simple relationship, particularly adaptable to closed-loop systems; it can, in fact, be considered the equation that defines the stability criterion for simple feedback systems. The equation (and its development) implies a closed-loop system can be considered marginally stable if it can be shown the loop product (KGH) of the system can have unity gain and an associated phase shift of 180° at some excitation frequency. This statement in the form of a set of equations is written as follows: If

(10.20a)
$$|KGH|_{s=j\omega} = 1$$

and

(10.20b)
$$\underline{/(KGH)}_{s=j\omega} = 180°$$

then

(10.20c)
$$\sigma_{out} \to \infty$$

and the system may be considered at least on the verge of instability. To use this criterion, we plot the gain vs. frequency curve for KGH and the phase vs. frequency curve for this product (i.e., the Bode plots for KGH; see Sec. 8.9). We examine these curves to determine whether the gain of KGH is ever unity when the phase angle is 180°; if this condition is somewhere satisfied, the closed-loop system would be considered marginally stable.*

The criterion established by the sequence of Eqs. (10.20a) to (10.20c) is meaningful from both an analytic and an experimental point of view when a single-loop system is under investigation. Based on the foregoing discussion, the analytic procedure suggested by the equations should now be self-evident: The loop product is formed, the Bode plots are constructed mathematically, and the results are interpreted as outlined. It should be noted that this procedure is not altered if the loop blocks include a time-delay function. Since Bode plots can be constructed even if there are transport lags in the system, this technique may also be used to study systems that have inherent transmission delays. The effect of the delay is included in the phase vs. frequency characteristics of the loop product (a review of Sec. 8.9 will serve to recall the relationship between phase and time delay); the gain characteristics are not affected by the lag functions. The Bode plots constructed with transport delay are inter-

---

* As a corollary, it may be stated, that, if the gain exceeds unity while the phase angle is 180°, the system is probably unstable. The use of the word "probably" may be somewhat surprising, but, although the criterion established is true for most practical uses, it is not universally true. Some system configurations have erratic-looking Bode plots where gain can exceed unity at 180° phase angle for a limited range of frequencies and yet the system can still be stable. For our purposes these may be considered unusual cases not generally encountered within the framework of the analytic work attempted here. However, the criterion established in this discussion errs on the safe side. If instability is indicated by the test proposed and there is some ambiguity, it should prompt additional investigation to verify or refute the conclusion.

preted in the same manner as previously outlined: Unity gain (or greater) with an associated 180° phase shift is interpreted as marginal or unstable closed-loop operation.

From a "real-system" point of view, KGH has physical significance which can be exploited for stability investigations. As a "real-world" interpretation of KGH we could consider this product to be the operational effect of the system on the signal transmission around the loop. (The loop product KGH is thus often described as the loop transmittance.) Examination of Fig. 10.18 indicates the loop product may thus be interpreted as transmission around the loop when the feedback path is interrupted (as by a switch); in an alternative phraseology, KGH could be considered simply the open-loop transmittance of the control system. In Fig. 10.18 this would imply KGH is the operational effect of a signal delivered at A and received at R when the switch sw is opened. This concept suggests a laboratory procedure which might be used to study the stability behavior of some closed-loop systems. If the loop of the system can be opened, the Bode plots for the loop transmittance can be constructed from experimental data by delivering a sinusoidal signal at one point (such as points A and B in Fig. 10.18 when the switch sw is opened). These Bode plots may then be used to evaluate the stability characteristics of the closed-loop system in the same way one would use Bode plots constructed from an operational description of KGH.

We can easily demonstrate the applicability of this newly developed viewpoint in a practical, realistic situation. In an earlier discussion (Sec. 8.9) it was shown the open-loop frequency characteristic of the pupillary control system can be obtained by special optical techniques. Sinusoidal excitations of light at different frequencies were beamed into the eye; the pupillary response at each frequency was recorded. From these data frequency-response curves were constructed. These curves were given in Fig. 8.39b and are repeated in Fig. 10.19. The gain curve was defined by the ratio of the percentage change in area to the percentage change in flux for each excitation frequency. This ratio of percents is a unitless number and reflects the transmittance around the system loop; it there-

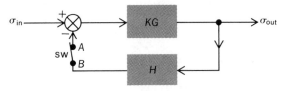

FIG. 10.18 *The standard representation of a linear single-loop feedback system.*

fore has the properties of loop transmittance KGH.* We can therefore accept that the gain vs. frequency curve and the phase vs. frequency curve of Fig. 10.19b may be used to assess the closed-loop stability of the pupillary system.

If we examine the Bode plots for the open-loop transmittance of the pupillary system we note the gain is about 0.1 when the phase angle is 180°. For these operating conditions the closed-loop system should show absolutely stable characteristics, and so it normally does. It is of interest to note that, if it is possible to increase the gain of the system by a factor of 10, we should then find the system gain is unity at a phase angle of 180°. This would indicate the system is at least marginally stable and has a resonant frequency of about 1.2 c/s. The implications of this statement are that the pupillary system can be expected to manifest sustained oscillation (even at constant excitation) at a frequency of approximately 1 Hz if the system gain is somehow increased to a high level. The possibility of a pupil oscillation was mentioned earlier; at the end of Sec. 10.3 it was pointed out oscillations can be induced when a circle of light is located at the edge of the pupil (as shown in Fig. 10.14). Stark and Cornsweet† did some studies in this area and found by this experimental technique that the mean frequency of oscillation for 10 normal subjects was 1.3 Hz. Such excellent correspondence between the predicted and measured values provides confidence in the validity of the proposed model.

---

* We can support this viewpoint in a somewhat more analytic fashion if we examine a block diagram of the experimental procedure. This diagram was presented in Fig. 8.38b and is repeated in Fig. 10.19. It is noted the excitation is supplied at B and the response is measured at A. If we consider σ_B the input and σ_A the output, the open-loop transmittance is given by

$$\frac{\sigma_A}{\sigma_B} = \text{open-loop transmittance} = KGH$$

Since σ_A is the delivered flux we may write $\sigma_B = \mathfrak{F}_{Rv}$; from the diagram in Fig. 10.19a we can also write $\sigma_A = E_i A_{pv}$. E_i is interpreted as the effective or equivalent illuminance that will deliver the quiescent flux level to the eye when the pupil area is at its quiescent value. In short, E_i is interpreted so that $E_i A_{pq} = \mathfrak{F}_{Rq}$. The signal at A can now be written

$$\sigma_A = \frac{\mathfrak{F}_{Rq}}{A_{pq}} A_{pv}$$

If we replace σ_A and σ_B by their equivalent expressions, we have

$$KGH = \frac{A_{pv}(\mathfrak{F}_{Rq}/A_{pq})}{\mathfrak{F}_{Rv}} = \frac{A_{pv}/A_{pq}}{\mathfrak{F}_{Rv}/\mathfrak{F}_{Rq}}$$

This ratio has, however, been described as the open-loop gain of the pupillary system and is plotted vs. frequency in Fig. 10.19b.

† L. Stark and Tom N. Cornsweet, Testing a Servoanalytic Hypothesis for Pupil Oscillations, *Science*, **127**(3298):588 (Mar. 14, 1958).

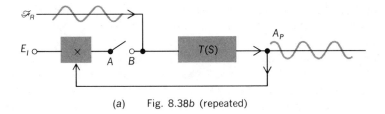

(a)   Fig. 8.38b (repeated)

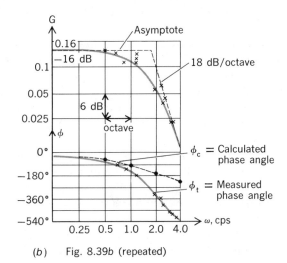

(b)   Fig. 8.39b (repeated)

FIG. 10.19   *Open-loop excitation and Bode plots of pupillary control system.*

Before we leave this example of the pupillary system it may be well to ask why a light imaged on the edge of the pupil induces pupillary hippus. From the presentation it was implied this procedure must have somehow increased the gain of the system which caused the system to approach the marginally stable condition. To examine this gain increase in more detail, we note from Fig. 10.13 that changes in incident illuminance do not cause the loop gain to be increased in a proportional or obvious fashion. The muscle block contains the factor $1/E_{iq}$; when the loop product is formed, the E_{iq} of the muscle block is canceled by the same term in the feedback block. The direct effect of illuminance on the system is thus canceled. Furthermore, a check of the system geometry will show that imagery on the pupil's edge actually causes a decrease in the system gain rather than an increase. The overall gain increase in the system must therefore be in the intrinsic (nonlinear) amplification capabilities of the sensory mechanism in the eye; this mechanism appears to be high gain when little flux is incident in the retina.

As an additional point, an examination of the Bode plots in Fig. 10.19 indicates destabilization must be accompanied by a gain increase, but it also indicates this may not be a complete answer. It is also possible that the change in mode of excitation has caused both the system transmission lag and the system time-constant to be increased. Both these changes have destabilizing effects and could very well have contributed to the marginally stable behavior induced by the edge illumination of the pupil.

We cannot close this introductory discussion of stability tests without some mention of the Nyquist criterion. If the discussion of the stability tests had been presented in a more formal fashion the Nyquist test would certainly have preceded the Bode test as the Nyquist method provides the more general and more complete criterion for a test of stability. The Nyquist test, in fact, can be used to justify and validate the Bode-test procedure.

In spite of its generality, however, the Nyquist method is not as universally used as the other tests. The difficulty probably lies in the need for polar plots, the relatively complex theoretical explanation needed to develop the test, and the unfamiliar and unusual-looking curves that must often be interpreted.

The level of background established here is not sufficient to develop the reasoning behind the Nyquist test. We shall, however, indicate briefly how the test is used by utilizing the Bode plots for reference. In a crude sense, Nyquist diagrams might be described as Bode plots on polar-coordinate paper. The Bode plots define a gain and phase value for each frequency; these numerical values for gain and phase define radial points in polar coordinates. A sequence of such points (which represent gain and phase) for all frequencies define the Nyquist plot.

To illustrate the transfer of points we can use the Bode plots in Fig. 10.19. We note, for example, at 0.25 Hz the system gain is about 0.16 unit and the phase angle is about 45°. On the polar plot this point would be located 0.16 unit from the origin at an angle of 45° to the horizontal (Fig. 10.20a). In a similar fashion, from Fig. 10.19 we see at 0.5 Hz the gain of 0.15 is accompanied by a phase angle of about 90°; this is shown in Fig. 10.20b. If we continue this procedure we generate a curve that represents the information from the Bode plot in polar coordinates. The heavy line in Fig. 10.20a shows the constructed curve.

The curve thus far generated represents one half of the actual Nyquist plot. Nyquist plots include positive and negative frequencies and therefore include positive and negative phase angles for each value of

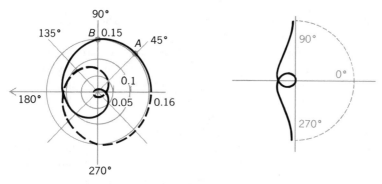

FIG. 10.20   *Nyquist plots.*

gain. This is a mathematical technicality, however, and merely indicates
that every point on a Nyquist plot has a mirror image about the horizontal axis. The total Nyquist plot is thus composed of the heavy curve
shown in Fig. 10.20a and its mirror image (the dashed curve in this same
figure).

To show how the Nyquist plot can be used to gauge stability we
recall that the criterion for marginal stability in the Bode diagram was
unity gain at a phase shift of 180°. The corresponding criterion in the
Nyquist plot is a polar curve that passes through the −1 point on the
horizontal axis. From Fig. 10.20 it is evident the pupillary control system
is stable at normal operating conditions.

Instability shows up in a more definitive manner in the Nyquist
plot than it does in the Bode diagram. The criterion for instability in the
Nyquist diagram is that the polar curve surround the −1 point (or that
the −1 point fall within the closed Nyquist curve). A complete statement
is as follows: If a Nyquist plot of the open-loop transmittance encloses
the −1 point, the closed-loop system will be unstable. If, therefore, the
open-loop gain of the pupillary control system increases (or other changes
occur) so that the polar curve in Fig. 10.20 tends to encompass the −1
point, such changes would tend to destabilize the closed-loop system. We
reached this same conclusion when the Bode diagrams were used to
examine the effects of gain changes on the stability of the closed-loop
system.

The "enclosure criterion" has not been developed by any discussion
here (except possibly on an intuitive basis); it can, however, be established
by formal mathematical analysis as shown first by Nyquist in about 1932.
It is this criterion, incidentally, that indicates why a closed-loop system
should be considered unstable if the Bode diagram of its loop trans-

mittance shows a gain greater than unity for a 180° phase shift. If this condition appears on the Bode diagram, the Nyquist plot will enclose the -1 point. As a final comment, it must be realized that the Nyquist plot is completely independent of the Bode diagram and vice versa. Neither is needed to obtain the other; both can be constructed directly from the mathematical representation of *KGH*. The only reason these two viewpoints were developed interdependently was for ease of presentation.

We close this section (and the book) with a recapitulation of comments made at the beginning of this chapter and elsewhere in the text. The information and ideas presented here should not be considered ends in themselves; instead they should be considered introductory viewpoints designed to indicate to the reader the possibilities and potentialities of such material for research work in the life sciences. For interested readers there are available tremendous sources of information on any and all topics included here, and these should be consulted for more detailed and expanded coverage.

Problems

10.1   The system shown in Fig. P10.1 is driven by a step-function $4u(t)$. Show that its response is given by $(10/3)u(t) + 2\epsilon^{-4t} - (16/3)\epsilon^{-3t}$.

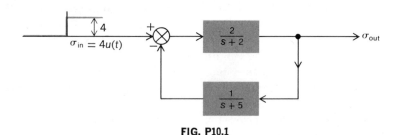

FIG. P10.1

10.2   One can obtain an analog simulation of a first-order system by using feedback around a simple integrator (see Prob. 5.4 and Appendix A, step 35). Thus,

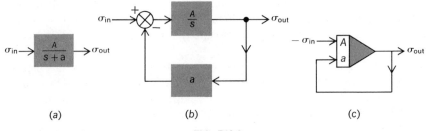

FIG. P10.2

if a system is represented by the block diagram of Fig. P10.2a, the same system can also be represented by the block diagram of Fig. P10.2b and by the analog simulation of Fig. P10.2c. Draw an analog simulation of the system shown in Fig. P10.1. Include numerical values as needed.

10.3 If a feedback control system is driven by an excitation σ_{in} (Fig. P10.3), show that the output cannot exactly follow step changes in the input unless the loop product (KGH) contains at least a single integration $(1/s)$. Show that the output cannot follow ramp changes in the input unless the product contains a double integration $(1/s^2)$, etc.

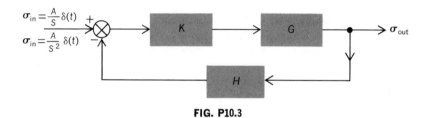

FIG. P10.3

10.4 Show that the diode circuit in Fig. P10.4 has dead-zone characteristics by plotting the e versus i characteristics of the circuit.

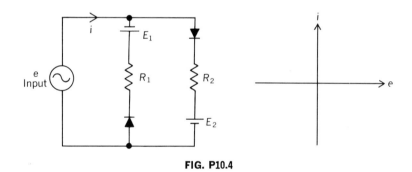

FIG. P10.4

    (a) What governs the width of the dead zone and the degree of symmetry of the dead zone?

    (b) What governs the slope of curve in each quadrant—during operation—when the input signal exceeds the dead-zone level?

    (c) If the input signal were oscillatory (say a triangular wave), sketch the possible output waveforms (i versus time).

10.5 The diagram of Fig. 10.6b (redrawn in Fig. P10.5) is assumed to be representative of the pituitary-adrenal system; the blocks are *assumed* to have the mathematical properties indicated.

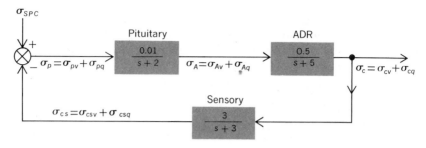

FIG. P10.5

(a) If the quiescent level of the cortisol, σ_{cq}, at a particular time is 10 $\mu g/100$ ml blood, what is the quiescent level of ACTH in the system?

(b) Since the result of (a) indicates (for the configuration proposed) that a steady delivery of ACTH is needed if σ_{cq} is to remain at a steady level, this implies a steady signal must be delivered to the pituitary block. What is the magnitude of the signal that must be delivered to this block?

(c) On the basis of the results obtained in (a) and (b), at what level must the cortisol set point be fixed to maintain the quiescent level indicated?

(d) From the previous answers it should be evident (for the configuration proposed) the set point (σ_{SPC}) and actual level of the controlled variable (σ_c) cannot be the same if the system is to maintain a quiescent cortisol level. (Then $\sigma_p = 0$ and $\sigma_A = 0$.) The control level and the set-point level may, however, approach each other if the blocks have different numerical values. Show σ_p ($= \sigma_{SPC} - \sigma_{cs}$) is reduced when $T(s)$ of the pituitary block is given by $1/s + 0.2$ and further reduced when $T(s)$ is $10/s + 0.02$.

(e) From the discussion in (d) it should be obvious the set-point level and the quiescent level can be identical if the pituitary block is of the form $T(s) = A/s$. Will the same statement be true if $T(s)$ is $A/s(s + a)$?

(f) The general conclusion from e is that the set-point level and quiescent level will be identical if the pituitary block contains an integral operator in its transfer function. Is this comment restricted to the pituitary block only, or is it applicable to other or all blocks in the closed loop?

10.6 The hypothetical configuration in Fig. P10.5 represents the quiescent operating condition of the pituitary-adrenal system. If we wish to perturb the system about the quiescent point, we can inject a signal anywhere within the loop path. Figure P10.6a shows how a pulse signal of ACTH might be introduced into the system.* To analyze the response, we note that if the system is not

---

* This is easier said than done. The adrenal gland is fed by numerous arterial branches from the phrenic artery, aorta, renal artery, etc. A true pulse signal to the adrenal cortex may not be possible. One might try to circumvent this problem by introducing a constant rate of ACTH (i.e., a step-rate in terms of micrograms per second) into the aortic trunk above the suprarenal artery. If the aortic flow can be considered constant during the injection, the ACTH concentration in the plasma rises suddenly to a constant level and then returns to zero after the injection. We can now assume that some percentage of the pulse concentration reaches the gland; this percentage value would be the percent of aortic blood delivered to the gland. Such information is available in the literature.

significantly altered by the additional signal we can assume a linear analysis would be valid. We need therefore only calculate the effect of σ_{Av} (the variable component of the ACTH signal reaching the adrenal cortex) and simply add these results to the known quiescent levels. This implies we may discount all quiescent levels in the system and redraw the system as if σ_{SPC} did not exist; the result is shown in Fig. P10-6b.

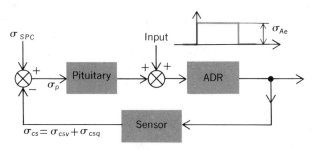

(a) Set point included

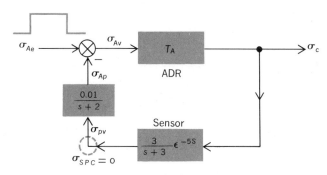

(b) Set point excluded

FIG. P10.6

(a) Suppose the pulse duration is 2 s and after 0.1 s the signal from the adrenal block (again assuming such a signal may be obtained) is found to be $0.5\epsilon^{-5t} - 0.5\epsilon^{-5(t-2)}u(t - 2)$. If the transmission lag from the ADR block to the sensor block is 5 s (the block thus includes the time-delay term ϵ^{-5s}) and the sensor and pituitary blocks have the operational forms shown in Fig. P10.6b, sketch the waveshape of the signal σ_A (the input to the ADR block) during the pulse excitation.

(b) On the basis of the conclusion derived in (a), what would be the transfer function of the ADR block?

(c) If there is no transmission lag in the system, the diagram of Fig. P10.6b can be considered a simple, single-loop control system. The overall transfer function for such a system is given by

$$\frac{\sigma_c}{\sigma_{Ae}} = \frac{T_A}{1 + T_A T_s T_p}$$

where T_A, T_s, and T_p are the transfer functions of the ADR, sensor, and pituitary blocks, respectively. If again the excitation (σ_{Ae}) is a 2-s pulse and the response (σ_c) is $0.5\epsilon^{-5t} - 0.5\epsilon^{-5(t-2)}u(t-2)$, how must the transfer function of the ADR block now be defined?

(*d*) If the transmission lag from the ADR block to the sensor block is 1 s, sketch the waveshape of the signal delivered to the ADR block. What procedures can now be used to define the transfer function of the ADR block?

10.7   Figure P10.7 shows a plot of pupil diameter vs. illuminance (in milli-lamberts, where 1 mL is equivalent to 1.076 mlm/cm²).

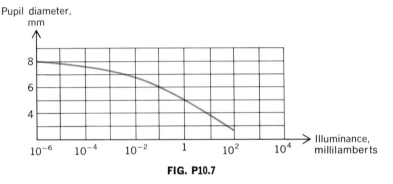

FIG. P10.7

(*a*) From these static characteristics, obtain the static characteristics that relate the pupil area (square millimeters) to the flux entering the eye (in milli-lumens). Note the log scale.

(*b*) If the pupil quiescent area is 20 mm², what is the quiescent flux entering the eye?

(*c*) If the muscle block has a transfer function $K/(s + a)^3$, can the set point of the system be anywhere along the flux vs. area curve? Where? Or why not?

(*d*) What is the small signal gain $\Delta\mathcal{F}/\Delta A$ about the quiescent area of 20 mm²? How should this number show up in the transfer function that relates flux ($\Delta\mathcal{F}$) as an input and area (ΔA) as an output?

10.8   If we interpret s as a differential operator, the flux signal ($\mathcal{F}$) and the aperture signal (A_p) in Fig. 10.12a may be related by

$$160(\mathcal{F}_{RO} - \mathcal{F}_R) = (s + a)^3 A_p = \frac{d^3 A_p}{dt^3} + 3a\frac{d^2 A_p}{dt^2} + 3a^2\frac{dA_p}{dt} + a^3 A_p$$

(*a*) Write the differential equation relating A_p and E_i and show it is a nonlinear equation.

(b) If E_i is considered to be a quiescent value with a variable component $(E_i = E_{iq} + E_{iv})$ and if A_p is similarly described $(A_p = A_{pq} + A_{pv})$, show that the differential equation derived in (a) reduces to a linear equation if the product of the variable terms can be neglected.

10.9 The denominator of the transfer function given by Eq. (10.7a) may be factored into $s + 15.4$ and $s^2 + 14.6s + 75$. The latter factor gives rise to two complex factors; it will therefore give rise to an oscillatory component in a transient response.

(a) Show the quadratic factor $s^2 + 14.6s + 75$ will yield two complex factors by showing the damping factor is less than unity. [See Sec. 9.3 and Eqs. (9.14) to (9.16).]

(b) The natural frequency of the system will evidently be $\sqrt{75}$. What is the *damped* natural frequency in radians per second and in cycles per second?

(c) If the excitation to the eye is a flicker of light described by $E_{iv} = 2 \times 10^{-4} \, \delta(t)$ mlm/mm², sketch the two components of the transient response and the total response. What is the maximum value of the diametral change of the pupil?

10.10 If the gain KK_f in the system of Fig. 10.15 is 3,375, it was noted in the text the system has a single time-constant of 25 s, an oscillating transient behavior with a frequency 12.9 rad/s, and a damping factor 0.2.

(a) Sketch roughly the response that may be expected if the system is driven with a step excitation.

(b) Suppose the gain of the system is increased so that KK_f is now 8,000. What will the time-constant, frequency, and damping now be for the closed-loop system? Sketch the impulse-response of this system.

(c) Sketch the impulse-response when KK_f is 15,600 (see Table 10.1).

10.11 Verify that the plots shown in Fig. P10.11b represent the root-locus plots for the system shown in Fig. P10.11a.

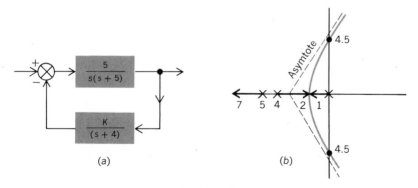

(a)

(b)

FIG. P10.11

10.12 In Chap. 5 it was shown the forward-loop transmittance of a dye-dilution system (to study circulation) can be determined by direct measurements because of the relatively long transport around the closed loop (see Sec. 5.8). The total closed-loop system might be represented by the diagram shown in Fig. P10.12. $\dot{W}$ would represent the rate of dye delivery to the central volume; K would be the concentration of dye delivered from the central volume. It is assumed the transfer function of the central volume was evaluated by the technique outlined in Chap. 5 and was found to be $(\frac{1}{150})(s + \frac{1}{3})$.

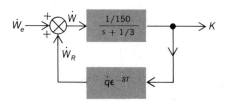

FIG. P10.12

Dye-laden blood is delivered from the central volume and is returned to the excitation site. The dye in the blood (W_R) mixes with the excitation signal (W_e) to form the input signal to the central volume (that is, $\dot{W}_e + \dot{W}_R = \dot{W}$). The rate of dye returned is the product of the dye concentration (K) and the blood flow; thus, $\dot{W}_R = K\dot{q}$.

Since it takes a finite amount of time for the blood to make a complete circuit the feedback includes a term to represent the transport lag (ϵ^{-st}). There are feedback loops (collateral circulation paths) at different flow rates and different transport lags. As an average value for these terms, use $\dot{q} = 50 \text{ cm}^3/\text{s}$ and $T = 3$ s.

(a) If there were no time lag, what would be the output concentration for a pulse-input which delivers 2 mg of dye in 1 s at a constant rate?

(b) If the time delay is not neglectable, it must be included in any model representation. On a mathematical basis ϵ^{-st} cannot be used directly as it leads to unwieldy and generally unsolvable differential equations. However, one can get a linear representation with time delay if an approximate form is used for the delay term. One such approximation is the Padé approximation which utilizes a form of series expansion for the delay term.* By using one form of this approximation one may write

$$\epsilon^{-30s} = \frac{(1 - 30s)/3}{[1 + 2(30s)]/3 + (30s)^2/(3)(2!)} = \frac{1 - 10s}{1 + 20s + 150s^2}$$

$$\epsilon^{-30s} = -\frac{(s - 10)/15}{s^2 + 0.133s + 0.0067}$$

* A Padé table may be found in J. G. Truxal, "Automatic Feedback Control System Synthesis," p. 550, McGraw-Hill Book Company, New York, 1955.

If this approximation for the delay term is used, the block diagram of Fig. P10.12 will then be a linearized representation of the system and linear analysis can be used to estimate performance. Will the transient response of the total system (including the delay approximation) be oscillatory?

(c) Can this system ever be unstable? Check with the Routh criterion.

(d) Can this system ever be unstable? Check with Bode plots using ϵ^{-st} directly.

REFERENCES

1. Langley, L.: "Homeostasis," Reinhold Publishing Corporation, New York, 1965.
2. Bernard, C.: "An Introduction to the Study of Experimental Medicine," The Macmillan Company, New York, 1957.
3. Cannon, W.: "The Wisdom of the Body," W. W. Norton & Company, Inc., New York, 1932.
4. "Homeostasis and Feedback Mechanisms," Symposia of the Society for Experimental Biology, Cambridge University Press, New York, 1964.
5. Guyton, Arthur C.: "Textbook of Medical Physiology," 3d ed., W. B. Saunders Company, Philadelphia, 1966.
6. Williams, Robert H. (ed.): "Textbook of Endocrinology," W. B. Saunders Company, Philadelphia, 1968.
7. Turner, C. Donnell: "General Endocrinology," W. B. Saunders Company, Philadelphia, 1966.
8. Yamamoto, William S., and John R. Brobeck (eds.): "Physiological Controls and Regulations," W. B. Saunders Company, Philadelphia, 1965.
9. Milhorn, Howard T., Jr.: "The Application of Control Theory to Physiological Systems," W. B. Saunders Company, Philadelphia, 1968.

APPENDIX A

NODAL ANALYSIS AND COMPUTER ELEMENTS

When a circuit diagram is used to represent a physical system, current is often used to represent one variable (e.g., flow) and voltage another variable (e.g., pressure). If an actual network is then constructed from the diagram, voltage and current measurements can be used to determine pertinent relationships between the variables. This step is not always necessary, and sometimes the relationships between the variables can be determined by direct analysis of the circuit diagram. This procedure is more general than the former as the results do not depend on numerical values. The technique of analyzing circuit diagrams and deriving relationships from such diagrams is the subject matter of network analysis.

There is much excellent literature on the subject of circuit analysis (two references are mentioned at the end of Chap. 2), and there is little point in repeating in detail what is readily available elsewhere. Here, therefore, is offered a very elementary programmed sequence which can serve either as a review of nodal analysis or as an introduction to this concept. For more detailed studies in this area the references and other literature should be consulted.

1. The program is designed to introduce the development of nodal equations based upon Kirchhoff's current law which states: *The total current leaving a junction is equal to the current entering the junction.* For the diagram in Fig. A.1a this is written algebraically as $i_1 = i_2 + i_3$. What is the corresponding equation for Fig. A.1b (assuming flow is of an incompressible fluid through a rigid pipe)?

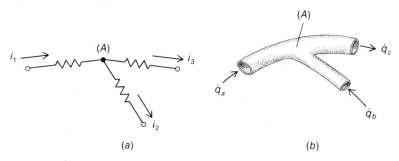

FIG. A.1

$$\dot{q}_a + \dot{q}_b = \dot{q}_c$$

2. If current to a junction (or node, as it is often called) is considered positive and flow from the node is considered negative, Kirchhoff's law may be stated: *The sum of the currents entering a node is zero.* In step 1, which currents or flows can be considered negative?

i_2, i_3, and $\dot{q}_c$

3. If we used the idea presented in step 2, the nodal equation would be $i_1 - i_2 - i_3 = 0$. Is this the same current equation as in step 1? What would be the corresponding equation for Fig. A.1b?

Yes
$$\dot{q}_a + \dot{q}_b - \dot{q}_c = 0$$

4. In step 1, if i_1 were 2 A and i_2 were 6 A, what would be the value of i_3?

$i_3 = -4$

5. The solution for i_3 in step 4 was negative. What can be concluded regarding the direction of the flow of i_3?

Current flows toward the node.

6. The convention chosen in step 2 was arbitrary. We could just as well have used a reverse definition: Current to a node is negative; current from the node is positive. With this new convention, write a nodal equation for point A in Fig. A.1b. How does this equation compare with that of step 3?

$$\dot{q}_c - \dot{q}_a - \dot{q}_b = 0$$

7. For the tubular network shown in Fig. A.7a, prove $\dot{q}_a$ must be the same as $\dot{q}_d$ by writing nodal equations at N_1 and N_2. If there is a cross branch between tubes b and c as shown by the phantom lines, will this affect the relationship $\dot{q}_a = \dot{q}_d$? With the experience now developed, it should be very easy to evaluate the flows i_3, i_4, i_6, and i_7. Show directions of flow by including an arrow.

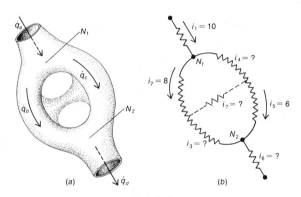

FIG. A.7

$$i_3 = 4, \ i_4 = 2$$
$$i_6 = 10, \ i_7 = 4$$

8. Flow through a resistor is governed by the voltage across the resistor. The relationship is $e = iR$ (Fig. A.8a). What is the value of the circuit i_1 in Fig. A.8b?

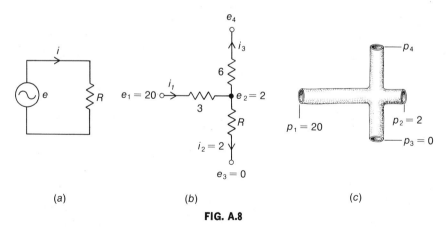

(a)   (b)   (c)

FIG. A.8

$$i_1 = 6$$

9. If $i_2 = 2$ in Fig. A.8*b*, how much is i_3?

$$i_3 = 4$$

10. If $i_2 = 2$ in Fig. A.8*b*, how much is R, and how much is e_4?

$$R = 1$$
$$e_4 = -22$$

11. If the electrical circuit of Fig. A.8*b* is the analog of the tube circuit in Fig. A.8*c*, how would you interpret the result $e_4 = -22$?

$$p_4 \text{ is suction.}$$

12. The electrical circuit in Fig. A.12*b* is assumed to be a representation of the capillary network shown in Fig. A.12*a*. If p_A, p_E, p_B, and p_F are given as shown, there are two unknown nodes. Write the nodal equations for the unknown nodes.

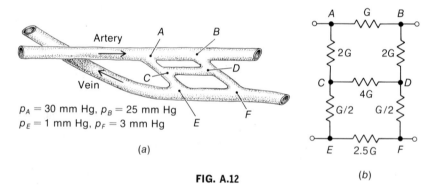

$p_A = 30$ mm Hg, $p_B = 25$ mm Hg
$p_E = 1$ mm Hg, $p_F = 3$ mm Hg

(a)

FIG. A.12

(b)

$$(30 - p_C)2G = (p_C - p_D)4G + (p_C - 1)\frac{G}{2}$$
$$(25 - p_D)2G = (p_D - p_C)4G + (p_D - 3)\frac{G}{2}$$

13. The two equations developed in step 12 have two unknowns (the pressures at the nodes). Determine the unknown node pressures by simultaneous solution of the two nodal equations.

$$p_C = 22.7 \text{ mmHg}$$
$$p_D = 21.8 \text{ mmHg}$$

14. It is noted in step 13 that the node pressures at C and D (Fig. A.12*b*) are almost the same value. If they were the same value, what would be the flow from C to D?

$$\text{Zero}$$

15. For the no-flow condition to occur in Fig. A.12b, p_C must be the same as p_D. Is there any possible adjustment of the resistances between A and C and B and D such that the capillary from C to D will be starved of blood flow? (Consider the conductances from A to C and from B to D to be nG, where n is some number; all other conductances can remain as shown. Solve for n when $p_C = p_D$.)

> Yes, when the conductances between A and C and B and D are $G/5$.

16. If the meter (M) which measures e_2 in the circuit in Fig. A.16 draws no current, then all the current that flows in R_1 must also enter R_2. Write a nodal equation for the node between the resistors.

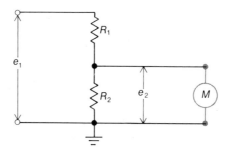

FIG. A.16

> $$\frac{e_1 - e_2}{R_1} = \frac{e_2}{R_2}$$

17. On the basis of the nodal equation developed in step 16, determine the meter reading if e_1 is 10 v, $R_1 = 2$, and $R_2 = 18$.

> $$e_2 = 9\text{V}$$

18. The circuit of step 16 is sometimes described as a voltage divider. Show e_2 is always a fractional part of e_1, where the fraction is governed by R_2 and the total circuit resistance $R_1 + R_2$.

> $$e_2 = \frac{R_2}{R_1 + R_2} e_1$$

19. A variable-voltage divider is often used in analog circuits to establish gain-constants. The divider in this case consists of a rheostat (or potentiometer), as shown in Fig. A.19. A contact arm slides across a resistive element; this essentially provides for a variable node between two resistive elements. Show e_2 is related to e_1 by a constant of proportionality (A) which depends upon the position of the slide wire. (The resistance R is proportional to its length.)

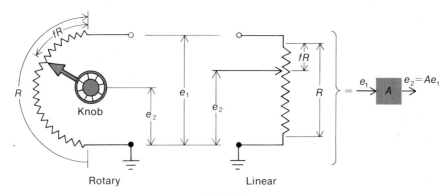

FIG. A.19

$$e_2 = (1 - f)e_1$$

20. The gain-constants established by potentiometric means cannot exceed unity; furthermore, these constants cannot be accurately maintained if variable currents are drawn from the circuit. For these reasons, amplifiers must also be used to set gain-constants in analog-computer circuits. An amplifier is a device that delivers a magnified version of an input signal; the magnification is described as the gain of the amplifier. If the gain of the amplifier in Fig. A.20 is 500 and the input from a phonograph cartridge is 100 mV, how much output does the amplifier deliver to the speaker?

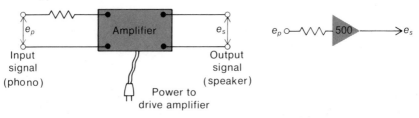

(*a*) Amplifier Schematic   (*b*) Equivalent representation

FIG. A.20

$$e_s = 500 \times 100 \times 10^{-3} \text{ V}$$
$$e_s = 50 \text{ V}$$

21. Setting a gain-constant is essentially the same as multiplying a variable by a constant. This is sometimes described as constant-multiplication. For versatility it should be possible to change this multiplicative constant at will. In a potentiometric multiplier this is accomplished by changing the resistance ratio; in a variable-gain amplifier this is accomplished by simply changing the gain. A more effective way to provide for constant-multiplication utilizes a high, constant-gain (or operational) amplifier. A schematic diagram of this multiplier is shown in Fig. A.21a. A is a very-high-gain amplifier. Resistors are connected around the amplifier as shown. Symbolically, the operation is designated as shown in Fig. A.21b. Write a nodal equation for the node at e_a in Fig. A.21a.

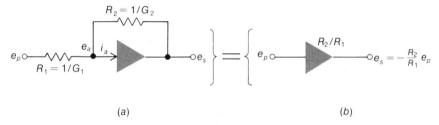

(a)   (b)

FIG. A.21

$$(e_p - e_a)G_1 + (e_s - e_a)G_2 = i_a$$

22. If the operational amplifier in step 21 has a gain of 10^4, the output of the system will be $e_s = 10^4 e_a$. With this new information, solve the nodal equation in step 21 for e_s.

$$e_s = \frac{-e_p G_1 + i_a}{G_2 - (G_1 + G_2)/10^4}$$

23. If R_1 in step 21 is 10^6 Ω and R_2 is 5×10^6 Ω, what is the equation for e_s?

$$e_s = \frac{-10^{-6} e_p + i_a}{(0.2 \times 10^{-6}) - (0.00012 \times 10^{-6})}$$

24. If the denominator of the solution for e_s (in step 23) is examined, we note the second term can be neglected compared with the first. We then have

$$e_s = \frac{-10^{-6} e_p + i_a}{0.2 \times 10^{-6}} = -5e_p + (5 \times 10^6) i_a$$

For ordinary operation the input-signal voltage e_p is of the order of 5 V; the input amplifier current (for a fair operational amplifier) would be 50 nanoamperes (nA) or 50×10^{-9} A. Based on these figures, approximately what is the output voltage?

$$e_s = (-25 + 0.25) \text{ V} \approx -25 \text{ V}$$

25. Good operational amplifiers draw practically no current ($i_a \approx 50 \times 10^{-12}$ A). Based upon this figure and the calculations in step 24, i_a can be considered neglectable as a node current. Furthermore, if e_s is approximately 25 V, e_a must have been about 25×10^{-4} V or 0.0025 V; since e_s is of the order of 5 V, e_a can be neglected (or considered zero) compared with e_s or e_p. In effect, therefore, e_a can be considered ground or reference and i_a can be considered zero. Using these approximations, write the nodal equation at e_a in Fig. A.21a and solve for e_s.

$$e_s = -\frac{G_1}{G_2} e_p$$
$$= -\frac{R_2}{R_1} e_p$$

26. From the solution of step 25 (if A is very high and $i_a \approx 0$) the configuration shown in Fig. A.21 may be used as a multiplier and a sign changer. Show a configuration to obtain $e_{\text{out}} = +26e_{\text{in}}$ (use two operational amplifiers).

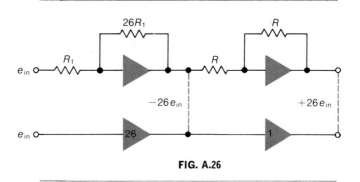

FIG. A.26

27. The resistive network shown in Fig. A.27 can serve as an adder if the meter current i_m is practically zero. Write a nodal equation for point A.

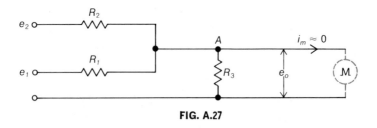

FIG. A.27

$$(e_2 - e_0)G_2 + (e_1 - e_0)G_1 = e_0G_3$$

28. If the resistors in Fig. A.27 are related so that $R_1 = R_2 = R_3$, write the solution for e_0.

$$e_0 = \tfrac{1}{3}(e_1 + e_2)$$

29. From the solution in step 28 it is seen the circuit of Fig. A.27 can be used as an adder to within a multiplicative constant. The constant could, of course, be adjusted by using a multiplier, as shown in Fig. A.29a. An alternative construction for an adder is shown in Fig. A.29b. If e_a and i_a are small enough to be neglected, write a nodal equation at A and show the circuit will perform addition (with a sign change) if $R_1 = R_2 = R_3$.

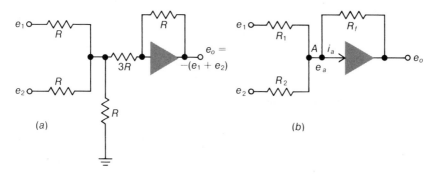

(a)

(b)

FIG. A.29

$$e_1 G_1 + e_2 G_2 + e_0 G_f = 0$$
$$e_1 \frac{G_1}{G_f} + e_2 \frac{G_2}{G_f} = -e_0$$

30. It is easy to see that the circuit of Fig. A.29b can be used to solve simple linear equations of the form $y = mx + b$. The voltage e_1 would be considered the variable x, e_2 would be the constant b; the slope m would be determined by properly adjusting R_f and R_1. Determine a set of values for the R's and the e's in Fig. A.29b to solve $y = 7x + 2$.

One possibility is
$e_1 = x; R_1 = 0.1$ MΩ
$R_f = 0.7$ M$\Omega; R_2 = 0.7$ MΩ
$e_2 = 2$ V

31. It is possible to use the same adder configuration to add the effects of two variables. If the total flow in a vessel is the sum of 80 percent of one flow input (see Fig. A.31a) and 50 percent of another, set up the circuit in Fig. A.31b so that the output voltage represents the total flow.

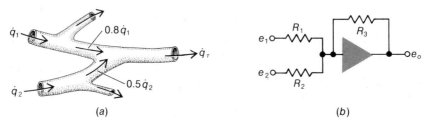

(a)   (b)

FIG. A.31

$$e_1 = \dot{q}_1; \ e_2 = \dot{q}_2$$
$$R_1 = 1.25 \ M\Omega; \ R_2 = 2 \ M\Omega$$
$$R_3 = 1 \ M\Omega$$

COMMENT: The subsequent material will be easier to follow after Chap. 7 has been studied.

32. In transform notation the impedance of a capacitor is $1/sC$. Mathematically, impedances are handled as resistances: impedances and resistances in series add; admittances (reciprocal impedance) in parallel add. What is the total impedance of the circuit in Fig. A.32a, and what is the total admittance of the circuit in Fig. A.32b?

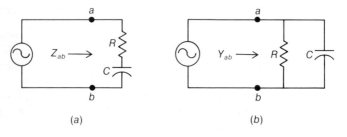

(a)   (b)

FIG. A.32

$$Z = R + \frac{1}{sC}$$
$$Y = \frac{1}{R} + sC$$

33. The technique for writing nodal equations does not change when impedances are present in the circuit. The sum of the currents entering a junction must still be zero. For the circuit in Fig. A.33, write the two nodal equations needed to solve for the unknown node voltages e_a and e_b.

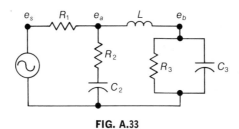

FIG. A.33

$$\text{Zero} = \frac{e_s - e_a}{R_1} + \frac{e_a}{R_2 + 1/sC_2} + \frac{e_a - e_b}{sL}$$

$$\text{Zero} = \frac{e_a - e_b}{sL} + e_b(G_3 + sC_3)$$

34. When capacitors are used with operational amplifiers, it is possible to assemble circuit configurations that will perform the mathematical operation of integration (and sometimes differentiation). The circuit shown in Fig. A.34a is such an integrator (integrators are often represented diagrammatically by the symbol shown in Fig. A.34b). By writing a nodal equation at A, show e_0 is the negative integral of e_1 to within a multiplicative constant if the operator $1/s$ is interpreted as integration. What values can be used for R and C to make the multiplicative constant unity?

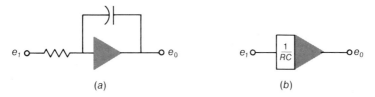

(a)   (b)

FIG. A.34

$$e_1 G + e_0 sC = 0$$

$$e_0 = -\frac{1}{RC}\left(\frac{1}{s}e_1\right)$$

35. The integral of a sum can be simulated if more than one input is applied to an integrator. Figure A.35 shows an arrangement where the output is the integral of the sum of input variables. What values should be used for the R's and C's to perform the operation $q = \int(0.8\dot{q}_1 + 0.5\dot{q}_2)\, dt$?

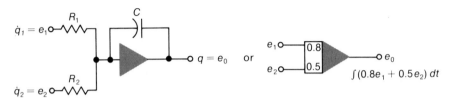

FIG. A.35

$$R_1 = 1.25 \text{ M}\Omega$$
$$C = 1 \ \mu\text{F}$$
$$R_2 = 2 \text{ M}\Omega$$

It is possible, of course, to continue to develop more complex circuits to do more complex operations. As pointed out earlier, however, there are many texts designed to do a more complete job of teaching circuit equations and analog-computer simulations. The intent here is simply to introduce this material and to provide an understanding of the subject matter. For more intensive studies, the texts mentioned in the reference lists of Chaps. 1 and 2 should be consulted.

APPENDIX B

SOURCE REPRESENTATIONS

In the early sections of this text, voltage generators were used almost exclusively to represent energy sources. The use of a voltage source to represent temperature and pressure seems entirely reasonable from an analog point of view: Temperature and pressure have been described as across-variables; voltage has been described as an across-variable; therefore, a voltage source may be used to represent a temperature or pressure source. Fortunately, voltage sources are a familiar part of everyday life; the 110-V outlet plugs in almost all homes serve potentially as voltage sources. It is easy to accept, therefore, that a voltage source can be used to represent an energy source which can supply an across-variable excitation. This serves to relate the idea of a source to our everyday experiences; it does not, however, really explain what a voltage source is.

Although the operating principle of the 110-V outlet as a voltage source may not be generally understood, the general features or properties of this source are well known by most users: Any one outlet can deliver current to a number of appliances (within limits, of course) without altering the delivered voltage. Thus an outlet may deliver current to a lamp, an amplifier, a turntable, and a clock— all at the same time or one at a time; in either case the voltages across the appliances remain very close to 110 V. As a comparable illustration, the ordinary 12-V

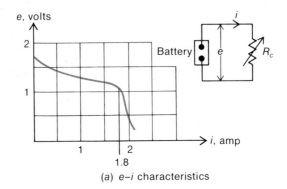

(a) e–i characteristics

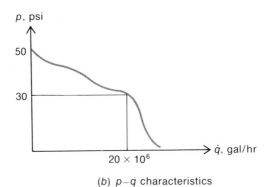

(b) p–q characteristics

FIG. B.1

battery in a car delivers 12 V to one or to many circuits. The voltage level of the source remains relatively constant even though variable amounts of current are drawn from the battery.

The examples given in the preceding paragraph illustrate the basic feature of the *constant*-voltage source; such sources deliver a constant voltage to externally connected devices even though the current demands on the source may vary. A plot of such characteristics appears as shown in Fig. B.1; when current flow is low (i.e., when the resistance R_c across the battery is high; see circuit inset) the battery voltage is 1.6 V. As the current flow increases (R_c is reduced) the battery voltage decreases to about 1.1 V; at high current drain the battery voltage is practically zero. If the current demands on the battery are in the range of $0.5 < i < 1.5$ the voltage variation is about 15 percent. For some applications a 15 percent variation can be neglected, and the battery can be said to act as a constant-voltage source.

The constant-voltage supply represents one type of source which delivers a constant level of the across-variable even though the delivery rate of the through-

variable can vary. Another type of source with these same operating characteristics is a city water-supply system: When there are few demands on the system
(i.e., few people use the water, and so delivery rate is low) city pressure is about
50 psi; as more water is drawn, the supply pressure reduces until, at very high
delivery rates, the pressure is practically zero. A typical plot of a water-supply
system might appear as shown in Fig. B.1b. For a wide range of delivery rates
the driving pressure remains relatively constant. The water system thus acts as a
source that delivers a constant level of an across-variable excitation, while the
through-variable may vary. Still another example of this same operating characteristic is a thermostatically controlled room. A well-made control system regulates a room temperature very closely in spite of varying heat loads.

The sources described above represent supplies that deliver relatively
constant levels of across-variable excitation although through-variable demands
may vary. In contrast to such sources, many supply generators have the reverse
characteristic. Such sources deliver a fixed level of the through-variable although
across-variable requirements may vary. An illustration of such a source is a
piston-type water pump; the fluid delivered depends primarily on the stroke of
the piston. For a fixed rate of piston motion the flow will be relatively independent
of the variable pressure levels on the pump. Another illustration of a constant
through-variable source is a burning lump of coal or an ignited gas jet or a diathermy unit. Each of these delivers radiated heat at a level relatively independent
of the temperature of the system under excitation. Electrically such sources are
also available as constant-current sources; these sources deliver fixed levels of
current although voltage demands may vary.

The descriptions of sources in the foregoing paragraphs seem to imply
there are two different types of excitation supplies. This is only partly true.
Although the inherent characteristics of a supply may make it better suited for
one purpose than another, it is almost always possible to convert a source from
one form of operation to another. The modification depends upon the graphical
characteristics of the source and the range of operation required. These ideas
are more easily understood if we develop and use analytic models to represent
excitation sources.

We can develop mathematical representations for sources in the same way
mathematical models were developed for system elements. The graphical model
can serve as a starting point; a mathematical equation that describes the overall
graphical characteristics is often impractical, and a piecewise-linearized representation is generally used instead. This procedure is particularly appropriate if
the source is to be used for a limited range of operations only. If, for example,
the operating range in Fig. B.1a is limited to $i < 1.8$ and in Fig. B.1b the range
is $\dot{q} < 20 \times 10^6$, both sources have approximately straight-line characteristics
and can be represented by an expression of the form $y = mx + b$. For the battery
characteristics we can write

(B.1) $\qquad e = -(\frac{1}{6})i + 1.5$

(B.1a) $\qquad e = -m_e i + E_0 \qquad m_e = \frac{1}{6} \qquad E_0 = 1.5 \text{ V}$

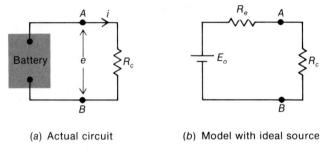

(a) Actual circuit   (b) Model with ideal source

FIG. B.2   *Model of a voltage source*

and for the water-supply source we can write

(B.2) $p = -10^{-6}\dot{q} + 50$

(B.2a) $p = -m_H\dot{q} + P_0$ $m_H = 10^{-6}$ $P_0 = 50$

Equations (B.1) and (B.2) can serve as mathematical models of the source characteristics shown in Fig. B.1 if operation is limited to the ranges specified.

　　The mathematical models are of interest as they provide an interpretation of the operating characteristics of the source not readily perceived from the graphical model alone. If, for example, we write Eq. (B.1a) as $e = E_0 - im_e$ and Eq. (B.2a) as $p = P_0 - im_H$ we see the actual sources may be considered devices that deliver fixed levels of excitations (E_0 or P_0) diminished by a term proportional to the flow from the source (im_e and $\dot{q}m_H$). The proportionality constant is determined from the slope of the characteristic curve for the source; for the situations shown in Fig. B.1 these slopes evidently have the units of resistance ($m_e = \Delta e/\Delta i = R_e$; $m_H = \Delta p/\Delta\dot{q} = R_H$). This viewpoint has interesting implications; it permits expressing a true source in terms of an ideal generator and an internal lossy effect. For example, the battery in Fig. B.2 (which is represented by the graphical model in Fig. B.1a) can be interpreted as an ideal constant-voltage source (E_0) driving through an internal resistive element (R_e). If we use the symbol $\pm$ to represent an *ideal* constant-voltage source, the actual battery circuit of Fig. B.2a can be redrawn as in Fig. B.2b. The actual battery to the left of points AB has been replaced by an ideal source E_0 and an internal resistance R_e. To show that the voltage e determined from the hypothetical model correctly describes the true voltage delivered from the actual battery, we write a nodal equation at A in the model circuit: $E_0 - e = iR_e$. This equation is identical to Eq. (B.1a) when $m_e = R_e$.

　　We can now appreciate how the battery serves as a voltage source. If we use the model representation in Fig. B.2b, there will be little voltage loss within the source if the source resistance (R_e) is very low, or if the current drawn is very low. These conditions are both satisfied to some extent if the driven system has high resistance. In Fig. B.2b, therefore, if the circuit resistance R_c were very much larger than the source internal resistance R_e (i.e., if $R_c \gg R_e$) the battery voltage delivered (e) would be very closely equal to E_0 in spite of variations of current drawn. These are the characteristics of a voltage source.

We can express these ideas analytically if we use the model configuration in Fig. B.2b. The circuit is a simple voltage divider (Appendix A, step 16); e can therefore be immediately expressed as

(B.3)
$$e = \frac{R_c}{R_e + R_c} E_0$$

From Eq. (B.3) we can readily see e is constant at E_0 if $R_c \gg R_e$.

By using this same procedure we can show the source represented by the characteristics in Fig. B.1a may also serve as a current source if the voltage requirements do not exceed $e = 1.0$ V. From the curve we can see that in this range the variation of i is "small" though e may vary over a "wide" range. This is the general feature of a current source. From an analytic point of view we note when $e \leq 1.0$ V ($i \geq 1.8$ A) the e-i characteristics may be approximately represented by

(B.4)
$$e = -m'_e i + E_0$$

where m'_e is the slope of the straight-line approximation (see Fig. B.1a). From the previous discussion we know that m'_e may be interpreted as an internal resistance; in this case, however, the resistance (R'_e) is relatively high. To examine the relationship for the current drawn from the battery, we rewrite Eq. (B.4)

(B.5)
$$\frac{E_0}{R'_e} - \frac{e}{R'_e} = i = I_0 - G'_e e$$

In this form we see the battery can be considered an ideal current source with an internal lossy effect given by G'_e. If we use the symbol ϕ to represent a current source, then the battery may be represented by the model in Fig. B.3. In this form we can easily see how the circuit in Fig. B.3 acts as a current source. If $R_c \ll R_e$, then little current will flow in R_e and practically all the I_0 will be delivered to the external circuit in spite of possible voltage variations across R_c.

As an analytic justification of this conclusion, we note Fig. B.3 represents

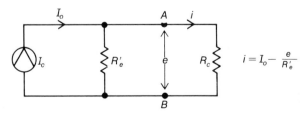

FIG. B.3

a current divider; the current i is given by

(B.6)
$$i = I_0 \frac{G_c}{G'_e + G_c}$$

If $G_c \gg G'_e$ (or $R_c \ll R'_e$), i remains approximately constant at $i = I_0$.

The results of the foregoing analysis are not necessarily restricted to constant-supply generators. Equation (B.3), for example, points out that e is equal to E_0 if the load resistor (R_c) is much larger than the internal resistance of the source (R_e). This conclusion is true even if E_0 is not constant but varies in some controllable fashion. Thus, if the e-i characteristics of Fig. B.1 shifted up and down with time as shown in Fig. B.4a, E_0 would vary with time; the voltage in Eq. (B.3) would therefore also vary with time in the same fashion. Under these circumstances we would have a variable-voltage supply that delivers a controllable voltage signal independent of the delivered current. The symbol for an ideal variable-voltage source is ⊖ . The circuit model for a variable-voltage supply driving an external circuit is thus the same as Fig. B.2 except the ideal constant source is replaced by the symbol for a variable supply.

In a similar fashion Eq. (B.6) indicates $i = I_0$ when the load resistance is less than the source resistance. This conclusion is true even if I_0 varies with time (or any other control) as shown in Fig. B.4b. Under these circumstances we would have a variable-current supply that delivers a current independent of the load voltage. The symbol for an ideal variable-current source is the same as that for a fixed supply. The diagram of Fig. B.3 would thus be used regardless of whether the source were fixed or variable.

The principles outlined in the foregoing paragraphs indicate to some extent how voltage sources or current sources may be devised from a given generator even though the generator characteristics are not known. A voltage source, for example, will deliver a controllable voltage to an external circuit or load regardless of the current delivered to the load. Thus in B.5a if the source indicated is to be a voltage source, e (the voltage across the load R_L) should be independent of i (the current delivered to the load R_L), even though the load may vary slightly.

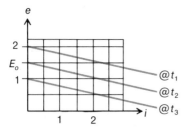

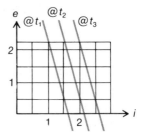

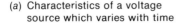

(a) Characteristics of a voltage source which varies with time

(b) Characteristics of a current source which varies with time

FIG. B.4

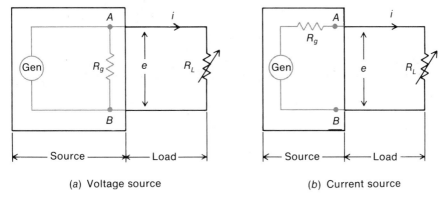

(a) Voltage source   (b) Current source

FIG. B.5 *Sources from energy supplies.*

To ensure this condition exists, the available generator is shorted with a low-resistance shunt ($R_g \ll R_L$), as shown in phantom view in Fig. B.5a. Almost all the generator current will now flow continuously through R_g and only a little to R_L. This condition will exist (even though R_L may vary) so long as $R_g \ll R_L$. In effect, then, variations in R_L will not affect the current flow to R_g; the voltage across R_g (that is, e) will thus not vary with R_L; the voltage across the load R_L is therefore independent of current delivered to R_L.

In a comparable fashion a current source is constructed from a generator by inserting a series resistor between the generator and load (see phantom view in Fig. B.5b). If the inserted resistor R_g has a very high resistance compared with the load resistor ($R_g \gg R_L$), almost all the generator voltage is lost across R_g and only a little across R_L. This condition will exist (even though R_L may vary) so long as $R_g \gg R_L$. In effect, then, variations in R_L will not affect the drop across R_g. The current through R_g (that is, i) will thus not vary with R_L; the current delivered to the load will thus be independent of the voltage across the load.

PARTIAL-FRACTION EXPANSION

As mentioned in the text, there are available many references that provide a detailed explanation of the expansion of algebraic fractions into partial fractions. The discussion here, therefore, will be no more than a simple introduction to the concept and mechanics of partial-fraction expansion. Mathematical justification for the procedures outlined will not be included. (The techniques to be described, however, can be justified mathematically.) For more complete treatments, other texts including the subject of partial-fraction expansion should be consulted.

The basic premise of partial-fraction expansion is: An algebraic fraction with a numerator of less degree than the denominator may be expanded into a sum of single-degree fractional terms; the number of fractional terms will be equal to the degree of the denominator. In mathematical terms, if

(C.1)
$$F(s) = \frac{s^n + a_1 s^{n-1} + a_2 s^{n-2} + \cdots + a_n}{s^m + b_1 s^{m-1} b_2 s^{m-2} \cdots b_m}$$

$$= \frac{(s + \alpha_1)(s + \alpha_2) \cdots (s + \alpha_n)}{(s + \beta_1)(s + \beta_2) \cdots (s + \beta_m)}$$

and $m > n$, then $F(s)$ may be written

(C.2)
$$F(s) = \frac{A_1}{s + \beta_1} + \frac{A_2}{s + \beta_2} + \cdots + \frac{A_m}{s + \beta_m}$$

As a numerical illustration of the general mathematical statements, if

(C.3) $$F(s) = \frac{s^2 + 6s + 5}{s^3 + 11s^2 + 36s + 36} = \frac{(s + 5)(s + 1)}{(s + 2)(s + 3)(s + 6)}$$

then, since the degree of the denominator (3) is greater than that of the numerator (2), $F(s)$ may be written

(C.4) $$F(s) = \frac{A_1}{s + 2} + \frac{A_2}{s + 3} + \frac{A_3}{s + 6}$$

The question now to be resolved is: How may the numerator terms (A_n) of the fractions be evaluated? In effect, we are asking how shall we choose A_1, A_2, and A_3 to ensure Eq. (C.5) is an equality:

(C.5) $$\frac{(s + 5)(s + 1)}{(s + 2)(s + 3)(s + 6)} = \frac{A_1}{s + 2} + \frac{A_2}{s + 3} + \frac{A_3}{s + 6}$$

One possible way to determine appropriate values for the numerator is to combine the right side of Eq. (C.5) and to match the s coefficients of the resultant numerator with the given numerator. Thus

(C.6) $$\frac{s^2 + 6s + 5}{(s + 2)(s + 3)(s + 6)}$$
$$= \frac{(A_1 + A_2 + A_a)s^2 + (9A_1 + 8A_2 + 5A_3)s + (18A_1 + 12A_2 + 6A_3)}{(s + 2)(s + 3)(s + 6)}$$

If this is to be an equality, the coefficients of s^2, s^1, and $s^0(= 1)$ on the right side of the equation must equal the coefficients of equivalent s terms on the left. We have, therefore, a set of three equations and three unknowns:

(C.7)
$$A_1 + A_2 + A_3 = 1$$
$$9A_1 + 8A_2 + 5A_3 = 6$$
$$18A_1 + 12A_2 + 6A_3 = 5$$

A simultaneous solution will yield the values of the A terms needed to ensure Eq. (C.5) is an equality. (The solution is $A_1 = -\frac{3}{4}$, $A_2 = +\frac{4}{3}$, $A_3 = +\frac{5}{12}$.)

A somewhat simpler procedure which can be used to evaluate the numerator constants of the expanded fractions is to separate and evaluate each constant in turn by appropriate numerical substitutions. The first step in the process is to transform the equality represented by Eq. (C.2) [and Eq. (C.5)] into one in which a numerator constant appears alone. In Eq. (C.5), for example, we can make A_1 stand alone if both sides of the equation are multiplied through by $s + 2$. We would then have

(C.8) $$\frac{(s + 5)(s + 1)}{(s + 3)(s + 6)} = A_1 + A_2\frac{s + 2}{s + 3} + A_3\frac{s + 2}{s + 6}$$

Equation (C.8) is an equality for all values of s. If we let s approach -2, the left side of Eq. (C.8) approaches $-\frac{3}{4}$; the coefficients of A_2 and A_3, however, approach zero [i.e., if $s \to -2$, $(s + 2)/(s + 3)$ approaches zero as does $(s + 2)/$

$(s + 6)$]. For the particular case where s approaches -2, Eq. (C.8) therefore reduces to

(C.9) $$\frac{(+3)(-1)}{(+1)(+4)} = \frac{-3}{4} = A_1 + (0)A_2 + (0)A_3 = A_1$$

This provides a solution for the numerator constant A_1.

We can now use this same technique to determine both numerator constants A_2 and A_3. For the A_2 term we multiply Eq. (C.5) through by $s + 3$; the result is

(C.10) $$\frac{(s + 5)(s + 1)}{(s + 2)(s + 6)} = \frac{s + 3}{s + 2} A_1 + A_2 + \frac{s + 3}{s + 6} A_3$$

Equation (C.10) is an equality for all s. If we let s approach -3, the coefficients of the A_1 and A_3 terms approach zero, and Eq. (C.10) becomes

(C.11) $$\frac{(+2)(-2)}{(-1)(+3)} = \frac{4}{3} = A_2$$

which can be taken as a solution for A_2. In a similar manner, the coefficient A_3 may be determined as

(C.12) $$\frac{(s + 5)(s + 1)}{(s + 2)(s + 3)} = \frac{s + 6}{s + 2} A_1 + \frac{s + 6}{s + 3} A_2 + A_3$$

$$\frac{(s + 5)(s + 1)}{(s + 2)(s + 3)} = \frac{(-1)(-5)}{(-4)(-3)} = \frac{5}{12} = A_3$$

The procedure outlined above is relatively simple and fairly general. It can be used for most cases of practical interest in systems studies. However, one situation often occurs where the simple technique just described is not sufficient to evaluate all numerator constants. This arises when the denominator contains factors of higher order than the first. To illustrate this point, we examine the fraction in Eq. (C.13):

(C.13) $$F(s) = \frac{s + 3}{s^4 + 3s^3 + 2s^2} = \frac{s + 3}{s^2(s + 2)(s + 1)}$$

As the denominator of this fraction is of fourth degree, the partial-fraction expansion should contain four terms. The denominators of three of the partial fractions are evident by inspection; they are s^2, $s + 2$, and $s + 1$. The denominator of the fourth term is simply s; this is not obvious by inspection but it can be shown that if the denominator of a fraction contains a factor of the form $(s + a)^n$ this factor will give rise to n terms in the partial-fraction expansion. These terms will be of the form $A_1/(s + a)^n$, $A_2/(s + a)^{n-1}$, . . . , $A_n/(s + a)$.

If we carry out the expansion suggested in the preceding paragraph, Eq. (C.13) may be written

(C.14) $$\frac{s + 3}{s^2(s + 2)(s + 1)} = \frac{A_1}{s^2} + \frac{A_2}{s} + \frac{A_3}{s + 2} + \frac{A_4}{s + 1}$$

where once again the A terms must be evaluated to satisfy the condition imposed by the equal sign. A_1, A_3, and A_4 are easily determined in the manner outlined previously.

(C.15)
$$\lim_{s \to 0} \frac{s + 3}{(s + 2)(s + 1)} = A_1 = \frac{3}{2}$$

(C.15a)
$$\lim_{s \to -2} \frac{s + 3}{s^2(s + 1)} = A_3 = -\frac{1}{4}$$

(C.15b)
$$\lim_{s \to -1} \frac{s + 3}{s^2(s + 2)} = A_4 = 2$$

If we attempt to use this same technique to evaluate A_2 we find both sides of Eq. (C.14) become infinite as we allow s to approach zero. This is a valid result as the equality is maintained, but it does not help to determine A_2. One possible way to handle this situation is to use the first procedure outlined: We can recombine the right side of Eq. (C.14) and match numerator coefficients. As A_2 is the only unknown in this case, only one equation is required and the solution is fairly simple.

A somewhat simpler procedure in this case is to set up an equation for A_2 by using any numerical value for s that does not cause any terms in Eq. (C.14) to become infinite. Since Eq. (C.14) is an equality for all values of s, the equality is maintained even if s takes on numerical values. Thus if s is taken as -3, we can write

(C.16)
$$0 = \frac{A_1}{9} + \frac{A_2}{-3} + \frac{A_3}{-1} + \frac{A_4}{-2}$$

As A_1, A_3, and A_4 have already been determined [Eqs. (C.15), (C.15a), and (C.15b)], A_2 can be evaluated immediately to be $A_2 = -\frac{1}{4}$.

The last technique which we shall describe to evaluate the numerator constants depends upon the use of differentiation to eliminate constants and isolate constant coefficients. We note in particular that differentiation of a constant (for example, dK/dx) removes the constant ($dK/dx = 0$) whereas differentiation of a constant and a variable $[d(Kx)/dx]$ causes a constant to stand alone $[d(Kx)/dx = K]$. To show how these ideas may be used, we use Eq. (C.14) once again for illustration. When we modify Eq. (C.14) in the manner suggested earlier, we can write

(C.17)
$$\frac{s + 3}{(s + 2)(s + 1)} = A_1 + A_2 s + \frac{A_3}{s + 2} + \frac{A_4}{s + 1}$$

This relationship was used to evaluate A_1, for when $s \to 0$ the right side of the equation reduces to A_1 and the left side reduces to a numerical value. We can also use this same expression to develop a relationship from which A_2 may be determined. If, for example, we differentiate Eq. (C.17), A_1 is removed from the equation, and A_2 is retained as a term. In particular, by differentiation we have

(C.18)
$$\frac{(s + 2)(s + 1) - (s + 3)(2s + 3)}{(s + 2)^2(s + 1)^2}$$
$$= A_2 + A_3 \frac{(s + 2)(2s) - s^2}{(s + 2)^2} + A_4 \frac{(s + 1)2s - s^2}{(s + 1)^2}$$

We can now evaluate A_2 in the same way A_1 was evaluated from Eq. (C.17). Specifically, if we let s approach zero, the right side of the equation reduces to A_2 and the whole equation reduces to a solution for A_2:

(C.19) $\lim\limits_{s \to 0} \dfrac{(s+2)(s+1) - (s+3)(2s+3)}{[(s+2)(s+1)]^2} = \dfrac{2-9}{4} = A_2 + 0 + 0 = -\dfrac{7}{4}$

To describe the latter technique in general mathematical terms we can state that if a fraction is of the form

(C.20) $F(s) = \dfrac{(s+a_1) \cdots (s+a_{n-1})}{(s+b)^n} = \dfrac{A_1}{(s+b)^n} + \dfrac{A_2}{(s+b)^{n-1}}$

$$+ \cdots + \dfrac{A_n}{s+b}$$

the numerator constants (A) of the partial-fraction expansion may be evaluated by

(C.21) $$A_1 = \lim\limits_{s \to -b} F(s)(s+b)^n$$

(C.22) $$A_2 = \lim\limits_{s \to -b} \dfrac{d}{ds}[F(s)(s+b)^{n-1}]$$

(C.23) $$A_n = \lim\limits_{s \to -b} \dfrac{d^{n-1}}{ds^{n-1}}[F(s)(s+b)]$$

TABLE OF
LAPLACE TRANSFORMS

Laplace transform $f(s)$	Time function $f(t) \quad t > 0$
1	$\delta(t)$ unit-impulse function
$\dfrac{1}{s}$	$u(t)$ unit-step function
$\dfrac{1}{s^2}$	t unit-ramp function
$\dfrac{1}{s^{n+1}}$	$\dfrac{1}{n!} t^n$ $n = $ positive integer
$\dfrac{1}{s+a}$	e^{-at}
$\dfrac{1}{(s+a)^2}$	te^{-at}
$\dfrac{1}{(s+a)(s+b)}$	$\dfrac{e^{-at} - e^{-bt}}{b - a}$
$\dfrac{1}{(s+a)^n}$	$\dfrac{1}{(n-1)!} t^{n-1} e^{-at}$ $n = $ positive integer
$\dfrac{1}{s(s+a)}$	$\dfrac{1}{a}(1 - e^{-at})$
$\dfrac{1}{s(s+a)(s+b)}$	$\dfrac{1}{ab}\left(1 - \dfrac{b}{b-a} e^{-at} + \dfrac{a}{b-a} e^{-bt}\right) \quad a \neq b$
$\dfrac{1}{s(s+a)^2}$	$\dfrac{1}{a^2}[1 - (1 + at)e^{-at}]$
$\dfrac{\omega}{s^2 + \omega^2}$	$\sin \omega t$
$\dfrac{s}{s^2 + \omega^2}$	$\cos \omega t$
$\dfrac{\omega_n^2}{s^2 + 2\zeta\omega_n s + \omega_n^2}$	$\dfrac{\omega_n^2}{\sqrt{1 - \zeta^2}} e^{-\zeta\omega_n t} \sin \omega_n \sqrt{1 - \zeta^2}\, t$
$\dfrac{\omega_n^2}{s(s^2 + 2\zeta\omega_n s + \omega_n^2)}$	$1 + \dfrac{1}{\sqrt{1 - \zeta^2}} e^{-\zeta\omega_n t} \sin(\omega_n \sqrt{1 - \zeta^2}\, t - \phi)$ where $\phi = \tan^{-1} \dfrac{\sqrt{1 - \zeta^2}}{-\zeta}$
$\dfrac{\omega_n^2}{s(s^2 + \omega_n^2)}$	$1 - \cos \omega_n t$
$\dfrac{s}{(s^2 + \omega_n^2)^2}$	$\dfrac{1}{2\omega_n} t \sin \omega_n t$
$\dfrac{1}{s^2(s+a)}$	$\dfrac{1}{a^2}(at - 1 + e^{-at})$

GLOSSARY OF TERMS

alveolus	An air sac of the lungs formed by the terminal dilatations of a bronchiole.
ampere	The practical mks unit of electric current that is equivalent to a flow of one coulomb per second or to the steady current produced by one volt applied across a resistance of one ohm.
anesthetic	Relating to drugs or methods which produce insensitivity to touch or pain.
antagonist	A muscle that acts in opposition to the action of another muscle, its agonist.
aorta	The great trunk artery that carries blood from the heart to be distributed by branch arteries through the body.
arteriole	One of the small terminal branches of an artery that ends in capillaries.
artery	One of the tubular branching muscular and elastic-walled vessels that carry blood from the heart to the body.
atrium, atria	The upper chambers on either side of the heart.

basal metabolism	The turnover of energy in a fasting and resting organism using energy solely to maintain the vital cellular activity, respiration, and circulation as measured by the basal metabolic rate.
bladder	Any sac or receptacle for a secretion.
BTSC	Body temperature standard conditions.
bronchiole	The last and tiniest subdivision of the bronchial tree opening into the alveoli.
calorie	The amount of heat required at a pressure of one atmosphere to raise the temperature of one gram of water one degree centigrade.
cardiac	Pertaining to the heart.
catheter	A tubular surgical instrument for withdrawing fluids from a cavity of the body.
clo	A unit of measurement representing approximately the insulation provided by $\frac{1}{4}$-in. thickness of wool.
compliance	A measure of the elastic properties of volumetric storage elements such as a balloon, the bladder, the lungs, etc.; quantitatively defined by the volume of fluid stored per unit pressure.
concentration	The ratio of the mass or volume of a solute to the mass or volume of the solution.
conductance	The readiness with which a conductor transmits an electric current; the reciprocal of electrical resistance.
coulomb	The unit of quantity in current electricity; the quantity afforded by an ampere of current in one second flowing against one ohm of resistance with a force of one volt.
cytoplasm	The protoplasm of a cell exclusive of that of the nucleus.
diaphragm	The musculomembranous partition that separates the abdomen from the thorax and provides part of the bellows action necessary for respiration.
diastole	The stage of dilatation or relaxation of the heart muscles, especially of the ventricles.
dicrotic	Pertaining to a double beat, as of the pulse. Generally refers to the notch in the pressure-pulse waveform distal to the aortic valve; the notch is due to the abrupt closure of the valve.
diffusion	The process whereby particles intermingle as the result of their spontaneous movement and in dissolved

substances move from a region of high to one of lower concentration.

diffusivity

A measure of the capability of materials to pass from one medium to another (or to become distributed throughout a medium) by virtue of spontaneous molecular motion; e.g., some gases diffuse into and through glass, and sodium and chloride ions (salt) become distributed throughout a salt and water mixture. Quantitatively defined by the amount of material transferred per unit time over a unit cross section due to a unit concentration gradient.

dynamic resistance

In electrical systems: the reciprocal of the slope of the voltage-current characteristics of a material. In generalized systems: the reciprocal of the slope of the across-through characteristics of a material when the slope includes time units to an odd power.

endothelium, endothelial

The layer of simple squamous cells which line the inner surface of the circulatory organs.

energy

A measure of work or the capability of performing work. Has units of foot-pounds, joules, ergs, watt-seconds, etc.

epithelium, epithelial

The covering of internal and external surfaces of the body, including the lining of vessels and other small cavities.

esophagus, esophageal

A musculomembranous canal extending from the pharynx to the stomach.

expirate
exhalate

A fraction or a portion of the gas discharge from the lungs. One of the constituents of lung discharge.

friction

Opposition to motion due to rubbing action between particles or surfaces.

hematocrit

An instrument for determining the relative amounts of plasma and corpuscles in blood; usually utilizes centrifugation for the determination.

impulse function

An excitation that has high amplitude but exists only for a brief instant.

inspirate
inhalate

A fraction or portion of the gas drawn into the lungs. One of the constituents of the intake.

insulator

Any substance or appliance of nonconducting properties to prevent the transfer of electricity, heat, or sound.

insulin

A protein hormone formed by the islet cells of Langerhans in the pancreas and secreted into the blood, where it regulates carbohydrate (sugar) metabolism.

intercellular	Situated between the cells of any structure.
intercostal	Situated between the ribs.
interpleural	Between two layers of the pleura, as between the visceral and the parietal pleura.
interstitial	Pertaining to or situated in the interstices (spaces or gaps) of a tissue.
intracellular	Situated or occurring within a cell or cells.
intrapleural	Situated within the space between the lungs and a thin tissue covering the lungs (the pleura).
inulin	A tasteless white polysaccharide found especially dissolved in the sap of the roots and rhizomes of composite plants. It is used in a test to determine renal functions.
in vitro	Within a glass; observable in a test tube.
in vivo	Within the living body.
ligament	Any tough fibrous band that connects bones or supports viscera.
metabolism	The sum of all the physical and chemical processes by which living organized substance is produced and maintained; also the transformation by which energy is made available for the uses of the organism.
mitral valve	A cardiac valve guarding the orifice between the left auricle and ventricle and preventing the return of blood to the auricle.
myocardial	Pertaining to the muscular tissue of the heart.
ohm	The practical mks unit of electric resistance equal to the resistance of a circuit in which a potential difference of one volt produces a current of one ampere.
ordinate	Vertical lines used as a base of reference in graphs. (See abscissa.)
osmosis	The passage of pure solvent from the lesser to the greater concentration when two solutions are separated by a membrane which selectively prevents the passage of solute molecules but is permeable to the solvent.
perfusion	The act of forcing a fluid through (an organ or tissue), especially by way of the blood vessels.
permeability	The property of a magnetizable substance that determines the degree in which it modifies the magnetic flux in the regions occupied by it in a magnetic field. The quality of state of being penetrable as by diffusion of material through pores or interstices.

plasma	The fluid part of blood, lymph, or milk as distinguished from suspended material.
pleural space	See intrapleural.
pneumotachograph	An instrument for recording the velocity of the respired air.
power	Capability; potency; the ability to act; mathematically described by the product of force and velocity.
pressure	Stress or strain, whether by compression, pull, thrust, or shear.
pulmonary	Pertaining to the lungs.
pulse function	An excitation signal that rises suddenly from one fixed level to another fixed level and after a finite period returns to its initial valve.
pupil	The opening at the center of the iris of the eye for the transmission of light.
quiescence	The stable unexcited state of a system.
resistance	The opposition by a conductor to the passage of an electric current.
respiration	The act or function of breathing.
solute	A substance dissolved in a solution. A solution consists of a solute and a solvent.
solvent	Dissolving, effecting a solution; a liquid that is capable of dissolving.
spirometer	An instrument for measuring the air taken into and exhaled from the lungs.
static resistance	The reciprocal slope of a line from the origin to some point on the across (voltage) vs. through (current) characteristics of a material.
step-function	An excitation signal that rises suddenly from one level to another level.
synergist	(1) A medicine that aids or cooperates with another; an adjuvant. (2) An organ that acts in concert with another.
systole	The period of the heart's contraction.
tendon	The fibrous cord of connective tissue in which the fibers of a muscle end and by which a muscle is attached to a bone or other structure.
thoracic cage	The cavity in which the heart and lungs lie.
thrust	The sideways force or pressure of one part of a structure against another part.

trachea	The windpipe; the cartilaginous and membranous tube descending from the larynx to the bronchi.
transient	Passing quickly into and out of existence; short-lived.
transmural	The differential measurements between the inside and outside of a vessel; e.g., the transmural pressure of a blood vessel is the pressure difference between the inside and outside of the vessel.
vascular	Relating to channels that convey body fluids or to a system of subchannels. A system of blood vessels might thus be described as a vascular bed.
vein	A vessel that conveys the blood to or toward the heart.
vena cava	One of the large veins by which in air-breathing vertebrates the blood is returned to the right atrium of the heart.
ventricle	Any small cavity, especially either one of the lower cavities of the heart, or one of several cavities of the brain.
viscosity	(1) The property of a fluid that resists internal flow. (2) The capability possessed by a solid of yielding continually under stress.
volt	The unit of electrical pressure or electromotive force, being the force necessary to cause one ampere of current to flow against one ohm of resistance.

INDEX